Advances in Quantum Science I

Taku Onishi
Editor

Advances in Quantum Science I

Theoretical and Methodological Approaches

Springer

Editor
Taku Onishi
Graduate School of Engineering
Mie University
Tsu, Japan

Department of Chemistry
University of Oslo
Oslo, Norway

ISBN 978-981-95-9716-1 ISBN 978-981-95-9717-8 (eBook)
https://doi.org/10.1007/978-981-95-9717-8

This Springer imprint is published by the registered company Springer Nature Singapore Pte Ltd.
The registered company address is: 152 Beach Road, #21-01/04 Gateway East, Singapore 189721, Singapore

Preface

Scientific phenomena related to electrons are based on the principle of quantum mechanics. Because electrons are theoretically treated as quantum particles. Recent research findings in quantum science are introduced from the wide aspects of theoretical development, computational method, Artificial Intelligence (AI), machine learning, computation, and experiment. Researchers who deal with electrons are able to find tips to pioneer new researches.

Volume 1 consists of two parts. Part I highlights theoretical challenges to be addressed: quantum electrodynamics, relativistic quantum chemistry, quantum fluctuation, and quantum entanglement. Part II introduces recent methodological development in quantum chemistry are in the spotlight.

Part I: Quantum Theoretical Physics

Theoretical developments related to relativistic field theory and quantum electrodynamics (QED) are explained in Chapters 1 to 4. Not only electrons but also positrons are theoretically handled. The Eriksen transformation and Wegner's flow equation approach are introduced in Chapter 1. From a quantum particle number conserving Hamiltonian, Schwinger's anomalous magnetic moment and the Lamb shift are obtained in Chapter 2. An effective nonrelativistic Hamiltonian for interacting electrons and/or positrons is derived in Chapter 3. In the application of a flow equation approach for Dirac Hamiltonian, a unitary equivalent and energy-separating Hamiltonian is obtained in Chapter 4. In Chapter 5, after explaining theoretical backgrounds of density matrix, density matrix is applied for a paramagnetic system, magnetic moment precession, and Kubo's formula for conductivity. Quantum fluctuations are also considered for physical quantities. Finally, the equation of density matrix is expressed in Wigner representation. In Chapter 6, S-matrix theory is applied to formulate quantum entanglement entropy. For elastic and inelastic scatterings, quantum entanglement entropies are evaluated using developed formulae. Here quantum particles are generally discussed.

Part II: Methodological Development in Quantum Chemistry

After ultrafast multistage electron transfer is explained, the theoretical approach is introduced in Chapter 7. The approach was applied for a molecular triad model (donor–acceptor–acceptor). Hot electron transfer was pointed out. Quantum battery is the device of charging and discharging electronic energy is based on the principles of quantum mechanics. In Chapter 8, the theoretical models of quantum battery are introduced. For organic molecular systems, it is recognised that coupled cluster singles and doubles, which is denoted as CCSD, reproduces electronic structure with high precision. In Chapter 9, the new coupled cluster method for relativistic systems is introduced.

Finally, I would like to thank Dr. Shinichi Koizumi and Springer staffs. I hope this book stimulates readers to come up with new scientific ideas.

For quantum science in the AI era, mathematics is entrance, and exit as well

Taku Onishi
Japan
A Happy Day 2026

Acknowledgements

The Editor, Taku Onishi, would like to express his special thanks to T.O.

Abbreviations

Eq.	Equation
Eqs.	Equations
Fml.	Formula
Fmls.	Formulas
Fig.	Figure
Figs.	Figures
QED	Quantum Electrodynamics
AI	Artificial Intelligence
DFT	Density Functional Theory
HF	Hartree-Fock
ET	Electron Transfer
QB	Quantum Battery
CCSD	Coupled Cluster Singles Doubles
2D	Two-Dimensional
CM	Centre of Mass
CMs	Centres of Mass
MD	Molecular Dynamics
CNT	Carbon Nanotube
GNR	Graphene Nanoribbon
CMO	Canonical Molecular Orbital
2DEG	Two-Dimensional Electron Gas
STM	Scanning Tunnelling Microscopy
TERS	Tip-Enhanced Raman Spectroscopy
XPS	X-ray Photoelectron Spectroscopy
3DOM	Three-Dimensionally Ordered Macroporous
PBG	Photonic Band Gap

About the Editor

Taku Onishi is a quantum scientist who graduated from Department of Chemistry, Osaka University, Japan in 1998. He gained Master degree in 2000, and Ph.D. in 2003, both from Department of Chemistry, Osaka University, Japan. In 2003, he took up academic tenure as assistant professor in Graduate School of Engineering, Mie University, Japan. During 2010–2011, he stayed at Centre for Theoretical and Computational Chemistry (CTCC), University of Norway as Excellent Young Researcher Overseas Visit Program, Japan Society for the Promotion of Science. Since 2010, he has been a guest researcher of Department of Chemistry, University of Oslo, Norway. In 2021, the title of guest professor was given from University of the Ryukyus, Japan. He has served on international scientific activities: a Member of Royal Society of Chemistry; Chair of the Next-Generation Science (NGS) Symposium; a position on the science committee of the International Conference of Computational Methods in Sciences and Engineering (ICCMSE); a member of the editorial board of Cogent Engineering, the Journal of Computational Methods in Sciences and Engineering (JCMSE). He has reviewed many international proceedings, books, and journals in various research fields of chemistry and physics. For example, AIP conference proceedings, Progress in Theoretical Chemistry and Physics, Cogent Chemistry, Physical Chemistry Chemical Physics, Molecular Physics, Dalton Transaction, The Journal of Physical Chemistry Letters, Journal of Computational Chemistry, International Journal of Quantum Chemistry, Journal of Solid State Chemistry, Solid State Ionics, Chemistry of Materials, Materials Chemistry and Physics, Chemical Engineering Journal, Chemical Physics, ACS catalysis, Applied Surface Science, Chemical Physics, Chemical Physics Letters, Chirality, Applied Surface Science, ACS catalysis, Biochimica et Biophysica Acta—Proteins and Proteomics, AIP advances, The Journal of Organic Chemistry, New Journal of Chemistry, etc.

About the Corresponding Authors

Nils Schopohl received his degree as a certified physicist (Dipl.-Phys.) in 1978 and received his doctorate (Dr. rer. nat.) in theoretical physics at the University of Hamburg in 1979. He was a postdoctoral researcher at IFF Jülich (Germany), Low Temperature Lab at TKK Otaniemi (Finland) and also at ILL Grenoble (France). He received his habilitation in theoretical physics from the University of Hamburg in 1989. In 1991 he became a member of CNRS at CRTB Grenoble in the position of a director of research (DR2), and in 1994 he was appointed Professor of Theoretical Physics at Ruhr-University Bochum. In 1997, he accepted a position as Full Professor of Theoretical Physics at the department of Physics at the Eberhard Karl University of Tübingen. He served as a Dean of the Faculty of Mathematics and Physics and also he was director of the Institute for Theoretical Physics for many years. Since his retirement in 2019 he is currently working as an appointed Senior Professor of Theoretical Physics. Besides his interest in fundamental physics research he is very interested in applications and holds various patents.

Elena V. Orlenko Dr. Sc. (habilitation), is a professor for the Theoretical Physics Department, in the High School Fundamental Physics Researches of Peter the Great St.Petersburg Polytechnic University. For many years, she taught third- and fourth-year students, giving them courses in quantum mechanics, statistical physics, and physical kinetics. Author of a new method of exchange perturbation theory taking into account interatomic electron permutations. This formalism has an invariant form and takes into account interatomic electron permutations among overlapping non-orthogonal state vectors. This method allows one to consider the time-dependent case for which the scattering matrix S has a general form that takes into account the principle of indistinguishability of identical particles; Theoretical study of magnetic ordering in low-dimensional correlated electron systems; A new approach to the theory of Non-Anderson weak localization of particles with respect to inelastic and exchange interactions; Spin ordering in cold atomic systems (Bose–Einstein condensation and Fermi condensate) and general correlated high-spin systems; Theory of superradiance phenomena in multilevel atomic systems.

Shigenori Seki Ph.D., is an associate professor at Setsunan University, Japan, specializing in superstring theory and theoretical particle physics. He earned his doctorate from Kyoto University, focusing on the fundamental mathematical structures of string theory. His recent research activities are primarily centered on two key pillars: the development of superstring theory and the exploration of quantum entanglement. He investigates these subjects within their respective theoretical contexts to uncover the intricate laws governing particles and fields. With an extensive international career, he has held research appointments at Tel Aviv University in Israel, the Institut des Hautes Études Scientifiques (IHÉS) in France, and the Asia Pacific Center for Theoretical Physics (APCTP) and Hanyang University in South Korea. He remains actively engaged in the global theoretical physics community, driven by a deep commitment to expanding the boundaries of human knowledge through meticulous mathematical analysis.

S. V. Feskov is a professor and leading researcher at Volgograd State University (VolSU), where he received his Ph.D. (2000) and D.Sc. (2012) in Chemical Physics. He began his scientific career in 1995 as a member of Prof. A. I. Ivanov's group, studying magnetic field effects in spin-selective radical reactions. From 2006 to 2009, he worked as a visiting researcher in the group of Prof. A. I. Burshtein at the Weizmann Institute of Science (Israel). His research is focused on theoretical and numerical modeling of ultrafast charge transfer processes in complex molecular systems, with particular interest in the role of environmental relaxation, magnetic field effects, and molecular diffusion in liquids.

A. I. Ivanov is a professor of Physics at Volgograd State University (VolSU). He graduated from the Department of Physics at Bashkir State University (Ufa, Russia) in 1973. Before joining VolSU in 1982, he was a researcher at the Institute of Chemistry, Russian Academy of Sciences (Ufa), where he also earned his Ph.D. in Physics. He later obtained his D.Sc. (habilitation) in Chemical Physics from the Physical Technical Institute of the Russian Academy of Sciences. His early research was devoted to the theory of elementary chemical reactions, nonradiative transitions in polyatomic molecules, and electron capture in the gas phase. Since the mid-1990s, his work has centered on the theory of ultrafast charge transfer in solutions. Since 2016, his primary research interest has been symmetry breaking charge transfer in electronically excited molecular aggregates.

Gabriel Hanna is a professor of Chemistry at the University of Alberta. He earned his PhD from the University of Toronto in 2006, where he developed mixed quantum-classical methods for simulating proton transfer in condensed phases, and completed postdoctoral work at the University of Michigan (2006–2009) on mixed quantum-classical approaches to multidimensional infrared spectroscopy. Hanna leads a theory and simulation program focused on open quantum systems, mixed quantum-classical dynamics, spectroscopy, quantum transport, and materials modeling. His group develops computationally tractable methods to simulate many-body quantum dynamics in realistic environments and applies them to problems including spectroscopic line shapes, energy transfer dynamics, and emergent quantum phenomena. A major emphasis of his recent work is the theoretical and computational design of quantum batteries, notably the symmetry-protected excitonic quantum battery. His simulations quantify charging and discharging times, storage efficiency, robustness to decoherence, and design principles for controlled energy release in experimentally relevant architectures. Hanna also advances materials science through computational studies of phosphonate metal-organic frameworks, combining molecular dynamics, density functional theory, and data-driven tools to predict and optimize stability, selectivity, and functionality for gas storage, remediation, optoelectronics, magnetism, and energy storage.

Sourav Pal is currently a Professor at Ashoka University, India, and was former Director of the Indian Institute of Science Education and Research (IISER) Kolkata and CSIR National Chemical Laboratory (NCL), Pune. He received Ph D degree from Calcutta University in 1985 and was a post-doctoral fellow in the University of Florida and Alexander von Humboldt Fellow. He has been a visiting Professor at the University of Arizona, Tucson, USA, the University of Torun, Poland, and the Institute for Molecular Sciences, Okazaki, Japan. He is a theoretical chemist and has contributed to methodological and conceptual developments on many-body electronic structure theory and properties using coupled-cluster theory, both at single reference and multi reference level. He also contributed to the area of density-based chemical reactivity as well as to catalytic and hydrogen storage materials using computational material science. He has been recognized by the highest award of Shanti Swarup Bhatnagar Prize in India and is a Fellow of Royal Society of Chemistry, UK and national academies of Science in India. He has published about 320 publications in high impact journals. He has played an important role in many scientific societies, including being the Founding member of the Board of Commonwealth Chemistry. He is presently Board member of Asia Pacific Association of Theoretical and Computation and in the Presidential Advisory committee of International Society of Theoretical Chemical Physics.

Contents

Part I
Quantum Theoretical Physics

Chapter 1
From the Relativistic Field Theory of Quantum Electrodynamics to a Hamiltonian for Electrons and Positrons as Point Particles with Attributes Mass, Charge and Spin

Nils Schopohl and Nadine Suzan Cetin

Abstract We consider the fundamental field theory Hamiltonian of Quantum Electrodynamics (QED) in Coulomb gauge. As an alternative to S-matrix theory and Feynman diagrams a new unitary transformation is unveiled, that brings with help of Wegner's flow equation method the original QED-Hamiltonian to a block-diagonal form, each block labeled by an eigenvalue of the operator $\hat{N} = \hat{N}_F + \hat{N}_P$, with $\hat{N}_F$ counting the number of occupied Dirac-modes, and $\hat{N}_P$ counting the number of occupied high energy photon modes in the Fock space of QED. A representation of that block-diagonal Hamiltonian as a perturbation series with regard to the coupling charge e establishes to first order the well known particle number conserving light-matter interaction term of quantum optics. To second order arise not only the specific energy shift terms adding to the bare Dirac Hamiltonian, as there are the anomalous magnetic moment of Schwinger and the Lamb shift terms, but in addition the effective two-body interactions between electrons and/or positrons representing the relativistic corrections to the Coulomb interaction. The other interaction terms obtained to second order in the perturbation theory comprise the Compton interaction. As a result there emerges a physical picture of electrons and positrons as if they were enumerable point particles with attributes mass, charge and spin, this way enabling for the relativistic quantum fields a corpuscular interpretation.

Keywords Quantum electrodynamics · Wegner flow equation method · Unitary transformation · Fock-space · Quantum field · Dirac Hamiltonian · Anomalous magnetic moment · Lamb-shift · Coulomb gauge · Relativistic corrections to the Coulomb interaction

---

N. Schopohl (✉)
Center for Quantum Science, Eberhard-Karls-Universität Tübingen, Tübingen, Germany
e-mail: nils.schopohl@uni-tuebingen.de

N. S. Cetin
Mathematical Institute, Charles University Prague, Prague, Czech Republic
e-mail: nadine-suzan.cetin@matfyz.cuni.cz

T. Onishi (ed.), *Advances in Quantum Science I*,
https://doi.org/10.1007/978-981-95-9717-8_1

1.1 Introduction

Non relativistic quantum mechanics was first formulated for a single particle. But from the very beginning nonrelativistic quantum mechanics also was able to cover interacting many particle systems, for instance the plurality of electrons bound to atoms or molecules. In full analogy to the many body problem posed in classical physics (astronomy), a generic nonrelativistic many particle Hamiltonian for a number of N electrons usually consists of two contributions: the one-body term counting altogether N individual contributions of the kinetic energy together with the potential energy of each particle, and the two-body term counting all contributions of different pairs of interacting particles. The extension of the methods of nonrelativistic quantum mechanics for one particle to problems containing a plurality $N > 1$ of particles, requires a straightforward extension of the concept of a probability amplitude, appendant to one particle moving in three-dimensional space, now to a $3N$-dimensional configuration space.

As it happens, the idea to extend Dirac's four-component relativistic single particle wavefunction to a relativistic many particle wavefunction depending on the positions of $N > 1$ interacting electrons rises severe problems. For instance, Lorentz-invariance of the envisaged relativistic N-electron wavefunction necessarily entangles time and space coordinates of each particle separately, thus requiring the introduction of multi-time variables [1], in effect suggesting in the relativistic case to imagine a $4N$-dimensional configuration space [2, 3]. And because both, positive and negative energies are required to span a complete system of eigenmodes already for the single particle Dirac Hamiltonian, it is not at all clear how to handle this feature when generalizing to a relativistic N-electron wavefunction.

Actually, an antecedent to this way of proceeding towards a relativistic N-electron wavefunction is the early 1929 work of Breit [4]. Guided by a close analogy to Dirac's single electron approach and relying on the physical picture, that two electrons would interact with the electromagnetic fields generated by the other (moving!) electron, Breit suggested for the description of *two* interacting electrons a new (effective) Hamiltonian, that would operate on the quantum level now on a sixteen-component two-particle wavefunction. From Breit's (not fully covariant) Hamiltonian emerges in the nonrelativistic limit indeed the well known Schrödinger-Pauli Hamiltonian for two interacting electrons, see e.g. [5]. However, the lowest order relativistic corrections to the particles dominant Coulomb interaction would lead Breit not only to physical interaction terms, for instance a magnetic dipole-dipole interaction as a natural consequence of each electron carrying spin (and thus magnetic moment), but also to an unexpected interaction term proportional to e^4 independent on $\hbar$ which contradicts experiment [6].

A covariant derivation of the results of Breit was established by Bethe and Salpeter, again restricting to *two* electrons, on the basis of S-matrix calculations, thus inferring (in a frame of reference with the total center of mass momentum being zero) a correct version of the Hamiltonian of Breit without the unphysical e^4 term [7, 8]. Unfortunately, an extension of the fully relativistic Bethe-Salpeter approach to a plurality of $N > 2$ interacting fermions seems to be a complicated undertaking, one of the problems being the adequate normalization of the many body wave function, e.g. [7, 9–11].

A better way of justifying a many-body Hamiltonian, that governs the interaction processes for electrons and/or positrons, is to dispense any explicit consideration of N-particle wave functions. Such an idea was first discussed by Itoh [12], who proceeded (in an intuitive way) directly from the field theory Hamiltonian of QED in Coulomb gauge. Subsequently Bialynicki-Birula [13] advanced the argument, that because the coupling between states far below the electron-positron production threshold is weak compared to the electrons rest energy mc^2, then this coupling could be treated in perturbation theory.

Expanding to quadratic order in the coupling charge e, then for non relativistic velocities of electrons a particle number conserving many body Hamiltonian was derived by Cohen-Tannoudji et al. in [14], comprising besides the Coulomb interaction between electrons the additional kinetic terms of Pauli-Schrödinger quantum mechanics. As derived from QED this approach provides compelling evidence why electrons (positrons) should be fermions carrying spin $s_z = \pm\frac{\hbar}{2}$ and obey to Fermi-Dirac statistics, thus connecting to the foundations of non relativistic quantum mechanics.

However, in [12–14] no reference to any details of other renowned QED-effects is found: neither mass nor charge renormalization is treated, nothing is said regarding Schwinger's g-2 correction, and for that matter the Lamb shift terms remain altogether unmentioned. In fact, from behind the reader's perspective, these authors rise a lasting impression: namely that for the comprehensive derivation of these genuine QED-effects indeed covariant S-matrix theory (Feynman diagrams) was indispensable.

In the ensuing considerations we construct, following the seminal work of Wegner on flow equations [15], with help of a suitable generator, a unitary transformation, that provides the passage from the minimally coupled relativistic fermion-field theory Hamiltonian of QED in Coulomb gauge, now to a block-diagonal unitary equivalent (still relativistic) many body Hamiltonian, which in point of fact *conserves* particle number. In effect the Fock space of the original Dirac-modes is decomposed by this unitary transformation into a direct sum of orthogonal subspaces, each subspace being labeled by a specific particle number, and this separately for electrons and positrons, hereby providing for the quantized matter fields of QED a corpuscular interpretation.

That way there emerges naturally a meaningful transition back to a conventional picture with electrons and/or positrons interacting via effective two-body forces and moving as point particles (or for that matter wave packets) with attributes mass m_e, charge $\pm |e|$ and spin $s_z = \pm\frac{\hbar}{2}$, eventually leaving behind, say in a Wilson chamber, their individual (different) traces.

From the solution to the Hamiltonian flow equations as a perturbation series advancing in powers of the coupling e, always treating electron and positron quantum fields as being coequal, we approve and sanction as well the specific energy shift terms adding to the bare Dirac Hamiltonian, in actual fact comprising to order e^2 the renormalization of mass, and to order e^3 the renormalization of charge, the anomalous magnetic moment term $g-2$ of Schwinger and last not least the Lamb-shift terms. A detailed discussion of these effects we postpone to our article [16], where a *gauge invariant* derivation of the Lamb shift terms to order e^3 is given as well.

A detailed derivation of the effective two-body interactions describing the relativistic corrections adding to the predominant Coulomb interaction between electrons and/or positrons we postpone to our chapter [17].

1.2 The Hamiltonian of Quantum Electrodynamics

The relativistic field theory Hamiltonian of Quantum Electrodynamics (QED) in Coulomb-gauge reads[1]:

$$\hat{\mathcal{H}}_{QED} = \hat{\mathcal{H}}_{D,0} + \hat{\mathcal{H}}_{\text{rad}} + \hat{\mathcal{H}}_{\perp} + \hat{\mathcal{V}}_C \tag{1.1}$$

Each term here enables comparison with a classical interpretation. The term $\hat{\mathcal{H}}_{D,0}$ comprises the relativistic ensemble of fermions carrying mass m_0 and charge e_0 as described by the Dirac Hamiltonian with minimal coupling to external (static) electromagnetic source fields. The second term $\hat{\mathcal{H}}_{\text{rad}}$ represents the radiation field Hamiltonian of Maxwell Electrodynamics (photons). The third term $\hat{\mathcal{H}}_{\perp}$ retains the coupling between the transversal electromagnetic vector potential field and the electric current density of the fermions, the last term $\hat{\mathcal{V}}_C$ describes the fermions mutual (electrostatic) Coulomb interactions.

$$\hat{\mathcal{H}}_{D,0} = \int_{V_P} d^3r \sum_{\nu,\nu' \in \{1,2,3,4\}} \hat{\Psi}^{\dagger}_{\nu}(\mathbf{r})\, \mathsf{H}^{(D,0)}_{\nu,\nu'}\, \hat{\Psi}_{\nu'}(\mathbf{r}) \tag{1.2}$$

$$\hat{\mathcal{H}}_{\text{rad}} = \frac{1}{2}\int_{V_P} d^3r' \sum_{b \in \{x,y,z\}} \left(\hat{\mathcal{E}}^{(T)}_b(\mathbf{r}')\, \varepsilon_0 \hat{\mathcal{E}}^{(T)}_b(\mathbf{r}') + \hat{\mathcal{B}}_b(\mathbf{r}') \frac{1}{\mu_0} \hat{\mathcal{B}}_b(\mathbf{r}') \right) \tag{1.3}$$

$$\hat{\mathcal{H}}_{\perp} = -\int_{V_P} d^3r \sum_{b \in \{x,y,z\}} \hat{j}_b(\mathbf{r})\, \hat{A}^{(T)}_b(\mathbf{r}) \tag{1.4}$$

$$\hat{\mathcal{V}}_C = \frac{1}{8\pi\varepsilon_0} \int_{V_P} d^3r \int_{V_P} d^3r' \frac{\hat{\varrho}(\mathbf{r})\, \hat{\varrho}(\mathbf{r}')}{|\mathbf{r}-\mathbf{r}'|} \tag{1.5}$$

Proficient summaries of the relativistic QED field theory Hamiltonian can be found, for instance, in [18–21]. In depth treatments of the physics of the radiation field can be found in [22, 23]. Notably, the Hamiltonian (1.1) is unitary equivalent to the manifestly gauge invariant Hamiltonian of QED in the *temporal* gauge, provided the action of (1.1) is restricted to the subspace of gauge invariant state vectors [13].

[1] Readers without pre-knowledge of relativistic quantum mechanics we refer to the appendix where some supplemental information and also further references to the literature are provided.

Last not least, a profound critics of the use of the Schrödinger picture in QED has been stated by P.A.M. Dirac [24], who pioneered in 1928 relativistic quantum mechanics and minimal coupling to external electromagnetic source fields, introducing his Hamiltonian for the relativistic electron in the guise [25]

$$\mathsf{H}^{(D,0)} = m_0 c^2 \beta + \sum_{b \in \{x,y,z\}} c\alpha_b \left(\mathsf{p}_b - e_0 A_b^{(\text{ext})}(\mathsf{x})\right) + e_0 \Phi^{(\text{ext})}(\mathsf{x})\, \mathsf{1}_{4\times 4} \tag{1.6}$$

whereas due to Born, Heisenberg and Jordan [26]

$$\left[\mathsf{p}_b, \mathsf{x}_a\right] = \frac{\hbar}{i}\delta_{a,b}\mathsf{1} \tag{1.7}$$

Here and in what follows we conform to Dirac's original representation of the 4×4-matrices α_b and β. Also we agree as of now on the summation convention for Cartesian indices $a, b, c \in \{x, y, z\}$.

Note that the bare mass m_0 and the bare charge e_0 entering the QED Dirac Hamiltonian (1.2) via (1.6), are actually *not* the observable mass m and *not* the observable charge e either, because as a result of the coupling of the Dirac quantum field to electromagnetic quantum fields the bare parameters m_0 and e_0 get renormalized into the *measurable* parameters m and e. How such a renormalization is accomplished for an effective Hamiltonian governing the low energy dynamics of electrons (positrons) as *particles* carrying charge e, mass m and magnetic moment, actually represents one of the central themes of Quantum Electrodynamics.

The procedure we propose in the ensuing is a *perturbative* flow equation approach by splitting up the bare parameters according to $m_0 = m + \delta m$ and $e_0 = e + \delta e$, and to consider the variation of mass δm along with the variation of charge $\frac{\delta e}{e}$ as being small perturbations of *second* order, that way determining δm and δe unambiguously from the requirement the limiting value of the flow should be independent of the high energy cutoff $\hbar\Omega_{max}$ of QED. As for δm and δe indeed only a (mild) dependence on this cutoff is coming into the theory, the claim both these quantities are small leads to the requirement $\frac{\alpha_{FS}}{\pi} \ln\left(\frac{\hbar\Omega_{max}}{mc^2}\right) \ll 1$, so that with $\alpha_{FS} \approx \frac{1}{137}$ a wide spectrum of high energy modes up to an extremely large value $\hbar\Omega_{max}$ is nonetheless covered with this approach. Details are expanded further below in subsequent sections.

The suggested splitting of the bare parameters according to $m_0 = m + \delta m$ and $e_0 = e + \delta e$ leads for now to the decomposition

$$\begin{aligned} \mathsf{H}^{(D,0)} &= \mathsf{H}^{(D)} + \delta\mathsf{H}^{(D)} \\ \mathsf{H}^{(D)} &= mc^2\beta + c\alpha_b\left(\mathsf{p}_b - eA_b^{(\text{ext})}(\mathsf{x})\right) + e\Phi^{(\text{ext})}(\mathsf{x})\,\mathsf{1}_{4\times 4} \\ \delta\mathsf{H}^{(D)} &= \delta m\, c^2\beta + \delta e\left(\Phi^{(\text{ext})}(\mathsf{x})\,\mathsf{1}_{4\times 4} - c\alpha_b A_b^{(\text{ext})}(\mathsf{x})\right), \end{aligned} \tag{1.8}$$

and correspondingly the field Hamiltonian (1.2) is decomposed

$$
\begin{aligned}
\hat{\mathcal{H}}_{D,0} &= \hat{\mathcal{H}}_D + \delta\hat{\mathcal{H}}_D \\
\hat{\mathcal{H}}_D &= \int_{V_P} d^3r \sum_{\nu,\nu' \in \{1,2,3,4\}} \hat{\Psi}^\dagger_\nu(\mathbf{r})\, \mathsf{H}^{(D)}_{\nu,\nu'} \hat{\Psi}_{\nu'}(\mathbf{r}) \\
\delta\hat{\mathcal{H}}_D &= \int_{V_P} d^3r \sum_{\nu,\nu' \in \{1,2,3,4\}} \hat{\Psi}^\dagger_\nu(\mathbf{r})\, \delta\mathsf{H}^{(D)}_{\nu,\nu'} \hat{\Psi}_{\nu'}(\mathbf{r})
\end{aligned}
\tag{1.9}
$$

In the end in (1.9) only the field Hamiltonian $\hat{\mathcal{H}}_D$ of the physical Dirac fermions matters, the effect of the variation $\delta\hat{\mathcal{H}}_D$ being effectively annulled by counter terms generated with the indicated perturbation theory approach. That way the physically observed mass m and the physically observed charge e enter the theory, thus causing Quantum Electrodynamics no longer being a complete theory in itself but rather an *effective* (i.e. phenomenological) theory, that enables nevertheless superprecise predictions for Dirac quasi particles (electrons, positrons) interacting with photons for all energies up to, say $\hbar\Omega_{max} < 1\,\text{GeV}$, without having to know values of their bare mass m_0 and their bare charge e_0 either.

It is convenient practice to consider the field modes as being confined in a large box of volume $V_P = L^3$ and obeying to periodic boundary conditions. At the end of the calculations then $L \to \infty$. The advantage of this procedure is that one considers discrete sums over mode labels k rather than integrals.

Introducing now the four-component Dirac modes $\phi_\nu(\mathbf{r}; k)$ with $\nu \in \{1, 2, 3, 4\}$ solving the eigenvalue problem

$$
\sum_{\nu'} \mathsf{H}^{(D)}_{\nu,\nu'} \phi_{\nu'}(\mathbf{r}; k) = E_k \phi_\nu(\mathbf{r}; k) \tag{1.10}
$$

we assume without reservation these eigenfunctions being ortho-normal and spanning a complete basis system

$$
\begin{aligned}
\int_{V_P} d^3r \sum_\nu \phi^\star_\nu(\mathbf{r}; k)\, \phi_\nu(\mathbf{r}; k') &= \delta_{k,k'} \\
\sum_k \phi_\nu(\mathbf{r}; k)\, \phi^\star_{\nu'}(\mathbf{r}'; k) &= \delta_{\nu,\nu'} \delta^3(\mathbf{r} - \mathbf{r}')
\end{aligned}
\tag{1.11}
$$

The index k, here serving to *count* the different modes, should be understood as a multi-index so that degenerate modes are not overlooked.

The associated four-component fermion quantum fields $\hat{\Psi}_\nu(\mathbf{r})$ and $\hat{\Psi}^\dagger_\nu(\mathbf{r})$ comprising coherently the degrees of freedom of matter and anti-matter [27] may be represented in the compact guise

$$
\begin{aligned}
\hat{\Psi}_\nu(\mathbf{r}) &= \sum_k \phi_\nu(\mathbf{r}; k)\, \hat{f}_k \\
\hat{\Psi}^\dagger_\nu(\mathbf{r}) &= \sum_k \phi^\star_\nu(\mathbf{r}; k)\, \hat{f}^\dagger_k
\end{aligned}
\tag{1.12}
$$

whereas

$$\{\hat{f}_k, \hat{f}_{k'}\} = \hat{0} = \{\hat{f}_k^{\dagger}, \hat{f}_{k'}^{\dagger}\} \qquad (1.13)$$
$$\{\hat{f}_k, \hat{f}_{k'}^{\dagger}\} = \delta_{k,k'}\hat{1}$$

are fermion creation and annihilation operators.

Regarding the operators of charge density and electrical current density in QED taking into account minimal coupling to external (static) electromagnetic source fields we adhere to a point splitting definition, as for instance advocated in the Pauli lectures [28]

$$\hat{\varrho}(\mathbf{r}) = \frac{e}{4}\lim_{\mathbf{r}'\to\mathbf{r}}\sum_{\nu}\left(\left[\hat{\Psi}_{\nu}^{\dagger}(\mathbf{r}), \hat{\Psi}_{\nu}(\mathbf{r}')\right] + \left[\hat{\Psi}_{\nu}^{\dagger}(\mathbf{r}'), \hat{\Psi}_{\nu}(\mathbf{r})\right]\right) \qquad (1.14)$$

$$\hat{j}_b(\mathbf{r}) = \frac{e}{4}\lim_{\mathbf{r}'\to\mathbf{r}}\sum_{\nu,\nu'}(c\alpha_b)_{\nu,\nu'}\left(\left[\hat{\Psi}_{\nu}^{\dagger}(\mathbf{r}), \hat{\Psi}_{\nu'}(\mathbf{r}')\right] + \left[\hat{\Psi}_{\nu}^{\dagger}(\mathbf{r}'), \hat{\Psi}_{\nu'}(\mathbf{r})\right]\right) \qquad (1.15)$$

The widely used textbook definitions of relativistic charge and current densities as normal ordered bilinear forms, for instance [29], actually don't apply in the presence of external source fields. For a comprehensive review see [30].

As detailed in the appendix, in the translational invariant case (no coupling at all to any external source fields) the energy spectrum of $\mathsf{H}_0^{(D)} \equiv \mathsf{H}^{(D)}|_{e=0}$ is made up of positive-energy and negative-energy eigenvalues

$$E_k \in \left\{-E_k^{(0)}, E_k^{(0)}\right\} \qquad (1.16)$$
$$E_k^{(0)} = \sqrt{(mc^2)^2 + (c\hbar\,|\mathbf{k}|)^2},$$

thus leaving a gap of width $2mc^2$ in that spectrum. This characteristic spectral gap still persists taking into account coupling to an external magnetostatic field only. But permitting a (weak) coupling to an external *electrostatic* field, say the Coulomb field of an atomic nucleus at a fixed position, the charge conjugation symmetry of the spectrum of $\mathsf{H}^{(D)}$ is broken. In view of positrons being repelled but electrons being attracted by a (positively charged) atomic nucleus, there exist for the electrons additional (bound) states with discrete positive eigenvalues below mc^2 whereas none exist in this case for positrons, always assuming the amplitude of the electrostatic field being *weak*, that is to say according to Weisskopf [31]

$$|\,\mathcal{E}^{ext}\,| \ll \frac{mc^2}{e\lambda_C} \qquad (1.17)$$
$$|\,\nabla\mathcal{E}^{ext}\,| \ll \frac{|\,\mathcal{E}^{ext}\,|}{\lambda_C}$$

with $\lambda_C = \frac{h}{mc^2}$ denoting the Compton wave length and m the electron mass. With the restrictions (1.17) then Schwinger pair creation processes are out of the question [32].

Being therefore able to discern with these restrictions between positive and negative energy eigenvalues E_k also in the presence of external fields, one agrees on the usual interpretation of $\hat{f}_k$ being a fermionic *annihilation* operator $\hat{c}_k$ for matter, $E_k > 0$, respectively $\hat{f}_k$ being a fermionic *creation* operator $\hat{b}_k^\dagger$ for anti-matter, $E_k < 0$, this interpretation being equivalent to the Bogoliubov transformation

$$\begin{aligned}\hat{f}_k &= \Theta_H(E_k)\,\hat{c}_k + \Theta_H(-E_k)\,\hat{b}_k^\dagger \\ \hat{f}_k^\dagger &= \Theta_H(E_k)\,\hat{c}_k^\dagger + \Theta_H(-E_k)\,\hat{b}_k\end{aligned} \tag{1.18}$$

whereas $\Theta_H(E_k)$ denotes the Heaviside distribution

$$\Theta_H(E_k) = \frac{1+z_k}{2} \tag{1.19}$$

$$z_k = \frac{E_k}{|\,E_k\,|} \tag{1.20}$$

Indeed, insertion of (1.12) into (1.9) leads with help of (1.11) and (1.18) at once to the general result

$$\begin{aligned}\hat{\mathcal{H}}_D &= \sum_k E_k \hat{f}_k^\dagger \hat{f}_k \\ &= \sum_{k\in I_+} |E_k|\,\hat{c}_k^\dagger \hat{c}_k + \sum_{k\in I_-} |E_k|\,\hat{b}_k^\dagger \hat{b}_k - \left(\sum_{k\in I_-} |E_k|\right)\hat{1}\end{aligned} \tag{1.21}$$

Here we introduced the index sets $I_\pm$ labeling the Dirac modes $\phi_\mu(\mathbf{r};k)$ assigned to positive and negative energy eigenvalues E_k , respectively

$$I_+ = \{k \mid E_k > 0\} \tag{1.22}$$

$$I_- = \{k \mid E_k < 0\} \tag{1.23}$$

As is evident from (1.21) the lowest possible eigenvalue of $\hat{\mathcal{H}}_D$ is the one with *zero* occupation number for modes with $k \in I_\pm$, thus implying the excitation energy above that ground state is by construction *positive*, regardless of matter modes ($E_k > 0$) or anti-matter modes ($E_k < 0$) being excited. The negative energy shift term in (1.21) represents the vacuum energy of Dirac fermions. Taking into account coupling to external electromagnetic source fields that vacuum energy acquires a physical meaning though, in actual fact contributing quantum corrections adding to the electromagnetic field energy, see for instance [19, 23, 31].

Likewise, introducing a complete set of photon mode creation and annihilation operators labeled by wavevector $\mathbf{q}$ and polarization indices $\nu \in \{I, II\}$, and obeying to Bose commutator rules

$$\left[\hat{a}_{\mathbf{q},\nu}, \hat{a}^{\dagger}_{\mathbf{q}',\nu'}\right] = \delta_{\nu,\nu'}\delta_{\mathbf{q},\mathbf{q}'} \tag{1.24}$$

$$\left[\hat{a}^{\dagger}_{\mathbf{q},\nu}, \hat{a}^{\dagger}_{\mathbf{q}',\nu'}\right] = \hat{0} = \left[\hat{a}_{\mathbf{q},\nu}, \hat{a}_{\mathbf{q}',\nu'}\right] \tag{1.25}$$

the (transversal) electromagnetic field operators that solve the homogeneous Maxwell equations can be represented as [14]

$$\begin{aligned}
\hat{\mathcal{A}}_b(\mathbf{r}) &= \frac{1}{\sqrt{V_P}} \sum_{\mathbf{q}} \sum_{\nu\in\{I,II\}} \tilde{\mathcal{A}}_b(\mathbf{q},\nu)\left(e^{i\mathbf{q}\cdot\mathbf{r}}\hat{a}_{\mathbf{q},\nu} + e^{-i\mathbf{q}\cdot\mathbf{r}}\hat{a}^{\dagger}_{\mathbf{q},\nu}\right) \\
\hat{\mathcal{E}}_b(\mathbf{r}) &= \frac{i}{\sqrt{V_P}} \sum_{\mathbf{q}} \sum_{\nu\in\{I,II\}} \tilde{\mathcal{E}}_b(\mathbf{q},\nu)\left(e^{i\mathbf{q}\cdot\mathbf{r}}\hat{a}_{\mathbf{q},\nu} - e^{-i\mathbf{q}\cdot\mathbf{r}}\hat{a}^{\dagger}_{\mathbf{q},\nu}\right) \\
\hat{\mathcal{B}}_b(\mathbf{r}) &= \sum_{b',b''\in\{x,y,z\}} \epsilon_{b,b',b''}\nabla_{b'}\hat{\mathcal{A}}_{b''}(\mathbf{r}),
\end{aligned} \tag{1.26}$$

with amplitudes

$$\begin{aligned}
\tilde{\mathcal{A}}_b(\mathbf{q},\nu) &= \sqrt{\frac{\hbar}{2\varepsilon_0\omega(\mathbf{q})}}\,u_b(\mathbf{q},\nu) \\
\tilde{\mathcal{E}}_b(\mathbf{q},\nu) &= \sqrt{\frac{\hbar\omega(\mathbf{q})}{2\varepsilon_0}}\,u_b(\mathbf{q},\nu)
\end{aligned} \tag{1.27}$$

determined by the vacuum dispersion relation $\omega(\mathbf{q}) = c\,|\mathbf{q}|$ of photons. The two transversal polarization vectors $u_b(\mathbf{q},\nu)$ of the photon, being normal to one another and being both located in the plane *perpendicular* to $\mathbf{q}$, thus obey to

$$\begin{aligned}
\sum_{\nu\in\{I,II\}} u_b(\mathbf{q},\nu)u^{\star}_{b'}(\mathbf{q},\nu) &= \delta_{b,b'} - \frac{q_b q_{b'}}{|\,\mathbf{q}\,|^2} \\
\sum_{b\in\{x,y,z\}} u^{\star}_b(\mathbf{q},\nu)u_b(\mathbf{q},\nu') &= \delta_{\nu,\nu'}
\end{aligned} \tag{1.28}$$

Here the $u_b(\mathbf{q},\nu)$ are chosen so that $u_b(\mathbf{q},\nu) = u_b(-\mathbf{q},\nu)$.

So, insertion of the expansions (1.26) into (1.3) leads together with (1.28) to the electromagnetic radiation Hamiltonian of QED in standard guise

$$\hat{\mathcal{H}}_{\text{rad}} = \sum_{\mathbf{q}} \hbar\omega_{\mathbf{q}} \sum_{\nu\in\{I,II\}} (\hat{a}^{\dagger}_{\mathbf{q},\nu}\hat{a}_{\mathbf{q},\nu} + \frac{1}{2}) \tag{1.29}$$

A useful compact notation for the electromagnetic quantum fields is [24]

$$\hat{\mathcal{A}}_b(\mathbf{r}) = \frac{1}{\sqrt{V_P}} \sum_{\mathbf{q}} e^{i\mathbf{q}\cdot\mathbf{r}} \sum_{\sigma\in\{+1,-1\}} \hat{\mathsf{A}}^{(\sigma)}_{b,\mathbf{q}} \tag{1.30}$$

$$\hat{\mathcal{E}}_b(\mathbf{r}) = \frac{1}{i\hbar}\left[\hat{\mathcal{A}}_b(\mathbf{r}), \hat{\mathcal{H}}_{\text{rad}}\right]$$

whereas

$$\begin{aligned} \hat{\mathsf{A}}^{(+1)}_{b,\mathbf{q}} &= \sum_{\nu\in\{I,II\}} \tilde{\mathcal{A}}_b(\mathbf{q},\nu)\,\hat{a}_{\mathbf{q},\nu} \\ \hat{\mathsf{A}}^{(-1)}_{b,\mathbf{q}} &= \sum_{\nu\in\{I,II\}} \tilde{\mathcal{A}}_b(-\mathbf{q},\nu)\,\hat{a}^{\dagger}_{-\mathbf{q},\nu} \end{aligned} \tag{1.31}$$

One readily confirms with help of (1.28) these linear combinations of photon creation- and annihilation operators obey to the commutator rules

$$\begin{aligned} \left[\hat{\mathsf{A}}^{(+1)}_{b,\mathbf{q}}, \hat{\mathsf{A}}^{(-1)}_{b',\mathbf{q}'}\right] &= \frac{\hbar}{2\varepsilon_0\omega(\mathbf{q})}\left(\delta_{b.b'} - \frac{q_b q_{b'}}{|\mathbf{q}|^2}\right)\delta_{\mathbf{q}+\mathbf{q}',\mathbf{0}}\cdot\hat{1} = \left[\hat{\mathsf{A}}^{(+1)}_{b',\mathbf{q}'}, \hat{\mathsf{A}}^{(-1)}_{b,\mathbf{q}}\right] \\ \left[\hat{\mathsf{A}}^{(+1)}_{b,\mathbf{q}}, \hat{\mathsf{A}}^{(+1)}_{b',\mathbf{q}'}\right] &= \hat{0} = \left[\hat{\mathsf{A}}^{(-1)}_{b,\mathbf{q}}, \hat{\mathsf{A}}^{(-1)}_{b',\mathbf{q}'}\right] \end{aligned} \tag{1.32}$$

If the QED-interaction terms (1.4) and (1.5) were ranked according to the contributions of electron and/or positron creation and annihilation operators separately, so that electrons and/or positrons were also interacting with photons of either polarization, the explicit writing of all these elementary acts gives quite lengthy expressions that would be laboriously to handle. The introduced notations (1.12) and (1.30) for the QED-quantum fields provide a compact style represention of the QED-interactions that suits appreciably better for the ensuing presentation of the calculations.

Inserting the mode expansions (1.12) and (1.30) into (1.4) then

$$\hat{\mathcal{H}}_{\perp} = -ec\frac{1}{\sqrt{V_P}}\sum_{\mathbf{q}}\sum_{\sigma\in\{+1,-1\}}\sum_{k,k'}\langle\phi_k|\,\alpha_b e^{iq_a x_a}\,|\phi_{k'}\rangle\,\frac{1}{2}\left[\hat{f}^{\dagger}_k, \hat{f}_{k'}\right]\hat{\mathsf{A}}^{(\sigma)}_{b,\mathbf{q}} \tag{1.33}$$

whereby

$$\langle\phi_k|\,\alpha_b e^{iq_a x_a}\,|\phi_{k'}\rangle = \int_{V_P} d^3r \sum_{\nu',\nu}\phi^{\star}_{\nu'}(\mathbf{r};k')\left(\alpha_b e^{i\mathbf{q}\cdot\mathbf{r}}\right)_{\nu',\nu}\phi_{\nu}(\mathbf{r};k) \tag{1.34}$$

Likewise, the QED-Coulomb interaction (1.5) assumes with the Fourier representation

$$\frac{1}{4\pi\,|\mathbf{r}-\mathbf{r}'|} = \int\frac{d^3q}{(2\pi)^3}\frac{e^{i\mathbf{q}\cdot(\mathbf{r}-\mathbf{r}')}}{|\mathbf{q}|^2} \tag{1.35}$$

the compact guise

$$\hat{V}_C = \frac{e^2}{2\varepsilon_0} \frac{1}{V_P} \sum_{\mathbf{q}} \frac{1}{|\mathbf{q}|^2} \sum_{k,k'} \sum_{K,K'} \langle \phi_k | e^{-iq_a x_a} | \phi_{k'} \rangle \langle \phi_K | e^{iq_a x_a} | \phi_{K'} \rangle \frac{1}{4} \left[\hat{f}_k^\dagger, f_{k'} \right] \left[\hat{f}_K^\dagger, f_{K'} \right] \tag{1.36}$$

whereby

$$\langle \phi_k | e^{iq_a x_a} | \phi_{k'} \rangle = \int_{V_P} d^3 r \sum_{\nu} \phi_\nu^\star \left(\mathbf{r}; k' \right) e^{i\mathbf{q}\cdot\mathbf{r}} \phi_\nu \left(\mathbf{r}; k \right) \tag{1.37}$$

Notably, for reasons of consistency with the use of periodic boundary conditions, the replacement

$$\int \frac{d^3 q}{(2\pi)^3} \frac{e^{i\mathbf{q}\cdot(\mathbf{r}-\mathbf{r}')}}{|\mathbf{q}|^2} \to \frac{1}{V_P} \sum_{\mathbf{q}} \frac{e^{i\mathbf{q}\cdot(\mathbf{r}-\mathbf{r}')}}{|\mathbf{q}|^2} \tag{1.38}$$

has been made in (1.36), with the error of this replacement in deed being only discernible near to the boundary of the large box volume V_P , and actually vanishing for $V_P \to \infty$.

1.3 Lifting the Lid on QED's Charge and Current Densities

If we take into account the effects of external electromagnetic source fields, then the point-splitting definitions of the charge and currrent density operators according to (1.14) lead not only to the established *normal ordered* bilinear forms constructed from the Dirac fermion field operators, for instance [29], but also additional c-number contributions emerge known as vacuum polarization charge density $\rho^{(\mathrm{pol})}(\mathbf{r})$ and vacuum polarization current density $j_b^{(\mathrm{pol})}(\mathbf{r})$

$$\hat{\varrho}(\mathbf{r}) = \mathcal{N}\left(e \sum_{\nu} \hat{\Psi}_\nu^\dagger(\mathbf{r}) \hat{\Psi}_\nu(\mathbf{r}) \right) + \rho^{(\mathrm{pol})}(\mathbf{r}) \cdot \hat{1} \tag{1.39}$$

$$\hat{j}_b(\mathbf{r}) = \mathcal{N}\left(e \sum_{\nu,\nu'} (c\alpha_b)_{\nu,\nu'} \hat{\Psi}_\nu^\dagger(\mathbf{r}) \hat{\Psi}_{\nu'}(\mathbf{r}) \right) + j_b^{(\mathrm{pol})}(\mathbf{r}) \cdot \hat{1} \tag{1.40}$$

Indeed, the QED-charge density (1.39) looks only deceptively real to a corresponding expression of the charge density common in nonrelativistic quantum mechanics. That is, because discriminating between matter and antimatter wave functions writing

$$\phi_\nu \left(\mathbf{r}; k \right) = U_\nu \left(\mathbf{r}; k \right) \Theta_H \left(E_k \right) + V_\nu \left(\mathbf{r}; k \right) \Theta_H \left(-E_k \right) \tag{1.41}$$

this engenders for the charge and current density operators (1.39), (1.40) of QED the decomposition (here we restrict to *static* source fields)

$$\hat{\varrho}\left(\mathbf{r}\right) = \hat{\varrho}_0\left(\mathbf{r}\right) + \hat{\varrho}_+\left(\mathbf{r}\right) + \hat{\varrho}_-\left(\mathbf{r}\right) + \rho^{(\mathrm{pol})}(\mathbf{r}) \cdot \hat{1} \tag{1.42}$$
$$\hat{j}_b(\mathbf{r}) = \hat{j}_{b,0}(\mathbf{r}) + \hat{j}_{b,+}(\mathbf{r}) + \hat{j}_{b,-}(\mathbf{r}) + j_b^{(\mathrm{pol})}(\mathbf{r}) \cdot \hat{1},$$

whereas upon insertion of (1.18) into (1.12) then (1.15) or else (1.42) imply

$$\hat{j}_{b,0}\left(\mathbf{r}\right) = e \begin{pmatrix} \sum_{k,k' \in I_+} \left(\sum_{\nu,\nu'} U_\nu^\star\left(\mathbf{r};k\right) \left(c\alpha_b\right)_{\nu,\nu'} U_{\nu'}\left(\mathbf{r};k'\right) \right) \hat{c}_k^\dagger \hat{c}_{k'} \\ - \sum_{k,k' \in I_-} \left(\sum_{\nu,\nu'} V_{\nu'}^\star\left(\mathbf{r};k\right) \left(c\alpha_b\right)_{\nu,\nu'} V_\nu\left(\mathbf{r};k'\right) \right) \hat{b}_{k'}^\dagger \hat{b}_k \end{pmatrix} \tag{1.43}$$

$$\hat{j}_{b,+}\left(\mathbf{r}\right) = e \sum_{k \in I_+} \sum_{k' \in I_-} \sum_{\nu,\nu'} U_\nu^\star\left(\mathbf{r};k\right) \left(c\alpha_b\right)_{\nu,\nu'} V_{\nu'}\left(\mathbf{r};k'\right) \hat{c}_k^\dagger \hat{b}_{k'}^\dagger = \left(\hat{j}_{b,-}\left(\mathbf{r}\right)\right)^\dagger \tag{1.44}$$

$$j_b^{(\mathrm{pol})}\left(\mathbf{r}\right) = -\frac{e}{2} \sum_k \sum_{\nu,\nu'} \phi_\nu^\star\left(\mathbf{r};k\right) \left(c\alpha_b\right)_{\nu,\nu'} z_k \phi_{\nu'}\left(\mathbf{r};k\right), \tag{1.45}$$

respectively

$$\hat{\varrho}_0\left(\mathbf{r}\right) = e \begin{pmatrix} \sum_{k,k' \in I_+} \left(\sum_\nu U_\nu^\star\left(\mathbf{r};k\right) U_\nu\left(\mathbf{r};k'\right) \right) \hat{c}_k^\dagger \hat{c}_{k'} \\ - \sum_{k,k' \in I_-} \left(\sum_\nu V_\nu^\star\left(\mathbf{r};k\right) V_\nu\left(\mathbf{r};k'\right) \right) \hat{b}_{k'}^\dagger \hat{b}_k \end{pmatrix} \tag{1.46}$$

$$\hat{\varrho}_+\left(\mathbf{r}\right) = e \sum_{k \in I_+} \sum_{k' \in I_-} \sum_\nu U_\nu^\star\left(\mathbf{r};k\right) V_\nu\left(\mathbf{r};k'\right) \hat{c}_k^\dagger \hat{b}_{k'}^\dagger = \left(\hat{\varrho}_-\left(\mathbf{r}\right)\right)^\dagger \tag{1.47}$$

$$\varrho^{(\mathrm{pol})}(\mathbf{r}) = -\frac{e}{2} \sum_k \sum_\nu \phi_\nu^\star\left(\mathbf{r};k\right) z_k \phi_\nu\left(\mathbf{r};k\right) \tag{1.48}$$

The physical interpretation of the c-number shifts $\varrho^{(\mathrm{pol})}(\mathbf{r})$ and $j_b^{(\mathrm{pol})}(\mathbf{r})$ being, that a so-called vacuum polarization $\mathbf{P}(\mathbf{r})$ and possibly as well a vacuum magnetization $\mathbf{M}(\mathbf{r})$ is generated by (static) external electromagnetic fields in the background, akin to the well known (static) electromagnetic response of a dielectric medium

$$\varrho^{(\mathrm{pol})}(\mathbf{r}) = -\nabla \cdot \mathbf{P}(\mathbf{r}) \tag{1.49}$$
$$\mathbf{j}^{(\mathrm{pol})}(\mathbf{r}) = \mathbf{rot}\,\mathbf{M}(\mathbf{r})$$

Note, that in the *absence* of an external *electrostatic* source field, $\Phi^{(\mathrm{ext})}\left(\mathbf{x}\right) \equiv 0$, there applies the charge conjugation symmetry, implicating immediately $\varrho^{(\mathrm{pol})}(\mathbf{r}) \equiv 0$ and $j_b^{(\mathrm{pol})}(\mathbf{r}) \equiv 0$, of course.

On physical grounds we expect the effect of external background fields being small of order *two* in the coupling e, as there should exist no difference replacing e by $-e$. Therefore we anticipate $\delta m = O\left(e^2\right)$ and also $\frac{\delta e}{e} = O\left(e^2\right)$.

The explicit calculation of the induced polarization charge density $\varrho^{(\mathrm{pol})}(\mathbf{r})$ and the induced polarization current $j_b^{(\mathrm{pol})}(\mathbf{r})$ we postpone to our chapter [16].

Introducing

$$\hat{N}_F = \sum_k \left(\Theta_H(E_k)\hat{c}_k^\dagger \hat{c}_k + \Theta_H(-E_k)\hat{b}_k^\dagger \hat{b}_k \right) \tag{1.50}$$

$$\hat{Q} = e \sum_k \left(\Theta_H(E_k)\hat{c}_k^\dagger \hat{c}_k - \Theta_H(-E_k)\hat{b}_k^\dagger \hat{b}_k \right) \tag{1.51}$$

the operator $\hat{N}_F$ counts the sum of the occupied matter and antimatter fermion modes in Fock space, while the operator $\hat{Q}$ determines their difference and thus the charge carried by the occupied fermion modes. Notably, $\hat{Q}$ is conserved even in the presence of interactions with the electromagnetic field.

For $\hat{N}_F$ one readily verifies the characteristics

$$\begin{aligned} \left[\hat{N}_F, \hat{\varrho}_0(\mathbf{r})\right] &= \hat{0} = \left[\hat{N}_F, \hat{j}_{b,0}(\mathbf{r})\right] \\ \left[\hat{N}_F, \hat{\varrho}_\pm(\mathbf{r})\right] &= \pm 2\hat{\varrho}_\pm(\mathbf{r}) \\ \left[\hat{N}_F, \hat{j}_{b,\pm}(\mathbf{r})\right] &= \pm 2\hat{j}_{b,\pm}(\mathbf{r}) \end{aligned} \tag{1.52}$$

As the pair terms $\hat{\varrho}_\pm(\mathbf{r})$ and $\hat{j}_{b,\pm}(\mathbf{r})$, see (1.47) and (1.44), cannot be decoded back in first quantization, owing to this a physical interpretation of the operator $\hat{\varrho}(\mathbf{r})$ as describing a *point cloud* comprising *particles* with attributes charge, mass and spin $\pm\frac{\hbar}{2}$, turns out being impossible that way, even though the orthonormality of the eigenfunctions of the Dirac Hamiltonian (1.8) implies

$$\begin{aligned} &\int d^3r \hat{\varrho}_\pm(\mathbf{r}) = \hat{0} \\ &\int d^3r \hat{\varrho}_0(\mathbf{r}) = \hat{Q} \equiv e \sum_k \left(\Theta_H(E_k)\hat{c}_k^\dagger \hat{c}_k - \Theta_H(-E_k)\hat{b}_k^\dagger \hat{b}_k \right) \end{aligned} \tag{1.53}$$

It being true that $\hat{Q}$ is determined solely by the partial charge density $\hat{\varrho}_0(\mathbf{r})$, it is yet incorrect to identify that quantity taken by itself as a physical charge density, as the complementing parts $\varrho_\pm(\mathbf{r})$ are deemed necessary for $\hat{\varrho}(\mathbf{r})$ together with $\hat{\mathbf{j}}(\mathbf{r})$ to comply with the fundamental continuity equation (time dependence in Heisenberg picture)

$$\frac{\partial \hat{\varrho}(\mathbf{r}, t)}{\partial t} + \nabla \cdot \hat{\mathbf{j}}(\mathbf{r}, t) = \hat{0} \tag{1.54}$$

So the appearance of the *pair* terms $\hat{\varrho}_\pm(\mathbf{r})$ and $\hat{j}_{b,\pm}(\mathbf{r})$ indeed elucidates the very nature of QED, insofar the QED Hamiltonian (1.1) conserves charge but violates conservation of total fermion number

$$\left[\hat{N}_F, \hat{\mathcal{H}}_{QED}\right] \neq \hat{0} \tag{1.55}$$

$$\left[\hat{Q}, \hat{\mathcal{H}}_{QED}\right] = \hat{0} \tag{1.56}$$

Introducing *normal* order, see for instance [33], by agreeing on a ground state $|G\rangle$ (vacuum) so that with respect to the fermion annihilation operators $\hat{c}_k$ and $\hat{b}_k$ there holds $\hat{c}_k |G\rangle = 0 = \hat{b}_k |G\rangle$, the operator (1.51) counting total charge can be rewritten as

$$\hat{Q} = \mathcal{N}\Big(e \sum_{k\in I} \hat{f}_k^{\dagger} \hat{f}_k\Big) = \mathcal{N}\Big(e \int d^3r \sum_{\nu,\nu'} \hat{\Psi}_\nu^{\dagger}(\mathbf{r})\hat{\Psi}_\nu(\mathbf{r})\Big), \tag{1.57}$$

whereas the operator (1.50) counting the occupied fermion states assumes the guise

$$\hat{N}_F = \mathcal{N}\Big(\sum_{k\in I} \hat{f}_k^{\dagger} z_k \hat{f}_k\Big) = \mathcal{N}\Big(\int d^3r \sum_{\nu,\nu'} \hat{\Psi}_\nu^{\dagger}(\mathbf{r})\Lambda_{\nu,\nu'}^{(D)}\hat{\Psi}_{\nu'}(\mathbf{r})\Big) \tag{1.58}$$

Here

$$\Lambda^{(D)} = \frac{\mathsf{H}^{(D)}}{\sqrt{\mathsf{H}^{(D)} \circ \mathsf{H}^{(D)}}} \tag{1.59}$$

denotes the energy sign operator solving the eigenvalue problem

$$\sum_{\nu'} \Lambda_{\nu,\nu'}^{(D)} \phi_{\nu'}(\mathbf{r}; k) = \frac{E_k}{|E_k|} \phi_\nu(\mathbf{r}; k) = z_k \phi_\nu(\mathbf{r}; k) \tag{1.60}$$

That way the right hand side in (1.58) provides an integral representation for $\hat{N}_F$. Alas, the obtained integrand in (1.57) is not anywhere near being *local*, thus prohibiting a straightforward physical interpretation as a particle density operator.

The proposal to find a local *particle* density operator $\hat{n}^{(e)}(\mathbf{r})$ for electrons, and distinguished from that a local *antiparticle* density operator $\hat{n}^{(p)}(\mathbf{r})$ for positrons, so that in line with physical intuition there existed a representation that completely disentangles matter and antimatter densities,

$$\hat{N}_F = \int d^3r (\hat{n}^{(e)}(\mathbf{r}) + \hat{n}^{(p)}(\mathbf{r})) \tag{1.61}$$

$$\hat{Q} = e \int d^3r (\hat{n}^{(e)}(\mathbf{r}) - \hat{n}^{(p)}(\mathbf{r})), \tag{1.62}$$

indeed appears in view of (1.53) and (1.57) for the moment being beyond all hope.

1.4 The Eriksen Transformation

Progress comes introducing the energy-separating unitary transformation T of Eriksen [34, 35], which transforms the 4×4-matrix Hamiltonian of the Dirac fermions $\mathsf{H}^{(D)}$ in (1.8) to a 2×2-block-diagonal matrix guise, the most important properties of the transformation T being closely linked to its generator Ω_u via

$$\begin{aligned} \mathsf{T} &= e^{-\Omega_u} \\ \beta\Omega_u &= -\Omega_u\beta \\ \Omega_u^{\dagger} &= -\Omega_u \\ \mathsf{T}^{\dagger}\mathsf{T} &= 1 = \mathsf{T}\mathsf{T}^{\dagger} \\ \Lambda^{(D)} &= \mathsf{T}^{\dagger}\,\beta\,\mathsf{T} \end{aligned} \tag{1.63}$$

With help of these relations it is readily confirmed that the unitary transformed Dirac Hamiltonian,

$$\mathsf{H}^{(NW)} = \mathsf{T}\,\mathsf{H}^{(D)}\mathsf{T}^{\dagger}, \tag{1.64}$$

obeys to the identity

$$\frac{\mathsf{H}^{(NW)}}{\sqrt{\mathsf{H}^{(NW)} \circ \mathsf{H}^{(NW)}}} = \beta \tag{1.65}$$

In [35] the operator $\mathsf{H}^{(NW)}$ (for good reason) is referred to as the Newton-Wigner (NW) Hamiltonian. Consequently, the unitary transformed Dirac eigenfunctions, see (1.41), as defined by

$$\begin{aligned} \phi_{\nu}^{(NW)}(\mathbf{r};k) &= \sum_{\nu'} \mathsf{T}_{\nu,\nu'}\phi_{\nu'}(\mathbf{r};k) \\ U_{\nu}^{(NW)}(\mathbf{r};k) &= \sum_{\nu'} \mathsf{T}_{\nu,\nu'}U_{\nu'}(\mathbf{r};k) \\ V_{\nu}^{(NW)}(\mathbf{r};k) &= \sum_{\nu'} \mathsf{T}_{\nu,\nu'}V_{\nu'}(\mathbf{r};k), \end{aligned} \tag{1.66}$$

are concurrent eigenfunctions of $\hat{\mathsf{H}}^{(NW)}$ *and* of the Dirac β matrix:

$$\sum_{\nu'} \hat{\mathsf{H}}_{\nu,\nu'}^{(NW)}\phi_{\nu'}^{(NW)}(\mathbf{r};k) = E_k\phi_{\nu}^{(NW)}(\mathbf{r};k) \tag{1.67}$$

$$\sum_{\nu'} (\beta)_{\nu,\nu'}\phi_{\nu'}^{(NW)}(\mathbf{r};k) = z_k\phi_{\nu}^{(NW)}(\mathbf{r};k) \tag{1.68}$$

This distinguishing mark immediately engenders for the eigenfunctions $U_{\nu}^{(NW)}(\mathbf{r};k)$ and $V_{\nu}^{(NW)}(\mathbf{r};k)$ of $\mathsf{H}^{(NW)}$ the expedient virtue of *energy separation*. To elucidate that important concept let us apply the Eriksen transformation T to a Dirac field operator (1.12), thus producing with the decompositions (1.18), (1.41) a field operator

$$\begin{aligned}\hat{\Phi}_{\nu}(\mathbf{r}) &\equiv \sum_{\nu'} \mathsf{T}_{\nu,\nu'} \hat{\Psi}_{\nu'}(\mathbf{r}) \\ &= \sum_{\nu'} \mathsf{T}_{\nu,\nu'} \sum_{k} \phi_{\nu'}(\mathbf{r}; k)\, \hat{f}_k \\ &= \sum_{k} \phi_{\nu}^{(NW)}(\mathbf{r}; k)\, \hat{f}_k\end{aligned} \tag{1.69}$$

$$\begin{aligned}&\equiv \sum_{k} \phi_{\nu}^{(NW)}(\mathbf{r}; k)\Big(\Theta_H(E_k)\hat{c}_k + \Theta_H(-E_k)\hat{b}_k^{\dagger}\Big) \\ &= \sum_{k} \left(\frac{1+z_k}{2} U_{\nu}^{(NW)}(\mathbf{r}; k)\hat{c}_k + \frac{1-z_k}{2} V_{\nu}^{(NW)}(\mathbf{r}; k)\hat{b}_k^{\dagger}\right) \\ &= \sum_{k\in I_+}\sum_{\nu'} \Big(\frac{\mathbb{1}_{4\times 4}+\beta}{2}\Big)_{\nu,\nu'} U_{\nu'}^{(NW)}(\mathbf{r}; k)\hat{c}_k + \sum_{k\in I_-}\sum_{\nu'} \Big(\frac{\mathbb{1}_{4\times 4}-\beta}{2}\Big)_{\nu,\nu'} V_{\nu'}^{(NW)}(\mathbf{r}; k)\hat{b}_k^{\dagger}\end{aligned}$$

Agreeing on the 4×4-matrices β, α_b and σ_b be given in Dirac representation, i.e.

$$\beta = \begin{pmatrix} \mathbb{1}_{2\times 2}, & \mathbb{0}_{2\times 2} \\ \mathbb{0}_{2\times 2}, & -\mathbb{1}_{2\times 2} \end{pmatrix}, \quad \alpha_b = \begin{pmatrix} \mathbb{0}_{2\times 2}, & \sigma_b^{(P)} \\ \sigma_b^{(P)}, & \mathbb{0}_{2\times 2} \end{pmatrix}, \quad \sigma_b = \begin{pmatrix} \sigma_b^{(P)}, & \mathbb{0}_{2\times 2} \\ \mathbb{0}_{2\times 2}, & \sigma_b^{(P)} \end{pmatrix} \tag{1.70}$$

then (1.69) is equivalent to

$$\hat{\Phi}_{\nu}(\mathbf{r}) = \sum_{k\in I_+} \begin{pmatrix} U_1^{(NW)}(\mathbf{r}, k) \\ U_2^{(NW)}(\mathbf{r}, k) \\ 0 \\ 0 \end{pmatrix}_{\nu} \hat{c}_k + \sum_{k\in I_-} \begin{pmatrix} 0 \\ 0 \\ V_3^{(NW)}(\mathbf{r}; k) \\ V_4^{(NW)}(\mathbf{r}; k) \end{pmatrix}_{\nu} \hat{b}_k^{\dagger} \equiv \begin{pmatrix} \hat{\psi}_1(\mathbf{r}) \\ \hat{\psi}_2(\mathbf{r}) \\ \hat{\chi}_1^{\dagger}(\mathbf{r}) \\ \hat{\chi}_2^{\dagger}(\mathbf{r}) \end{pmatrix}_{\nu} \tag{1.71}$$

The eigenfunctions (1.41) of the Dirac Hamiltonian (1.6) spanning a complete orthonormal basis this property deems valid as well for the eigenfunctions (1.66) of the Newton-Wigner Hamiltonian (1.64). So, the eigenfunctions $U_{\mu}^{(NW)}(\mathbf{r}, k)$ with $k \in I_+$ and $\mu \in \{1, 2\}$ span a complete orthonormal basis in the *matter* section of the QED Fock space, whereas the eigenfunctions $V_{\mu}^{(NW)}(\mathbf{r}, k)$ with $k \in I_-$ and $\mu \in \{3, 4\}$ span independently a complete orthonormal basis in the corresponding *antimatter* section, always assuming external electric source fields being weak, of course.

Thus the following linear combinations of the fermionic matter and antimatter creation and annihilation operators, as itentified from (1.71), namely

$$\hat{\psi}_1(\mathbf{r}) = \sum_{k\in I_+} U_1^{(NW)}(\mathbf{r}, k)\, \hat{c}_k \tag{1.72}$$

$$\hat{\psi}_2(\mathbf{r}) = \sum_{k \in I_+} U_2^{(NW)}(\mathbf{r}, k)\, \hat{c}_k$$
$$\hat{\chi}_1^\dagger(\mathbf{r}) = \sum_{k \in I_-} V_3^{(NW)}(\mathbf{r}; k)\, \hat{b}_k^\dagger \tag{1.73}$$
$$\hat{\chi}_2^\dagger(\mathbf{r}) = \sum_{k \in I_-} V_4^{(NW)}(\mathbf{r}; k)\, \hat{b}_k^\dagger,$$

in fact comprise along with their hermitean adjoints *genuine* fermion field operators, whereby these field operators comprehensively describe the quantum fields of electrons and positrons as *separated* entities. Notably, consistent with the definition (1.73) the *two*-component positron field operators $\hat{\chi}_\mu(\mathbf{r})$ are determined by the *complex conjugated* (!) eigenmodes $V_{2+\mu}^{\star(NW)}(\mathbf{r}, k)$ of the unitary transformed Dirac Hamiltonian (1.64), a feature strongly linked to charge conjugation and time reversal.

Based on the completeness and orthonormality of the respective eigenfunctions of $\hat{\mathsf{H}}^{(NW)}$ in the matter respectively antimatter sector of the QED-Fock space there holds indeed for $\mu, \mu' \in \{1, 2\}$

$$\left\{\hat{\psi}_\mu(\mathbf{r}), \hat{\psi}_{\mu'}(\mathbf{r}')\right\} = \hat{0} = \left\{\hat{\psi}_\mu^\dagger(\mathbf{r}), \hat{\psi}_{\mu'}^\dagger(\mathbf{r}')\right\} \tag{1.74}$$
$$\left\{\hat{\psi}_\mu(\mathbf{r}), \hat{\psi}_{\mu'}^\dagger(\mathbf{r}')\right\} = \delta_{\mu,\mu'}\delta^{(3)}(\mathbf{r}-\mathbf{r}')\,\hat{1}$$
$$\left\{\hat{\chi}_{\mu'}(\mathbf{r}'), \hat{\chi}_\mu(\mathbf{r})\right\} = \hat{0} = \left\{\hat{\chi}_{\mu'}^\dagger(\mathbf{r}'), \hat{\chi}_\mu^\dagger(\mathbf{r})\right\}$$
$$\left\{\hat{\chi}_{\mu'}(\mathbf{r}'), \hat{\chi}_\mu^\dagger(\mathbf{r})\right\} = \delta_{\mu,\mu'}\delta^{(3)}(\mathbf{r}-\mathbf{r}')\,\hat{1},$$

in addition all electron field operators anti-commuting by construction with all positron field operators.

In contrast to the indices $\nu, \nu' \in \{1, 2, 3, 4\}$ labeling the *four-component* Dirac quantum fields (1.12), here the indices $\mu, \mu' \in \{1, 2\}$ designate *two-component* fermion quantum fields suitable for the *individual* description of matter (electrons) and antimatter (positrons). In general $\mu, \mu' \in \{1, 2\}$ cannot be identified with the quantum labels $\pm\frac{\hbar}{2}$ of the spin operator $\hat{S}_z = \frac{\hbar}{2}\sigma_z$, but of course in the special case of a *constant* external magnetostatic field a determined linear combination may qualify as an eigenfunction of $\hat{S}_z$.

Next introducing *particle density* operators for electrons (e) and positrons (p) with these two-component fermion quantum fields, given by

$$\hat{n}^{(e)}(\mathbf{r}) = \sum_\mu \hat{\psi}_\mu^\dagger(\mathbf{r})\, \hat{\psi}_\mu(\mathbf{r}) \tag{1.75}$$

$$\hat{n}^{(p)}(\mathbf{r}) = \sum_\mu \hat{\chi}_\mu^\dagger(\mathbf{r})\, \hat{\chi}_\mu(\mathbf{r}), \tag{1.76}$$

one obtains directly from (1.57) in terms of the unitary transformed four-component Dirac field operators (1.71) for the operator of total fermion number $\hat{N}_F$ and the operator $\hat{Q}$ of total fermion charge (carefully respecting the rules for the *normal ordering* operator $\mathcal{N}$ applied to fermions) a validation of the proposed integral representations (1.61) and (1.62):

$$\begin{aligned}\hat{N}_F &= \mathcal{N}\Big(\int d^3r \sum_{\nu,\nu'} \hat{\Psi}_\nu^\dagger(\mathbf{r})\Lambda_{\nu,\nu'}^{(D)}\hat{\Psi}_{\nu'}(\mathbf{r})\Big) = \mathcal{N}\Big(\int d^3r \sum_{\nu,\nu'} \hat{\Psi}_\nu^\dagger(\mathbf{r})(\mathsf{T}^\dagger\beta\mathsf{T})_{\nu,\nu'}\hat{\Psi}_{\nu'}(\mathbf{r})\Big) \\ &= \mathcal{N}\Big(\int d^3r \sum_{\nu,\nu'} \hat{\Phi}_\nu^\dagger(\mathbf{r})(\beta)_{\nu,\nu'}\hat{\Phi}_{\nu'}(\mathbf{r})\Big) \\ &= \int d^3r \mathcal{N}\Big(\hat{\psi}_1^\dagger(\mathbf{r})\hat{\psi}_1(\mathbf{r}) + \hat{\psi}_2^\dagger(\mathbf{r})\hat{\psi}_2(\mathbf{r}) - \hat{\chi}_2(\mathbf{r})\hat{\chi}_2^\dagger(\mathbf{r}) - \hat{\chi}_1(\mathbf{r})\hat{\chi}_1^\dagger(\mathbf{r})\Big) \\ &= \int d^3r \Big(\hat{n}^{(e)}(\mathbf{r}) + \hat{n}^{(p)}(\mathbf{r})\Big), \end{aligned} \tag{1.77}$$

and by the same token

$$\begin{aligned}\hat{Q} &= \mathcal{N}\Big(e\int d^3r \sum_{\nu} \hat{\Psi}_\nu^\dagger(\mathbf{r})\hat{\Psi}_\nu(\mathbf{r})\Big) = \mathcal{N}\Big(e\int d^3r \sum_{\nu,\nu'} \hat{\Psi}_\nu^\dagger(\mathbf{r})(\mathsf{T}^\dagger\mathsf{T})_{\nu,\nu'}\hat{\Psi}_{\nu'}(\mathbf{r})\Big) \\ &= \mathcal{N}\Big(e\int d^3r \sum_{\nu} \hat{\Phi}_\nu^\dagger(\mathbf{r})\hat{\Phi}_\nu(\mathbf{r})\Big) \\ &= e\int d^3r \mathcal{N}\Big(\hat{\psi}_1^\dagger(\mathbf{r})\hat{\psi}_1(\mathbf{r}) + \hat{\psi}_2^\dagger(\mathbf{r})\hat{\psi}_2(\mathbf{r}) + \hat{\chi}_2(\mathbf{r})\hat{\chi}_2^\dagger(\mathbf{r}) + \hat{\chi}_1(\mathbf{r})\hat{\chi}_1^\dagger(\mathbf{r})\Big) \\ &= e\int d^3r \Big(\hat{n}^{(e)}(\mathbf{r}) - \hat{n}^{(p)}(\mathbf{r})\Big) \end{aligned} \tag{1.78}$$

Based on the insight imparted by these representations of the operators $\hat{N}_F$ and $\hat{Q}$ as integrals over local density distributions $\hat{n}^{(e)}(\mathbf{r})$ and $\hat{n}^{(p)}(\mathbf{r})$, allocatable to electrons (e) and positrons (p) separately, see (1.61) and (1.62) together with (1.75) and (1.76), our strategy to derive from the relativistic quantum field theory Hamiltonian $\hat{\mathcal{H}}_{QED}$ stated in (1.1) it's nonrelativistic limit coinciding with the Bethe-Salpeter Hamiltonian $\hat{\mathcal{H}}_{BS}$, see for instance [8, 12], can be seen as a three step procedure.

The first step aims at finding a field theory Hamiltonian $\hat{\mathcal{H}}_U$ being unitary equivalent to $\hat{\mathcal{H}}_{QED}$, so that by construction $\hat{\mathcal{H}}_U$ commutes with the operator of total fermion number $\hat{N}_F$, concurrently also eliminating emission and absorption processes with the high energy photons.

The second step consists in decoding $\hat{\mathcal{H}}_U$ back to first quantization. This step requires a normal ordering procedure, thus creating besides the electron-electron, electron-positron, positron-positron interaction terms additional energy shifts for the Dirac fermions, induced in the presence of external electromagnetic source fields by the QED-interactions (1.4) and (1.5). These energy shift terms comprise in the Hamiltonian (1.2) the usual renormalization of mass, and in the presence of external electromagnetic source fields the renormalization of charge, the Lamb-shift and the anomalous magnetic moment of Schwinger. As an additional result of the normal ordering procedure a c-number spectral shift term is obtained adding to the Dirac-fermions vacuum energy in (1.21), that cannot be ignored, as it acquires a physical meaning in the context of vacuum polarization insofar it depends on the strength of external source fields.

The third step disentangles with help of the energy separating Eriksen transformation thereafter all the matrix elements originally built with Dirac eigenfunctions, see for instance (1.37) and (1.34), eventually into contributions allocatable to electrons and positrons separately. In marked contrast to derivations given in [8, 12] the above described method enables an entirely symmetrical treatment of electrons and positrons.

Proceeding in this manner a physical picture of a plurality of point particles with attributes mass, charge and spin is obtained , these particles interacting pairwise, and concurrently, interacting as well with photons (light). Restricting the particles to move at a speed much smaller than light, eventually the Hamiltonian of non relativistic many-body Schrödinger-Pauli quantum mechanics emerges.

1.5 The Flow Equation Method

The flow equation method, originally developed by Wegner for problems in nonrelativistic many body physics [15, 36], pertains to a continuous unitary transformation of a given Hamiltonian $\hat{H}$ via

$$\begin{aligned} \hat{H}(s) &= \hat{U}(s) \circ \hat{H} \circ \hat{U}^{\dagger}(s) \\ \frac{d}{ds}\hat{U}(s) &= \hat{\eta}(s) \circ \hat{U}(s) \\ \hat{U}(0) &= \hat{1}_{4\times 4} \\ \hat{\eta}^{\dagger}(s) &= -\hat{\eta}(s) \end{aligned} \tag{1.79}$$

Given an antisymmetric generator $\hat{\eta}(s)$ for the flow, the unitary transformation of an hermitean operator $\hat{H}$ is determined solving an initial value problem

$$\begin{aligned} \frac{d}{ds}\hat{H}(s) &= \left[\hat{\eta}(s), \hat{H}(s)\right] \\ \hat{H}(0) &= \hat{H} \end{aligned} \tag{1.80}$$

Of course, the crux of the method is the choice of the generator $\hat{\eta}(s)$, with the essential point being, that $\hat{\eta}(s)$ controls the physical properties of the limiting value $\hat{H}(\infty) = \lim_{s\to\infty} \hat{H}(s)$.

For a comprehensive review discussing various generators with a multitude of applications of the flow equation method see [37]. A detailed discussion of the flow equation method for constructing an energy-separating Hamiltonian unitary equivalent to the minimally coupled Dirac Hamiltonian (1.6) can be found in [35].

Guided by Wegner's insight [38] let us consider, in order to find a suitable generator $\eta(s)$ of the wanted flow, the functional

$$\Phi(s) \equiv \frac{1}{2}\mathrm{tr}\left(\left[\hat{N}_F, \hat{H}(s)\right] \circ \left[\hat{H}(s), \hat{N}_F\right]\right) + \frac{1}{2}\mathrm{tr}\left(\left[\hat{N}_P, \hat{H}(s)\right] \circ \left[\hat{H}(s), \hat{N}_P\right]\right), \tag{1.81}$$

that we specify for our purpose to depend on the hermitian operator $\hat{N}_F$ introduced afore in (1.57) counting the number of occupied Dirac fermion modes,

$$\hat{N}_F = \sum_k z_k \mathcal{N}\left(f_k^\dagger f_k\right) \tag{1.82}$$

$$z_k = \begin{cases} 1, & \text{if } E_k = |E_k| \\ -1, & \text{if } E_k = -|E_k| \end{cases}$$

Concurrently the functional depends on the hermitian operator $\hat{N}_P$ counting the number of high energy photon modes with wavelength *smaller* than for instance a Bohr radius a_B

$$\hat{N}_P = \sum_q \kappa_q a_q^\dagger a_q \tag{1.83}$$

$$\kappa_q = \begin{cases} 1, & \text{if } |\mathbf{q}| > \frac{2\pi}{a_B} \\ 0, & \text{if } |\mathbf{q}| \leq \frac{2\pi}{a_B} \end{cases}$$

Here q denotes a multi-index $q = (\mathbf{q}, \nu)$ specifying wave vectors $\mathbf{q}$ and indices $\nu \in \{I, II\}$ to discriminate the polarization of photons. Notably, the interaction with long wave length photons provoking for instance the emission and absorption of visible light by electrons bound to atoms, remain. This is because the generator of the flow eliminates only the particle number violating interaction processes with the high energy photons.

Then, provided $\hat{H}(s)$ obeys to the flow equation

$$\frac{d}{ds}\hat{H}(s) = \left[\hat{\eta}^{(F,P)}(s), \hat{H}(s)\right] \tag{1.84}$$

$$\hat{H}(s=0) = \frac{1}{mc^2}\left(\hat{\mathcal{H}}_D + \delta\hat{\mathcal{H}}_D + \hat{\mathcal{H}}_{rad} + \hat{\mathcal{H}}_\perp + \hat{\mathcal{V}}_C\right)$$

with the specific generator of the flow chosen as

$$\hat{\eta}^{(F,P)}(s) = \left[\hat{H}(s), \left(\left[\hat{N}_F, \left[\hat{N}_F, \hat{H}(s)\right]\right] + \left[\hat{N}_P, \left[\hat{N}_P, \hat{H}(s)\right]\right]\right)\right], \tag{1.85}$$

there holds by construction

$$\Phi(s) \geq 0 \tag{1.86}$$

$$\frac{d}{ds}\Phi(s) = -\mathrm{tr}\left(\hat{\eta}^{(F,P)}(s) \circ \left(\hat{\eta}^{(F,P)}(s)\right)^{\dagger}\right) \leq 0 \tag{1.87}$$

Together (1.86) and (1.87) imply at once

$$\lim_{s\to\infty} \hat{\eta}^{(F,P)}(s) = \hat{0} \tag{1.88}$$

Evidently, the feature (1.88) is compatibel with the limiting value $\lim_{s\to\infty} \hat{H}(s)$ of that flow *conserving* all at once total fermion number *and* total number of high energy photons.

$$\left[\hat{N}_F, \hat{H}(\infty)\right] = \hat{0} = \left[\hat{N}_P, \hat{H}(\infty)\right] \tag{1.89}$$

The effectiveness of Wegner's Hamiltonian flow equation method [38] is indeed unparalleled. No other method enables with comparable easiness to attain a particle number conserving Hamiltonian $\hat{H}(s \to \infty)$ being unitary equivalent to the original field theory Hamiltonian $\hat{H}(s = 0)$.

Notably, the proposed flow equation method (1.84) with the specific generator (1.85) in effect yields a unitary equivalent Hamiltonian for electrons and/or positrons as particles, where the relevant eigenvalues of $\hat{H}(\infty)$ eventually concentrate into well separated *clusters* discriminated by the numbers N_P of high energy photons. In that respect the proposed flow equation method is a close relative to the effective Hamiltonian method conceived by Avan et al. in [39–41].

1.6 Perturbative Solution of Hamiltonian Flow

As the exact solution to the Hamiltonian flow (1.84) is not available, we present in the ensuing a perturbative solution of that flow

$$\hat{H}(s) = \sum_{j=1}^{\infty} \hat{H}^{(j)}(s), \tag{1.90}$$

progressing in different orders $\hat{H}^{(j)}(s)$ of smallness in the coupling e, directly related to the dimensionless fine structure parameter $\alpha_{FS} = \frac{e^2}{4\pi\epsilon_0\hbar c} \simeq \frac{1}{137}$.

Insertion of the ansatz (1.90) into (1.84) and (1.85) leads to a set of coupled ordinary differential equations determining the operators $\hat{H}^{(j)}(s)$ recursively for $j = 1, 2, \ldots$, once the zero order $\hat{H}_0(s)$ term was available

$$\frac{d\hat{H}^{(j)}(s)}{ds} = \sum_{\substack{j',j''=0 \\ j'+j''\leq j}}^{\infty} \left[C^{(F,P)}\left(\hat{H}^{(j')}(s), \hat{H}^{(j'')}(s)\right), \hat{H}^{(j-j'-j'')}(s)\right] \tag{1.91}$$

Here the symbol (F=Fermion)

$$C^{(F,P)}(\hat{X},\hat{Y}) \equiv \left[\hat{X},\left[\hat{N}_F,\left[\hat{N}_F,\hat{Y}\right]\right]\right]+\left[\hat{X},\left[\hat{N}_P,\left[\hat{N}_P,\hat{Y}\right]\right]\right] \tag{1.92}$$

shortens in what follows the writing (a lot). For instance, according to (1.85) then

$$\hat{\eta}^{(F,P)}(s) = C^{(F,P)}\left(\hat{H}(s),\hat{H}(s)\right) \tag{1.93}$$

Note the bilinearity property. Given a complex number z and hermitean operators $\hat{X}_1,\hat{X}_2,\hat{Y}_1,\hat{Y}_2$ then

$$\begin{aligned} C^{(F,P)}\left(z\,\hat{X}_1,\hat{Y}_1\right) &= z\,C^{(F,P)}\left(\hat{X}_1,\hat{Y}_1\right) = C^{(F,P)}\left(\hat{X}_1,z\,\hat{Y}_1\right) \\ C^{(F,P)}\left(\hat{X}_1+\hat{X}_2,\hat{Y}_1\right) &= C^{(F,P)}\left(\hat{X}_1,\hat{Y}_1\right)+C^{(F,P)}\left(\hat{X}_2,\hat{Y}_1\right) \\ C^{(F,P)}\left(\hat{X}_1,\hat{Y}_1+\hat{Y}_2\right) &= C^{(F,P)}\left(\hat{X}_1,\hat{Y}_1\right)+C^{(F,P)}\left(\hat{X}_1,\hat{Y}_2\right) \end{aligned} \tag{1.94}$$

1.6.1 Initial Data of Hamiltonian Flow

For the envisaged perturbation theory approach it is vital to pay attention to the different orders of smallness in the coupling e upon insertion of the decompositions (1.42) of charge density and current density into the interaction terms (1.4) and (1.5). In view of an arrangement according to the coupling strength e then a *consistent* choice of the initial data at $s=0$ for the flow (1.84) reads

$$\begin{aligned} \hat{H}^{(0)}(0) &= \frac{1}{mc^2}\left(\hat{\mathcal{H}}_D+\hat{\mathcal{H}}_{\mathrm{rad}}\right) \\ \hat{H}^{(1)}(0) &= \frac{1}{mc^2}\hat{\mathcal{H}}_\perp^{(1)} \\ \hat{H}^{(2)}(0) &= \frac{1}{mc^2}\left(\hat{\mathcal{V}}_C^{(2)}+\delta\hat{\mathcal{H}}_D^{(2)}+\hat{\mathcal{H}}_\perp^{(2)}\right) \\ \hat{H}^{(3)}(0) &= \frac{1}{mc^2}\left(\hat{\mathcal{V}}_C^{(3)}+\delta\hat{\mathcal{H}}_D^{(3)}\right) \\ \hat{H}^{(4)}(0) &= \frac{1}{mc^2}\hat{\mathcal{V}}_C^{(4)} \\ \hat{H}^{(j)}(0) &= 0 \text{ if } j \geq 5 \end{aligned} \tag{1.95}$$

So, the representation of charge density and current density according to (1.39) and (1.40) induces upon insertion into the original interaction terms (1.4) and (1.5) for the light-matter interaction term (1.33) of QED the decomposition

$$\hat{\mathcal{H}}_\perp = \hat{\mathcal{H}}_\perp^{(1)}+\hat{\mathcal{H}}_\perp^{(2)}$$

$$\hat{\mathcal{H}}_{\perp}^{(1)} = -ec\frac{1}{\sqrt{V_P}}\sum_{\mathbf{q}}\sum_{k,k'}\langle\phi_k|\,\alpha_b e^{iq_a\mathsf{x}_a}\,|\phi_{k'}\rangle\,\mathcal{N}\left(\hat{f}_k^\dagger\hat{f}_{k'}\right)\left(\mathsf{A}_{b,\mathbf{q}}^{(+1)}+\mathsf{A}_{b,\mathbf{q}}^{(-1)}\right)$$

$$\hat{\mathcal{H}}_{\perp}^{(2)} = \frac{ec}{2}\frac{1}{\sqrt{V_P}}\sum_{\mathbf{q}}\sum_{k}\langle\phi_k|\,\alpha_b e^{iq_a\mathsf{x}_a}z_k\,|\phi_k\rangle\left(\mathsf{A}_{b,\mathbf{q}}^{(+1)}+\mathsf{A}_{b,\mathbf{q}}^{(-1)}\right) \tag{1.96}$$

Here we used the identities

$$\frac{1}{2}\left[\hat{f}_k^\dagger,\hat{f}_{k'}\right] = \mathcal{N}\left(f_k^\dagger f_{k'}\right) - \frac{z_k}{2}\delta_{k,k'} \tag{1.97}$$

and

$$\mathcal{N}\left(\hat{f}_k^\dagger\hat{f}_{k'}\right) = \hat{f}_k^\dagger\hat{f}_{k'} - \frac{1-z_k}{2}\delta_{k,k'}\hat{1} \tag{1.98}$$

with z_k as introduced afore in (1.20).

Likewise one obtains for the Coulomb interaction term (1.36) the decomposition

$$\hat{\mathcal{V}}_C = \hat{\mathcal{V}}_C^{(2)} + \hat{\mathcal{V}}_C^{(3)} + \hat{\mathcal{V}}_C^{(4)}$$

$$\hat{\mathcal{V}}_C^{(2)} = \frac{e^2}{2\varepsilon_0}\frac{1}{V_P}\sum_{\mathbf{q}}\frac{1}{|\mathbf{q}|^2}\sum_{k,k'}\sum_{K,K'}\langle\phi_k|\,e^{-iq_a\mathsf{x}_a}\,|\phi_{k'}\rangle\langle\phi_K|\,e^{iq_a\mathsf{x}_a}\,|\phi_{K'}\rangle\mathcal{N}\left(\hat{f}_k^\dagger f_{k'}\right)\mathcal{N}\left(\hat{f}_K^\dagger f_{K'}\right)$$

$$\hat{\mathcal{V}}_C^{(3)} = -\frac{e^2}{2\varepsilon_0}\frac{1}{V_P}\sum_{\mathbf{q}}\frac{1}{|\mathbf{q}|^2}\sum_{k}\langle\phi_k|\,e^{-iq_a\mathsf{x}_a}z_k\,|\phi_k\rangle\sum_{K,K'}\langle\phi_K|\,e^{iq_a\mathsf{x}_a}\,|\phi_{K'}\rangle\mathcal{N}\left(\hat{f}_K^\dagger f_{K'}\right)$$

$$\hat{\mathcal{V}}_C^{(4)} = \frac{e^2}{2\varepsilon_0}\frac{1}{V_P}\sum_{\mathbf{q}}\frac{1}{|\mathbf{q}|^2}\sum_{k}\langle\phi_k|\,e^{-iq_a\mathsf{x}_a}z_k\,|\phi_k\rangle\sum_{K}\langle\phi_K|\,e^{iq_a\mathsf{x}_a}z_K\,|\phi_K\rangle\cdot\hat{1} \tag{1.99}$$

The reason for such an arrangement according to size being that the *diagonal* sums, $\sum_k\langle\phi_k|\,e^{-iq_a\mathsf{x}_a}z_k\,|\phi_k\rangle$ and $\sum_k\langle\phi_k|\,\alpha_b e^{-iq_a\mathsf{x}_a}z_k\,|\phi_k\rangle$, actually themselves add up in the presence of external source fields to a result proportional to the coupling e.

Correspondingly, the correction term dealing with the renormalization of the free Dirac fermions in (1.9) decomposes into a second order and a third order contribution

$$\delta\hat{\mathcal{H}}_D = \delta\hat{\mathcal{H}}_D^{(2)} + \delta\hat{\mathcal{H}}_D^{(3)}$$

$$\delta\hat{\mathcal{H}}_D^{(2)} = \delta m\,c^2\sum_{k,k'}\langle\phi_k|\,\beta\,|\phi_{k'}\rangle\,\mathcal{N}\left(\hat{f}_k^\dagger f_{k'}\right)$$

$$\delta\hat{\mathcal{H}}_D^{(3)} = \delta e\sum_{k,k'}\langle\phi_k|\left(\Phi^{(ext)}(\mathsf{x})\,1_{4\times4} - c\alpha_b\mathsf{A}_b^{(ext)}(\mathsf{x})\right)|\phi_{k'}\rangle\,\mathcal{N}\left(\hat{f}_k^\dagger f_{k'}\right) \tag{1.100}$$

1.6.2 The Zero Order Term Term $\hat{H}^{(0)}(s)$

The ordinary differential equation (ode) determining the zero order term $\hat{H}_0(s)$ is nonlinear

$$\frac{d}{ds}\hat{H}^{(0)}(s) = \left[C^{(F,P)}\left(\hat{H}^{(0)}(s), \hat{H}^{0)}(s)\right), \hat{H}^{(0)}(s)\right] \tag{1.101}$$

But because

$$\left[\hat{N}_F, \hat{H}^{(0)}(0)\right] = \hat{0} = \left[\hat{N}_P, \hat{H}^{(0)}(0)\right] \tag{1.102}$$

then for $j = 0, 1, 2, ...$

$$C^{(F)}\left(\hat{H}^{(j)}(s), \hat{H}^{0)}(s)\right) \equiv \hat{0} \tag{1.103}$$

Hence the only solution to (1.101) conforming with the posed initial data (1.95) is

$$\hat{H}^{(0)}(s) = \hat{H}^{(0)} = \frac{1}{mc^2}\left(\hat{\mathcal{H}}_D + \hat{\mathcal{H}}_{\text{rad}}\right) \equiv \hat{H}_D + \hat{H}_{\text{rad}} \tag{1.104}$$

whereby with *scaled* variables $\tilde{E}_k = \frac{E_k}{mc^2}$, $\tilde{\omega}_q = \frac{\hbar\omega_{\mathbf{q}}}{mc^2}$, and a multi-index $q \equiv (\mathbf{q}, \nu)$ we write

$$\hat{H}_D = \sum_k \tilde{E}_k \hat{f}_k^\dagger \hat{f}_k \tag{1.105}$$

$$\hat{H}_{\text{rad}} = \sum_q \tilde{\omega}_q \left(\hat{a}_q^\dagger \hat{a}_q + \frac{1}{2}\right)$$

With $\hat{H}^{(0)}(s)$ being determined now all the higher order operators $\hat{H}^{(j)}(s)$ for $j = 1, 2, ...$ in the perturbation series (1.90) follow recursively solving *linear* ode's only, with the corresponding initial values given in (1.95).

1.6.3 The First Order Term $\hat{H}^{(1)}(s)$

Setting $j = 1$ in the recursion (1.91) then with (1.103) one needs to solve the initial value problem

$$\frac{d\hat{H}^{(1)}(s)}{ds} = -\left[\hat{H}^{(0)}, C^{(F,P)}\left(\hat{H}^{(0)}, \hat{H}^{(1)}(s)\right)\right] \tag{1.106}$$

$$\hat{H}^{(1)}(0) = \frac{1}{mc^2}\hat{\mathcal{H}}_\perp^{(1)}$$

Notably there holds

$$
\begin{aligned}
\left[\hat{N}_F, \mathcal{N}\left(\hat{f}_k^\dagger \hat{f}_{k'}\right)\right] &= (z_k - z_{k'})\, \mathcal{N}\left(\hat{f}_k^\dagger \hat{f}_{k'}\right) \\
\left[\hat{H}_D, \mathcal{N}\left(\hat{f}_k^\dagger \hat{f}_{k'}\right)\right] &= \left(\tilde{E}_k - \tilde{E}_{k'}\right) \mathcal{N}\left(\hat{f}_k^\dagger \hat{f}_{k'}\right) \\
\left[\hat{N}_P, \hat{\mathsf{A}}^{(\sigma)}_{b,\mathbf{q}}\right] &= -\sigma \kappa_q \hat{\mathsf{A}}^{(\sigma)}_{b,\mathbf{q}} \\
\left[\hat{H}_{\text{rad}}, \hat{\mathsf{A}}^{(\sigma)}_{b,\mathbf{q}}\right] &= -\sigma \tilde{\omega}_q \hat{\mathsf{A}}^{(\sigma)}_{b,\mathbf{q}} \\
\left[\hat{H}^{(0)}, C^{(F)}(\hat{H}^{(0)}, \mathcal{N}\left(\hat{f}_k^\dagger \hat{f}_{k'}\right) \hat{\mathsf{A}}^{(\sigma)}_{b,\mathbf{q}})\right] &= \left((z_k - z_{k'})^2 + \kappa_q^2\right) \left(\tilde{E}_k - \tilde{E}_{k'} - \sigma \tilde{\omega}_q\right)^2 \mathcal{N}\left(\hat{f}_k^\dagger \hat{f}_{k'}\right) \hat{\mathsf{A}}^{(\sigma)}_{b,\mathbf{q}}
\end{aligned}
\tag{1.107}
$$

Conforming with the posed initial value at $s = 0$, see (1.95), (1.96) these commutators engender the first order perturbation operator $\hat{H}^{(1)}(s)$ solving (1.106) is

$$
\begin{aligned}
\hat{H}^{(1)}(s) = (-\frac{e}{mc}) \frac{1}{\sqrt{V_P}} \sum_{\mathbf{q},\sigma} \sum_{k,k'} \langle \phi_k | \, \alpha_b e^{i q_a \mathrm{x}_a} \, | \phi_{k'} \rangle \, \mathcal{N}\left(\hat{f}_k^\dagger \hat{f}_{k'}\right) \hat{\mathsf{A}}^{(\sigma)}_{b,\mathbf{q}} \times \\
\times \exp\left[-s \left((z_k - z_{k'})^2 + \kappa_q^2\right) \left(\tilde{E}_k - \tilde{E}_{k'} - \sigma \tilde{\omega}_q\right)^2\right]
\end{aligned}
\tag{1.108}
$$

Indeed, this assertion is readily confirmed calculating the derivative $\frac{d\hat{H}^{(1)}(s)}{ds}$ and comparing with the commutator on the right hand side of (1.106).

Taking the limit $s \to \infty$ then only terms with $z_k = z_{k'}$ *and* $\kappa_q = 0$ can contribute in (1.108). Even though there exists yet another contribution from high energy modes constrained by $\tilde{E}_k - \tilde{E}_{k'} - \sigma \tilde{\omega}_q = 0$, this particular contribution is subdominant as it has an extra factor $\frac{1}{V_P}$ without summation over a corresponding wave number and thus may be safely ignored for $V_P \to \infty$. So there holds

$$
\begin{aligned}
\hat{H}^{(1)}(\infty) &= \lim_{s \to \infty} \hat{H}^{(1)}(s) = \hat{H}^{(1,0)} + O(\frac{1}{V_P}) \\
\hat{H}^{(1,0)} &= -\frac{e}{mc} \frac{1}{\sqrt{V_P}} \sum_{\mathbf{q}} (1 - \kappa_q) \sum_{k,k'} \frac{1 + z_k z_{k'}}{2} \langle \phi_k | \, \alpha_b e^{i q_a \mathrm{x}_a} \, | \phi_{k'} \rangle \, \mathcal{N}\left(\hat{f}_k^\dagger \hat{f}_{k'}\right) \left(\hat{\mathsf{A}}^{(+1)}_{b,\mathbf{q}} + \hat{\mathsf{A}}^{(-1)}_{b,\mathbf{q}}\right)
\end{aligned}
\tag{1.109}
$$

Note the limiting value $\hat{H}^{(1,0)}$ is void of any pair creation or pair annihilation processes, as there holds by construction

$$
\left[\hat{N}_F, \hat{H}^{(1,0)}\right] = \hat{0} \tag{1.110}
$$

By this means $\hat{H}^{(1,0)}$ governs the usual particle number conserving interaction of charge carriers with the quanta of electromagnetic radiation. More specifically, in view of the factor $1 - \kappa_q$ constraining the wavenumbers of photons in (1.109) it is about emission and/or absorption of light up to the energy scale of hard X-rays.

1.6.4 The Second Order Term $\hat{H}^{(2)}(s)$

Setting $j = 2$ in the recursion (1.91) then with (1.103) one obtains after elementary rearrangements

$$\frac{d\hat{H}^{(2)}(s)}{ds} = \begin{pmatrix} \left[C^{(F,P)}\left(\hat{H}^{(0)}, \hat{H}^{(2)}(s)\right), \hat{H}^{(0)}\right] + \\ +\left[C^{(F,P)}\left(\hat{H}^{(0)}, \hat{H}^{(1)}(s)\right), \hat{H}^{(1)}(s)\right] + \\ +\left[C^{(F,P)}\left(\hat{H}^{(1)}(s), \hat{H}^{(1)}(s)\right), \hat{H}^{(0)}\right] \end{pmatrix}, \tag{1.111}$$

with the inhomogeneous term, i.e. the last two lines in (1.111), now being fixed by the previous result (1.108) obtained for $\hat{H}^{(1)}(s)$. Apparently the task to solve the ode (1.111) for the second order perturbation term $\hat{H}^{(2)}(s)$ is more laborious, but it can with help of the superposition principle be split into two separate tasks writing

$$\hat{H}^{(2)}(s) = \hat{H}^{(2,h)}(s) + \hat{H}^{(2,i)}(s), \tag{1.112}$$

whereas $\hat{H}^{(2,h)}(s)$ solves the *homogeneous* initial value problem

$$\begin{aligned} \frac{d\hat{H}^{(2,h)}(s)}{ds} &= -\left[\hat{H}^{(0)}, C^{(F,P)}(\hat{H}^{(0)}, \hat{H}^{(2,h)}(s))\right] \\ \hat{H}^{(2,h)}(s=0) &= \frac{1}{mc^2}\left(\hat{\mathcal{V}}_C^{(2)} + \delta\hat{\mathcal{H}}_D^{(2)} + \hat{\mathcal{H}}_\perp^{(2)}\right), \end{aligned} \tag{1.113}$$

and the term $\hat{H}^{(2,i)}(s)$ designates the *special* solution to the *inhomogeneous* ode (1.111) with *homogeneous* initial data, that is $\hat{H}^{(2,i)}(s=0) = \hat{0}$.

Determination of $\hat{H}^{(2,h)}(s)$

The readily confirmed operator identities

$$\begin{aligned} \left[\hat{H}^{(0)}, C^{(F,P)}(\hat{H}^{(0)}, \hat{\mathsf{A}}_{b,\mathbf{q}}^{(\sigma)})\right] &= \kappa_q^2\tilde{\omega}_q^2\hat{\mathsf{A}}_{b,\mathbf{q}}^{(\sigma)} \\ \left[\hat{H}^{(0)}, C^{(F,P)}(\hat{H}^{(0)}, \mathcal{N}\left(\hat{f}_k^\dagger\hat{f}_{k'}\right))\right] &= (z_k - z_{k'})^2\left(\tilde{E}_k - \tilde{E}_{k'}\right)^2\mathcal{N}\left(\hat{f}_k^\dagger\hat{f}_{k'}\right) \\ \left[\hat{H}^{(0)}, C^{(F,P)}\left(\hat{H}^{(0)}, \mathcal{N}\left(\hat{f}_k^\dagger\hat{f}_{k'}\right)\mathcal{N}\left(\hat{f}_K^\dagger\hat{f}_{K'}\right)\right)\right] &= (z_k - z_{k'} + z_K - z_{K'})^2(\tilde{E}_k - \tilde{E}_{k'} + \tilde{E}_K - \tilde{E}_{K'})^2 \times \\ &\quad \times \mathcal{N}\left(\hat{f}_k^\dagger\hat{f}_{k'}\right)\mathcal{N}\left(\hat{f}_K^\dagger\hat{f}_{K'}\right) \end{aligned} \tag{1.114}$$

suggest the solution of the homogeneous ode (1.113) may be decomposed as

$$\hat{H}^{(2,h)}(s) = \hat{H}^{(2,h,F)}(s) + \hat{H}^{(2,h,P)}(s) \tag{1.115}$$

whereby

$$\hat{H}^{(2,h,P)}(s) = \frac{e}{2mc}\frac{1}{\sqrt{V_P}}\sum_{\mathbf{q}',\sigma'}\sum_{k''}\langle\phi_{k''}|\,\alpha_{b'}e^{i\,q'_a\mathrm{x}_a}z_{k''}\,|\phi_{k''}\rangle\exp\left(-s\kappa_{q'}^2\tilde{\omega}_{q'}^2\right)\hat{\mathsf{A}}_{b',\mathbf{q}'}^{(\sigma')} \tag{1.116}$$

and

$$\begin{aligned}&\hat{H}^{(2,h,F)}(s) \\ &= \frac{1}{mc^2}\left(\begin{array}{c}\frac{e^2}{2\varepsilon_0}\frac{1}{V_P}\sum_{\mathbf{q}}\frac{1}{|\mathbf{q}|^2}\sum_{k,k'}\sum_{K,K'}\langle\phi_k|\,e^{-iq_a\mathrm{x}_a}\,|\phi_{k'}\rangle\langle\phi_K|\,e^{iq_a\mathrm{x}_a}\,|\phi_{K'}\rangle\times \\ \times\exp\left[-s(z_k-z_{k'}+z_K-z_{K'})^2(\tilde{E}_k-\tilde{E}_{k'}+\tilde{E}_K-\tilde{E}_{K'})^2\right]\times\mathcal{N}\left(\hat{f}_k^\dagger\hat{f}_{k'}\right)\mathcal{N}\left(\hat{f}_K^\dagger\hat{f}_{K'}\right) \\ +\sum_{k,k'}\langle\phi_k|\,\delta m\,c^2\beta\,|\phi_{k'}\rangle\exp\left[-s(z_k-z_{k'})^2(\tilde{E}_k-\tilde{E}_{k'})^2\right]\mathcal{N}\left(\hat{f}_k^\dagger\hat{f}_{k'}\right)\end{array}\right)\end{aligned} \tag{1.117}$$

Indeed there holds

$$\hat{H}^{(2,h)}(s=0) = \frac{1}{mc^2}\left(\hat{\mathcal{V}}_C^{(2)} + \delta\hat{\mathcal{H}}_F^{(2)} + \hat{\mathcal{H}}_\perp^{(2)}\right), \tag{1.118}$$

and calculating $\frac{d\hat{H}^{(2,h)}(s)}{ds}$ the right hand side of the homogenous ode (1.113) is reproduced.

The Limiting Value $\hat{H}^{(2,h)}(\infty)$

Apparently there exists according to (1.117) no contribution to $\hat{H}^{(2,h,F)}(s\to\infty)$ *unless* $z_k - z_{k'} + z_K - z_{K'} = 0$, respectively $z_k - z_{k'} = 0$. Note that a possible contribution of terms that originates from a constraint $\tilde{E}_k - \tilde{E}_{k'} + \tilde{E}_K - \tilde{E}_{K'} = 0$, can be ignored for $V_P\to\infty$, see the reasoning leading to (1.109).

Regarding the normal ordering of fermion operators (1.18) there holds as an identity

$$\mathcal{N}\left(\hat{f}_k^\dagger\hat{f}_{k'}\right)\mathcal{N}\left(\hat{f}_K^\dagger\hat{f}_{K'}\right) = \begin{cases}\mathcal{N}\left(\hat{f}_k^\dagger\hat{f}_K^\dagger\hat{f}_{K'}\hat{f}_{k'}\right) \\ +\frac{1+z_K}{2}\delta_{k',K}\mathcal{N}\left(\hat{f}_k^\dagger\hat{f}_{K'}\right) - \frac{1-z_k}{2}\delta_{k,K'}\mathcal{N}\left(\hat{f}_K^\dagger\hat{f}_{k'}\right) \\ +\frac{1-z_k}{2}\frac{1+z_K}{2}\delta_{k',K}\delta_{k,K'}\hat{1}\end{cases} \tag{1.119}$$

By inserting (1.119) into (1.117) eventually

$$\hat{H}^{(2,h)}(\infty) = \lim_{s\to\infty}\hat{H}^{(2,h)}(s) \tag{1.120}$$

$$= \frac{1}{mc^2}\left\{\begin{array}{l} \frac{e^2}{2\varepsilon_0}\frac{1}{V_P}\sum_{\mathbf{q}}\frac{1}{|\mathbf{q}|^2}\sum_{k,k',K,K'\in I}\langle\phi_k|\,e^{-iq_a\mathrm{x}_a}\,|\phi_{k'}\rangle\,\langle\phi_K|\,e^{iq_a\mathrm{x}_a}\,|\phi_{K'}\rangle\times \\ \times\delta_{z_k-z_{k'}+z_K-z_{K'},0}\left(\begin{array}{c}\mathcal{N}\left(\hat{f}_k^\dagger\hat{f}_K^\dagger\hat{f}_{K'}\hat{f}_{k'}\right) \\ +\delta_{k',K}\frac{1+z_K}{2}\mathcal{N}\left(\hat{f}_k^\dagger\hat{f}_{K'}\right)-\delta_{k,K'}\frac{1-z_k}{2}\mathcal{N}\left(\hat{f}_K^\dagger\hat{f}_{k'}\right) \\ +\frac{1-z_k}{2}\frac{1+z_K}{2}\delta_{k,K'}\delta_{k',K}\hat{1}\end{array}\right) \\ +\sum_{k,k'\in I}\frac{1+z_kz_{k'}}{2}\langle\phi_k|\,\delta m\,c^2\beta\,|\phi_{k'}\rangle\,\mathcal{N}\left(\hat{f}_k^\dagger\hat{f}_{k'}\right) \\ +\frac{ec}{2}\frac{1}{\sqrt{V_P}}\sum_{\mathbf{q}}\left(1-\kappa_q\right)\sum_{k\in I}\langle\phi_k|\,\alpha_b e^{iq_a\mathrm{x}_a}z_k\,|\phi_k\rangle\left(\hat{\mathsf{A}}_{b,\mathbf{q}}^{(+1)}+\hat{\mathsf{A}}_{b,\mathbf{q}}^{(-1)}\right)\end{array}\right.$$

Determination of $\hat{H}^{(2,i)}(s)$

The inhomogenous term in the ode (1.111) reads

$$I^{(2)}(s)\equiv\left[C^{(F,P)}\left(\hat{H}^{(0)},\hat{H}^{(1)}(s)\right),\hat{H}^{(1)}(s)\right]+\left[C^{(F,P)}\left(\hat{H}^{(1)}(s),\hat{H}^{(1)}(s)\right),\hat{H}^{(0)}\right] \tag{1.121}$$

So, in order to find $\hat{H}^{(2,i)}(s)$ we thus need to solve the initial value problem

$$\begin{aligned}\frac{d\hat{H}^{(2,i)}(s)}{ds}&=-\left[\hat{H}^{(0)},C^{(F)}\left(\hat{H}^{(0)},\hat{H}^{(2)}(s)\right)\right]+I^{(2)}(s) \\ \hat{H}^{(2,i)}(s)&=0\end{aligned} \tag{1.122}$$

Taking advantage of the features (1.94) of the symbol $C^{(F,P)}\left(\hat{X},\hat{Y}\right)$ introduced in (1.92) leads upon insertion of (1.104) and (1.108) straightforwardly to the following (lengthy though) result

$$\begin{aligned}\hat{I}^{(2)}(s)=&\left(\frac{e}{mc}\right)^2\sum_{k,k'\in I}\sum_{K,K'\in I}\frac{1}{V_P}\sum_{\mathbf{q},\mathbf{q}'}\sum_{\sigma,\sigma'}\langle\phi_k|\,\alpha_b e^{iq_a\mathrm{x}_a}\,|\phi_{k'}\rangle\,\langle\phi_K|\,\alpha_{b'}e^{iq'_a\mathrm{x}_a}\,|\phi_{K'}\rangle\times \\ &\times\exp\left[-s\left(\begin{array}{c}\left(\kappa_q^2+(z_k-z_{k'})^2\right)\cdot\left(\tilde{E}_k-\tilde{E}_{k'}-\sigma\tilde{\omega}_q\right)^2+ \\ +\left((\kappa_{q'}^2+(z_K-z_{K'})^2)\right)\cdot\left(\tilde{E}_K-\tilde{E}_{K'}-\sigma'\tilde{\omega}_{q'}\right)^2\end{array}\right)\right]\times \\ &\times\left(\kappa_q^2+(z_k-z_{k'})^2\right)\left(\left(\tilde{E}_K-\tilde{E}_{K'}-\sigma'\tilde{\omega}_{q'}\right)+2\left(\tilde{E}_k-\tilde{E}_{k'}-\sigma\tilde{\omega}_q\right)\right)\times \\ &\times\left[\left(\mathcal{N}\left(\hat{f}_k^\dagger\hat{f}_{k'}\right)\mathsf{A}_{b,\mathbf{q}}^{(\sigma)}\right),\left(\mathcal{N}\left(\hat{f}_K^\dagger\hat{f}_{K'}\right)\mathsf{A}_{b',\mathbf{q}'}^{(\sigma')}\right)\right]\end{aligned} \tag{1.123}$$

Next, using the basic commutators (1.107) the validity of the operator identity

$$\left[\hat{H}^{(0)}, C^{(F,P)}\left(\hat{H}^{(0)}, \left(\hat{f}_k^\dagger \hat{f}_{k'} \mathsf{A}_{b,\mathbf{q}}^{(\sigma)} \hat{f}_K^\dagger \hat{f}_{K'} \mathsf{A}_{b',\mathbf{q}'}^{(\sigma')}\right)\right)\right] \tag{1.124}$$
$$= \left((z_k - z_{k'} + z_K - z_{K'})^2 + \left(-\sigma\kappa_q - \sigma'\kappa_{q'}\right)^2\right) \times$$
$$\times \left(\tilde{E}_k - \tilde{E}_{k'} + \tilde{E}_K - \tilde{E}_{K'} - \sigma\tilde{\omega}_q - \sigma'\tilde{\omega}_{q'}\right)^2 \left(\hat{f}_k^\dagger \hat{f}_{k'} \mathsf{A}_{b,\mathbf{q}}^{(\sigma)} \hat{f}_K^\dagger \hat{f}_{K'} \mathsf{A}_{b',\mathbf{q}'}^{(\sigma')}\right)$$

is confirmed without difficulty. Also, the coefficients in (1.124) remain invariant under a permutation of the sequence of fermion and/or photon operators.

This way, the seeked for special solution to the *inhomogeneous* ode (1.111) may be represented on the basis of Duhamel's principle eventually as

$$\hat{H}^{(2,i)}(s) = \left(\frac{e}{mc}\right)^2 \sum_{k,k'\in I} \sum_{K,K'\in I} \frac{1}{V_P} \sum_{\mathbf{q},\mathbf{q}'} \sum_{\sigma,\sigma'} \langle\phi_k|\,\alpha_b e^{iq_a x_a}\,|\phi_{k'}\rangle \langle\phi_K|\,\alpha_{b'} e^{iq'_a x_a}\,|\phi_{K'}\rangle \times$$
$$\times \int_0^s ds' \exp\left[-(s-s')\left(\begin{array}{c}\left(\left(-\sigma\kappa_q - \sigma'\kappa_{q'}\right)^2 + (z_k - z_{k'} + z_K - z_{K'})^2\right)\times \\ \times\left(\tilde{E}_k - \tilde{E}_{k'} + \tilde{E}_K - \tilde{E}_{K'} - \sigma\tilde{\omega}_q - \sigma'\tilde{\omega}_{q'}\right)^2\end{array}\right)\right] \times$$
$$\times \exp\left[-s'\left(\begin{array}{c}\left(\kappa_q^2 + (z_k - z_{k'})^2\right)\cdot\left(\tilde{E}_k - \tilde{E}_{k'} - \sigma\tilde{\omega}_q\right)^2 + \\ +\left(\kappa_{q'}^2 + (z_K - z_{K'})^2\right)\cdot\left(\tilde{E}_K - \tilde{E}_{K'} - \sigma'\tilde{\omega}_{q'}\right)^2\end{array}\right)\right] \times$$
$$\times \left(\kappa_q^2 + (z_k - z_{k'})^2\right)\left(\left(\tilde{E}_K - \tilde{E}_{K'} - \sigma'\tilde{\omega}_{q'}\right) + 2\left(\tilde{E}_k - \tilde{E}_{k'} - \sigma\tilde{\omega}_q\right)\right) \times$$
$$\times \left[\left(\mathcal{N}\left(\hat{f}_k^\dagger \hat{f}_{k'}\right)\mathsf{A}_{b,\mathbf{q}}^{(\sigma)}\right), \left(\mathcal{N}\left(\hat{f}_K^\dagger \hat{f}_{K'}\right)\mathsf{A}_{b',\mathbf{q}'}^{(\sigma')}\right)\right] \tag{1.125}$$

Indeed there holds $\hat{H}^{(2,i)}(s=0) = \hat{0}$ and calculating the derivative $\frac{d\hat{H}^{(2,i)}(s)}{ds}$ reproduces the right hand side of the inhomogeneous ode (1.111).

The Limiting Value $\hat{H}^{(2,i)}(\infty)$

With help of the integral

$$X \geq 0, Y > 0$$
$$\lim_{s\to\infty} \int_0^s ds' e^{-(s-s')X} e^{-s'Y} = -\lim_{s\to\infty} \frac{e^{-sX}}{X-Y} = \frac{1}{Y}\delta_{X,0} \tag{1.126}$$

there follows for $s \to \infty$ with similar reasoning as in (1.120) the limiting value

$$\hat{H}^{(2,i)}(\infty) = \left(\frac{e}{mc}\right)^2 \sum_{k,k'} \sum_{K,K'} \frac{1}{V_P} \sum_{\mathbf{q},\mathbf{q}'} \kappa_q \kappa_{q'} \sum_{\sigma,\sigma'} \langle \phi_k | \alpha_b e^{iq_a x_a} | \phi_{k'} \rangle \langle \phi_K | \alpha_{b'} e^{iq'_a x_a} | \phi_{K'} \rangle \times$$
$$\times \frac{\left(\tilde{E}_K - \tilde{E}_{K'} - \sigma' \tilde{\omega}_{q'}\right) + 2\left(\tilde{E}_k - \tilde{E}_{k'} - \sigma \tilde{\omega}_q\right)}{\left(\tilde{E}_k - \tilde{E}_{k'} - \sigma \tilde{\omega}_q\right)^2 + \left(\tilde{E}_K - \tilde{E}_{K'} - \sigma' \tilde{\omega}_{q'}\right)^2} \cdot \delta_{\sigma\kappa_q + \sigma'\kappa_{q'},0} \cdot \delta_{z_k - z_{k'} + z_K - z_{K'},0} \times$$
$$\times \left[\left(\mathcal{N}\left(\hat{f}_k^\dagger \hat{f}_{k'}\right) \mathsf{A}_{b,\mathbf{q}}^{(\sigma)}\right), \left(\mathcal{N}\left(\hat{f}_K^\dagger \hat{f}_{K'}\right) \mathsf{A}_{b',\mathbf{q}'}^{(\sigma')}\right)\right] \tag{1.127}$$

Here we restricted the photon wave vectors to the region $\kappa_q = 1$ as the contribution of long wavelength photons arising out of the region $\kappa_q = 0$ in (1.127) is negligible.

Under an index transformation

$$\begin{aligned} &\left\{\left(k, k', b, q, \sigma\right), \left(K, K', b', q', \sigma'\right)\right\} \\ \to &\left\{\left(K, K', b', q', \sigma'\right), \left(k, k', b, q, \sigma\right)\right\} \end{aligned} \tag{1.128}$$

the commutator in (1.127) changes sign. This observation thus enables to reshape $\hat{H}^{(2,i)}(\infty)$ in a symmetrized guise

$$\hat{H}^{(2,i)}(\infty) = \left(\frac{e}{mc}\right)^2 \sum_{k,k'} \sum_{K,K'} \frac{1}{V_P} \sum_{\mathbf{q},\mathbf{q}'} \kappa_q \kappa_{q'} \langle \phi_k | \alpha_b e^{iq_a x_a} | \phi_{k'} \rangle \langle \phi_K | \alpha_{b'} e^{iq'_a x_a} | \phi_{K'} \rangle \times$$
$$\times \frac{1}{2} \sum_{\sigma,\sigma'} \delta_{\sigma+\sigma',0} \cdot \delta_{z_k - z_{k'} + z_K - z_{K'},0} \cdot \frac{\left(\tilde{E}_k - \tilde{E}_{k'} - \sigma \tilde{\omega}_q\right) - \left(\tilde{E}_K - \tilde{E}_{K'} - \sigma' \tilde{\omega}_{q'}\right)}{\left(\tilde{E}_k - \tilde{E}_{k'} - \sigma \tilde{\omega}_q\right)^2 + \left(\tilde{E}_K - \tilde{E}_{K'} - \sigma' \tilde{\omega}_{q'}\right)^2} \times$$
$$\times \left[\left(\mathcal{N}\left(\hat{f}_k^\dagger \hat{f}_{k'}\right) \mathsf{A}_{b,\mathbf{q}}^{(\sigma)}\right), \left(\mathcal{N}\left(\hat{f}_K^\dagger \hat{f}_{K'}\right) \mathsf{A}_{b',\mathbf{q}'}^{(\sigma')}\right)\right] \tag{1.129}$$

$$= \left(\frac{e}{mc}\right)^2 \sum_{k,k'} \sum_{K,K'} \frac{1}{V_P} \sum_{\mathbf{q},\mathbf{q}'} \kappa_q \kappa_{q'} \langle \phi_k | \alpha_b e^{iq_a x_a} | \phi_{k'} \rangle \langle \phi_K | \alpha_{b'} e^{iq'_a x_a} | \phi_{K'} \rangle \times$$
$$\times \delta_{z_k - z_{k'} + z_K - z_{K'},0} \cdot \frac{\left(\tilde{E}_k - \tilde{E}_{k'} - \tilde{\omega}_q\right) - \left(\tilde{E}_K - \tilde{E}_{K'} + \tilde{\omega}_{q'}\right)}{\left(\tilde{E}_k - \tilde{E}_{k'} - \tilde{\omega}_q\right)^2 + \left(\tilde{E}_K - \tilde{E}_{K'} + \tilde{\omega}_{q'}\right)^2} \times$$
$$\times \left[\left(\mathcal{N}\left(\hat{f}_k^\dagger \hat{f}_{k'}\right) \mathsf{A}_{b,\mathbf{q}}^{(+)}\right), \left(\mathcal{N}\left(\hat{f}_K^\dagger \hat{f}_{K'}\right) \mathsf{A}_{b',\mathbf{q}'}^{(-)}\right)\right] \tag{1.130}$$

Exploting the summation with regard to the photon labels $\sigma, \sigma' \in \{+, -\}$, also once more using the symmetry under (1.128) and thereafter rearranging

$$\left[\left(\mathcal{N}\left(\hat{f}_k^\dagger \hat{f}_{k'}\right) \mathsf{A}_{b,\mathbf{q}}^{(+)}\right), \left(\mathcal{N}\left(\hat{f}_K^\dagger \hat{f}_{K'}\right) \mathsf{A}_{b',\mathbf{q}'}^{(-)}\right)\right] = \left\{ \begin{aligned} &\left[\mathcal{N}\left(\hat{f}_k^\dagger \hat{f}_{k'}\right), \mathcal{N}\left(\hat{f}_K^\dagger \hat{f}_{K'}\right)\right] \mathsf{A}_{b',\mathbf{q}'}^{(-)} \mathsf{A}_{b,\mathbf{q}}^{(+)} \\ &+ \mathcal{N}\left(\hat{f}_k^\dagger \hat{f}_{k'}\right) \mathcal{N}\left(\hat{f}_K^\dagger \hat{f}_{K'}\right) \left[\mathsf{A}_{b,\mathbf{q}}^{(+)}, \mathsf{A}_{b',\mathbf{q}'}^{(-)}\right], \end{aligned} \right. \tag{1.131}$$

there follows with help of (1.119) for the searched for limiting value (1.130) eventually

$$
\begin{aligned}
\hat{H}^{(2,i)}(\infty) &= \left(\frac{e}{mc}\right)^2 \sum_{k,k'} \sum_{K,K'} \frac{1}{V_P} \sum_{\mathbf{q},\mathbf{q}'} \kappa_q \kappa_{q'} \langle \phi_k | \alpha_b e^{i q_a x_a} | \phi_{k'} \rangle \langle \phi_K | \alpha_{b'} e^{i q'_a x_a} | \phi_{K'} \rangle \times \\
&\times \delta_{z_k - z_{k'} + z_K - z_{K'},0} \frac{\left(\tilde{E}_k - \tilde{E}_{k'} - \tilde{\omega}_q\right) - \left(\tilde{E}_K - \tilde{E}_{K'} + \tilde{\omega}_{q'}\right)}{\left(\tilde{E}_k - \tilde{E}_{k'} - \tilde{\omega}_q\right)^2 + \left(\tilde{E}_K - \tilde{E}_{K'} + \tilde{\omega}_{q'}\right)^2} \times \\
&\times \left(\begin{array}{l} \left(\delta_{k',K} \mathcal{N}\left(\hat{f}_k^\dagger \hat{f}_{K'}\right) - \delta_{k,K'} \mathcal{N}\left(\hat{f}_K^\dagger \hat{f}_{k'}\right) + \delta_{k,K'} \delta_{k',K} \frac{z_K - z_k}{2}\right) \hat{\mathsf{A}}^{(-)}_{b',\mathbf{q}'} \hat{\mathsf{A}}^{(+)}_{b,\mathbf{q}} + \\
+ \left[\hat{\mathsf{A}}^{(+)}_{b,\mathbf{q}} \,,\, \hat{\mathsf{A}}^{(-)}_{b',\mathbf{q}'}\right] \left\{ \begin{array}{l} \mathcal{N}\left(\hat{f}_k^\dagger \hat{f}_K^\dagger \hat{f}_{K'} \hat{f}_{k'}\right) \\ + \delta_{k',K} \frac{1+z_K}{2} \mathcal{N}\left(\hat{f}_k^\dagger \hat{f}_{K'}\right) - \delta_{k,K'} \frac{1-z_k}{2} \mathcal{N}\left(\hat{f}_K^\dagger \hat{f}_{k'}\right) \\ + \delta_{k,K'} \delta_{k',K} \frac{1-z_k}{2} \frac{1+z_K}{2} \end{array} \right. \end{array} \right)
\end{aligned} \tag{1.132}
$$

By construction both terms, (1.120) and (1.132), conserve fermion number and photon number. So, the limiting value of the second order term

$$
\hat{H}^{(2)}(\infty) = \hat{H}^{(2,h)}(\infty) + \hat{H}^{(2,i)}(\infty) \tag{1.133}
$$

indeed governs the effective particle-particle interactions to order e^2, and by implication the QED-interaction induced energy changes (self-energy) of the Dirac particles in the presence of *external* source fields as well.

Notably the effective interaction terms in Eq. (1.132) are distinctly less singular than expected in view of results obtainable from second-order perturbation theory [41]. This feature is in full agreement with early findings of Lenz and Wegner [36], who discovered with their flow equation approach, as compared to the original unitary transformation method of Fröhlich [42], a better result for the effective electron-electron interaction in superconductors.

As there exist altogether *six* possibilities to choose the energy signs in (1.120) and in (1.132) conforming with the constraint $z_k - z_{k'} + z_K - z_{K'} = 0$, see table (1.1), the derived expressions for the limiting value $\hat{H}^{(2,0)}(\infty)$ indeed require a detailed inspection to capture the essence of the involved physics. The concomitant discussions we postpone to Sect. 1.7.

1.6.5 The Third Order Term $\hat{H}^{(3)}(s)$

Setting $j = 3$ in the recursion (1.91) and minding the identity (1.103), there remain altogether six commutators to be analyzed

Table 1.1 Allowed energy sign combinations

$(z_k, z_{k'}, z_K, z_{K'})$	$z_k - z_{k'} + z_K - z_{K'} = 0$	$\hat{f}_k^\dagger \hat{f}_K^\dagger \hat{f}_{K'} \hat{f}_{k'}$	$\mathcal{N}\left(\hat{f}_k^\dagger \hat{f}_K^\dagger \hat{f}_{K'} \hat{f}_{k'}\right)$
$(1, 1, 1, 1)$	1-1+1-1=0	$\hat{c}_k^\dagger \hat{c}_K^\dagger \hat{c}_{K'} \hat{c}_{k'}$	$\hat{c}_k^\dagger \hat{c}_K^\dagger \hat{c}_{K'} \hat{c}_{k'}$
$(-1, -1, -1, -1)$	-1+1-1+1=0	$\hat{b}_k \hat{b}_K \hat{b}_{K'}^\dagger \hat{b}_{k'}^\dagger$	$\hat{b}_{K'}^\dagger \hat{b}_{k'}^\dagger \hat{b}_k \hat{b}_K$
$(1, 1, -1, -1)$	1-1-1+1=0	$\hat{c}_k^\dagger \hat{b}_K \hat{b}_{K'}^\dagger \hat{c}_{k'}$	$-\hat{c}_k^\dagger \hat{c}_{k'} \hat{b}_{K'}^\dagger \hat{b}_K$
$(-1, -1, 1, 1)$	-1+1+1-1=0	$\hat{b}_k \hat{c}_K^\dagger \hat{c}_{K'} \hat{b}_{k'}^\dagger$	$-\hat{c}_K^\dagger \hat{c}_{K'} \hat{b}_{k'}^\dagger \hat{b}_k$
$(1, -1, -1, 1)$	1+1-1-1=0	$\hat{c}_k^\dagger \hat{b}_K \hat{c}_{K'} \hat{b}_{k'}^\dagger$	$\hat{c}_k^\dagger \hat{c}_{K'} \hat{b}_{k'}^\dagger \hat{b}_K$
$(-1, 1, 1, -1)$	-1-1+1+1=0	$\hat{b}_k \hat{c}_K^\dagger \hat{b}_{K'}^\dagger \hat{c}_{k'}$	$\hat{c}_K^\dagger \hat{c}_{k'} \hat{b}_{K'}^\dagger \hat{b}_k$

$$\frac{d\hat{H}^{(3)}(s)}{ds} = \begin{cases} \left[C^{(F,P)}\left(\hat{H}^{(0)}, \hat{H}^{(3)}(s)\right), \hat{H}^{(0)}\right] \\ + \left[C^{(F,P)}\left(\hat{H}^{(0)}, \hat{H}^{(1)}(s)\right), \hat{H}^{(2)}(s)\right] \\ + \left[C^{(F,P)}\left(\hat{H}^{(0)}, \hat{H}^{(2)}(s)\right), \hat{H}^{(1)}(s)\right] \\ + \left[C^{(F,P)}\left(\hat{H}^{(1)}(s), \hat{H}^{(2)}(s)\right), \hat{H}^{(0)}\right] \\ + \left[C^{(F,P)}\left(\hat{H}^{(2)}(s), \hat{H}^{(1)}(s)\right), \hat{H}^{(0)}\right] \\ + \left[C^{(F,P)}\left(\hat{H}^{(1)}(s), \hat{H}^{(1)}(s)\right), \hat{H}^{(1)}(s)\right] \end{cases} \tag{1.134}$$

Here the last five lines in (1.134) comprise in total the *inhomogeneous* term, its constituent parts being composed of the previously obtained results for $\hat{H}^{(1)}(s)$ and $\hat{H}^{(2)}(s)$. As before, the superposition principle enables to represent the searched for solution $\hat{H}^{(3)}(s)$ in the guise

$$\hat{H}^{(3)}(s) = \hat{H}^{(3,h)}(s) + \hat{H}^{(3,i)}(s) \tag{1.135}$$

whereby $\hat{H}^{(3,h)}(s)$ solves the homogeneous ode corresponding to inhomogeneous initial data

$$\begin{aligned} \frac{d\hat{H}^{(3,h)}(s)}{ds} &= -\left[\hat{H}^{(0)}, C^{(F,P)}\left(\hat{H}^{(0)}, \hat{H}^{(3)}(s)\right)\right] \\ \hat{H}^{(3,h)}(0) &= \frac{1}{mc^2}\left(\hat{\mathcal{V}}_C^{(3)} + \delta\hat{\mathcal{H}}_D^{(3)}\right) \end{aligned} \tag{1.136}$$

and $\hat{H}^{(3,i)}(s)$ solves the inhomogeneous ode (1.134) corresponding to the *homogeneous* initial data $\hat{H}^{(3,i)}(0) = \hat{0}$ compliant with the arrangements (1.95) and (1.99).

The Homogeneous Solution $\hat{H}^{(3,h)}(s)$

Proceeding in full analogy to the construction of $\hat{H}^{(2,h)}$ in (1.115) the solution to the homogeneous ode (1.136) is found in the guise

$$\hat{H}^{(3,h)}(s) = \frac{1}{mc^2} \begin{cases} -\frac{e^2}{2\varepsilon_0}\frac{1}{V_P}\sum_{\mathbf{q}}\frac{1}{|\mathbf{q}|^2}\sum_{k''}\langle\phi_{k''}|\, e^{-iq_a \mathsf{x}_a} z_{k''}\, |\phi_{k''}\rangle \sum_{k,k'}\langle\phi_k|\, e^{iq_a\mathsf{x}_a}\, |\phi_{k'}\rangle \times \\ \times \exp\left[-s(z_k - z_{k'})^2(\tilde{E}_k - \tilde{E}_{k'})^2\right] \mathcal{N}\left(\hat{f}_k^\dagger \hat{f}_{k'}\right) \\ \\ +\delta e \sum_{k,k'}\langle\phi_k|\left(\Phi^{(ext)}(\mathsf{x})\, 1_{4\times 4} - c\alpha_b \mathsf{A}_b^{(ext)}(\mathsf{x})\right)|\phi_{k'}\rangle \times \\ \times \exp\left[-s(z_k - z_{k'})^2(\tilde{E}_k - \tilde{E}_{k'})^2\right] \mathcal{N}\left(\hat{f}_k^\dagger \hat{f}_{k'}\right) \end{cases} \tag{1.137}$$

Evidently $\hat{H}^{(3,h)}(s=0)$ certifies the posed initial data and calculating $\frac{d\hat{H}^{(3,h)}(s)}{ds}$ the right hand side of the homogenous ode (1.136) is reproduced.

The Limiting Value $\hat{H}^{(3,h)}(\infty)$

In the limit $s \to \infty$ only terms with $z_k = z_{k'}$ contribute to the operator $\hat{H}^{(3,h)}(s)$ in (1.137), thus obtaining

$$\hat{H}^{(3,h)}(\infty) = \frac{1}{mc^2}\left(\sum_{k,k'\in I_+} + \sum_{k,k'\in I_-}\right) \times \tag{1.138}$$

$$\times \begin{pmatrix} -\frac{e^2}{2\varepsilon_0}\frac{1}{V_P}\sum_{\mathbf{q}}\frac{1}{|\mathbf{q}|^2}\sum_{k''}\langle\phi_{k''}|\, e^{-iq_a\mathsf{x}_a} z_{k''}\, |\phi_{k''}\rangle\langle\phi_k|\, e^{iq_a\mathsf{x}_a}\, |\phi_{k'}\rangle + \\ + \delta e \cdot \langle\phi_k|\left(\Phi^{(ext)}(\mathsf{x})\, 1_{4\times 4} - c\alpha_b\mathsf{A}_b^{(ext)}(\mathsf{x})\right)|\phi_{k'}\rangle \end{pmatrix} \mathcal{N}\left(\hat{f}_k^\dagger \hat{f}_{k'}\right)$$

Apparently, the external potential term $\Phi^{(ext)}(\mathsf{x})$ in the second line of (1.138) can be grouped together with the first line there, thus providing an equation to determine the charge shift parameter δe. Yet, as the external vector potential term $\mathsf{A}_b^{(ext)}(\mathsf{x})$ is standing all alone in (1.138), now the next thing to do is to inspect the *inhomogeneous* solution $\hat{H}^{(3,i)}(s)$ for a corresponding match.

The Inhomogeneous Solution $\hat{H}^{(3,i)}(s)$

To facilitate our analysis it proves beneficial to employ the superposition principle and to insert $\hat{H}^{(2)}(s) = \hat{H}^{(2,h,P)}(s) + \hat{H}^{(2,h,F}(s) + \hat{H}^{(2,i)}(s)$ into the inhomogeneous term of the ode (1.134), the details listed in (1.116), (1.117) and (1.125). Making use of the bi-linearity property (1.94) of the symbol $C^{(F)}$ then a corresponding decomposition of the solution $\hat{H}^{(3,i)}(s)$ of the inhomogeneous ode (1.134) is found

$$\hat{H}^{(3,i)}(s) = \hat{H}^{(3,i)}_{1,1,1}(s) + \hat{H}^{(3,i)}_{2,i}(s) + \hat{H}^{(3,i)}_{2,h,F}(s) + \hat{H}^{(3,i)}_{2,h,P}(s)\,, \tag{1.139}$$

whereby the individual terms in (1.139) solve the associated inhomogeneous ode's conforming with the choice of homogeneous initial data, i.e.

$$\frac{d\hat{H}^{(3,i)}_{1,1,1}(s)}{ds} = \begin{cases} -\left[\hat{H}^{(0)}(s), C^{(F,P)}\left(\hat{H}^{(0)}(s), \hat{H}^{(3,i)}_{1,1,1}(s)\right)\right] \\ +\left[C^{(F,P)}\left(\hat{H}^{(1)}(s), \hat{H}^{(1)}(s)\right), \hat{H}^{(1)}(s)\right] \end{cases} \tag{1.140}$$

and choosing a label $L \in \{(2, h, P), (2, h, F), (2, i)\}$ the other terms in (1.139) solve

$$\frac{d\hat{H}^{(3,i)}_{L}(s)}{ds} = \begin{cases} -\left[\hat{H}^{(0)}(s), C^{(F,P)}\left(\hat{H}^{(0)}(s), \hat{H}^{(3,i)}_{L}(s)\right)\right] \\ +\left[C^{(F,P)}\left(\hat{H}^{(0)}(s), \hat{H}^{(1)}(s)\right), \hat{H}^{(L)}(s)\right] + \left[C^{(F,P)}\left(\hat{H}^{(0)}(s), \hat{H}^{(L)}(s)\right), \hat{H}^{(1)}(s)\right] \\ +\left[C^{(F,P)}\left(\hat{H}^{(1)}(s), \hat{H}^{(L)}(s)\right), \hat{H}^{(0)}(s)\right] + \left[C^{(F,P)}\left(\hat{H}^{(L)}(s), \hat{H}^{(1)}(s)\right), \hat{H}^{(0)}(s)\right] \end{cases} \tag{1.141}$$

A survey of the inhomogeneous terms in (1.140) and (1.141) reveals the operators $\hat{H}^{(3,i)}_{1,1,1}(s)$, $\hat{H}^{(3,i)}_{2,i}(s)$ and $\hat{H}^{(3,i)}_{2,h,F}(s)$ describe (highly complex) interactions between *three* particles and what's more, photon stimulated interactions between two particles etc., that for now we don't want to follow at in detail, because we are here not interested in applying the theory to systems with a *high* particle density, where such terms certainly take on greater significance as building blocks for the construction of the fourth order term $\hat{H}^{(4)}(s)$.

Actually, the less complex patterned operator $\hat{H}^{(3,i)}_{2,h,P}(s)$ is the only one term in (1.139) conserving particle number and thus is the only one term of third order with a finite limiting value for $s \to \infty$. In addition, because $\hat{H}^{(3,i)}_{2,h,P}(s)$ manifestly depends on the Fourier transform of the polarization current (1.45), this term sticks out being capable of neutralizing the external vector potential term in (1.138).

With the basic rules (1.107) a straightforward evaluation of the four commutators, representing in total the inhomogeneous terms in (1.141) for $L = (2, h, P)$ gives on the basis of Duhamel's principle the searched for solution of (1.141) in the guise

$$
\hat{H}^{(3,i)}_{2,h,P}(s) = -\left(\frac{e}{mc}\right)^2 \frac{1}{2V_P} \sum_{\mathbf{q},\mathbf{q}'} \sum_{k,k'} \sum_{\sigma,\sigma'} \begin{pmatrix} \left(\kappa_q^2 + (z_k - z_{k'})^2\right)\left(\tilde{E}_k - \tilde{E}_{k'} - \sigma\tilde{\omega}_q\right) + \kappa_{q'}^2 \sigma' \tilde{\omega}_{q'} + \\ -\kappa_{q'}^2 \left(\tilde{E}_k - \tilde{E}_{k'}\right) + \left(\kappa_q^2 + (z_k - z_{k'})^2\right)\left(\tilde{E}_k - \tilde{E}_{k'}\right) \end{pmatrix} \times \tag{1.142}
$$
$$
\times \int_0^s ds' \exp\left[-(s-s')\left(\tilde{E}_k - \tilde{E}_{k'}\right)^2 (z_k - z_{k'})^2\right] \exp\left[-s'\left(\kappa_q^2 + (z_k - z_{k'})^2\right)(\tilde{E}_k - \tilde{E}_{k'} - \sigma\tilde{\omega}_q)^2 + \kappa_{q'}^2 \tilde{\omega}_{q'}^2\right] \times
$$
$$
\times \sum_{k''} \langle \phi_{k''} | \alpha_{b'} e^{-iq_a \mathsf{x}_a} z_{k''} | \phi_{k''} \rangle \langle \phi_k | \alpha_b e^{iq_a \mathsf{x}_a} | \phi_{k'} \rangle \mathcal{N}\left(\hat{f}_k^\dagger \hat{f}_{k'}\right) \left[\mathsf{A}^{(\sigma)}_{b,\mathbf{q}}, \mathsf{A}^{(\sigma')}_{b',\mathbf{q}'}\right]
$$

The Limiting Value $\hat{H}^{(3,i)}(\infty)$

With help of the integral (1.126) one finds in analogy to (1.127) for the operator $\hat{H}^{(3,i)}_{2,h,P}(s)$ in (1.142), that again only terms with $z_k = z_{k'}$ contribute for $s \to \infty$, thus obtaining

$$
\hat{H}^{(3,i)}_{2,h,P}(\infty) = -\left(\frac{e}{mc}\right)^2 \frac{1}{2V_P} \sum_{\mathbf{q},\mathbf{q}'} \kappa_q \kappa_{q'} \sum_{\sigma,\sigma'} \left[\mathsf{A}^{(\sigma)}_{b,\mathbf{q}}, \mathsf{A}^{(\sigma')}_{b',\mathbf{q}'}\right] \times \tag{1.143}
$$
$$
\times \sum_{k,k'} \delta_{z_k - z_{k'},0} \begin{pmatrix} \kappa_q^2 \left(\tilde{E}_k - \tilde{E}_{k'} - \sigma\tilde{\omega}_q\right) + \kappa_{q'}^2 \sigma' \tilde{\omega}_{q'} + \\ + \left(\kappa_q^2 - \kappa_{q'}^2\right)\left(\tilde{E}_k - \tilde{E}_{k'}\right) \end{pmatrix} \frac{1}{\kappa_q^2 \left(\tilde{E}_k - \tilde{E}_{k'} - \sigma\tilde{\omega}_q\right)^2 + \kappa_{q'}^2 \tilde{\omega}_{q'}^2} \times
$$
$$
\times \sum_{k''} \langle \phi_{k''} | \alpha_{b'} e^{-iq_a \mathsf{x}_a} z_{k''} | \phi_{k''} \rangle \langle \phi_k | \alpha_b e^{iq_a \mathsf{x}_a} | \phi_{k'} \rangle \mathcal{N}\left(\hat{f}_k^\dagger \hat{f}_{k'}\right)
$$

Taking into account (1.32) the evaluation of the photon commutator gives

$$
\left[\mathsf{A}^{(\sigma)}_{b,\mathbf{q}}, \mathsf{A}^{(\sigma')}_{b',\mathbf{q}'}\right] = \sigma \cdot \delta_{\sigma',-\sigma} \cdot \delta_{\mathbf{q}',-\mathbf{q}} \cdot \frac{\hbar}{2\varepsilon_0 \omega(\mathbf{q})} \left(\delta_{b.b'} - \frac{q_b q_{b'}}{|\mathbf{q}|^2}\right) \tag{1.144}
$$

Being interested only in the contributions of the high energy photons $\kappa_q = 1$ then

$$
\hat{H}^{(3,i)}_{2,h,P}(\infty) = -\left(\frac{e}{mc}\right)^2 \sum_{\mathbf{q}} \kappa_q \frac{\hbar}{2\varepsilon_0 \omega(\mathbf{q})} \left(\delta_{b.b'} - \frac{q_b q_{b'}}{|\mathbf{q}|^2}\right) \times \tag{1.145}
$$
$$
\times \frac{1}{2V_P} \left(\sum_{k,k' \in I_+} + \sum_{k,k' \in I_-}\right) \sum_{\sigma \in \{+,-\}} \frac{\sigma\left(\tilde{E}_k - \tilde{E}_{k'}\right) - 2\tilde{\omega}_q}{\left(\tilde{E}_k - \tilde{E}_{k'} - \sigma\tilde{\omega}_q\right)^2 + \tilde{\omega}_q^2} \times
$$
$$
\times \sum_{k''} \langle \phi_{k''} | \alpha_{b'} e^{-iq_a \mathsf{x}_a} z_{k''} | \phi_{k''} \rangle \langle \phi_k | \alpha_b e^{iq_a \mathsf{x}_a} | \phi_{k'} \rangle \mathcal{N}\left(\hat{f}_k^\dagger \hat{f}_{k'}\right)
$$

Eventually minding $\tilde{\omega}_q \equiv \frac{\hbar c |\mathbf{q}|}{mc^2}$ and realizing

$$\sum_{\sigma\in\{+,-\}} \frac{\sigma\left(\tilde{E}_k - \tilde{E}_{k'}\right) - 2\tilde{\omega}_q}{\left(\tilde{E}_k - \tilde{E}_{k'} - \sigma\tilde{\omega}_q\right)^2 + \tilde{\omega}_q^2} = -\frac{2}{\tilde{\omega}_q}\left(1 - \frac{\left(\tilde{E}_k - \tilde{E}_{k'}\right)^4}{4\tilde{\omega}_q^4 + \left(\tilde{E}_k - \tilde{E}_{k'}\right)^4}\right), \tag{1.146}$$

there follows with neglect of small terms $O\left(\frac{\tilde{E}_k - \tilde{E}_{k'}}{\tilde{\omega}_q}\right)^4$ the result

$$\hat{H}^{(3,i)}(\infty) \simeq \hat{H}^{(3,i)}_{2,h,P}(\infty) = \frac{1}{mc^2}\frac{e^2}{2\varepsilon_0}\frac{1}{V_P}\sum_{\mathbf{q}} \kappa_q \frac{1}{|\mathbf{q}|^2}\left(\delta_{b.b'} - \frac{q_b q_{b'}}{|\mathbf{q}|^2}\right)\times \tag{1.147}$$
$$\times \sum_{k''} \langle\phi_{k''}|\,\alpha_{b'} e^{-iq_a x_a} z_{k''}\,|\phi_{k''}\rangle \sum_{k,k'} \frac{1 + z_k z_{k'}}{2}\langle\phi_k|\,\alpha_b e^{iq_a x_a}\,|\phi_{k'}\rangle \mathcal{N}\left(\hat{f}_k^\dagger \hat{f}_{k'}\right)$$

1.7 The Unitary Transformed Hamiltonian of QED

Because we focus in the ensuing on systems with a *low* particle density, assuming the mean distance between particles being (much) greater than a Compton wave length, the contributions of the more complex patterned higher order operators $\hat{H}^{(n)}(s)$ in the expansion (1.90) for $n \geq 4$ can be safely neglected for $s \to \infty$.

With help of Table 1.1, there follow to first order in the expansion parameter e from (1.109) explicit expressions for the respective light-electron and light-positron interaction terms, and to second order the effective two-body interactions together with additional self-energy terms as encoded in the limiting value $\hat{H}^{(2)}(\infty)$. The third order term $\hat{H}^{(3)}(\infty)$ comprises the effects of vacuum polarization on the respective energy shifts of electrons (positrons) in the presence of external electromagnetic fields.

Adding the results of the perturbation theory obtained in the previous Sect. 1.6 the resulting approximation for the unitary transformed QED-Hamiltonian reads

$$\begin{aligned}\hat{\mathcal{H}}_U &= mc^2 \lim_{s\to\infty}\left(\hat{H}^{(0)}(s) + \hat{H}^{(1)}(s) + \hat{H}^{(2)}(s) + \hat{H}^{(3)}(s) + \cdots\right) \\ &= mc^2\left(\hat{H}^{(0)} + \hat{H}^{(1)}(\infty) + \hat{H}^{(2,h)}(\infty) + \hat{H}^{(2,i)}(\infty) + \hat{H}^{(3,h)}(\infty) + \hat{H}^{(3,i)}(\infty) + \cdots\right)\end{aligned} \tag{1.148}$$

We emphasize all the terms in (1.148) apply *without* restricting any particles to move at slow speed compared to light

$$mc^2\left(\hat{H}^{(0)} + \hat{H}^{(1)}(\infty)\right) = \hat{\mathcal{H}}_D + \hat{\mathcal{H}}_{rad} + \underbrace{\hat{\mathcal{H}}^{(e,\text{rad})} + \hat{\mathcal{H}}^{(p,\text{rad})}}_{\text{photon emission/absorption}} \tag{1.149}$$

$$mc^2 \hat{H}^{(2,h)}(\infty) = \underbrace{\hat{\mathcal{U}}_C^{(e,e)} + \hat{\mathcal{U}}_C^{(p,p)} + \hat{\mathcal{U}}_C^{(e,p)}}_{\text{two-body interactions}} + \underbrace{\hat{\mathcal{M}}_C^{(e)} + \hat{\mathcal{M}}_C^{(p)}}_{\text{fermion self-energy}} + \underbrace{\mathcal{K}_C}_{\text{vacuum energy}} \quad (1.150)$$

$$mc^2 \hat{H}^{(2,i)}(\infty) = \begin{cases} \overbrace{\hat{\mathcal{U}}_\perp^{(e,e)} + \hat{\mathcal{U}}_\perp^{(p,p)} + \hat{\mathcal{U}}_\perp^{(e,p)}}^{\text{two-body interactions}} + \overbrace{\hat{\mathcal{M}}_\perp^{(e)} + \hat{\mathcal{M}}_\perp^{(p)}}^{\text{fermion self-energy}} \\ + \underbrace{\hat{\mathcal{V}}_\perp^{(e,ph)} + \hat{\mathcal{V}}_\perp^{(p,ph)}}_{\text{Compton interaction}} + \underbrace{\hat{\Pi}^{(ph)}}_{\text{photon self-energy}} + \underbrace{\mathcal{K}_\perp}_{\text{vacuum energy change}} \end{cases} \quad (1.151)$$

$$mc^2 \hat{H}^{(3)}(\infty) = mc^2 \left(\hat{H}^{(3,h)}(\infty) + \hat{H}^{(3,i)}(\infty) \right) \equiv \underbrace{\hat{\mathcal{M}}_{VP}^{(e)} + \hat{\mathcal{M}}_{VP}^{(p)}}_{\text{fermion self-energy vacuum polarization}} \quad (1.152)$$

1.7.1 The Interaction of Electrons and Positrons with Low Energy Photons

To save writing we introduce abstract bra- and ket-notation, so that the decomposition (1.41) reads now

$$|\phi_k\rangle = \Theta_H(E_k)|U_k\rangle + \Theta_H(-E_k)|V_k\rangle \quad (1.153)$$

The common interaction processes of light with the charge carriers of matter, regarding absorption- and/or emission of *single* photons by electrons (positrons), are now readily obtained from the limiting value (1.109) after bringing the units in as

$$\begin{aligned} \hat{\mathcal{H}}_{\text{low}}^{(e,\text{rad})} &= -\frac{1}{\sqrt{V_P}} \sum_{\mathbf{q}} \left(1 - \kappa_q\right) \sum_{k,k' \in I_+} \langle U_k | c\alpha_b e^{iq_a x_a} | U_{k'} \rangle \, \hat{c}_k^\dagger \hat{c}_{k'} \left(e\hat{\mathsf{A}}_{b,\mathbf{q}}^{(-1)} + e\hat{\mathsf{A}}_{b,\mathbf{q}}^{(+1)} \right) \\ \hat{\mathcal{H}}_{\text{low}}^{(p,\text{rad})} &= +\frac{1}{\sqrt{V_P}} \sum_{\mathbf{q}} \left(1 - \kappa_q\right) \sum_{K,K' \in I_-} \langle V_K | c\alpha_b e^{iq_a x_a} | V_{K'} \rangle \, \hat{b}_{K'}^\dagger \hat{b}_K \left(e\hat{\mathsf{A}}_{b,\mathbf{q}}^{(-1)} + e\hat{\mathsf{A}}_{b,\mathbf{q}}^{(+1)} \right) \end{aligned} \quad (1.154)$$

Here and everywhere else in what follows attention should be paid to the reversed order of quantum labels (K', K) for the positron mode creation- and annihilation-operators $\hat{b}_{K'}^\dagger \hat{b}_K$ as compared to the corresponding order of quantum labels (k, k') for the electron mode creation- and annihilation-operators $\hat{c}_k^\dagger \hat{c}_{k'}$. Actually, this feature being directly related to Feynman's interpretation of positrons being electrons propagating "backwards" in time [43].

Note that the particle number conserving interaction of the charge carriers with the quanta of electromagnetic radiation as stated in (1.154) coincides with the first order light-matter interaction operator common in the literature on atomic physics, quantum optics, solid state et cetera, where it is used to describe the interaction of electrons (positrons) with photons, possibly up to the energy scale of hard X-rays.

1.7.2 The Two-Body Interactions

The two-body interactions (e-e), (p-p) and (e-p) between electrons and positrons coming from the QED Coulomb interaction (1.5) assume according to the result (1.120) the guise

$$\hat{\mathcal{U}}_C^{(e,e)} = \frac{e^2}{2\varepsilon_0} \sum_{k,k',K,K' \in I_+} \int \frac{d^3q}{(2\pi)^3} \frac{1}{|\mathbf{q}|^2} \langle U_k | e^{-iq_a x_a} | U_{k'} \rangle \langle U_K | e^{iq_a x_a} | U_{K'} \rangle \hat{c}_k^\dagger \hat{c}_K^\dagger \hat{c}_{K'} \hat{c}_{k'} \tag{1.155}$$

$$\hat{\mathcal{U}}_C^{(p,p)} = \frac{e^2}{2\varepsilon_0} \sum_{k,k',K,K' \in I_-} \int \frac{d^3q}{(2\pi)^3} \frac{1}{|\mathbf{q}|^2} \langle V_k | e^{-iq_a x_a} | V_{k'} \rangle \langle V_K | e^{iq_a x_a} | V_{K'} \rangle \hat{b}_{K'}^\dagger \hat{b}_{k'}^\dagger \hat{b}_k \hat{b}_K \tag{1.156}$$

$$\hat{\mathcal{U}}_C^{(e,p)} = \frac{e^2}{\varepsilon_0} \sum_{k,k' \in I_+} \sum_{K,K' \in I_-} \int \frac{d^3q}{(2\pi)^3} \frac{1}{|\mathbf{q}|^2} \begin{pmatrix} -\langle U_k | e^{-iq_a x_a} | U_{k'} \rangle \langle V_K | e^{iq_a x_a} | V_{K'} \rangle \\ +\langle U_k | e^{-iq_a x_a} | V_{K'} \rangle \langle V_K | e^{iq_a x_a} | U_{k'} \rangle \end{pmatrix} \hat{c}_k^\dagger \hat{c}_{k'} \hat{b}_{K'}^\dagger \hat{b}_K, \tag{1.157}$$

whereas according to the result (1.132) the corresponding effective two-body interactions arising from the transversal QED-interaction (1.4) are with help of Table 1.1 readily identified as

$$\hat{\mathcal{U}}_\perp^{(e,e)} = \frac{e^2}{m} \sum_{k,k' \in I_+} \sum_{K,K' \in I_+} \int \frac{d^3q}{(2\pi)^3} \frac{\hbar}{2\varepsilon_0 \omega(\mathbf{q})} \left(\delta_{b,b'} - \frac{q_b q_{b'}}{|\mathbf{q}|^2} \right) \times \tag{1.158}$$

$$\times \frac{\left(\tilde{E}_k - \tilde{E}_{k'} - \tilde{\omega}_q\right) - \left(\tilde{E}_K - \tilde{E}_{K'} + \tilde{\omega}_q\right)}{\left(\tilde{E}_k - \tilde{E}_{k'} - \tilde{\omega}_q\right)^2 + \left(\tilde{E}_K - \tilde{E}_{K'} + \tilde{\omega}_q\right)^2} \langle U_k | \alpha_b e^{iq_a x_a} | U_{k'} \rangle \langle U_K | \alpha_{b'} e^{-iq_a x_a} | U_{K'} \rangle \hat{c}_k^\dagger \hat{c}_K^\dagger \hat{c}_{K'} \hat{c}_{k'}$$

$$\hat{\mathcal{U}}_\perp^{(p,p)} = \frac{e^2}{m} \sum_{k,k' \in I_-} \sum_{K,K' \in I_-} \int \frac{d^3q}{(2\pi)^3} \frac{\hbar}{2\varepsilon_0 \omega(\mathbf{q})} \left(\delta_{b,b'} - \frac{q_b q_{b'}}{|\mathbf{q}|^2} \right) \times \tag{1.159}$$

$$\times \frac{\left(\tilde{E}_{K'} - \tilde{E}_K - \tilde{\omega}_q\right) - \left(\tilde{E}_{k'} - \tilde{E}_k + \tilde{\omega}_q\right)}{\left(\tilde{E}_{K'} - \tilde{E}_K - \tilde{\omega}_q\right)^2 + \left(\tilde{E}_{k'} - \tilde{E}_k + \tilde{\omega}_q\right)^2} \langle V_k | \alpha_b e^{iq_a x_a} | V_{k'} \rangle \langle V_K | \alpha_{b'} e^{-iq_a x_a} | V_{K'} \rangle \hat{b}_{K'}^\dagger \hat{b}_{k'}^\dagger \hat{b}_k \hat{b}_K$$

and

$$\hat{\mathcal{U}}_\perp^{(e,p)} = \frac{e^2}{m} \sum_{k,k' \in I_+} \sum_{K,K' \in I_-} \int \frac{d^3q}{(2\pi)^3} \frac{\hbar}{2\varepsilon_0 \omega(\mathbf{q})} \left(\delta_{b,b'} - \frac{q_b q_{b'}}{|\mathbf{q}|^2} \right) \times \tag{1.160}$$

$$
\times \begin{cases} - \left(\frac{(\tilde{E}_k - \tilde{E}_{k'} - \tilde{\omega}_q) - (\tilde{E}_K - \tilde{E}_{K'} + \tilde{\omega}_q)}{(\tilde{E}_k - \tilde{E}_{k'} - \tilde{\omega}_q)^2 + (\tilde{E}_K - \tilde{E}_{K'} + \tilde{\omega}_q)^2} + \frac{(\tilde{E}_K - \tilde{E}_{K'} - \tilde{\omega}_q) - (\tilde{E}_k - \tilde{E}_{k'} + \tilde{\omega}_q)}{(\tilde{E}_K - \tilde{E}_{K'} - \tilde{\omega}_q)^2 + (\tilde{E}_k - \tilde{E}_{k'} + \tilde{\omega}_q)^2} \right) \times \\ \times \langle U_k | \, \alpha_b e^{i q_a \mathsf{x}_a} \, | U_{k'} \rangle \langle V_K | \, \alpha_{b'} e^{-i q_a \mathsf{x}_a} \, | V_{K'} \rangle \, \hat{c}_k^\dagger \hat{c}_{k'} \hat{b}_{K'}^\dagger \hat{b}_K \\ \\ + \left(\frac{(\tilde{E}_k - \tilde{E}_{K'} - \tilde{\omega}_q) - (\tilde{E}_K - \tilde{E}_{k'} + \tilde{\omega}_q)}{(\tilde{E}_k - \tilde{E}_{K'} - \tilde{\omega}_q)^2 + (\tilde{E}_K - \tilde{E}_{k'} + \tilde{\omega}_q)^2} + \frac{(\tilde{E}_K - \tilde{E}_{k'} - \tilde{\omega}_q) - (\tilde{E}_k - \tilde{E}_{K'} + \tilde{\omega}_q)}{(\tilde{E}_K - \tilde{E}_{k'} - \tilde{\omega}_q)^2 + (\tilde{E}_k - \tilde{E}_{K'} + \tilde{\omega}_q)^2} \right) \times \\ \times \langle U_k | \, \alpha_b e^{i q_a \mathsf{x}_a} \, | V_{K'} \rangle \langle V_K | \, \alpha_{b'} e^{-i q_a \mathsf{x}_a} \, | U_{k'} \rangle \, \hat{c}_k^\dagger \hat{c}_{k'} \hat{b}_{K'}^\dagger \hat{b}_K \end{cases}
$$

Actually the first line in the electron-positron interaction terms (1.157), (1.160) resembles the terms (1.158), (1.159) describing two-body interaction processes with the particles involved keeping their identity, while the second line in (1.157), (1.160) is different and comprises the so called *annihilation* processes [21, 44–46]. A discussion of further details as regards positronium we postpone to the ensuing chapter III.

On a final note, because the above derived effective two-body interactions between Dirac fermions are represented with fully relativistic matrix elements, these interactions determine as well the amplitudes for Møller scattering ($e + e \rightarrow e + e$) and also Bhabha-scattering ($e + p \rightarrow e + p$). For a beautiful résumé see [21].

1.7.3 The Compton Interaction

Besides the effective two-body interaction terms between electrons and/or positrons enlisted in the unitary transformed QED-Hamiltonian (1.148), regard must be given yet to another term in (1.151), actually this term describing the interaction between *photons* and electrons (or else positrons). Because photon number is conserved in these specific interaction processes with electrons (positrons), justifiably here the photons as well manifest their *corpuscular* nature, a feature discovered by Compton in the scattering of X-rays from electrons [47]. Eliminating in (1.132) a rational function $\mathsf{R}(\tilde{E}_{k''})$ in front of the intermediate states $|\phi_{k''}\rangle \langle \phi_{k''}|$ to be summed over inserting a spectral representation $\mathsf{R}(\mathsf{H}^{(D)}) = \sum_{k'' \in I} \mathsf{R}(\tilde{E}_{k''}) \, |\phi_K\rangle \langle \phi_{k''}|$, this way an operational form of the matrix element contributing to the Compton interaction results. Restricting the fermion mode labels $k, k' \in I_+$, eventually the electron-photon part of the Compton interaction emerges in the guise

$$
\hat{\mathcal{V}}_{\perp}^{(e,ph)} = -\frac{e^2}{m} \frac{1}{V_P} \sum_{\mathbf{q},\mathbf{q}'} \kappa_q \kappa_{q'} \hat{\mathsf{A}}_{b',\mathbf{q}'}^{(-1)} \hat{\mathsf{A}}_{b,\mathbf{q}}^{(1)} \times \tag{1.161}
$$

$$
\times \sum_{k,k' \in I_+} \langle U_k | \begin{pmatrix} \alpha_b e^{i q_a \mathsf{x}_a} \frac{2\mathsf{H}^{(D)} - \tilde{E}_k - \tilde{E}_{k'} + \tilde{\omega}_q + \tilde{\omega}_{q'}}{(\mathsf{H}^{(D)} - \tilde{E}_k + \tilde{\omega}_q)^2 + (\mathsf{H}^{(D)} - \tilde{E}_{k'} + \tilde{\omega}_{q'})^2} e^{i q'_a \mathsf{x}_a} \alpha_{b'} + \\ \\ + \alpha_{b'} e^{i q'_a \mathsf{x}_a} \frac{2\mathsf{H}^{(D)} - \tilde{E}_{k'} - \tilde{E}_k - \tilde{\omega}_q - \tilde{\omega}_{q'}}{(\mathsf{H}^{(D)} - \tilde{E}_{k'} - \tilde{\omega}_q)^2 + (\mathsf{H}^{(D)} - \tilde{E}_k - \tilde{\omega}_{q'})^2} e^{i q_a \mathsf{x}_a} \alpha_b \end{pmatrix} | U_{k'} \rangle \, \hat{c}_k^\dagger \hat{c}_{k'}
$$

Here the elementary photon mode creation and annihilation operators are encoded according to (1.31) in the definitions of the operators $\hat{\mathsf{A}}^{(\pm 1)}_{b',\mathbf{q}'}$, so that in effect photons contact electrons and are scattered. Of course, an analogous expression is found from (1.132) for the positron-photon part of the Compton interaction, and so we desist to state it here explicitly.

Restricting for simplicity the considerations to electrons (positrons) propagating in free space (absence of external source fields), then an intuitively accessible interpretation of a collision between a *corpuscular* massless photon with an electron (positron) can be given, whereby after the collision a new state of individual momentum and individual energy for both, photon and electron (positron), is produced in accord with conservation of energy and momentum, akin to collisions between billiards. By this means a complete derivation of the full relativistic Compton process is enabled. Compared with results for the matrix elements of Compton scattering obtained from S-matrix perturbation theory, for instance [46, 48], the denominators in (1.161) are yet less singular, a distinctive feature of the flow equation approach [36].

1.7.4 The Self-Energy of Dirac Fermions

A fundamental feature of Quantum Electrodynamics is, that the characteristic energy change of a *single* emitted and/or absorbed particles is not solely determined by the spectrum of the single particle Dirac Hamiltonian $\hat{\mathcal{H}}_D$, but as well by the self-energy terms, denoted in (1.150), (1.151) and (1.152) as $\hat{\mathcal{M}}^{(e)}_C$, $\hat{\mathcal{M}}^{(e)}_\perp$ and $\hat{\mathcal{M}}^{(e)}_{VP}$ for electrons, respectively as $\hat{\mathcal{M}}^{(p)}_C$, $\hat{\mathcal{M}}^{(p)}_\perp$ and $\hat{\mathcal{M}}^{(p)}_{VP}$ for positrons. Explicit expressions for the self-energy terms of *electrons*, see (1.150), (1.151), (1.152), are then readily identified from the single particle terms enlisted in (1.120), (1.132), (1.138), (1.145) as

$$\hat{\mathcal{M}}^{(e)}_C = \sum_{k,k'\in I_+} \left(\begin{array}{c} \delta m\, c^2 \, \langle U_k|\, \beta\, |U_{k'}\rangle + \\ + \frac{e^2}{2\varepsilon_0} \frac{1}{V_P} \sum_{\mathbf{q}} \frac{1}{|\mathbf{q}|^2} \langle U_k|\, e^{-iq_a \mathsf{x}_a} \sum_{k''\in I} z_{k''} \, |\phi_{k''}\rangle \, \langle \phi_{k''}|\, e^{iq_a \mathsf{x}_a} \, |U_{k'}\rangle \end{array} \right) \hat{c}^\dagger_k \hat{c}_{k'} \tag{1.162}$$

$$\hat{\mathcal{M}}^{(e)}_\perp = \frac{e^2}{m} \frac{1}{V_P} \sum_{\mathbf{q}} \frac{\hbar}{2\varepsilon_0 \omega(\mathbf{q})} \left(\delta_{b.b'} - \frac{q_b q_{b'}}{|\mathbf{q}|^2} \right) \sum_{k,k'\in I_+} \times$$
$$\times \left(\langle U_k|\, \alpha_b e^{-iq_a \mathsf{x}_a} \sum_{k''\in I} \frac{\tilde{E}_k + \tilde{E}_{k'} - 2\tilde{E}_{k''} - 2\tilde{\omega}_q z_{k''}}{\left(\tilde{E}_{k''} + \tilde{\omega}_q z_{k''} - \tilde{E}_k\right)^2 + \left(\tilde{E}_{k''} + \tilde{\omega}_q z_{k''} - \tilde{E}_{k'}\right)^2} |\phi_{k''}\rangle \, \langle \phi_{k''}|\, \alpha_{b'} e^{iq_a \mathsf{x}_a} \, |U_{k'}\rangle \right) \hat{c}^\dagger_k \hat{c}_{k'} \tag{1.163}$$

(1.164)

$$\hat{\mathcal{M}}_{VP}^{(e)} = \sum_{k,k'\in I_+} \left(\begin{array}{c} -\frac{e^2}{2\varepsilon_0}\frac{1}{V_P}\sum_{\mathbf{q}}\frac{1}{|\mathbf{q}|^2}\sum_{k''} \langle \phi_{k''} | \, e^{-iq_a x_a} z_{k''} \, |\phi_{k''}\rangle \langle U_k | \, e^{iq_a x_a} \, |U_{k'}\rangle + \\ + \langle U_k | \, \delta e \left(\Phi^{(ext)}(\mathsf{x}) \, 1_{4\times4} - c\alpha_b \mathsf{A}_b^{(ext)}(\mathsf{x}) \right) |U_{k'}\rangle + \\ + \frac{e^2}{2\varepsilon_0}\frac{1}{V_P}\sum_{\mathbf{q}}\frac{1}{|\mathbf{q}|^2}\left(\delta_{b.b'} - \frac{q_b q_{b'}}{|\mathbf{q}|^2}\right) \sum_{k''\in I} \langle \phi_{k''} | \, \alpha_{b'} e^{-iq_a x_a} z_{k''} \, |\phi_{k''}\rangle \langle U_k | \, \alpha_b e^{iq_a x_a} \, |U_{k'}\rangle \end{array} \right) \hat{c}_k^\dagger \hat{c}_{k'}$$

The terms comprising the self-energy of *positrons* are identified correspondingly

$$\hat{\mathcal{M}}_{C,Y}^{(p)} = -\sum_{K,K'\in I_-} \left(\begin{array}{c} \delta m \, c^2 \, \langle V_K | \, \beta \, | V_{K'}\rangle + \\ + \frac{e^2}{2\varepsilon_0}\frac{1}{V_P}\sum_{\mathbf{q}}\frac{1}{|\mathbf{q}|^2} \langle V_K | \, e^{-iq_a x_a} \sum_{k''\in I} z_{k''} \, |\phi_{k''}\rangle \langle \phi_{k''} | \, e^{iq_a x_a} \, | V_{K'}\rangle \end{array} \right) \hat{b}_{K'}^\dagger \hat{b}_K \tag{1.165}$$

$$\hat{\mathcal{M}}_{\perp,Y}^{(p)} = -\frac{e^2}{m}\frac{1}{V_P}\sum_{\mathbf{q}}\frac{\hbar}{2\varepsilon_0\omega(\mathbf{q})}\left(\delta_{b.b'} - \frac{q_b q_{b'}}{|\mathbf{q}|^2}\right)\sum_{K,K'\in I_-} \times$$
$$\times \left(\langle V_K | \, \alpha_b e^{-iq_a x_a} \sum_{k''\in I} \frac{\tilde{E}_K + \tilde{E}_{K'} - 2\tilde{E}_{k''} - 2\tilde{\omega}_q z_{k''}}{\left(\tilde{E}_{k''} + \tilde{\omega}_q z_{k''} - \tilde{E}_K\right)^2 + \left(\tilde{E}_{k''} + \tilde{\omega}_q z_{k''} - \tilde{E}_{K'}\right)^2} |\phi_{k''}\rangle \langle \phi_{k''} | \, \alpha_{b'} e^{iq_a x_a} \, | V_{K'}\rangle \right) \hat{b}_{K'}^\dagger \hat{b}_K \tag{1.166}$$

$$\hat{\mathcal{M}}_{VP}^{(p)} = -\sum_{K,K'\in I_-} \left(\begin{array}{c} -\frac{e^2}{2\varepsilon_0}\frac{1}{V_P}\sum_{\mathbf{q}}\frac{1}{|\mathbf{q}|^2}\sum_{k''\in I} \langle \phi_{k''} | \, e^{-iq_a x_a} z_{k''} \, |\phi_{k''}\rangle \langle V_K | \, e^{iq_a x_a} \, |V_{K'}\rangle + \\ + \langle V_K | \, \delta e \left(\Phi^{(ext)}(\mathsf{x}) \, 1_{4\times4} - c\alpha_b \mathsf{A}_b^{(ext)}(\mathsf{x}) \right) |V_{K'}\rangle + \\ + \frac{e^2}{2\varepsilon_0}\frac{1}{V_P}\sum_{\mathbf{q}}\frac{1}{|\mathbf{q}|^2}\left(\delta_{b.b'} - \frac{q_b q_{b'}}{|\mathbf{q}|^2}\right) \sum_{k''\in I} \langle \phi_{k''} | \, \alpha_{b'} e^{-iq_a x_a} z_{k''} \, |\phi_{k''}\rangle \langle V_K | \, \alpha_b e^{iq_a x_a} \, |V_{K'}\rangle \end{array} \right) \hat{b}_{K'}^\dagger \hat{b}_K \, . \tag{1.167}$$

The three self-energy terms enlisted in (1.162), (1.163), (1.164) for electrons, and in (1.165), (1.166), (1.167) for positrons , in actual fact comprise famous key effects of quantum electrodynamics, namely charge and mass renormalization, Schwinger's magnetic moment correction and last not least the terms required to explain the Lamb shift experiment.

In our ensuing chapter [16] we go into the details and show, how to derive these effects directly from the expressions $\hat{\mathcal{M}}_C^{(e)}$, $\hat{\mathcal{M}}_\perp^{(e)}$ and $\hat{\mathcal{M}}_{VP}^{(e)}$ in an accurate logical way with a complete relativistic *and* gauge invariant calculation to first order in the fine structure constant $\alpha_{FS} = \frac{e^2}{4\pi\epsilon_0\hbar c} \approx \frac{1}{137}$, making full use of a lesser known, but conclusively substantiated physical cutoff method invented by Dirac [24].

1.7.5 Comment on the Self-Energy of Photons

With the described unitary transformation method of the Hamiltonian of QED leading to the particle number conserving Hamiltonian (1.148), not only a self-energy for electrons and positrons is obtained, but (perhaps not surprisingly) there is identified from the limiting value (1.132) a term bilinear in the photon creation an annihilation operators

$$\hat{\Pi}^{(ph)} = \frac{e^2}{m}\frac{1}{V_P}\sum_{\mathbf{q},\mathbf{q}'}\kappa_q\kappa_{q'}\sum_{\nu,\nu'\in\{I,II\}}\tilde{\mathcal{A}}_{b'}\left(-\mathbf{q}',\nu'\right)\tilde{\mathcal{A}}_b\left(\mathbf{q},\nu\right)\hat{a}^{\dagger}_{-\mathbf{q}',\nu'}\hat{a}_{\mathbf{q},\nu}\times \tag{1.168}$$
$$\times\sum_{k,k'}\frac{z_{k'}-z_k}{2}\langle\phi_k|\,\alpha_b e^{iq_a\mathrm{x}_a}\,|\phi_{k'}\rangle\langle\phi_{k'}|\,\alpha_{b'}e^{iq'_a\mathrm{x}_a}\,|\phi_k\rangle\frac{\left(\tilde{E}_k-\tilde{E}_{k'}-\tilde{\omega}_q\right)+\left(\tilde{E}_k-\tilde{E}_{k'}-\tilde{\omega}_{q'}\right)}{\left(\tilde{E}_k-\tilde{E}_{k'}-\tilde{\omega}_q\right)^2+\left(\tilde{E}_k-\tilde{E}_{k'}-\tilde{\omega}_{q'}\right)^2}$$

As this term depends as well on the full spectrum of Dirac fermions, it is manifest, that photons may acquire concurrently *material-character* as well [49]. But interpreting a corpuscular photon as a *hybrid* that comprises electrodynamic *and* material degrees of freedom (both, electron and positron), prima facie contradicts the *experimentally* tested concept of photons propagating freely in outer space at the universal speed c of light in arbitrary direction.

For *zero* external electromagnetic fields the associated operator (1.168) of photons we denote as $\hat{\Pi}^{(0,ph)}$. In this case the Dirac fermions are governed by the translational invariant Hamiltonian $\mathsf{H}_0^{(D)} = mc^2\beta + c\alpha_b\mathsf{p}_b$. If then the concerning matrix elements in (1.168) are evaluated with free fermion states, and if in particular in the summation over fermion labels k, k' the gauge invariant regularization method of Pauli and Villars [50] is employed, a considerable simplification is obtained, namely $\hat{\Pi}^{(0,ph)} \equiv 0$. So, the propagation of photons in free space is found *unobstructed* in calculations with free Dirac fermions, an important and well known result [28, 51].

But that scenario changes for Dirac fermions moving in the presence of *strong* electromagnetic background fields. For instance, early on Delbrück predicted [52], that an external electromagnetic field, such as the Coulomb field *near* to a nucleus, may scatter incident photons *elastically* by a process in which electron-positron pairs would eventually act as a kind of catalytic agent and produce effects which otherwise cannot occur within linear electrodynamics. The underlying physical idea is, that the incident photon creates a (virtual) electron-positron pair, whereupon this pair undergoes scattering in the external Coulomb field. And the subsequent reconversion of that (virtual) pair of course leads then to a re-emission of a photon.

In actual fact the fermionic matrix elements stated in (1.132), as obtained from calculations with the Hamiltonian (1.8) by taking into account an external Coulomb field of a (nearby) nucleus, as well comprise the Delbrück scattering amplitude, albeit in Furry representation [53]. Actually, the Delbrück scattering effect has been experimentally verified, for a review see [54].

Another effect of the polarizable quantum vacuum, that can be ascribed to the photon self-energy-operator $\hat{\Pi}^{(ph)}$, regards the propagation of photons in outer space and is concerned with an external field induced bi-refringence (resemblant the Cotton-Mouton effect in gases), leading to a plane wave velocity of photons other than c, even though such a deviation (after bringing the units in) is indeed *miniscule* and should be discernible only with (extreme) external electromagnetic field strengths occurring in the context of astrophysics. For a comprehensive review of the experimental situation see [55].

1.7.6 Comment on the Vacuum Energy Change

The contribution of Dirac fermions to the zero-point energy is readily identified as the spectral shift term in (1.21), the *minus* sign being a consequence of the anti-commutator rules. Combined with the corresponding *positive* zero-point energy term of the photons in (1.29) there results to *zero* order in the expansion of the unitary transformed Hamiltonian of QED in (1.148) a vacuum energy [23]

$$\mathcal{K}_0 = \sum_{\mathbf{q}} \hbar\omega_{\mathbf{q}} - \sum_{k\in I_-} |E_k| \tag{1.169}$$

Now, considering the *second* order terms in the unitary transformed Hamiltonian (1.148), one obtains from (1.120) one more spectral shift term, referred to as $\mathcal{K}_C$ in (1.150), given by

$$\mathcal{K}_C = \frac{e^2}{2\varepsilon_0}\frac{1}{V_P}\sum_{\mathbf{q}}\sum_{k\in I_+}\sum_{K\in I_-}\frac{\Theta_{q,k,K}}{|\mathbf{q}|^2}\langle U_k|\, e^{iq_a x_a}\,|V_K\rangle\langle V_K|\, e^{-iq_a x_a}\,|U_k\rangle\,, \tag{1.170}$$

and from (1.132) a corresponding spectral shift term, referred to as $\mathcal{K}_\perp$ in (1.151), given by

$$\mathcal{K}_\perp = \frac{e^2}{2\varepsilon_0}\frac{\hbar}{mc}\frac{1}{V_P}\sum_{\mathbf{q}}\frac{1}{V_P}\sum_{\mathbf{q}}\sum_{k\in I_+}\sum_{K\in I_-}\left(\delta_{b,b'} - \frac{q_b q_{b'}}{|\mathbf{q}|^2}\right)\frac{\Theta_{q,k,K}}{|\mathbf{q}|}\times$$
$$\times\ \langle U_k|\,\alpha_b e^{iq_a x_a}\,|V_K\rangle\langle V_K|\,\alpha_{b'} e^{-iq_a x_a}\,|U_k\rangle\,\frac{1}{\tilde{E}_k - \tilde{E}_K - \tilde{\omega}_q} \tag{1.171}$$

The total $\mathcal{K}_0 + \mathcal{K}_C + \mathcal{K}_\perp$ contributes to the particle number conserving Hamiltonian (1.148) then a zero-point energy of the *polarizable* QED-vacuum, which acquires in the presence of external *spatially* varying fields a physical meaning [31]. Even the zero-point energy contribution $\mathcal{K}_0^{(ph)} = \sum_{\mathbf{q}} \hbar\omega_{\mathbf{q}}$ of the quantum vacuum contributed by mere photons leads, by taking the so called (electromagnetic) surface modes of a material body into account, then to the well known dispersive Casimir-Polder and van der Waals forces, which have been knowledgeably reviewed to perfection by Barash and Ginzburg [56]. The extra contribution obtained from (1.170),

(1.171) now will bring into this scenario to order α_{FS} additional vacuum polarization effects. It would be interesting to investigate what effect these extra terms could have on previous studies of intermolecular forces, for instance the energy to vaporize a unit mass of material (latent heat) [23].

1.8 Closing

We dispense for now a more in-depth discussion, as our principal focus in this article was, to establish directly from the relativistic quantum field theory of QED in Coulomb gauge, a logical deduction of an effective particle number conserving Hamiltonian $\hat{\mathcal{H}}_U$ for pairwise interacting fermions as resumed in (1.148), that enables a retranslation back to first quantization and allows this way to regard electrons and positrons as interacting point particles with attribute mass, charge and spin. In the non relativistic limit $\hat{\mathcal{H}}_U$ indeed coincides with the Hamiltonian of non-relativistic Schrödinger-Pauli quantum mechanics.

With regard to the details how to obtain from the self-energy terms of Dirac fermions, as enlisted in (1.162), (1.163), (1.164) and in (1.165), (1.166), (1.167) respectively, eventually the renormalization of charge and mass, we shall care about in our chapter [16], where we also show in detail with a gauge invariant computing procedure relying on Dirac's cutoff method, how to obtain Schwinger's $g-2$ correction of the magnetic moment of the electron, and what's more, *all* the terms required to explain the Lamb shift experiment to order α_{FS}.

Last not least we refer to our article [17], wherein our method of derivation of the relativistic corrections to the predominant Coulomb interaction between electrons and/or positrons by means of the energy separating Eriksen transformation is presented in detail.

From what has been said, the unitary transformation method of the QED-Hamiltonian as presented in this chapter, apparently opens an effective alternative to the renowned causal S-matrix theory and/or Feynman diagrams, respectively.

Actually, it was Dirac's considered opinion, that one should in Quantum Electrodynamics desist the Schrödinger picture, i.e. avoid the calculation of the time evolution of any *states* in QED-Fock space, but instead deal with the time dependence of *operators* in the Heisenberg picture only [24]. In that respect our Hamiltonian flow equation approach is near to Dirac's.

Acknowledgements We thank Daniel Braun for useful discussions.

Appendix

In this appendix we recapitulate a few distinguishing features of the Dirac equation and its momentous importance to the theory of the relativistic electron. We hereby hope to provide useful background information to an audience, which on one hand has some basic knowledge of states in Fock space and the method of second quantization, but on the other hand is not primarily specialized in *relativistic* field theory. We aim for this audience attaining a deeper understanding of our approach outlined in the section (1.2) of this chapter, where the Hamiltonian of Quantum Electrodynamics is introduced as a starting point for the relativistic field theory of the electron on the assumption the reader already had some pre-knowledge.

The literature on the subject is vast and there are many excellent clearly written text books, for instance [57]. Very carefully written *introductions* into Quantum Electrodynamics and the physics of the Quantum Theory of Radiation one finds in the classic books [14, 48]. For a particularly informative review of the nascence (and history) of Quantum Electrodynamics we refer to a book of P.W. Milonni [23].

The Dirac Equation

To start with, let us bring in mind the circumstance, that the Maxwell equations of electrodynamics are invariant under a Lorentz transformation, whereas the Schrödinger equation is not.

So, to describe an electron moving at relativistic speed, Dirac adhered to the idea of introducing a kind of wavefunction $\Psi(\mathbf{r}, t)$ for a particle carrying mass and charge, by describing its time evolution exactly as if it was a Schrödinger type equation with a Hamiltonian $\mathsf{H}^{(D)}$

$$(i\hbar\frac{\partial}{\partial t} - \mathsf{H}^{(D)})\Psi(\mathbf{r}, t) = 0 \tag{1.172}$$

But different from the familiar non relativistic Hamiltonian of the Schrödinger wave mechanics, describing for instance a freely moving particle with a Hamiltonian

$$\mathsf{H}_0^{(S)} = \frac{\mathbf{p}^2}{2m} \tag{1.173}$$

Dirac seeked a guise $\mathsf{H}_0^{(D)}$ for that Hamiltonian completely *different* from $\mathsf{H}_0^{(S)}$, in the main merely ensuring it has an eigenvalue equal to the familiar *relativistic* energy dispersion relation for a freely moving particle of mass m and momentum $\hbar\mathbf{k}$

$$E_{\mathbf{k}}^{(0)} = \sqrt{m^2c^4 + (c\hbar\mathbf{k})^2} \tag{1.174}$$

His disruptive idea was to accept for $\mathsf{H}_0^{(D)}$ exclusively a guise, where the quantum mechanical momentum operator $\mathsf{p}_b = \frac{\hbar}{i}\frac{\partial}{\partial r_b}$ and the time derivative $\mathsf{p}_0 = -i\hbar\frac{1}{c}\frac{\partial}{\partial t}$ were in a *linear* relation, just as position $\mathbf{r}$ and time t are connected in elementary special relativity.

So Dirac introduced four (constant dimensionless) parameters, β and $\alpha_x, \alpha_y, \alpha_z$ and wrote[2]

$$\mathsf{H}_0^{(D)} = \beta mc^2 + c\alpha_b \mathsf{p}_b \tag{1.175}$$

For reasons of compatibility he then required that the square $\mathsf{H}_0^{(D)}\mathsf{H}_0^{(D)}$, as readily obtained from (1.175) in the guise

$$\mathsf{H}_0^{(D)}\mathsf{H}_0^{(D)} = m^2c^4\beta\beta + mc^3(\beta\alpha_b + \alpha_b\beta)\mathsf{p}_b + \frac{c^2}{2}\left(\alpha_b\alpha_{b'} + \alpha_{b'}\alpha_b\right)\mathsf{p}_b\mathsf{p}_{b'}, \tag{1.176}$$

should have eigenvalues

$$\left(E_{\mathbf{k}}^{(0)}\right)^2 = m^2c^4 + (c\hbar\mathbf{k})^2 \tag{1.177}$$

As $\hbar k_b$ is an eigenvalue of the momentum operator p_b, one needs to infer the afore introduced parameters cannot be accepted as being real numbers. Instead β and α_b are subject to some conditions

$$\begin{aligned} \beta\beta &= \mathsf{1} \\ \beta\alpha_b + \alpha_b\beta &= \mathsf{0} \\ \alpha_b\alpha_{b'} + \alpha_{b'}\alpha_b &= 2\,\delta_{b,b'}\mathsf{1} \end{aligned} \tag{1.178}$$

These conditions can be fulfilled with β and α_b being square matrices. However, one cannot find matrices of smaller rank than 4 satisfying the required properties (1.178). This is, because in compliance with these three requirements there holds

$$\alpha_x\alpha_x = \alpha_y\alpha_y = \alpha_z\alpha_z = \beta\beta = \mathsf{1}, \tag{1.179}$$

and so α_b and β have eigenvalues ± 1. From the cyclic invariance of the trace of a matrix one readily gathers (summation convention suspended here)

$$\begin{aligned} \mathrm{tr}\,(\beta) &= \mathrm{tr}\,(\beta\alpha_b\alpha_b) = \mathrm{tr}\,(\alpha_b\beta\alpha_b) = -\mathrm{tr}\,(\beta\alpha_b\alpha_b) = -\mathrm{tr}\,(\beta) \equiv \mathsf{0} \\ \mathrm{tr}\,(\alpha_b) &= \mathrm{tr}\,(\beta\alpha_b\beta) = -\mathrm{tr}\,(\alpha_b\beta\beta) = -\mathrm{tr}\,(\alpha_b) \equiv \mathsf{0} \end{aligned}$$

In view of β and α_b being traceless and having eigenvalues ± 1, so then β and α_b must be *even*-dimensional matrices. But the smallest possible rank 2 cannot be accepted, because there exist only *three* traceless and anticommuting linear independent 2×2 matrices, which may be taken as the Pauli spin matrices

---

[2] To simplify notation we use from now on the summation convention.

$$\mathsf{1}_{2\times 2} = \begin{pmatrix} 1, 0 \\ 0, 1 \end{pmatrix}, \ \sigma_x^{(P)} = \begin{pmatrix} 0, 1 \\ 1, 0 \end{pmatrix}, \ \sigma_y^{(P)} = \begin{pmatrix} 0, -i \\ i, 0 \end{pmatrix}, \ \sigma_z^{(P)} = \begin{pmatrix} 1, 0 \\ 0, -1 \end{pmatrix} \tag{1.180}$$

These obey to the algebra

$$\sigma_b^{(P)} \sigma_{b'}^{(P)} = \delta_{b,b'} \mathsf{1}_{2\times 2} + i\epsilon_{bb'b''}\sigma_{b''}^{(P)} \tag{1.181}$$

The conditions (1.178) can be realized with 4×4 matrices, though. In terms of a convenient 2×2-block matrix notation with Pauli-spin matrices (1.180) one has in Dirac representation

$$\beta = \begin{pmatrix} \mathsf{1}_{2\times 2}, & \mathsf{0}_{2\times 2} \\ \mathsf{0}_{2\times 2}, & -\mathsf{1}_{2\times 2} \end{pmatrix}, \ \alpha_b = \begin{pmatrix} \mathsf{0}_{2\times 2}, & \sigma_b^{(P)} \\ \sigma_b^{(P)}, & \mathsf{0}_{2\times 2} \end{pmatrix}, \ \sigma_b = \begin{pmatrix} \sigma_b^{(P)}, & \mathsf{0}_{2\times 2} \\ \mathsf{0}_{2\times 2}, & \sigma_b^{(P)} \end{pmatrix} \tag{1.182}$$

Besides the relations (1.178) one readily confirms the multiplication rule

$$\alpha_b \alpha_{b'} = \delta_{b,b'} \mathsf{1}_{4\times 4} + i\epsilon_{bb'b''}\sigma_{b''} \tag{1.183}$$

So, because β, α_b and σ_b are 4×4-matrices, the wavefunction $\Psi(\mathbf{r}, t)$ envisaged by Dirac necessarily comprises *four* components $\Psi_\nu(\mathbf{r}, t)$, say labeled with an index $\nu \in \{1, 2, 3, 4\}$.

In the presence of electromagnetic fields, described by a vector potential A_b and a scalar potential Φ, one uses in quantum mechanics the substitutions

$$\begin{aligned} i\hbar\frac{\partial}{\partial t} &\rightarrow i\hbar\frac{\partial}{\partial t} - e\Phi \equiv -c\Pi_0 \\ \mathsf{p}_b &\rightarrow \mathsf{p}_b - e\mathsf{A}_b \equiv \Pi_b \end{aligned} \tag{1.184}$$

to ensure gauge invariance of the theory. Note the charge of the electron being $e = -|e|$. Compliant with the time evolution (1.172) then one promptly obtains the Dirac Hamiltonian of the relativistic electron with the minimal coupling to electromagnetic potentials in the guise

$$\mathsf{H}^{(D)} = mc^2\beta + e\Phi \cdot \mathsf{1}_{4\times 4} + c\alpha_b (\mathsf{p}_b - e\mathsf{A}_b) \tag{1.185}$$

From (1.172) and its adjoint one readily infers that the bilinear expressions

$$\begin{aligned} \varrho(\mathbf{r}, t) &= e \sum_\nu \Psi_\nu^\star(\mathbf{r}, t)\Psi_\nu(\mathbf{r}, t) \\ j_b(\mathbf{r}, t) &= e \sum_{\nu,\nu'} \Psi_\nu^\star(\mathbf{r}, t)\, (c\alpha_b)_{\nu,\nu'}\, \Psi_{\nu'}(\mathbf{r}, t) \end{aligned} \tag{1.186}$$

indeed obey to the continuity equation

$$\frac{\partial}{\partial t}\varrho(\mathbf{r},t)+\frac{\partial}{\partial r_b}j_b(\mathbf{r},t)=0 \tag{1.187}$$

So, from (1.187) the physical interpretations of ϱ and j_b in (1.186) is, that ϱ describes the charge density and j_b describes the current density. Notably, in the expression defining the current density in (1.186) no explicit coupling to the vector potential $e\mathsf{A}_b$ occurs. Apparently, the matrix $c\alpha_b$ has taken over a role as if it was a particle velocity.

On the Lorentz Invariance of the Dirac Equation

The equations describing physical variables must be the same for two observers in uniform relative motion. This is the principle of special relativity.

Say one observer was writing the Maxwell equations of electrodynamics in terms of a coordinate system $(ct, \mathbf{r})$, another observer instead writing them in terms of a coordinate system $(ct', \mathbf{r}')$. For simplicity let us assume the coordinate system $(ct', \mathbf{r}')$ is an exact imitation of the Cartesian coordinate system $(ct, \mathbf{r})$, but moving relative to it at velocity $\mathbf{v} = v\mathbf{n}$.

In this case, according to special relativity, both systems are connected by a Lorentz boost in the direction of the unit vector $\mathbf{n}$. Specifically with u denoting a *rapidity* parameter such a boost reads

$$\begin{aligned}\mathbf{n}\cdot\mathbf{r}' &= \cosh(u)\mathbf{n}\cdot\mathbf{r}-\sinh(u)ct\\ ct' &= \cosh(u)ct-\sinh(u)\mathbf{n}\cdot\mathbf{r}\\ \mathbf{n}\wedge\mathbf{r}' &= \mathbf{n}\wedge\mathbf{r}\end{aligned} \tag{1.188}$$

where $\cosh(u)=\frac{1}{\sqrt{1-\frac{v^2}{c^2}}}$, $\sinh(u)=\frac{\frac{v}{c}}{\sqrt{1-\frac{v^2}{c^2}}}$, and accordingly $\tanh(u)=\frac{v}{c}$.

All the other Lorentz transformations emerge from (1.188) combining a Lorentz boost with general orthogonal transformations of the space coordinates and possibly time inversion as well.

To transform more complicated objects from one coordinate system to another it is expedient to use tensor analysis. In special relativity certain physical quantities can be arranged as a *four*-tupel, with the elements of such four-tupels transforming between themselves according to (1.188) in the same way as $x = (ct, \mathbf{r})$ gets transformed into $x' = (ct', \mathbf{r}')$. By this means there emerges the notion of a *four-vector*.

For instance $(ct, \mathbf{r})$ is a four-vector x^μ, likewise $(\frac{E}{c}, \hbar\mathbf{k})$ is a four-vector p^μ. With the index $\mu \in \{0, 1, 2, 3\}$ indicating which of the four components is considered, we agree on using for the *contravariant* components of a four-vector the notation

$$x^0 = ct\ ,\ x^1 = r_x\ ,\ x^2 = r_y\ ,\ x^3 = r_z$$

$$p^0 = \frac{E}{c}\,,\ p^1 = \hbar k_x\,,\ p^2 = \hbar k_y\,,\ p^3 = \hbar k_z \tag{1.189}$$

The corresponding *covariant* components of a four-vector are then in terms of the (flat space) metric tensor $g_{\mu\mu'}$ defined as $x_\mu = g_{\mu\mu'}x^{\mu'}$, $p_\mu = g_{\mu\mu'}p^{\mu'}$. Accepting the metric of Minkowski space-time in the guise

$$g_{\mu\mu'} = \begin{pmatrix} 1,\ 0,\ 0,\ 0 \\ 0,\ -1,\ 0,\ 0 \\ 0,\ 0,\ -1,\ 0 \\ 0,\ 0,\ 0,\ -1 \end{pmatrix}_{\mu,\mu'} \tag{1.190}$$

we agree on the notation

$$\begin{aligned} x_0 &= ct\,,\ x_1 = -r_x\,,\ x_2 = -r_y\,,\ x_3 = -r_z \\ p_0 &= \frac{E}{c}\,,\ p_1 = -\hbar k_x\,,\ p_2 = -\hbar k_y\,,\ p_3 = -\hbar k_z \end{aligned} \tag{1.191}$$

Other important examples of four-vectors are the four-current density j^μ, the four potential A^μ and the four-dimensional gradient $\partial^\mu = g^{\mu\mu'}\frac{\partial}{\partial x^{\mu'}}$, the associated contravariant components of these four-vectors being

$$\begin{aligned} j^0 &= c\rho\,,\ j^1 = j_x\,,\ j^2 = j_y\,,\ j^3 = j_z \\ A^0 &= \frac{1}{c}\Phi\,,\ A^1 = A_x\,,\ A^2 = A_y\,,\ A^3 = A_z \\ \partial^0 &= \frac{1}{c}\frac{\partial}{\partial t}\,,\ \partial^1 = -\frac{\partial}{\partial r_x}\,,\ \partial^2 = -\frac{\partial}{\partial r_y}\,,\ \partial^3 = -\frac{\partial}{\partial r_z} \end{aligned} \tag{1.192}$$

Note that $g^{\mu\nu}g_{\nu\mu'} = \delta_{\mu,\mu'}$ is defining the inverse of $g_{\mu\mu'}$. With (1.190) being a diagonal matrix apparently there holds $g^{\mu\mu'} = g_{\mu\mu'}$, of course.

Scalar products of two four-vectors x^μ and y^μ are denoted as $x_\mu y^\mu$, usually referred to as a Lorentz-scalar They have the important property keeping their value fixed under any Lorentz transformation. Here are some examples

$$x_\mu x^\mu = (ct)^2 - r^2 = (ct')^2 - r'^2 = x'_\mu x'^\mu \tag{1.193}$$

$$p_\mu p^\mu = \frac{E^2}{c^2} - (\hbar k)^2 = \frac{E'^2}{c^2} - (\hbar k')^2 = p'_\mu p'^\mu \tag{1.194}$$

$$x_\mu p^\mu = Et - \hbar\mathbf{k}\cdot\mathbf{r} = E't' - \hbar\mathbf{k}'\cdot\mathbf{r}' = x'_\mu p'^\mu \tag{1.195}$$

$$j_\mu A^\mu = \rho\Phi - \mathbf{j}\cdot\mathbf{A} = \rho'\Phi' - \mathbf{j}'\cdot\mathbf{A}' = j'_\mu A'^\mu \tag{1.196}$$

$$\partial^\mu j_\mu = \frac{\partial}{\partial t}\rho + \frac{\partial}{\partial r_b}j_b = 0 = \frac{\partial}{\partial t'}\rho' + \frac{\partial}{\partial r'_b}j'_b = \partial'^\mu j'_\mu \tag{1.197}$$

For instance, (1.193) entails the light cone remaining invariant. Due to (1.195) the phase $\varphi = \frac{E}{\hbar}t - \mathbf{k} \cdot \mathbf{r}$ of a plane wave $e^{i\varphi}$ is an invariant too. Note as well (1.197) implies total charge being unchanged under a Lorentz transformation.

Another important conclusion follows from considering the rest frame $(ct'', \mathbf{r}'')$ of a massive particle. Then $E'' = mc^2, c\hbar\mathbf{k}'' = \mathbf{0}$ and (1.194) at once implies $c^2 p''_\mu p''^\mu = E''^2 = m^2c^4 = c^2 p_\mu p^\mu = E^2 - (c\hbar k)^2$, thus providing a stringent proof the rest energy mc^2 being Lorentz invariant.

In the frame $x = (ct, \mathbf{r})$, now combining the Hamiltonian (1.185) with the time evolution (1.172) and minding the definitions (1.184) for the gauge invariant derivatives, the Dirac equation reads

$$\left(c\Pi_0 + c\alpha_b\Pi_b + mc^2\beta\right)\Psi(x) = 0 \tag{1.198}$$

After multiplication on the left with β one obtains introducing

$$\begin{aligned} \gamma^0 &= \beta \\ \gamma^b &= \beta\alpha_b \text{ with } b \in \{x, y, z\} \end{aligned} \tag{1.199}$$

the Dirac equation in the familiar guise

$$\left(c\gamma^\mu\Pi_\mu + mc^2 \cdot 1_{4\times4}\right)\Psi(x) = 0 \tag{1.200}$$

Notably there holds

$$\gamma^\mu\gamma^{\mu'} + \gamma^{\mu'}\gamma^\mu = 2g^{\mu\mu'} \cdot 1_{4\times4} \tag{1.201}$$

Unfortunately, the peculiar Hermitean conjugation property

$$\begin{aligned} (\gamma^b)^\dagger &= -\gamma^b \\ (\gamma^0)^\dagger &= \gamma^0 \end{aligned} \tag{1.202}$$

(valid in any unitary equivalent representation) prevents the matrices γ^μ transforming like the components of a four-vector. For this reason the prevalent habit of taking $\gamma^\mu\Pi_\mu$ as a kind of Lorentz scalar is *incorrect*. Instead, as has been emphasized by Dirac [58], the Lorentz invariance of (1.200) emerges from the algebraic properties of the matrices γ^μ.

According to the principle of relativity it is required there holds in the frame $x' = (ct', \mathbf{r}')$ instead of (1.200) now the Dirac equation

$$\left(c\gamma^\mu\Pi'_\mu + mc^2 \cdot 1_{4\times4}\right)\Psi'(x') = 0 \tag{1.203}$$

Here the matrices γ^μ remain fixed. It is the four-vector Π_μ that is changed under a Lorentz transformation to $\Pi'_{\mu'}$ and accordingly the Dirac wave function $\Psi(x)$ is changed to $\Psi'(x')$. As the four components $\Psi_\nu(x)$ do *not* build a four-vector, a different transformation law is seeked for, so that an observer in the frame x' can reconstruct $\Psi'(x')$ if $\Psi(x)$ was known to him.

Progress comes assuming for every Lorentz-transformation $x'^{\mu'} = L^{\mu'}_{\ \mu} x^{\mu}$ there exists a *linear* transformation $\mathcal{L} = \mathcal{L}[L]$ operating on the components $\Psi_\nu(x)$ of the Dirac wave function so that

$$\Psi'_{\nu'}(x') = \sum_{\nu} \mathcal{L}_{\nu',\nu} \Psi_\nu(x) \tag{1.204}$$

To find that transformation $\mathcal{L}$ in the guise of a 4×4-matrix it suffices to restrict to an *infinitesimal* Lorentz transformation connecting x and x' by

$$\begin{aligned} x'^{\mu'} &= \frac{\partial x'^{\mu'}}{\partial x^{\mu}} x^{\mu} = L^{\mu'}_{\ \mu} x^{\mu} = \left(\delta^{\mu'}_{\mu} + \epsilon^{\mu'}_{\ \mu} + \cdots\right) x^{\mu} \\ \epsilon^{\mu}_{\ \mu'} &= -\epsilon^{\mu}_{\ \mu'} \\ \left|\epsilon^{\mu}_{\ \mu'}\right| &\ll 1 \end{aligned} \tag{1.205}$$

This applies, because any Lorentz transformation can be obtained from a succession of infinitesimal Lorentz transformations.

The associated infinitesimal Lorentz transformations of the four-potential and the four-momentum operator read correspondingly

$$\begin{aligned} A'^{\mu'}(x') &= \left(\delta^{\mu'}_{\mu} + \epsilon^{\mu'}_{\ \mu}\right) A^{\mu}(x) \\ \partial'^{\mu'} &= \left(\delta^{\mu'}_{\mu} + \epsilon^{\mu'}_{\ \mu}\right) \partial^{\mu} \\ \Pi'^{\mu'} &= \left(\delta^{\mu'}_{\mu} + \epsilon^{\mu'}_{\ \mu}\right) \Pi^{\mu} \end{aligned} \tag{1.206}$$

Being interested in the inverse Lorentz transformation then of course

$$\begin{aligned} \Pi^{\mu} &= \left(\delta^{\mu}_{\mu'} - \epsilon^{\mu}_{\ \mu'}\right) \Pi'^{\mu'} \\ \Pi_{\mu} &= g_{\mu\mu''} \left(\delta^{\mu''}_{\mu'} - \epsilon^{\mu''}_{\ \mu'}\right) g^{\mu'\mu'''} \Pi'_{\mu'''} = \left(\delta^{\ \mu'}_{\mu} - \epsilon_{\mu}^{\ \mu'}\right) \Pi'_{\mu'} \end{aligned} \tag{1.207}$$

Likewise, the representation $\mathcal{L} = \mathcal{L}[L]$ operating on the components $\Psi_\nu(x)$ for the case of an *infinitesimal* Lorentz transformation we may write as

$$\begin{aligned} \Psi'_{\nu'}(x') &= \sum_{\nu} \left(\delta_{\nu',\nu} + \Omega_{\nu',\nu}\right) \Psi_\nu(x) \\ \left|\Omega_{\nu',\nu}\right| &\ll 1 \end{aligned} \tag{1.208}$$

The task is now to establish a connection between the 4×4-matrix Ω and the generator $\epsilon^{\mu'}_{\mu}$ of the Lorentz transformation. To this end we express $\Psi_\nu(x)$ in terms of $\Psi'_{\nu'}(x')$ by inverting (1.208)

$$\Psi_\nu(x) = \sum_{\nu'} \left(\delta_{\nu,\nu'} - \Omega_{\nu,\nu'}\right) \Psi'_{\nu'}(x') \tag{1.209}$$

Upon insertion of (1.207) and (1.209) into the original Dirac equation (1.200) one finds

$$\begin{aligned} 0 &= \sum_{\nu'} \left(c\gamma^\mu \Pi_\mu + mc^2 \cdot 1_{4\times4}\right)_{\nu,\nu'} \Psi_{\nu'}(x) \\ &= \sum_{\nu',\nu''} \left(c\gamma^\mu \left(\delta^{\mu'}_{\mu} - \epsilon^{\mu'}_{\mu}\right) \Pi'_{\mu'} + mc^2 \cdot 1_{4\times4}\right)_{\nu,\nu'} \left(\delta_{\nu',\nu''} - \Omega_{\nu',\nu''}\right) \Psi'_{\nu''}(x') \\ &= \sum_{\nu',\nu''} \left(c\gamma^{\mu'} \Pi'_{\mu'} + mc^2 \cdot 1_{4\times4} - c\gamma^\mu \epsilon^{\mu'}_{\mu} \Pi'_{\mu'}\right)_{\nu,\nu'} \left(\delta_{\nu',\nu''} - \Omega_{\nu',\nu''}\right) \Psi'_{\nu''}(x') \end{aligned} \tag{1.210}$$

Multiplication from the left with $\left(\delta_{\nu''',\nu} + \Omega_{\nu''',\nu}\right)$ simplifies the analyis. With neglect of the higher order terms and relabeling summation indices $(\mu, \mu') \to (\mu', \mu)$

$$\begin{aligned} 0 &= \sum_{\nu',\nu'',\nu'''} \left(\delta_{\nu''',\nu} + \Omega_{\nu''',\nu}\right) \left(c\gamma^{\mu'} \Pi'_{\mu'} + mc^2 \cdot 1_{4\times4} - c\gamma^\mu \epsilon^{\mu'}_{\mu} \Pi'_{\mu'}\right)_{\nu,\nu'} \left(\delta_{\nu',\nu''} - \Omega_{\nu',\nu''}\right) \Psi_{\nu''}(x') \\ &= \sum_{\nu'} \left(c\gamma^{\mu'} \Pi'_{\mu'} + mc^2 \cdot 1_{4\times4} + \left[\Omega, \left(c\gamma^{\mu'} \Pi'_{\mu'} + mc^2 \cdot 1_{4\times4}\right)\right] - c\gamma^\mu \epsilon^{\mu'}_{\mu} \Pi'_{\mu'}\right)_{\nu,\nu'} \Psi'_{\nu'}(x') \\ &= \sum_{\nu'} \left(c\gamma^\mu \Pi'_\mu + mc^2 \cdot 1_{4\times4} + \left(\left[\Omega, \gamma^\mu\right] - \gamma^{\mu'} \epsilon^{\mu}_{\mu'}\right) c\Pi'_\mu\right)_{\nu,\nu'} \Psi'_{\nu'}(x') \end{aligned} \tag{1.211}$$

Apparently, if it was possible here to determine Ω from the requirement

$$\left[\Omega, \gamma^\mu\right] \stackrel{!}{=} \gamma^{\mu'} \epsilon^{\mu}_{\mu'}, \tag{1.212}$$

then by means of the identity (1.211) the relativistic invariance of the Dirac equation has been proven.

At first sight it appears problematic to find Ω concurrently meeting the four conditions posed in (1.212) for $\mu \in \{0, x, y, z\}$. It is the magic of the algebra of the Dirac matrices that Ω can in deed be found *independent* on the index μ and given solely in terms of the generators $\epsilon_{\kappa\kappa'} = \epsilon^{\mu'}_{\kappa} g_{\mu'\kappa'}$ of the infinitesimal Lorentz transformation and the matrices $\gamma^\kappa \gamma^{\kappa'}$, specifically

$$\Omega = \frac{1}{4} \epsilon_{\kappa\kappa'} \gamma^\kappa \gamma^{\kappa'} \tag{1.213}$$

The proof uses the multiplication rules (1.201) and the anti-symmetry $\epsilon_{\kappa\kappa'} = -\epsilon_{\kappa'\kappa}$.

$$
\begin{aligned}
\left[\Omega, \gamma^{\mu}\right] &= \frac{1}{4}\epsilon_{\kappa\kappa'}\left[\gamma^{\kappa}\gamma^{\kappa'}, \gamma^{\mu}\right] \\
&= \frac{1}{4}\epsilon_{\kappa\kappa'}\left(\gamma^{\kappa}\left(\gamma^{\kappa'}\gamma^{\mu} + \gamma^{\mu}\gamma^{\kappa'}\right) - \left(\gamma^{\kappa}\gamma^{\mu} + \gamma^{\mu}\gamma^{\kappa}\right)\gamma^{\kappa'}\right) \\
&= \frac{1}{4}\epsilon_{\kappa\kappa'}\left(\gamma^{\kappa}2g^{\kappa'\mu} - 2g^{\kappa\mu}\gamma^{\kappa'}\right) \\
&= \frac{1}{2}\epsilon_{\kappa\kappa'}g^{\kappa'\mu}\gamma^{\kappa} - \frac{1}{2}\epsilon_{\kappa'\kappa}g^{\kappa'\mu}\gamma^{\kappa} \\
&= \epsilon_{\kappa\kappa'}g^{\kappa'\mu}\gamma^{\kappa} \\
&= \gamma^{\kappa}\epsilon_{\kappa}^{\ \mu}
\end{aligned}
$$

So the result (1.213) together with (1.211) engenders $\Psi'_{\upsilon'}(x')$ being a solution of the Dirac equation in the frame x', this way establishing via the transformation law (1.208) the Lorentz invariance of the Dirac equation. Actually the derived transformation law (1.209) classifies $\Psi_{\upsilon}(x)$ not transforming as a four-vector at all, but instead transforming as the components of a four-dimensional object known as Dirac spinor.

Nonrelativistic Limit

To verify that the Dirac equation

$$
i\hbar\frac{\partial}{\partial t}\Psi(\mathbf{r}, t) = \left(\beta mc^2 + e\Phi + c\alpha_b\Pi_b\right)\Psi(\mathbf{r}, t) \tag{1.214}
$$

describes an electron interacting with external electromagnetic fields just like in Schrödinger-Pauli quantum mechanics, one restricts to the low energy sector, where kinetic and potential energies can be considered as being very much less than the rest energy mc^2. In this case it is expedient to search for a solution to (1.214) in the guise

$$
\Psi(\mathbf{r}, t) = e^{-it\frac{mc^2}{\hbar}}\begin{pmatrix}\psi_+(\mathbf{r}, t)\\ \psi_-(\mathbf{r}, t)\\ \chi_+(\mathbf{r}, t)\\ \chi_-(\mathbf{r}, t)\end{pmatrix}, \tag{1.215}
$$

whereby $\Psi(\mathbf{r}, t)$ is here represented with *two-component* envelope wavefunctions referred to as $\psi(\mathbf{r}, t)$ and $\chi(\mathbf{r}, t)$. Bearing in mind the definition of β and α_b in (1.182), one finds upon insertion of this ansatz into (1.214) the two coupled equations

$$
\begin{aligned}
\left(i\hbar\frac{\partial}{\partial t} - e\Phi\right)\psi(\mathbf{r}, t) &= c\sigma_b{}^{(P)}\Pi_b\chi(\mathbf{r}, t) \\
\left(i\hbar\frac{\partial}{\partial t} - e\Phi + 2mc^2\right)\chi(\mathbf{r}, t) &= c\sigma_b{}^{(P)}\Pi_b\psi(\mathbf{r}, t)
\end{aligned} \tag{1.216}
$$

As the envelope functions $\chi_\pm(\mathbf{r}, t)$ and $\psi_\pm(\mathbf{r}, t)$ can be assumed to vary only *slowly* on the time scale $\frac{\hbar}{mc^2}$, the equations determining these two component functions can be decoupled in the style of perturbation theory, thus obtaining to lowest order the equations

$$\chi(\mathbf{r}, t) = \frac{1}{2mc}\sigma_b^{(P)}\Pi_b\psi(\mathbf{r}, t) \tag{1.217}$$

$$\left(i\hbar\frac{\partial}{\partial t} - e\Phi\right)\psi(\mathbf{r}, t) = \frac{1}{2m}\left(\sigma_b^{(P)}\Pi_b\right)^2\psi(\mathbf{r}, t) \tag{1.218}$$

Apparently $\chi(\mathbf{r}, t)$ in (1.217) represents the so-called *small* components of the Dirac wavefunction. By contrast $\psi(\mathbf{r}, t)$ in (1.218) represents the so-called large components.

The kinetic term in (1.218) conceals the coupling of the magnetic moment of the electron to the magnetic induction field $\mathsf{B}_a = \epsilon_{bb'a}\frac{\partial}{\partial r_b}\mathsf{A}_{b'}$. In fact, with the algebra of Pauli spin-matrices (1.181) there follows

$$\begin{aligned}\left(\sigma_b^{(P)}\Pi_b\right)^2 &= \sigma_b^{(P)}\sigma_{b'}^{(P)}\Pi_b\Pi_{b'} \\ &= \Pi_b\Pi_b 1_{2\times2} + i\epsilon_{bb'a}\Pi_b\Pi_{b'}\sigma_a^{(P)} \\ &= \Pi_b\Pi_b 1_{2\times2} - i\epsilon_{bb'a}\left[\mathsf{p}_b, e\mathsf{A}_{b'}\right]\sigma_a^{(P)} \\ &= (\mathsf{p}_b - e\mathsf{A}_b)(\mathsf{p}_b - e\mathsf{A}_b)1_{2\times2} - e\hbar\mathsf{B}_a\sigma_a^{(P)}\end{aligned}$$

By this means Eq. (1.218) determining the time evolution of $\psi(\mathbf{r}, t)$ changes into the equivalent guise

$$i\hbar\frac{\partial}{\partial t}\psi(\mathbf{r}, t) = \left(e\Phi + \frac{(\mathsf{p} - e\mathsf{A})^2}{2m} - \frac{e\hbar}{2m}\sigma_a^{(P)}\mathsf{B}_a\right)\psi(\mathbf{r}, t) \tag{1.219}$$

Remarkably, Eq. (1.219) for the two-component envelope functions $\psi_\pm(\mathbf{r}, t)$ coincides in every detail with the *single-particle* Schrödinger-Pauli equation of quantum mechanics for the nonrelativistic spin-$\frac{1}{2}$ electron with minimal coupling to electromagnetic potentials. An extension of the scalar Schrödinger wave equation to a two component wave function was introduced by Pauli in an ad hoc manner in the year 1927 in order to describe the electron's spin. The deeper theoretical foundation of why the spin of the electron is by necessity a direct consequence of combining quantum mechanics with special relativity is due to P.A.M. Dirac, who introduced his fundamental equation only a year later in 1928.

For a uniform external magnetic induction field B there holds $\mathsf{A} = \frac{1}{2}\mathsf{B} \wedge \mathsf{r}$. Then one finds from (1.219) in terms of the operators of orbital angular momentum $\mathsf{L}_a = (\mathsf{r} \wedge \mathsf{p})_a$ and spin $\mathsf{S}_a = \frac{\hbar}{2}\sigma_a^{(P)}$

$$i\hbar\frac{\partial}{\partial t}\psi(\mathbf{r}, t) = \left(e\Phi + \frac{\mathsf{p}^2}{2m} - \frac{e}{2m}\left(\mathsf{L}_a + 2\mathsf{S}_a\right)\mathsf{B}_a + \frac{e^2}{2m}\mathsf{A}^2\right)\psi(\mathbf{r}, t) \tag{1.220}$$

Obviously a gyromagnetic ratio $g = 2$ for the electron is here *deduced* from the nonrelativistic limit of the Dirac equation, whereas in Schrödinger-Pauli quantum mechanics this value for g had to be taken from experiment.

Besides this great success of the Dirac equation, in addition it accounts for the observed *fine structure* of the spectrum of the hydrogen atom! A detailed description of these and several other major achievements of the Dirac equation can be found in advanced textbooks on quantum mechanics, for instance [18, 19, 21, 59, 60].

Plane Wave Solutions of the Free-Particle Dirac Equation

For a *freely* moving particle governed by the Hamiltonian (1.175) the Dirac equation reads

$$i\hbar\frac{\partial}{\partial t}\Psi(\mathbf{r},t) = \left(mc^2\beta + c\alpha_b \mathsf{p}_b\right)\Psi(\mathbf{r},t) \tag{1.221}$$

In this case, of course, there exist plane wave solutions. Making the ansatz

$$\Psi_\nu(\mathbf{r},t) = \mathsf{C}_\nu(\mathbf{k})\frac{1}{\sqrt{V_P}}\exp\left\{i\,(\mathbf{k}\cdot\mathbf{r} - \frac{E_k}{\hbar}t)\right\} \tag{1.222}$$

there follows from the requirement, that this ansatz should solve (1.221), an equivalent eigenvalue problem of linear algebra

$$\sum_{\nu'}(\mathcal{H}_\mathbf{k})_{\nu,\nu'}\,\mathsf{C}_{\nu'}(\mathbf{k}) = E_k\mathsf{C}_\nu(\mathbf{k}) \tag{1.223}$$

$$\mathcal{H}_\mathbf{k} = mc^2\beta + c\hbar\,k_b\alpha_b \tag{1.224}$$

To diagonalize the matrix $\mathcal{H}_\mathbf{k}$ in (1.224) let us look for a unitary transformation $\mathsf{T}_\mathbf{k}$ depending on a freely selectable c-number parameter $\vartheta_\mathbf{k}$ yet to be determined

$$\begin{aligned}\mathsf{T}_\mathbf{k} &\equiv \mathbb{1}_{4\times4}\cos\vartheta_\mathbf{k} + n_a\alpha_a\beta\sin\vartheta_\mathbf{k} \\ \mathsf{T}_\mathbf{k}^\dagger\mathsf{T}_\mathbf{k} &= \mathbb{1}_{4\times4} = \mathsf{T}_\mathbf{k}\mathsf{T}_\mathbf{k}^\dagger\end{aligned} \tag{1.225}$$

With help of the identity $(n_a\alpha_a)(n_b\alpha_b) \equiv \mathbb{1}_{4\times4}$ and minding elementary trigonometric formulas then with $k_b = |\mathbf{k}|\,n_b$ there holds [61]

$$\begin{aligned}&\mathsf{T}_\mathbf{k}^\dagger\,\mathcal{H}_\mathbf{k}\,\mathsf{T}_\mathbf{k} \\ &= \left(\mathbb{1}_{4\times4}\cos\vartheta_\mathbf{k} + \beta\alpha_{a'}n_{a'}\sin\vartheta_\mathbf{k}\right)\left(mc^2\beta + c\hbar\,|\mathbf{k}|\,n_b\alpha_b\right)\left(\mathbb{1}_{4\times4}\cos\vartheta_\mathbf{k} + n_a\alpha_a\beta\sin\vartheta_\mathbf{k}\right) \\ &= \left(\cos 2\vartheta_\mathbf{k} - \frac{mc^2}{c\hbar\,|\mathbf{k}|}\sin 2\vartheta_\mathbf{k}\right)c\hbar\,|\mathbf{k}|\,n_b\alpha_b + \left(\cos 2\vartheta_\mathbf{k} + \frac{c\hbar\,|\mathbf{k}|}{mc^2}\sin 2\vartheta_\mathbf{k}\right)mc^2\beta\end{aligned} \tag{1.226}$$

Choosing $\vartheta_{\mathbf{k}}$ such that the coefficient of the matrix $n_b\alpha_b$ *vanishes*, leads to

$$\tan 2\vartheta_{\mathbf{k}} = \frac{c\hbar\,|\mathbf{k}|}{mc^2} \tag{1.227}$$

Provided (1.227) holds true, then indeed $\mathsf{T}_{\mathbf{k}}$ diagonalizes the matrix $\mathcal{H}_{\mathbf{k}}$, i.e.

$$\mathsf{T}_{\mathbf{k}}^{\dagger}\,\mathcal{H}_{\mathbf{k}}\,\mathsf{T}_{\mathbf{k}} = \left(\cos 2\vartheta_{\mathbf{k}} + \frac{c\hbar\,|\mathbf{k}|}{mc^2}\sin 2\vartheta_{\mathbf{k}}\right) mc^2\beta = \beta E_{\mathbf{k}}^{(0)} \tag{1.228}$$

$$E_{\mathbf{k}}^{(0)} = \sqrt{\left(mc^2\right)^2 + (c\hbar\mathbf{k})^2}$$

Notably, the positive eigenvalues $E_k = E_{\mathbf{k}}^{(0)}$, respectively the negative eigenvalues $E_k = -E_{\mathbf{k}}^{(0)}$, if viewed per se, are twofold degenerate, as is immediately revealed inserting the diagonal matrix β, see (1.182). The deeper physical reason for this degeneracy being, that indeed $\mathcal{H}_{\mathbf{k}}$ commutes with the *helicity* operator $n_a\sigma_a$, i.e.

$$\begin{aligned}\left[n_a\sigma_a, \mathcal{H}_{\mathbf{k}}\right] &= c\hbar\,|\mathbf{k}|\,n_a n_b\left[\sigma_a, \alpha_b\right] \\ &= c\hbar\,|\mathbf{k}|\,\underbrace{n_a n_b \epsilon_{abb'}}_{\equiv 0} 2i\alpha_{b'} \equiv 0_{4\times 4}\end{aligned} \tag{1.229}$$

Because $\mathcal{H}_{\mathbf{k}}$ is a hermitean 4×4-matrix it has, of course, four mutually orthogonal eigenvectors $\mathsf{C}_\nu^{(j)}(\mathbf{k})$ and four associated eigenvalues $E_k^{(j)}$, that we distinguish here with a label $j \in \{1, 2, 3, 4\}$.

For $j = 1, 2$ the eigenvalues of $\mathcal{H}_{\mathbf{k}}$ are given by $E_k^{(j)} = E_{\mathbf{k}}^{(0)}$, exactly coinciding with the relativistic kinetic energy of a freely moving particle stated afore in (1.174). But for $j = 3, 4$ the eigenvalues of $\mathcal{H}_{\mathbf{k}}$ are *negative* and given by $E_k^{(j)} = -E_{\mathbf{k}}^{(0)}$.

Now, by construction the searched for eigenvectors $\mathsf{C}_\nu^{(j)}(\mathbf{k})$ conform with the jth-column of the unitary 4×4-matrix $\mathsf{T}_{\mathbf{k}}$. This is because (1.228) implies upon multiplication with the unitary matrix $\mathsf{T}_{\mathbf{k}}$ from the left

$$\mathcal{H}_{\mathbf{k}}\,\mathsf{T}_{\mathbf{k}} = \mathsf{T}_{\mathbf{k}}\,\beta E_{\mathbf{k}}^{(0)} \tag{1.230}$$

Obviously this is equivalent to (1.223). With that said there holds in view of (1.225)

$$\mathsf{T}_{\mathbf{k}} = \xi_k^{(+)} 1_{4\times 4} + \xi_k^{(-)} n_a\alpha_a\beta \tag{1.231}$$

$$= \begin{pmatrix} \xi_k^{(+)} & , 0 & , -\xi_k^{(-)} n_z & , -\xi_k^{(-)}\left(n_x - i\,n_y\right) \\ 0 & , \xi_k^{(+)} & , -\xi_k^{(-)}\left(n_x + i\,n_y\right) & , \xi_k^{(-)} n_z \\ \xi_k^{(-)} n_z & , \xi_k^{(-)}\left(n_x - i\,n_y\right) & , \xi_k^{(+)} & , 0 \\ \xi_k^{(-)}\left(n_x + i\,n_y\right) & , -\xi_k^{(-)} n_z & , 0 & , \xi_k^{(+)} \end{pmatrix}$$

where

$$\xi_k^{(\pm)} = \sqrt{\frac{1}{2}\left(1 \pm \cos 2\vartheta_{\mathbf{k}}\right)} = \sqrt{\frac{1}{2}\left(1 \pm \frac{mc^2}{E_{\mathbf{k}}^{(0)}}\right)} \tag{1.232}$$

Because $(n_a\sigma_a)^2 = \mathbb{1}_{4\times 4}$ the eigenvalues η of the helicity operator $n_a\sigma_a$ are $\eta \in \{+1, -1\}$. So η may serve as well as a label to count the two-fold degeneracy of the eigenvectors $\mathsf{C}_\nu^{(j)}(\mathbf{k})$ of $\mathcal{H}_{\mathbf{k}}$ corresponding to the eigenvalues $E_k^{(j)}$. Note these eigenvectors $\mathsf{C}_\nu^{(j)}(\mathbf{k})$ are readily identified being the jth-column vector of the matrix (1.231) for $j = 1, 2, 3, 4$, and due to the unitarity of $\mathsf{T}_{\mathbf{k}}$, see (1.225), these eigenvectors together span a complete orthonormal basis system.

So, *concurrent* eigenvectors of $\mathcal{H}_{\mathbf{k}}$ and of the helicity operator $n_a\sigma_a$ can be found as certain linear combinations built from these eigenvectors $\mathsf{C}_\nu^{(j)}(\mathbf{k})$.

Introducing the linear combinations

$$U_\nu(\mathbf{k}, \eta) = \sum_{j\in\{1,2\}} W_j^{(\eta)} \mathsf{C}_\nu^{(j)}(\mathbf{k}) \tag{1.233}$$
$$V_\nu(\mathbf{k}, \eta) = \sum_{j\in\{3,4\}} W_j^{(\eta)} \mathsf{C}_\nu^{(j)}(\mathbf{k})$$

then there holds by construction, of course

$$\sum_{\nu'} (\mathcal{H}_{\mathbf{k}})_{\nu,\nu'} U_{\nu'}(\mathbf{k}, \eta) = E_{\mathbf{k}}^{(0)}\, U_\nu(\mathbf{k}, \eta)$$
$$\sum_{\nu'} (\mathcal{H}_{\mathbf{k}})_{\nu,\nu'} V_{\nu'}(\mathbf{k}, \eta) = -E_{\mathbf{k}}^{(0)}\, V_\nu(\mathbf{k}, \eta) \tag{1.234}$$

But still we need to find the c-number coefficients $W_j^{(\eta)}$ so that there holds as well

$$\sum_{\nu'} (n_a\sigma_a)_{\nu,\nu'} U_{\nu'}(\mathbf{k}, \eta) = \eta\, U_\nu(\mathbf{k}, \eta)$$
$$\sum_{\nu'} (n_a\sigma_a)_{\nu,\nu'} V_{\nu'}(\mathbf{k}, \eta) = \eta\, V_\nu(\mathbf{k}, \eta) \tag{1.235}$$

To this end let us rewrite (1.233) in the guise

$$U_\nu(\mathbf{k}, \eta) = \sum_{\nu'} (\mathsf{T}_{\mathbf{k}})_{\nu,\nu'} \begin{pmatrix} W_1^{(\eta)} \\ W_2^{(\eta)} \\ 0 \\ 0 \end{pmatrix}_{\nu'} \quad \text{and} \quad V_\nu(\mathbf{k}, \eta) = \sum_{\nu'} (\mathsf{T}_{\mathbf{k}})_{\nu,\nu'} \begin{pmatrix} 0 \\ 0 \\ W_3^{(\eta)} \\ W_4^{(\eta)} \end{pmatrix}_{\nu'}$$

Then, making use of the unitarity relation (1.225), the eigenvalue equations (1.235) are readily converted into

$$\sum_{\nu'}\left(\mathsf{T}_{\mathbf{k}}^{\dagger}n_a\sigma_a\mathsf{T}_{\mathbf{k}}\right)_{\nu,\nu'}\begin{pmatrix}W_1^{(\eta)}\\W_2^{(\eta)}\\W_3^{(\eta)}\\W_4^{(\eta)}\end{pmatrix}_{\nu'}=\eta\begin{pmatrix}W_1^{(\eta)}\\W_2^{(\eta)}\\W_3^{(\eta)}\\W_4^{(\eta)}\end{pmatrix}_{\nu}$$

But the matrix product here simplifies upon insertion of (1.231) into

$$\mathsf{T}_{\mathbf{k}}^{\dagger}n_a\sigma_a\mathsf{T}_{\mathbf{k}}=n_a\sigma_a=\begin{pmatrix}n_z & ,n_x-in_y & ,0 & ,0\\ n_x+in_y & ,-n_z & ,0 & ,0\\ 0 & ,0 & ,n_z & ,n_x-in_y\\ 0 & ,0 & ,n_x+in_y & ,-n_z\end{pmatrix} \tag{1.236}$$

Thus, minding $n_a n_a=1$ and $\eta^2=1$ it is readily confirmed

$$\begin{aligned}W_1^{(\eta)}&=\frac{n_z+\eta}{\sqrt{2\,(1+\eta n_z)}}=W_3^{(\eta)}\\ W_2^{(\eta)}&=\frac{n_x+i\,n_y}{\sqrt{2\,(1+\eta n_z)}}=W_4^{(\eta)}\end{aligned} \tag{1.237}$$

Looking up the column vectors $\mathbf{C}_{\nu}^{(j)}(\mathbf{k})$ from (1.231) and inserting the weights (1.237) one obtains directly from (1.233)

$$U_{\nu}(\mathbf{k},\eta)=\frac{1}{\sqrt{2\,(1+\eta\,n_z)}}\begin{pmatrix}(n_z+\eta)\,\xi_k^{(+)}\\(n_x+i\,n_y)\,\xi_k^{(+)}\\(n_z+\eta)\,\eta\,\xi_k^{(-)}\\(n_x+i\,n_y)\,\eta\,\xi_k^{(-)}\end{pmatrix}_{\nu} \tag{1.238}$$

$$V_{\nu}(\mathbf{k},\eta)=\frac{1}{\sqrt{2\,(1+\eta\,n_z)}}\begin{pmatrix}-(n_z+\eta)\,\eta\,\xi_k^{(-)}\\-(n_x+i\,n_y)\,\eta\,\xi_k^{(-)}\\(n_z+\eta)\,\xi_k^{(+)}\\(n_x+i\,n_y)\,\xi_k^{(+)}\end{pmatrix}_{\nu} \tag{1.239}$$

$$n_b=\frac{k_b}{|\mathbf{k}|}\quad,\quad\eta\in\{+,-\} \tag{1.240}$$

These are indeed the searched for concurrent eigenstates of the Hamiltonian $\mathcal{H}_{\mathbf{k}}$ and the helicity operator $n_b\sigma_b$. Positive and negative energy eigenstates and/or positive and negative helicity eigenstates are mutually orthogonal. To check the eigenvalue equations (1.235) is easy. The probe of the original eigenvalue equations (1.234) is not difficult either, observing $\xi_k^{(\pm)}$ as defined in (1.232) obeys to

$$\frac{\xi_k^{(+)}}{\xi_k^{(-)}}=\frac{E_{\mathbf{k}}^{(0)}+mc^2}{c\hbar\,|\mathbf{k}|}\quad,\quad\frac{\xi_k^{(-)}}{\xi_k^{(+)}}=\frac{E_{\mathbf{k}}^{(0)}-mc^2}{c\hbar\,|\mathbf{k}|} \tag{1.241}$$

From what has been said eventually a complete system of solutions of the free-particle Dirac equation is obtained

$$
\begin{aligned}
\eta, \tau &\in \{+1, -1\} \\
k &= (\mathbf{k}, \eta, \tau) \\
\phi_\nu\,(\mathbf{k}, \eta, \tau) &= \frac{1+\tau}{2} U_\nu(\mathbf{k}, \eta) + \frac{1-\tau}{2} V_\nu(\mathbf{k}, \eta) \\
\Psi_\nu(\mathbf{r}, t; k) &= \phi_\nu\,(\mathbf{k}, \eta, \tau)\, \frac{1}{\sqrt{V_P}} \exp\left\{ i\,(\mathbf{k}\cdot\mathbf{r} - \tau\, \frac{E^{(0)}_{\mathbf{k}}}{\hbar} t) \right\}
\end{aligned}
\tag{1.242}
$$

The label τ specifies the positive-energy solutions ($\tau = +1$), respectively the negative-energy solutions ($\tau = -1$) with eigenvalue $\tau\, E^{(0)}_{k}$. The labels $\mathbf{k}$, η, τ serve to *count* all the modes. The multi-index k is introduced in (1.242) to simplify notation. So it follows from what has been said

$$
\begin{aligned}
(\mathcal{H}_{\mathbf{k}})_{\nu,\nu'} \phi_{\nu'}\,(k) &= E_k\, \phi_\nu\,(k) \\
E_k \equiv E_{\mathbf{k},\eta,\tau} &= \tau\, E^{(0)}_{\mathbf{k}} \\
\phi_{\nu'}\,(k) &= \phi_{\nu'}\,(\mathbf{k}, \eta, \tau)
\end{aligned}
$$

Actually the modes (1.242) are normalized here as plane waves subject to periodic boundary conditions in a box volume V_P. They satisfy the usual orthogonality and completeness relations with regard to the labels $\mathbf{k}$, η, τ. Notably, the solutions with a fixed wave vector $\mathbf{k}$ themselves already form a complete orthogonal system with regard to the labels η, τ. Consequently

$$
\begin{aligned}
\sum_{\tau,\eta\in\{+1,-1\}} \phi(k)\phi^\dagger(k) &= \mathbb{1}_{4\times4} \\
\sum_{\tau,\eta\in\{+1,-1\}} E_k \cdot \phi(k)\phi^\dagger(k) &= \mathcal{H}_{\mathbf{k}} \\
\sum_{\tau,\eta\in\{+1,-1\}} \tau \cdot \phi(k)\phi^\dagger(k) &= \Lambda_{\mathbf{k}} \equiv \frac{1}{E^{(0)}_{\mathbf{k}}} \mathcal{H}_{\mathbf{k}} \\
\sum_{\tau,\eta\in\{+1,-1\}} \frac{1\pm\tau}{2} \cdot \phi(k)\phi^\dagger(k) &= \frac{1}{2}\,(\mathbb{1}_{4\times4} \pm \Lambda_{\mathbf{k}})
\end{aligned}
\tag{1.243}
$$

Notice: Often one uses *other* spinors than (1.238), (1.239). These other spinors originate from a Lorentz boost applied to the four eigenstates of the Dirac Hamiltonian $mc^2\beta$ in the rest frame. But a Lorentz boost ist *not* a unitary transformation and so such spinors have a different normalization, for instance [19, 59].

Reframing the Negative Energy Solutions as Positrons

Even though the Dirac eigenvalue problem can be solved for the Coulomb potential exactly [19, 60] and thus explains very successfully the fine structure of the spectrum of the hydrogen atom by restricting to the positive-energy solutions, a general dispense of the negative-energy solutions is highly problematic. Not only exist valid objections to such an ad hoc truncation of a mathematical nature. There exist also deep *physical* reasons to take the negative-energy solutions seriously.

Regarding, for instance, the interaction of the relativistic electron with a rapidly varying electromagnetic radiation field, it is readily perceived, that an electron initially in a positive-energy state can by emission of a high energy photon transit to a negative-energy state! But as this process obviously may reiterate, this poses indeed a severe problem of stability.

This vexing difficulty led Dirac (again) to a new disruptive idea. He proposed that all *negative-energy* states of the free-particle Hamiltonian with energy ranging from $-\infty < -E_k^{(0)} < -mc^2$ were actually filled up with electrons obeying to the Pauli exclusion principle. This way the transit of a positive-energy electron to any of these negative-energy states should be prevented and so the *stability* of the positive energy states of the relativistic electron would be re-established.

In effect Dirac imagined the groundstate of the relativistic electron as a new kind of *many particle* state, in which all negative-energy states were occupied, but all positive-energy states, being separated by an energy gap $2mc^2$ from the filled "sea" of negative-energy states, were unoccupied. Such a concept of the ground state has actually a close analogy with the physics of an insulator, where a filled valence band is separated from an empty conduction band by an energy gap, the zero of the energy scale being in the middle of this gap.

If accepted as a logical possibility, then electrons in the filled negative-energy states (valence band) cannot change their states unless enough energy is absorbed, say from a rapidly changing external electromagnetic field, so that a transition from a negative-energy state (valence-band) across an energy gap $2mc^2$ to a positive-energy state (conduction band) was enabled. In effect this way a particle-hole pair gets created representing an excitation from that ground state.

To elucidate (for a free-particle Hamiltonian) the notion of a hole, consider *removing* an electron with momentum $\hbar\mathbf{k}$, helicity η and energy $-E_k^{(0)}$ from that valence band. Let us denote the total energy of the *filled* valence band as E_v, its total momentum as $\mathbf{P}_v$, its total helicity as η_v and its total charge as q_v. If that electron was taken out then these *conserved* quantities changed as follows

$$
\begin{aligned}
q_v &\to q_v' = q_v - e = q_v + |e| \\
\eta v &\to \eta_v' = \eta_v - \eta \\
\mathbf{P}_v &\to \mathbf{P}_v' = \mathbf{P}_v - \hbar\mathbf{k} \\
E_v &\to E_v' = E_v - (-E_k^{(0)}) = E_v + E_k^{(0)}
\end{aligned}
\tag{1.244}
$$

Naturally this process of creating a state with one negative-energy electron missing can be understood as being equivalent to the *creation* of a "hole" in the valence band. Notably, such a hole in the valence band now carries a *positive* charge $-e = |e|$. Also it has the opposite momentum $-\hbar\mathbf{k}$ and opposite helicity $-\eta$ compared to that of the removed electron, and last not least this hole has now a *positive* energy $E_k^{(0)}$. The latter feature enables the interpretation of a hole behaving like a real particle, analogous to the electron, as its rest energy is mc^2 and so it should be assigned a *positive* mass m identical to that of the removed electron!

In his article [62] from 1931 Dirac writes "A hole, if there were one, would be a new kind of particle, unknown to experimental physics, having the same mass and opposite charge of the electron".

In point of fact, this prediction of Dirac was confirmed experimentally a year later in 1932 with the discovery of the positron by C.D. Anderson [63].

Closing Remark

According to the basic idea of Dirac, the absorption of a photon of energy exceeding $2mc^2$ may cause a negative-energy electron to transit from a state with energy $-E_k$ to a positive-energy state with energy $E_{k'}$, thus leaving a hole of energy E_k behind in that "sea". In other words an electron-positron *pair* with specific momenta and energies $E_{k'} + E_k \geq 2mc^2$ was created this way. The inverse process is the annihilation of a pair, corresponding to the recombination of hole and electron with concurrent emission of a high energy photon.

Notably, in these processes the total number of particles is not conserved, whereas the total charge remains constant. Dirac's hole theory predicted the existence of the positron (anti-matter) and has given results in full agreement with experiments, even though the formulation of the theory, which is based on the assumption of an *infinite* number of *preexisting* electrons filling the "sea" of negative energy states, certainly rises questions.

But once the existence of the positron is accepted, a relativistic theory of light-matter interaction can be *directly* formulated with the method of second-quantization as a many-particle problem, where quantized fermion fields (comprising electrons and positrons as excitations of a Dirac quantum field) interact with quantized electromagnetic fields (photons) from the start. In this manner one gets (for ever) rid of the "sea" of filled negative-energy electrons. This major conceptional step forward in fact constitutes the commencement of modern quantum field theory.

In closing we hope that a reader acquainted with second quantization and nonrelativistic quantum mechanics has by now gathered some stock of information about relativistic quantum mechanics, so that the reading of Sect. 1.2 should be facilitated.

References

1. Tomonaga SI. On a relativistically invariant formulation of the quantum theory of wave fields. Progress Theoret Phys. 1946;1(2):27–41.
2. Dirac PAM, Fock V, Podolsky B. On quantum electrodynamics. In: Schwinger J, editors. Selected papers on quantum electrodynamics. Dover Publications Inc.; 1958.
3. Deckert DA, Nickel L. Multi-time dynamics of the Dirac-Fock-Podolsky model of QED. J Math Phys. 2019;60(7).
4. Breit G. The effect of retardation on the interaction of two electrons. Phys Rev. 1929;34(4).
5. Landau L. On the relativistic correction to the Schrödinger equation in many body systems. Phys Z Sovjet. 1935;8.
6. Breit G. Dirac's equation and the spin-spin interactions of two electrons. Phys Rev. 1932;39.
7. Sucher J. Relativistic many-electron Hamiltonians. Phys Scr. 1987;36:271.
8. Bethe HA, Salpeter EE. Quantum mechanics of one- and two-electron atoms. A Plenum/Rosetta edition; 1977.
9. Brown GE, Ravenhall DG. On the interaction of two electrons. Proc Roy Soc. 1951; (A208):552.
10. Mandelzweig VB, Wallace SJ. QED based two-body dirac equation. Phys Lett B. 1987;197:469–73.
11. Kutzelnigg W. Solved and unsolved problems in relativistic quantum chemistry. Chem Phys. 2012;395:16–34.
12. Itoh T. Derivation of nonrelativistic Hamiltonian for electrons from quantum electrodynamics. Rev Mod Phys. 1965;37(1).
13. Bialynicki-Birula I. The Hamiltonian of quantum electrodynamics. In: Barut AO, editor, Quantum electrodynamics and quantum optics, 1984.
14. Cohen-Tannoudji C, Dupon-Roc J, Grynberg G. Photons&Atoms. Introduction to quantum electrodynamics. WILEY-VCH; 2004.
15. Wegner F. Flow equations. Ann Phys. 1994;3:77.
16. Schopohl N, Cetin NS. Advances in Quantum Science I - Theoretical and Methodological Approaches, Chapter 2: Obtaining Schwinger's Magnetic Moment Anomaly and the Lamb Shift Terms from a Particle Number Conserving Hamiltonian being Unitary Equivalent to the Relativistic Hamiltonian of Quantum Electrodynamics. Springer; 2026.
17. Schopohl N, Cetin NS. Advances in Quantum Science I - Theoretical and Methodological Approaches, Chapter 3: The Nonrelativistic Hamiltonian of Electrons and Positrons: a Derivation from Quantum Electrodynamics. Springer Nature; 2026.
18. Zee A. Quantum field theory in a Nutshell, 2nd ed. Princeton University Press; 2010.
19. Itzykson C, Zuber JB. Quantum field theory. McGRAW HILL International Editions; 1985.
20. Schweber SS. Harper international student reprint: An introduction to relativistic quantum field theory; 1964.
21. Landau LD, Lifschitz JM. Quantenelektrodynamik. Verlag Harri Deutsch; 2009.
22. Cohen-Tannoudjii C. Atoms in electromagnetic fields. World Scientific; 1994.
23. Milonni PW. The quantum vacuum. An introduction to quantum electrodynamics. Academic Press Inc.; 1994.
24. Dirac PAM. Lectures on quantum field theory. Academic Press London; 1966.
25. Dirac PAM. The quantum theory of the electron. Proc Roy Soc A. 1928;117:610–24.
26. Born M, Heisenberg W, Jordan P. Zur Quantenmechanik II. Zeitschrift für Physik. 1925.
27. Majorana E. Teoria Simmetrica dell'Elettrone e del Positrone. Il Nuovo Cimento. 1937;14:171–84.
28. Pauli W. Pauli Lectures on Physics 6. Selected topics in field quantization. Cambridge MIT; 1973.
29. Bjorken JD, Drell SD. Relativistic quantum fields. McGraw-Hill; 1965.
30. Schlemmer J, Zahn J. The current density in quantum electrodynamics in external potentials. Ann Phys. 2015;359:31–45.
31. Viktor F, Weisskopf. Über die Elektrodynamik des Vakuums auf Grund der Quanten-Theorie des Elektrons. Mat Fys Medd. 1936;14.

32. Schwinger J. On gauge invariance and vacuum polarization. Phys Rev. 1950;82:664–79.
33. Fetter AL, Walecka JD. Quantum theory of many particle systems. Mc Graw Hill; 1971.
34. Eriksen E. Foldy-Wouthuysen-Tranformation. Exact solution with generalization to the two-particle problem. Phys. Rev. 1958;111(1011).
35. Schopohl N, Cetin NS. Advances in Quantum Science I - Theoretical and Methodological Approaches, Chapter 4: A Flow Equation Approach Transforming the Dirac Hamiltonian into a Unitary Equivalent and Energy-Separating Hamiltonian. Springer Nature; 2026.
36. Lenz P, Wegner F. Flow equations for electron-phonon interactions. Nucl Phys B. 1996;482:639–712.
37. Kehrein S. The Flow equation approach to many-particle systems. Springer; 2006.
38. Wegner F. Flow equations and normal ordering: A survey. J Phys A: Math Gen. 2006;39:8821–8230.
39. Avan P, Cohen-Tannoudji C, Dupont-Roc J, Fabre C. Effect of high frequency irradiation on the dynamical properties of weakly bound electrons. Journal de Physique. 1976;37(9).
40. Dupont-Roc J, Fabre C, Cohen-Tannoudji C. Physical interpretations for radiative corrections in the non-relativistic limit. J Phys BAtom Molec Phys. 1978;11(4).
41. Dupont Roc J, Cohen Tannoudji C. Tendances Actuelles En Physique Atomique, Les Houches, Ecoles d'Ete de Physique, Session 38, volume 1, chapter Effective Hamiltonian approach to g-2 relativistic calculation, p. 157–180. North-Holland Physics Publishing Amsterdam. Oxford. New York. Tokyo; 1982.
42. Fröhlich H. Proc Roy Soc A. 1952;215:291.
43. Feynman RP. The theory of positrons. Phys Rev. 1949;76(6):749–59.
44. Landau LD. On the electron-positron interaction. In Collected Papers of L. D. Landau, p. 532–539. Pergamon Press Ltd.;1965.
45. Schwinger J. Particles, sources and fields, vol. 2. Perseus; 1998.
46. Feynman RP. Quantum electrodynamics. Advanced book classics. Addison Wesley; 1961.
47. Compton AH. Recoil of electrons from scattered x-rays. Nature. 1923;112:435.
48. Heitler W. The quantum theory of radiation. 3rd ed. New York: Dover Publications Inc.; 1984.
49. Wentzel G. New aspects of the photon self-energy problem. Phys Rev. 1948;74:1070–5.
50. Pauli W, Villars F. On the invariant regularization in relativistic quantum theory. Rev Mod Phys. 1949;21:434–44.
51. Ma ST. Vacuum polarization in the positron theory. Lond Edinb Dublin Philos Mag J Sci. 1949;40(310):1112–28.
52. Delbrueck M. Z Physik. 1933;84:144.
53. Furry WH. On bound states and scattering in positron theory. Phys Rev. 1951;81:115–24.
54. Schumacher M. Delbrueck scattering. Radiat Phys Chem. 1999;56:101–11.
55. Battesti R, Rizzo C. Magnetic and electric properties of a quantum vacuum. Rep Prog Phys. 2013;76:016401 (23pp).
56. Barash YuS, Ginzburg VL. The dielectric function of condensed systems. Chapter 6: Electromagnetic Fluctuations and Molecular Forces in Condensed Matter. Modern Problems in Condensed Matter Sciences. Elsevier Science Publishers B.V. North Holland Physics; 1989.
57. Thaller B. The Dirac equation. Berlin Heidelberg: Springer; 1992.
58. Dirac PAM. Lectures on quantum mechanics and relativistic field theories. Martino Publishing; 2012.
59. Baym G. Lectures on quantum mechanics. The Benjamin/Cummings Publishing Company Inc; 1969.
60. Drell SD, Bjorken DJD. Relativistic quantum mechanics. McGraw-Hill; 1964.
61. Foldy L, Wouthuysen S. On the Dirac theory of spin 1/2 particles and its non-relativistic limit. Phys Rev. 1950;78(1):29–36.
62. Dirac PAM. Quantised singularities in the electromagnetic field. Proc R Soc Lond. 1931;A133:60–72.
63. Anderson CD. Cosmic-Ray positive and negative electrons. Phys Rev. 1933;43:405.

Chapter 2
Obtaining Schwinger's Magnetic Moment Anomaly and the Lamb Shift Terms from a Particle Number Conserving Hamiltonian $\hat{\mathscr{H}}_U$ Being Unitary Equivalent to the Hamiltonian of Quantum Electrodynamics

Nils Schopohl and Nadine Suzan Cetin

Abstract Based on a particle number conserving Hamiltonian $\hat{\mathscr{H}}_U$ derived afore in [1] with a flow equation approach from the fundamental field theory Hamiltonian of QED in Coulomb gauge, full relativistic calculations of the self-energy of electrons and positrons moving in the presence of external electromagnetic fields are presented to order 1 in the fine structure constant. It is shown with help of a gauge invariant cut off method due to Dirac, how the best known effects of QED, above all mass and charge renormalization, but also vacuum polarization, Schwinger's anomalous magnetic moment and what's more the Lamb shift terms, all together representing specific energy shift terms adding to the bare Dirac Hamiltonian, can be obtained with straightforward calculations, throughout respecting gauge invariance and retaining first and second order gradient terms only, but neglecting the third order terms as the particles are considered not being relativistic in the initial and final states. In contrast to previous inspiring work based on an effective Hamiltonian approach [2], notably our result for the self-energy of electrons (positrons) is devoid of gradient terms, neither "odd" terms $\alpha_a \Pi_a$ nor "even" gradient terms $\Pi_a \Pi_a$ are found, so that when adding these self-energy corrections to the Dirac Hamiltonian the relativistic invariance of the resulting renormalized Dirac Hamiltonian remains unchanged.

Keywords Quantum electrodynamics · Wegner flow equation method · Unitary transformation · Fock-space · Quantum field · Dirac Hamiltonian · Coulomb gauge · Self-energy · Anomalous magnetic moment · Lamb shift · Fine structure constant

---

N. Schopohl (✉)
Center for Quantum Science, Eberhard-Karls-Universität Tübingen, Tübingen, Germany
e-mail: nils.schopohl@uni-tuebingen.de

N. S. Cetin
Mathematical Institute, Charles University Prague, Prague, Czech Republic
e-mail: nadine-suzan.cetin@matfyz.cuni.cz

T. Onishi (ed.), *Advances in Quantum Science I*,
https://doi.org/10.1007/978-981-95-9717-8_2

2.1 Introduction

In chapter [1] we derived by means of a flow equation approach a second quantized, particle number conserving many body Hamiltonian $\hat{\mathscr{H}}_U$ for electrons and positrons in the guise of a perturbation series in the coupling e, that is by construction unitary equivalent to the fundamental field theory Hamiltonian $\hat{\mathscr{H}}_{QED}$ of relativistic quantum electrodynamics. If this many body Hamiltonian $\hat{\mathscr{H}}_U$ is narrowed down to the one-particle and to two-particle terms only, then

$$\hat{\mathcal{H}}_U = \hat{\mathcal{H}}_D^{(\mathrm{ren})} + \hat{\mathcal{U}}_{\mathrm{int}} + \cdots \tag{2.1}$$

The various effective two-particle interactions $\hat{\mathcal{U}}_{\mathrm{int}}$ between electrons and/or positrons we elaborate in detail in our chapter [3]. So we are concerned in what follows in the main with the effective *one-particle* terms $\hat{\mathcal{H}}_D^{(\mathrm{ren})}$ governing the motion of electrons and positrons in the presence of external (static) electromagnetic fields.

According to the results obtained afore in chapter [1], these effective one-particle terms assume the guise

$$\hat{\mathcal{H}}_D^{(\mathrm{ren})} = \hat{\mathscr{H}}_D + \hat{\mathcal{M}}^{(e)} + \hat{\mathcal{M}}^{(p)} \tag{2.2}$$

whereby

$$\hat{\mathscr{H}}_D = \sum_{k \in I} E_k \hat{f}_k^\dagger \hat{f}_k \tag{2.3}$$

Here we agree on the usual interpretation of $\hat{f}_k$ being an electron *annihilation* operator $\hat{c}_k$ for $E_k > 0$, respectively $\hat{f}_k$ being a positron *creation* operator $\hat{b}_k^\dagger$ for $E_k < 0$, this interpretation being equivalent to a Bogoliubov transformation

$$\begin{aligned} \hat{f}_k &= \Theta_H(E_k)\, \hat{c}_k + \Theta_H(-E_k)\, \hat{b}_k^\dagger \\ \hat{f}_k^\dagger &= \Theta_H(E_k)\, \hat{c}_k^\dagger + \Theta_H(-E_k)\, \hat{b}_k \end{aligned} \tag{2.4}$$

Due to the presence of the self-energy terms $\hat{\mathcal{M}}^{(e)} + \hat{\mathcal{M}}^{(p)}$ in (2.2) the energy of an emitted and/or absorbed electron (positron) is not solely determined by the eigenvalues E_k of the single particle Dirac Hamiltonian (2.3) , but undergoes characteristic modifications. In detail these self-energy terms decompose into

$$\begin{aligned} \hat{\mathcal{M}}^{(e)} &= \hat{\mathcal{M}}_C^{(e)} + \hat{\mathcal{M}}_\perp^{(e)} + \hat{\mathcal{M}}_{VP}^{(e)} \\ \hat{\mathcal{M}}^{(p)} &= \hat{\mathcal{M}}_C^{(p)} + \hat{\mathcal{M}}_\perp^{(p)} + \hat{\mathcal{M}}_{VP}^{(p)} \end{aligned} \tag{2.5}$$

In (2.5) the terms tagged with a letter C or a symbol $\perp$ result from normal ordering, whereas the terms tagged with a letter VP relate to the effects of vacuum polarization. The latter originate from the definition of charge and current densities entering the original QED-Hamiltonian, see [1, Sect. 3]. We desist here to list the corresponding expressions for the self-energies obtained in [1, Sect. 7.4], as they will be introduced separately in what follows as starting points of our calculations anyway.

The self-energy terms in (2.5) acquire in point of fact *after* the renormalization of charge and mass, a fundamental physical meaning, in as much these terms explain in the presence of external electromagnetic fields two key effects of quantum electrodynamics (QED), namely Schwinger's magnetic moment anomaly $g - 2 > 0$ and also the Lamb shift terms.

Saying this we allude to the fact, that *if* the Dirac equation for the bound electron in the hydrogen atom was solved *without* taking into account the self-energy corrections in (2.2), then one finds the spin magnetic moment of the electron being $\mathbf{M} = g\frac{e}{2m}\mathbf{S}$ with $g = 2$, and further one obtains the *fine structure* of the spectrum of the hydrogen atom, with the salient feature some of the energy levels of the bound electron still remaining degenerate, as regards for instance the energy levels $E_{2s_{1/2}}$ and $E_{2p_{1/2}}$.

But if the contribution of the self-energy terms is taken into account in the one-particle Hamiltonian (2.2), then $g - 2 = \frac{\alpha_{FS}}{\pi}$ in agreement with Schwinger's result [4, 5] and further it is found, that the degeneracy of the energy levels $E_{2s_{1/2}}$ and $E_{2p_{1/2}}$ in the hydrogen atom is lifted by a tiny amount $\Delta = E_{2s_{1/2}} - E_{2p_{1/2}} > 0$, in agreement with a value $\Delta \approx 1052\,\text{MHz}$ as detected unequivocally first in the experiments of Lamb and Retherford [6].

2.2 Dirac's Cutoff Procedure

As it turns out, the straightforward evaluation of the self-energy terms in (2.5) is indeed problematic, as one encounters, like always in QED, some of the integrals occurring are plagued by an infinity coming in, the reason being the integration intervals over the wave numbers of the fermion and/or photon modes under consideration extend to infinity.

So the idea is, just to put an upper limit to the wave numbers involved, as the correct matter-light interaction processes at ultra-high wave number (or ultrashort distance) are unknown anyway. Alas, upon introducing a *sharp* cutoff by simply limiting the magnitude of the wave numbers involved, and adding such regulated self-energy terms to the Hamiltonian (2.3) as indicated in (2.2), concurrently the relativistic invariance of $\hat{\mathcal{H}}_D^{(\text{ren})}$ gets broken too even for small energy processes [7].

By all means, if a cut-off is introduced to prevent an infinity coming in at ultra-high wave number into the theory of QED, it should be introduced such that the fundamental symmetries of the original Hamiltonian of QED remain intact.

A renowned cutoff procedure to circumvent the difficulty with a sharp cutoff, by introducing a soft cutoff, is due to Pauli and Villars [8, 9]. Alternate such mathematically founded methods are, to mention a few, dimensional regularization [10] or else Schwinger's truncation of his "time" parameter s in his renowned work on vacuum polarization and gauge invariance [5].

A lesser-known, yet *physically* substantiated method to introduce a suitable cutoff has been invented by P.A.M. Dirac [7], who suggests to delimit with a *high* energy cutoff $\hbar\Omega_{max}$ the *total* of the *kinetic* energies associated with the Dirac modes *and* the photon modes in the matrix elements to be summed over (including as well any other intermediate states), that way ensuring gauge invariance in an elementary way.

With a view to a quasiclassical picture, to determine the kinetic energy $E_k^{(\mathrm{kin})}$ of a Dirac fermion associated with an eigenstate $|\phi_k\rangle$ a reasonable approximation should be to exclude the potential energy contribution from the energy eigenvalue E_k *afterwards*, i.e., identifying

$$E_k^{(\mathrm{kin})} \equiv E_k - e\Phi^{(\mathrm{ext})} \tag{2.6}$$

The described cutoff procedure amounts to multiplying a matrix element, say in a term to be summed over a photon mode label q and fermion mode labels k', k now with a Heaviside step function [7]

$$\begin{aligned} \langle\phi_{k'}|\, \alpha_b e^{iq_a x_a}\, |\phi_k\rangle &\to \langle\phi_{k'}|\, \alpha_b e^{iq_a x_a}\, |\phi_k\rangle\, \Theta_{k',k}^{(q)} \\ \Theta_{k',k}^{(q)} &\equiv \Theta_H\left(\hbar\Omega_{\mathrm{max}} - \hbar\omega_q - \left|E_{k'}^{(\mathrm{kin})}\right| - \left|E_k^{(\mathrm{kin})}\right|\right) \end{aligned} \tag{2.7}$$

Of course, any mode labels *not* being summed over, even in case the associated states $|\phi_k\rangle$ building the matrix element have possibly a high kinetic energy $mc^2 \ll E_k^{(kin)}$, for physical reasons all such states are constrained by having their kinetic energies being much smaller than the high energy cutoff $\hbar\Omega_{\mathrm{max}}$. So the contribution of modes carrying labels *not* being summed over can be safely ignored in the argument of the step function in (2.7). For instance, in case a fermion state label k is not being summed over, the associated cutoff factor we shall indicate as $\Theta_{k'}^{(q)} \equiv \Theta_H\left(\hbar\Omega_{\mathrm{max}} - \hbar\omega_q - \left|E_{k'}^{(\mathrm{kin})}\right|\right)$. Likewise if a photon label q is not summed over we shall indicate the associated cutoff factor as $\Theta_{k',k} \equiv \Theta_H\left(\hbar\Omega_{\mathrm{max}} - \left|E_{k'}^{(\mathrm{kin})}\right| - \left|E_k^{(\mathrm{kin})}\right|\right)$.

As we substantiate in the ensuing, our perception is, that Dirac's cutoff method leads, in a *reasoned* and consistent way, to the elimination of the bare mass m_0 and the bare charge e_0 in favor of the observable mass m and the observable charge e, the latter parameters being *measurable*, say, in experiments with the Lorentz force or else in deflection experiments. Notably, results obtained with Dirac's method depend on the cutoff only mildly, specifically only a small logarithmic dependence on the cutoff is in the pertaining integrals coming in, being of order

$$\alpha_{FS} \ln\left(\frac{\hbar\Omega_{max}}{mc^2}\right) \ll 1 \tag{2.8}$$

actually this being in total agreement with results obtained by other regularization methods [8]. Since $\alpha_{FS} \approx \frac{1}{137}$ therefore the cutoff should obey to $1 \ll \frac{\hbar\Omega_{max}}{mc^2} \ll e^{137}$, thus still leaving an extremely wide spectrum of *relativistic* modes, all of them respecting the condition (2.8) to enable perturbation theory.

So, after the renormalization of charge and mass the remaining parts of the self-energy operator of electrons (positrons), thereupon acquire in terms of e and m a physical meaning, explaining in the presence of external electromagnetic fields Schwinger's spin anomaly and also the terms responsible for the energy shift detected in the Lamb experiment. Eventually a physically transparent and *logical* QED-theory is obtained this way, where one solves equations and comprehends, which terms in the theory are small to the point that their neglect is justifiable. In the words of P.M.A. Dirac [7]:

"QED cannot be a complete theory in itself as we cannot make QED consistently relativistically invariant for high-energy processes. Well, that does not really matter. We shall find that we can make the cut off at a fairly high energy value, of the order of several hundreds of millions of volts. It is reasonable to make the cutoff at an energy value somewhere there and we shall then have a theory that fails for processes involving energies of more than some hundreds of million of volts, but which we can hope will be accurate for lower energies. We may thus hope it will be accurate for calculating things like the anomalous magnetic moment and the Lamb shift, where there are no high-energy processes involved."

Be that as it may, QED certainly cannot claim being physically correct at really high energy $E_k \gg 1\,\text{GeV}$ anyway, as it ignores the possibility of electrons and positrons interacting with *other* emerging "particles," for instance muons, neutrinos, etc., all appearing in such high energy processes. So, QED should be considered being an *effective* theory, essentially a kind of "low" energy approximation to the Standard Model [10]. To quote once more P.M.A. Dirac [7]:

"I look upon it in this way: We have a quantum electrodynamics that cannot be accurate for high-energy processes, because we do not know how to take into account high-energy interactions, and our cutting-off the interaction for high energies is just an expression of our ignorance of how high-energy processes should be taken into account."

On a final note, what has been said with regard to the self-energy of electrons (positrons) concerns as well the self-energy of photons and the contributions to the vacuum energy [11].

2.3 Vacuum Polarization and the Renormalization of Charge

An elementary understanding of vacuum polarization in quantum electrodynamics is perhaps within reach following the example set by the physics of dielectric materials, where it is common to define the *induced* electric potential $\Phi^{(\text{ind})}(\mathbf{r})$ and the *induced* vector potential $\mathsf{A}^{(\text{ind})}(\mathbf{r})$, these quantities representing the *difference* between the physical potentials and the bare external potentials, so that

$$\begin{aligned} \Phi(\mathbf{r}) &= \Phi^{(\text{ext})}(\mathbf{r}) + \Phi^{(\text{ind})}(\mathbf{r}) \\ \mathsf{A}(\mathbf{r}) &= \mathsf{A}^{(\text{ext})}(\mathbf{r}) + \mathsf{A}^{(\text{ind})}(\mathbf{r}) \end{aligned} \tag{2.9}$$

By definition the physical potential $\Phi(\mathbf{r})$ generates the *local* electric field $\mathcal{E} = -\nabla\Phi$ that acts on a single (test) charge e under the *combined* influences of the polarizable charge carrier ensemble of the probe under consideration *and* the *fixed* external fields in the background, while the external potential $\Phi^{(\text{ext})}$ generates the external field $\mathcal{E}^{(\text{ext})} = -\nabla\Phi^{(\text{ext})}$ that would act on that test charge in a system void of any other charge carriers. Likewise the physical vector potential $\mathsf{A}(\mathbf{r})$ generates the *local* magnetic induction field $\mathcal{B} = \mathbf{rot}\,\mathsf{A}$, whereas the external vector potential $\mathsf{A}^{(\text{ext})}(\mathbf{r})$ generates the external magnetic induction field $\mathcal{B}^{(\text{ext})} = \mathbf{rot}\,\mathsf{A}^{(\text{ext})}$, that would act on a *moving* test charge in a system void of any other charge carriers.

So, while $\mathcal{E}^{(\text{ext})}$ and $\mathcal{B}^{(\text{ext})}$ are completely controlled by the external sources, it is the charge carrier ensemble of the probe itself, that decides on the value of the *local* fields $\mathcal{E}$ and $\mathcal{B}$ in response to applying background fields $\mathcal{E}^{(\text{ext})}$ and $\mathcal{B}^{(\text{ext})}$, just as a gas in equilibrium would for prescribed temperature and fixed volume then decide on its pressure, of course.

Conforming with the structure of the vacuum polarization terms, that we derived afore in [1, Eq. (159)] for electrons in the guise[1]

$$\hat{\mathcal{M}}_{VP}^{(e)} = \sum_{k,k' \in I_+} \begin{pmatrix} -\frac{e^2}{2\varepsilon_0}\frac{1}{V_P}\sum_{\mathbf{q}}\frac{1}{|\mathbf{q}|^2}\sum_{k'' \in I}\langle\phi_{k''}|\, e^{-iq_a \mathsf{x}_a} z_{k''}\,|\phi_{k''}\rangle\langle U_k|\, e^{iq_a \mathsf{x}_a}\,|U_{k'}\rangle + \\ +\langle U_k|\,\delta e\left(\Phi^{(ext)}(\mathsf{x})\,1_{4\times 4} - c\alpha_b \mathsf{A}_b^{(ext)}(\mathsf{x})\right)|U_{k'}\rangle + \\ +\frac{e^2}{2\varepsilon_0}\frac{1}{V_P}\sum_{\mathbf{q}}\frac{1}{|\mathbf{q}|^2}\left(\delta_{b.b'} - \frac{q_b q_{b'}}{|\mathbf{q}|^2}\right)\sum_{k'' \in I}\langle\phi_{k''}|\,\alpha_{b'} e^{-iq_a \mathsf{x}_a} z_{k''}\,|\phi_{k''}\rangle\langle U_k|\,\alpha_b e^{iq_a \mathsf{x}_a}\,|U_{k'}\rangle \end{pmatrix} \hat{c}_k^\dagger \hat{c}_{k'}, \tag{2.10}$$

and in [1, Eq. (162)] for positrons in the guise

$$\hat{\mathcal{M}}_{VP}^{(p)} = -\sum_{K,K' \in I_-} \begin{pmatrix} -\frac{e^2}{2\varepsilon_0}\frac{1}{V_P}\sum_{\mathbf{q}}\frac{1}{|\mathbf{q}|^2}\sum_{k'' \in I}\langle\phi_{k''}|\, e^{-iq_a \mathsf{x}_a} z_{k''}\,|\phi_{k''}\rangle\langle V_K|\, e^{iq_a \mathsf{x}_a}\,|V_{K'}\rangle + \\ +\langle V_K|\,\delta e\left(\Phi^{(ext)}(\mathsf{x})\,1_{4\times 4} - c\alpha_b \mathsf{A}_b^{(ext)}(\mathsf{x})\right)|V_{K'}\rangle + \\ +\frac{e^2}{2\varepsilon_0}\frac{1}{V_P}\sum_{\mathbf{q}}\frac{1}{|\mathbf{q}|^2}\left(\delta_{b.b'} - \frac{q_b q_{b'}}{|\mathbf{q}|^2}\right)\sum_{k'' \in I}\langle\phi_{k''}|\,\alpha_{b'} e^{-iq_a \mathsf{x}_a} z_{k''}\,|\phi_{k''}\rangle\langle V_K|\,\alpha_b e^{iq_a \mathsf{x}_a}\,|V_{K'}\rangle \end{pmatrix} \hat{b}_{K'}^\dagger \hat{b}_K, \tag{2.11}$$

it is now suggestive to conceive the vacuum of quantum electrodynamics as being a polarizable medium. Therefore let us identify the induced (electrostatic) potential, respectively the induced (magnetostatic) vector potential hereby as

$$\begin{aligned} \Phi^{(\text{ind})}(\mathbf{r}) &\equiv -\frac{e}{2\varepsilon_0}\frac{1}{V_P}\sum_{\mathbf{q}}\frac{1}{|\mathbf{q}|^2}\sum_{k',k'' \in I}\langle\phi_{k'}|\, e^{-iq_a \mathsf{x}_a}\,|\phi_{k''}\rangle\langle\phi_{k''}|\,\Lambda^{(D)}\,|\phi_{k'}\rangle \\ &\stackrel{!}{=} \sum_{\mathbf{q}} e^{i\mathbf{q}\cdot\mathbf{r}}\,\Phi_F^{(\text{ind})}(\mathbf{q}) \end{aligned} \tag{2.12}$$

$$\begin{aligned} \mathsf{A}_b^{(\text{ind})}(\mathbf{r}) &\equiv -\frac{e}{2\varepsilon_0 c^2}\frac{1}{V_P}\sum_{\mathbf{q}}\frac{1}{|\mathbf{q}|^2}\left(\delta_{b.b'} - \frac{q_b q_{b'}}{|\mathbf{q}|^2}\right)\sum_{k',k'' \in I}\langle\phi_{k'}|\, e^{-iq_a \mathsf{x}_a}\,|\phi_{k''}\rangle\langle\phi_{k''}|\, c\alpha_{b'}\Lambda^{(D)}\,|\phi_{k'}\rangle \\ &\stackrel{!}{=} \sum_{\mathbf{q}} e^{i\mathbf{q}\cdot\mathbf{r}}\,\mathsf{A}_{F,b}^{(\text{ind})}(\mathbf{q}) \end{aligned} \tag{2.13}$$

---

[1] We use notation such that $I = I_- \cup I_+$ with $I_+ \equiv \{k \mid E_k > 0\}$ and $I_- \equiv \{k \mid E_k < 0\}$.

By this means the vacuum polarization parts of the self-energy of electrons (positrons), detailed afore in (2.10), (2.11) can be re-expressed in the guise

$$\hat{\mathcal{M}}_{VP}^{(e)} = \sum_{k,k' \in I_+} \langle U_k | \begin{pmatrix} \delta e \left(\Phi^{(\text{ext})}(\mathsf{x})\, 1_{4\times 4} - c\alpha_b \mathsf{A}_b^{(\text{ext})}(\mathsf{x}) \right) \\ +e \left(\Phi^{(\text{ind})}(\mathsf{x})\, 1_{4\times 4} - c\alpha_b \mathsf{A}_b^{(\text{ind})}(\mathsf{x}) \right) \end{pmatrix} | U_{k'} \rangle\, \hat{c}_k^\dagger \hat{c}_{k'} \tag{2.14}$$

$$\hat{\mathcal{M}}_{VP}^{(p)} = -\sum_{K,K' \in I_-} \langle V_K | \begin{pmatrix} \delta e \left(\Phi^{(\text{ext})}(\mathsf{x})\, 1_{4\times 4} - c\alpha_b \mathsf{A}_b^{(\text{ext})}(\mathsf{x}) \right) \\ +e \left(\Phi^{(\text{ind})}(\mathsf{x})\, 1_{4\times 4} - c\alpha_b \mathsf{A}_b^{(\text{ind})}(\mathsf{x}) \right) \end{pmatrix} | V_{K'} \rangle\, \hat{b}_{K'}^\dagger \hat{b}_K$$

On physical grounds the induced electric potential $\Phi^{(\text{ind})}(\mathbf{r})$ is directly connected to the vacuum polarization charge density $\varrho^{(\text{pol})}(\mathbf{r})$ and by extension, the induced vector potential $\mathsf{A}_b^{(\text{ind})}(\mathbf{r})$ in Coulomb gauge is directly connected to the vacuum polarization current density $j_b^{(\text{pol})}(\mathbf{r})$ via the respective Poisson equations

$$-\nabla^2 \Phi^{(\text{ind})}(\mathbf{r}) = \frac{1}{\varepsilon_0} \varrho^{(\text{pol})}(\mathbf{r}) \tag{2.15}$$

$$-\nabla^2 \mathsf{A}_b^{(\text{ind})}(\mathbf{r}) = \frac{1}{\varepsilon_0 c^2} j_b^{(\text{pol})}(\mathbf{r}) \tag{2.16}$$

whereby we recapitulate according to what has been said in [1, Sect. 3]

$$\varrho^{(\text{pol})}(\mathbf{r}) = -\frac{e}{2} \sum_k \sum_\nu \phi_\nu^\star(\mathbf{r}; k)\, z_k \phi_\nu(\mathbf{r}; k) \tag{2.17}$$

$$j_b^{(\text{pol})}(\mathbf{r}) = -\frac{e}{2} \sum_k \sum_{\nu,\nu'} \phi_\nu^\star(\mathbf{r}; k)\, (c\alpha_b)_{\nu,\nu'}\, z_k \phi_{\nu'}(\mathbf{r}; k) \tag{2.18}$$

with $z_k = \frac{E_k}{|E_k|}$ denoting an eigenvalue of Dirac's energy-sign operator $\Lambda^{(D)}$, i.e.,

$$\sum_{\nu'} \Lambda_{\nu,\nu'}^{(D)} \phi_{\nu'}(\mathbf{r}; k) = z_k \phi_\nu(\mathbf{r}; k) \tag{2.19}$$

With a choice of periodic boundary conditions in a *large* box of volume V_P, the finite size corrections to the solutions $\Phi^{(\text{ind})}(\mathbf{r})$ and $A_b^{(\text{ind})}(\mathbf{r})$ of the Poisson equations (2.15) and (2.16) arising near to the boundary of that box can be neglected with confidence. In this case the Fourier coefficients $\Phi_F^{(\text{ind})}(\mathbf{q})$ and $A_{F,b}^{(\text{ind})}(\mathbf{q})$ of the induced potentials are linked to the Fourier coefficients $\varrho_F^{(\text{pol})}(\mathbf{q})$ and $j_{F,b}^{(\text{pol})}(\mathbf{q})$ associated with (2.17) and (2.18) simply by what is conforming with the Fourier transformation rules of the respective Poisson equations in the bulk

$$\Phi_F^{(\text{ind})}(\mathbf{q}) = \frac{\varrho_F^{(\text{pol})}(\mathbf{q})}{\varepsilon_0 |\mathbf{q}|^2} \tag{2.20}$$

$$\mathsf{A}_{F,b}^{(\text{ind})}(\mathbf{q}) = \frac{1}{\varepsilon_0 c^2} \frac{j_{F,b}^{(\text{pol})}(\mathbf{q})}{|\mathbf{q}|^2} \tag{2.21}$$

In line with (2.12), (2.13) and (2.20), (2.21) the task to determine the induced potentials $\Phi^{(\text{ind})}(\mathbf{r})$ and $A_b^{(\text{ind})}(\mathbf{r})$ by solving the Poisson equations (2.15) and (2.16) is now reduced to the calculation of the respective Fourier coefficients of the vacuum polarization charge density (2.17) and the vacuum polarization current density (2.18), given by

$$\begin{aligned} \varrho_F^{(\text{pol})}(\mathbf{q}) &= \frac{1}{V_P} \int d^3 r e^{-i\mathbf{q}\cdot\mathbf{r}} \varrho^{(\text{pol})}(\mathbf{r}) \\ &= -\frac{e}{2V_P} \sum_{k'} \langle \phi_{k'} | \, e^{-iq_a \mathsf{x}_a} \Lambda^{(D)} \, | \phi_{k'} \rangle \end{aligned} \tag{2.22}$$

$$\begin{aligned} j_{F,b}^{(\text{pol})}(\mathbf{q}) &= \frac{1}{V_P} \int d^3 r e^{-i\mathbf{q}\cdot\mathbf{r}} j_b^{(\text{pol})}(\mathbf{r}) \\ &= -\frac{ce}{2V_P} \sum_{k'} \langle \phi_{k'} | \, e^{-iq_a \mathsf{x}_a} \alpha_{b'} \Lambda^{(D)} \, | \phi_{k'} \rangle \end{aligned} \tag{2.23}$$

In the absence of external sources, then of course there exists no vacuum-polarization, i.e., $\varrho_F^{(\text{pol})}(\mathbf{q}) \equiv 0$ and $j_{F,b}^{(\text{pol})}(\mathbf{q}) \equiv 0$. So it is natural to expect these quantities are in the presence of external sources connected to the external potentials via a response kernel. The outcome of a calculation of these Fourier coefficients in line with Dirac's gauge invariant cutoff method is[2]

$$\begin{aligned} \varrho_F^{(\text{pol})}(\mathbf{q}) &= -I(\mathbf{q}) \, \Phi_F^{(\text{ext})}(\mathbf{q}) \\ j_{F,b}^{(\text{pol})}(\mathbf{q}) &= -I(\mathbf{q}) \, c^2 \mathsf{A}_{F,b}^{(\text{ext})}(\mathbf{q}) \end{aligned} \tag{2.24}$$

with the response kernel $I(\mathbf{q})$ given by

$$\frac{1}{\varepsilon_0} \frac{I(\mathbf{q})}{|\mathbf{q}|^2} = \frac{\alpha_{FS}}{\pi} \left(\frac{2}{3} \ln\left(\tilde{\Omega}_{max}\right) - \frac{5}{9} - \frac{|\tilde{\mathbf{q}}|^2}{4} \int_0^1 dy \frac{y^2 \left(1 - \frac{1}{3} y^2\right)}{1 + \frac{|\tilde{\mathbf{q}}|^2}{4} \left(1 - y^2\right)} \right) \tag{2.25}$$

Here $\tilde{\mathbf{q}} \equiv \frac{\hbar}{mc} \mathbf{q}$ and $\tilde{\Omega}_{max} \equiv \frac{\hbar \Omega_{max}}{mc^2} \gg 1$ denote a high energy cutoff (scaled units). A straightforward deduction of (2.24) and (2.25) we present in the subsequent section.

The next step regarding the energy shift of electrons (positrons) brought about by the self-energy terms (2.14) consists to insert the response kernel (2.25) into (2.24), so that in compliance with (2.20), (2.21) there follows

[2] Details of these calculations are presented in the Appendix A.

$$\delta e\ \Phi^{(\text{ext})}(\mathbf{r}) + e\Phi^{(\text{ind})}(\mathbf{r}) = \sum_{\mathbf{q}} e^{i\mathbf{q}\cdot\mathbf{r}} \left(\delta e\ \Phi_F^{(\text{ext})}(\mathbf{q}) + e\Phi_F^{(\text{ind})}(\mathbf{q})\right) \tag{2.26}$$
$$= \sum_{\mathbf{q}} e^{i\mathbf{q}\cdot\mathbf{r}} \left(\frac{\delta e}{e} - \frac{1}{\varepsilon_0}\frac{I(\mathbf{q})}{|\mathbf{q}|^2}\right) e\ \Phi_F^{(\text{ext})}(\mathbf{q})$$

$$\delta e\ \mathsf{A}_b^{(\text{ext})}(\mathbf{r}) + e\mathsf{A}_b^{(\text{ind})}(\mathbf{r}) = \sum_{\mathbf{q}} e^{i\mathbf{q}\cdot\mathbf{r}} \left(\delta e\ \mathsf{A}_{F,b}^{(\text{ext})}(\mathbf{r}) + e\mathsf{A}_{F,b}^{(\text{ind})}(\mathbf{r})\right) \tag{2.27}$$
$$= \sum_{\mathbf{q}} e^{i\mathbf{q}\cdot\mathbf{r}} \left(\frac{\delta e}{e} - \frac{1}{\varepsilon_0}\frac{I(\mathbf{q})}{|\mathbf{q}|^2}\right) e\ \mathsf{A}_{F,b}^{(\text{ext})}(\mathbf{q})$$

with these expressions still depending on the high energy cutoff parameter $\tilde{\Omega}_{max}$ via the response kernel (2.25).

The procedure to get rid of this problem with the high energy cutoff is *charge renormalization* [5, 12, 13]. This means the charge shift δe, as introduced into the Dirac Hamiltonian $\mathsf{H}^{(D)}$ in [1, Eq. (8)] to replace the *bare* charge $e_0 = e + \delta e$ by the observable charge e, will upon insertion of (2.25) into (2.26), (2.27) be eliminated with the requirement to compensate the constant term independent on $\mathbf{q}$ in (2.25), meaning that δe is determined by the requirement

$$\frac{\delta e}{e} \stackrel{!}{=} \frac{\alpha_{FS}}{\pi}\left(\frac{2}{3}\ln\left(\tilde{\Omega}_{max}\right) - \frac{5}{9}\right) \tag{2.28}$$

As a matter of fact, the *observable* charge $|e|$, as it emerges after the elimination of the shift parameter δe in line with (2.28), is *smaller* than the bare charge $|e_0| = |e|\left(1 + \frac{\delta e}{e}\right)$. Notably, the *bare* charge e_0 is unobservable, because in *experiment* it is the *observable* charge e that gives the force $\mathbf{F} = \pm|e|\mathscr{E}$ acting on electrons (positrons) in the presence of a *local* electrostatic field $\mathscr{E} = -\nabla\Phi$. Notably (2.28) concurs with Dirac in [7, pp. 125–126], even though his calculational method and his choice of gauge differs completely from ours.

Agreeing on the external sources not having Fourier components with very short wavelength, and with δe being fixed by (2.28) and removed from the kernels in (2.26), (2.27) the remaining measurable contribution to the self-energy terms (2.14) due to vacuum polarization is given by expanding the integral in (2.25) in powers of $|\tilde{\mathbf{q}}|^2$, thus leading in position space representation for the respective kernels in (2.26), (2.27) to

$$\delta e\Phi^{(\text{ext})}(\mathbf{r}) + e\Phi^{(\text{ind})}(\mathbf{r}) = -\frac{\alpha_{FS}}{\pi}\frac{1}{15}\left(\frac{\hbar}{mc}\right)^2 \nabla^2\, e\Phi^{(\text{ext})}(\mathbf{r}) + \cdots \tag{2.29}$$
$$\delta e\mathsf{A}_b^{(\text{ext})}(\mathsf{x}) + e\mathsf{A}_b^{(\text{ind})}(\mathsf{x}) = -\frac{\alpha_{FS}}{\pi}\frac{1}{15}\left(\frac{\hbar}{mc}\right)^2 \nabla^2\, e\mathsf{A}_b^{(\text{ext})}(\mathbf{r}) + \cdots$$

Eventually there follows for the energy shift of electrons (positrons) as induced by vacuum polarization in the presence of electromagnetic background fields

$$\hat{\mathcal{M}}_{VP}^{(e)} = \sum_{k,k'\in I_+} \langle U_k| \frac{\alpha_{FS}}{\pi}\frac{1}{15}\left(\frac{\hbar}{mc}\right)^2 \left(-e\,\nabla^2\Phi^{(\text{ext})}(\mathsf{x}) + c\alpha_b e\,\nabla^2\mathsf{A}_b^{(ext)}(\mathsf{x})\right)|U_{k'}\rangle\,\hat{c}_k^\dagger\hat{c}_{k'}$$

$$\hat{\mathcal{M}}_{VP}^{(p)} = \sum_{K,K'\in I_-} \langle V_K| \frac{\alpha_{FS}}{\pi}\frac{1}{15}\left(\frac{\hbar}{mc}\right)^2 \left(+e\,\nabla^2\Phi^{(\text{ext})}(\mathsf{x}) - c\alpha_b e\,\nabla^2\mathsf{A}_b^{(\text{ext})}(\mathsf{x})\right)|V_{K'}\rangle\,\hat{b}_{K'}^\dagger\hat{b}_K \quad (2.30)$$

For applications in atomic physics the gradients of the external magnetic induction field are completely insignificant and so the contribution of $-\nabla^2\mathbf{A}^{(ext)}(\mathsf{x}) = \mathbf{rot}\,\mathbf{B}^{(\text{ext})}$ to the self-energy terms (2.30) can be ignored.

Accordingly, an electron nearby an atomic nucleus experiences, with the self-energy correction (2.29) added to the external Coulomb potential $\Phi^{(\text{ext})}(\mathbf{r})$ of the nucleus, actually a (miniscule) stronger attraction than if only a pure Coulomb force was effective. This effect was discovered by Uehling [14], see also [15].

For a derivation of the Ueling potential with the methodology of relativistic S-matrix theory, one-loop renormalization, and Pauli-Villars regularization see [16, p. 327].

Even though the effect of the vacuum polarization is small for electrons, precision measurements of the Lamb shift (for instance in hydrogen) demonstrate its existence.

2.4 The Coulomb Self-Energy of Electrons

Our starting point in the ensuing is the Coulomb self-energy of electrons as presented afore in [1, Sect. 7.4]

$$\hat{\mathcal{M}}_C^{(e)} = \sum_{k,k'\in I_+} \left(\begin{array}{c} \delta m\, c^2\,\langle U_k|\,\beta\,|U_{k'}\rangle + \\ +\frac{e^2}{2\varepsilon_0}\int \frac{d^3q}{(2\pi)^3}\frac{1}{|\mathsf{q}|^2}\langle U_k|\,e^{-iq_a\mathsf{x}_a}\sum_{k''\in I} z_{k''}\,|\phi_{k''}\rangle\langle\phi_{k''}|\,e^{iq_a\mathsf{x}_a}\,|U_{k'}\rangle \end{array}\right)\hat{c}_k^\dagger\hat{c}_{k'} \quad (2.31)$$

Here the involved *intermediate* sum $\sum_{k''\in I}\ldots$ obviously requires a knowledge of *all* eigenvalues $\tilde{E}_{k''}$ and *all* eigenstates $|\phi_{k''}\rangle$ of the Dirac Hamiltonian $\tilde{\mathsf{H}}^{(D)}$, that is the summation over mode labels $k'' \in I$ encloses both, the positive *and* the negative energy states. Now, the task to determine the spectrum of $\tilde{\mathsf{H}}^{(D)}$ with the external electromagnetic fields taken into account is difficult and will in general be approachable only with numerical methods. Serendipitously though, such calculations can be prevented by re-arranging and simplifying some of the terms in (2.31) with help of the energy-sign operator $\Lambda^{(D)} = \sum_{k\in I} z_k\,|\phi_k\rangle\langle\phi_k|$ introduced in the appendix in (2.127). By this means then

$$\hat{\mathcal{M}}_C^{(e)} = mc^2\sum_{k,k'\in I_+}\langle U_k|\left(\frac{\delta m}{m}\beta + Y_C\right)|U_{k'}\rangle\;\hat{c}_k^\dagger\hat{c}_{k'}, \quad (2.32)$$

whereby in terms of $\tilde{q} = \lambda_C q$ and taking into account Dirac's high energy cut off

$$Y_C = \frac{\alpha_{FS}}{4\pi^2} \int d^3\tilde{q} \frac{\Theta_H\left(\tilde{\Omega}_{\max} - 2\tilde{q}\right)}{\tilde{q}^2} e^{-iq_a \mathsf{x}_a} \circ \Lambda^{(D)} \circ e^{iq_a \mathsf{x}_a} \tag{2.33}$$

with $\alpha_{FS} = \frac{e^2}{4\pi\varepsilon_0\hbar c}$ the fine structure constant and $\lambda_C = \frac{\hbar}{mc}$ the Compton length.

As the position operator x_a conjugate to the momentum operator p_a generates translations in reciprocal space, there holds

$$e^{-iq_a \mathsf{x}_a} \circ \mathsf{p}_b \circ e^{+iq_a \mathsf{x}_a} = \mathsf{p}_b + \hbar q_b \tag{2.34}$$

Bearing in mind the integral representation (2.127) for $\Lambda^{(D)}$, we have for a Dirac Hamiltonian in the presence of electromagnetic background fields, namely

$$\mathsf{H}^{(D)} = mc^2\beta + c\alpha_a\left(\mathsf{p}_a - e\mathsf{A}_a^{(\text{ext})}(\mathsf{x})\right) + e\Phi^{(\text{ext})}(\mathsf{x}) \tag{2.35}$$

accordingly a q-shifted energy-sign operator given by

$$\Lambda_q^{(D)} \equiv e^{-iq_a \mathsf{x}_a} \circ \Lambda^{(D)} \circ e^{iq_a \mathsf{x}_a} = \int_{-\infty}^{\infty} \frac{d\omega}{\pi} \frac{1_{4\times 4}}{-i\omega 1_{4\times 4} + \mathsf{H}^{(D)} + c\alpha_a \hbar q_a} \tag{2.36}$$

Restricting the considerations to the low-energy sector of QED, where the electron is not relativistic in the states $|U_k\rangle$ and $|U_{k'}\rangle$ in (2.32), the application of the operator $\Pi_a \equiv \mathsf{p}_a - e\mathsf{A}_a^{(\text{ext})}$ on those states produces a small result, so that we can expand the integrand in (2.36) in the quantity $c\alpha_a\Pi_a$, retaining second order terms $(c\alpha_a\Pi_a)^2$, but neglecting terms of third order. As shown in Appendix B the result of a manifest *gauge invariant* expansion of the q-shifted energy-sign operator in small quantities gives

$$\Lambda_q^{(D)} = \Lambda_q^{(0)} + \Lambda_q^{(1)} + \Lambda_q^{(2)} + \cdots, \tag{2.37}$$

whereby in terms of the definitions

$$w = \sqrt{1 + \tilde{q}^2} \tag{2.38}$$
$$n_b = \frac{q_b}{q}$$

the zeroth order term in the expansion is

$$\Lambda_q^{(0)} = \frac{1}{w}\left(\beta + \tilde{q} n_c \alpha_c\right) = \mathsf{P}_{q,+}^{(0)} - \mathsf{P}_{q,-}^{(0)} = \sum_{s\in\{+,-\}} s \cdot \mathsf{P}_{q,s}^{(0)} \tag{2.39}$$
$$\mathsf{P}_{q,\pm}^{(0)} = \frac{1}{2}\left(1_{4\times 4} \pm \Lambda_q^{(0)}\right)$$

Notably, the first order term $\Lambda_q^{(1)}$ in the expansion (2.37) now comprises as well the (scaled) electrostatic field, see (2.139), so that

$$\Lambda_q^{(1)} = \sum_{s\epsilon\{+,-\}} \left(\tfrac{1}{w}\tilde{\Pi}_a - \tfrac{1}{2}\tfrac{1}{w^2}i\tilde{\mathcal{E}}_a \cdot s\right) \mathsf{P}_{q,-s}^{(0)}\alpha_a \mathsf{P}_{q,s}^{(0)} \tag{2.40}$$

The second order term $\Lambda_q^{(2)}$ in the expansion (2.37) is already appreciably extended[3]

$$\Lambda_q^{(2)} = \begin{pmatrix} \left(\frac{1}{2w^2}\tilde{\Pi}_a\tilde{\Pi}_{a'} + \frac{1}{8w^4}\tilde{\mathcal{E}}_a\tilde{\mathcal{E}}_{a'}\right)\alpha_a\Lambda_q^{(0)}\alpha_{a'} + \frac{1}{4w^3}\begin{pmatrix} \left[\tilde{\Pi}_a, i\tilde{\mathcal{E}}_a\right]1_{4\times4} \\ +\left(\tilde{\mathcal{E}}_a\tilde{\Pi}_{a'} + \tilde{\Pi}_{a'}\tilde{\mathcal{E}}_a\right)\varepsilon_{aa'c}\sigma_c \end{pmatrix} + \\ +\left(-\frac{1}{w^2}\tilde{\Pi}_a\tilde{\Pi}_{a'} + \frac{1}{4\tilde{w}^4}\tilde{\mathcal{E}}_a\tilde{\mathcal{E}}_{a'}\right)\frac{\tilde{q}}{\tilde{w}}n_{a'}\alpha_a + \frac{3}{4w^3}\frac{\tilde{q}}{\tilde{w}}n_{a'}i\tilde{\mathcal{E}}_a\tilde{\Pi}_{a'}\alpha_a\Lambda_q^{(0)} + \\ +\left(-\frac{1}{w^2}\tilde{\Pi}_{a'}\tilde{\Pi}_a + \frac{1}{4w^4}\tilde{\mathcal{E}}_a\tilde{\mathcal{E}}_{a'}\right)\frac{\tilde{q}}{\tilde{w}}n_{a'}\alpha_a - \frac{3}{4\tilde{w}^3}\frac{\tilde{q}}{\tilde{w}}n_{a'}\tilde{\Pi}_{a'}i\tilde{\mathcal{E}}_a\Lambda_q^{(0)}\alpha_a + \\ +\left(\frac{3}{2w^2}\tilde{\Pi}_a\tilde{\Pi}_{a'} - \frac{5}{8\tilde{w}^4}\tilde{\mathcal{E}}_a\tilde{\mathcal{E}}_{a'}\right)\frac{\tilde{q}^2}{w^2}n_a n_{a'}\Lambda_q^{(0)} + \frac{\tilde{q}^2}{2w^5}n_a n_{a'}\left[\tilde{\Pi}_a, i\tilde{\mathcal{E}}_{a'}\right]1_{4\times4} \end{pmatrix} \tag{2.41}$$

For reasons of compatibility with the general property $\Lambda_q^2 = 1_{4\times4}$ we rechecked in Appendix B the following identities being valid

$$\begin{aligned} \Lambda_q^{(0)}\Lambda_q^{(0)} &= 1_{4\times4} \\ \Lambda_q^{(0)}\Lambda_q^{(1)} + \Lambda_q^{(1)}\Lambda_q^{(0)} &= 0_{4\times4} \\ \Lambda_q^{(0)}\Lambda_q^{(2)} + \Lambda_q^{(1)}\Lambda_q^{(1)} + \Lambda_q^{(2)}\Lambda_q^{(0)} &= 0_{4\times4} \end{aligned} \tag{2.42}$$

Upon insertion of the expansion (2.37) into (2.33) we have after passing to spherical coordinates

$$\mathsf{Y}_C = \frac{\alpha_{FS}}{\pi}\int_0^{\frac{\tilde{\Omega}_{\max}}{2}} d\tilde{q}\int\frac{d\Omega_n}{4\pi}\left(\Lambda_q^{(0)} + \Lambda_q^{(1)} + \Lambda_q^{(2)} + \cdots\right) \tag{2.43}$$

For the rest it is straightforward integration and exploiting the algebra of Dirac matrices. Averaging over the directions n_b gives

$$\begin{aligned} \int\frac{d\Omega_n}{4\pi}\Lambda_q^{(0)} &= \frac{1}{w}\beta \\ \int\frac{d\Omega_n}{4\pi}\Lambda_q^{(1)} &= \left(\frac{2}{3}\frac{1}{w} + \frac{1}{3}\frac{1}{w^3}\right)\alpha_a\tilde{\Pi}_a - \frac{1}{2}\frac{1}{w^3}i\tilde{\mathcal{E}}_a\alpha_a\beta \\ \int\frac{d\Omega_n}{4\pi}\Lambda_q^{(2)} &= \begin{pmatrix} -\frac{1}{2w^5}\tilde{\Pi}_a\tilde{\Pi}_a\beta + \frac{1}{2w^3}\tilde{\mathcal{B}}_c\sigma_c\beta \\ +\left(\frac{1}{6w^3} + \frac{1}{12w^5}\right)\left[\tilde{\Pi}_a, i\tilde{\mathcal{E}}_a\right]1_{4\times4} \\ +\frac{1}{4w^5}\left(\tilde{\mathcal{E}}_a\tilde{\Pi}_{a'} + \tilde{\Pi}_{a'}\tilde{\mathcal{E}}_a\right)\varepsilon_{aa'c}\sigma_c \\ +O\left(\tilde{\mathcal{E}}_a\tilde{\mathcal{E}}_a\beta\right) \end{pmatrix} \end{aligned} \tag{2.44}$$

and eventually

---

[3] Details of the derivation of (2.41) are presented in the Appendix B.

$$
\mathsf{Y}_C = \frac{\alpha_{FS}}{\pi}\int_0^{\frac{\tilde{\Omega}_{\max}}{2}} d\tilde{q} \begin{pmatrix} \frac{1}{w}\beta + \left(\frac{2}{3}\frac{1}{w} + \frac{1}{3}\frac{1}{w^3}\right)\alpha_a\tilde{\Pi}_a - \frac{1}{2}\frac{1}{w^3} i\tilde{\mathscr{E}}_a\alpha_a\beta \\ -\frac{1}{2w^5}\tilde{\Pi}_a\tilde{\Pi}_a\beta + \frac{1}{2w^3}\tilde{\mathscr{B}}_c\sigma_c\beta \\ +\left(\frac{1}{6w^3} + \frac{1}{12w^5}\right)\left[\tilde{\Pi}_a, i\tilde{\mathscr{E}}_a\right]1_{4\times 4} + \frac{1}{2\tilde{w}^5}\tilde{\mathscr{E}}_a\tilde{\Pi}_{a'}\varepsilon_{aa'c}\sigma_c \\ +O\left(\tilde{\mathscr{E}}_a\tilde{\mathscr{E}}_a\beta\right) \end{pmatrix} \tag{2.45}
$$

For the special case of zero electric field $\mathscr{E}_a \equiv 0$ the integrands enlisted in (2.45) proof (after elementary algebraic manipulations) all being identical to the ones obtained in the inspiring work [2] with different methods of derivation.

The calculation of the q-integrals in (2.45) is elementary and gives for $\tilde{\Omega}_{\max} \ggg 1$

$$
\int_0^{\frac{\tilde{\Omega}_{\max}}{2}} d\tilde{q}\left\{\frac{1}{w}, \frac{1}{w^3}, \frac{1}{w^5}\right\} = \left\{\ln\left(\tilde{\Omega}_{\max}\right), 1, \frac{2}{3}\right\} \tag{2.46}
$$

With the results (2.46) inserted into (2.45) then the self-energy (2.32) evaluates to

$$
\hat{\mathcal{M}}_C^{(e)} = mc^2 \sum_{k,k'\in I_+} \langle U_k| \begin{pmatrix} \left(\frac{\delta m}{m} + \frac{\alpha_{FS}}{\pi}\ln\tilde{\Omega}_{max}\right)\beta + \\ \\ +\frac{\alpha_{FS}}{\pi}\begin{pmatrix} \left(\frac{2}{3}\ln\tilde{\Omega}_{max} + \frac{1}{3}\right)\alpha_b\tilde{\Pi}_b - \frac{1}{2}i\tilde{\mathscr{E}}_b\alpha_b\beta - \\ -\frac{1}{3}\tilde{\Pi}_b\tilde{\Pi}_b\beta + \frac{1}{2}\tilde{\mathscr{B}}_c\sigma_c\beta + \\ +\frac{2}{9}\left[\tilde{\Pi}_c, i\tilde{\mathscr{E}}_c\right]1_{4\times 4} + \frac{1}{3}\tilde{\mathscr{E}}_a\tilde{\Pi}_{a'}\varepsilon_{aa'c}\sigma_c \\ +O\left(\tilde{\mathscr{E}}_a\tilde{\mathscr{E}}_a\beta\right) \end{pmatrix} \end{pmatrix} |U_{k'}\rangle\, \hat{c}_k^\dagger\hat{c}_{k'} \tag{2.47}
$$

Notably, in view of the dependence of $\hat{\mathcal{M}}_C^{(e)}$ on the cut off $\tilde{\Omega}_{\max}$, the terms proportional to β in (2.47), when taken *together* with a contribution of the transversal self-energy $\hat{\mathcal{M}}_\perp^{(e)}$, eventually will disappear with the procedure of mass renormalization. Actually, the gradient terms in (2.47) give at first sight more reason for concern, as they ruin the relativistic invariance of the renormalized Dirac Hamiltonian (2.2). Fortunately, in the end there is nothing to worry about, because the 1st-order gradient term in (2.47) being proportional to $\ln\tilde{\Omega}_{\max}$, when added to other gradient terms, that emerge evaluating the transversal self-energy $\hat{\mathcal{M}}_\perp^{(e)}$ with the gauge invariant cut off procedure of Dirac, disappears from (2.2) as well.

With regard to the remaining 1st- and 2nd-order gradients in (2.47), we now show that they annul each other in favor of electric field dependent terms. Because we restricted to eigenstates $|\phi_k\rangle$ and $|\phi_{k'}\rangle$ associated with non-relativistic eigenvalues E_k and $E_{k'}$ of the Dirac Hamiltonian $\mathsf{H}^{(D)}$, we have $E_k - E_{k'} = O(\text{small}^2)$ in case $z_k z_{k'} = 1$, i.e., there holds for the matrix elements of the operator $i\tilde{\mathscr{E}}_a\alpha_a\beta$ as a relation of equivalence with neglect of third order terms $O(\text{small}^3)$

$$
\begin{aligned}
\langle\phi_k|\,\frac{1}{2}i\,\tilde{\mathscr{E}}_a\alpha_a\beta\,|\phi_{k'}\rangle &= \langle\phi_k|\left(-\alpha_a\tilde{\Pi}_a + \tilde{\Pi}_a\tilde{\Pi}_a\beta - \tilde{\mathscr{B}}_c\sigma_c\beta\right)|\phi_{k'}\rangle \\
\langle\phi_k|\,\frac{1}{2}i\,\tilde{\mathscr{E}}_a\alpha_a\beta\,|\phi_{k'}\rangle &= \langle\phi_k|\left(\frac{1}{4}\left[\tilde{\Pi}_a, i\tilde{\mathscr{E}}_a\right]1_{4\times 4} + \frac{1}{2}\tilde{\mathscr{E}}_a\tilde{\Pi}_b\varepsilon_{abc}\sigma_c\right)|\phi_{k'}\rangle
\end{aligned} \tag{2.48}
$$

The proof of (2.48) is actually within easy reach, because there holds for $z_k z_{k'} = 1$

$$
\begin{aligned}
O(\text{small}^3) &= \langle \phi_k | \left(\tilde{E}_k - \tilde{E}_{k'} \right) \alpha_b \tilde{\Pi}_b \beta \, | \phi_{k'} \rangle \\
&= \langle \phi_k | \left[\tilde{\mathsf{H}}^{(D)}, \alpha_b \tilde{\Pi}_b \beta \right] | \phi_{k'} \rangle \\
&= \langle \phi_k | \left[\left(\beta + \alpha_a \tilde{\Pi}_a + \tilde{\Phi}^{(\text{ext})} \right), \alpha_b \tilde{\Pi}_b \beta \right] | \phi_{k'} \rangle \\
&= \langle \phi_k | \left(-2\alpha_a \tilde{\Pi}_a + 2\tilde{\Pi}_a \tilde{\Pi}_a \beta - 2\tilde{\mathcal{B}}_c \sigma_c \beta - i\tilde{\mathcal{E}}_a \alpha_a \beta \right) | \phi_{k'} \rangle
\end{aligned}
$$

and featuring a double commutator there also holds for $z_k z_{k'} = 1$

$$
\begin{aligned}
O(\text{small}^4) &= \langle \phi_k | \left(\tilde{E}_k - \tilde{E}_{k'} \right)^2 \tilde{\Phi}^{(\text{ext})} \, | \phi_{k'} \rangle \\
&= \langle \phi_k | \left[\tilde{\mathsf{H}}^{(D)}, \left[\tilde{\mathsf{H}}^{(D)}, \tilde{\Phi}^{(\text{ext})} \right] \right] | \phi_{k'} \rangle \\
&= \langle \phi_k | \left(\left[\tilde{\Pi}_a, i\tilde{\mathcal{E}}_a \right] 1_{4\times 4} + 2\tilde{\mathcal{E}}_a \tilde{\Pi}_b \varepsilon_{abc} \sigma_c - 2i\tilde{\mathcal{E}}_a \alpha_a \beta \right) | \phi_{k'} \rangle
\end{aligned}
$$

For $k = k'$ the error in (2.48) is exactly zero for *all* eigenstates of $\mathsf{H}^{(D)}$, of course.

With the equivalence relations (2.48) taken into account we finally obtain from (2.47) for the Coulomb self-energy of electrons

$$
\hat{\mathcal{M}}_C^{(e)} = mc^2 \sum_{k,k' \in I_+} \langle U_k | \left(\begin{array}{c} \left(\frac{\delta m}{m} + \frac{\alpha_{FS}}{\pi} \ln \tilde{\Omega}_{max} \right) \beta + \\ + \frac{\alpha_{FS}}{\pi} \left(\begin{array}{c} \left(\frac{2}{3} \ln \tilde{\Omega}_{max} \right) \alpha_b \tilde{\Pi}_b + \frac{1}{6} \tilde{\mathcal{B}}_c \sigma_c \beta - \\ - \frac{1}{9} \left[\tilde{\Pi}_c, i\tilde{\mathcal{E}}_c \right] 1_{4\times 4} - \frac{1}{3} \tilde{\mathcal{E}}_a \tilde{\Pi}_{a'} \varepsilon_{aa'c} \sigma_c \\ + O\left(\tilde{\mathcal{E}}_a \tilde{\mathcal{E}}_a \beta \right) \end{array} \right) \end{array} \right) | U_{k'} \rangle \, \hat{c}_k^{\dagger} \hat{c}_{k'} \tag{2.49}
$$

Note that here and in what follows we are not writing explicitly the coefficient of the *square* of the electric field amplitude in front of a factor α_{FS} anymore, because a contribution proportional to $\alpha_{FS} \tilde{\mathcal{E}}_a \tilde{\mathcal{E}}_a \beta$ is of fourth order in the coupling e. But such fourth order terms we have to neglect for consistency with the perturbation solution of the Hamiltonian flow equations presented in [1], where we maintained for the purpose of charge renormalization due to vacuum polarization the third order terms, see (2.28), but thereafter neglected the fourth order terms in the coupling e already.

2.5 The Transversal Self-energy of Electrons

The transversal self-energy of electrons is according to [1, Sect. 7.4] given by

$$
\hat{\mathcal{M}}_{\perp}^{(e)} = \frac{e^2}{m} \frac{1}{V_P} \sum_{\mathbf{q}} \frac{\hbar}{2\varepsilon_0 \omega(\mathbf{q})} \left(\delta_{b.b'} - \frac{q_b q_{b'}}{|\mathbf{q}|^2} \right) \sum_{k,k' \in I_+} \times \tag{2.50}
$$

$$
\times \left(\langle U_k | \alpha_b e^{-iq_a \mathsf{x}_a} \sum_{k'' \in I} \frac{\tilde{E}_k + \tilde{E}_{k'} - 2\tilde{E}_{k''} - 2\tilde{\omega}_q z_{k''}}{\left(\tilde{E}_{k''} + \tilde{\omega}_q z_{k''} - \tilde{E}_k \right)^2 + \left(\tilde{E}_{k''} + \tilde{\omega}_q z_{k''} - \tilde{E}_{k'} \right)^2} | \phi_{k''} \rangle \langle \phi_{k''} | e^{iq_a \mathsf{x}_a} \alpha_{b'} | U_{k'} \rangle \right) \hat{c}_k^{\dagger} \hat{c}_{k'}
$$

It is consistent to simplify the matrix element here in the following way

$$\sum_{k''\in I} \frac{\tilde{E}_k+\tilde{E}_{k'}-2\tilde{E}_{k''}-2\tilde{\omega}_q z_{k''}}{\left(\tilde{E}_{k''}+\tilde{\omega}_q z_{k''}-\tilde{E}_k\right)^2+\left(\tilde{E}_{k''}+\tilde{\omega}_q z_{k''}-\tilde{E}_{k'}\right)^2} |\phi_{k''}\rangle\langle\phi_{k''}| \tag{2.51}$$

$$= \sum_{k''\in I} \left(\frac{1}{\frac{\tilde{E}_k+\tilde{E}_{k'}}{2} - \tilde{E}_{k''} - \tilde{\omega}_q z_{k''}} - \frac{\left(\frac{\tilde{E}_k-\tilde{E}_{k'}}{2}\right)^2}{\left(\frac{\tilde{E}_k+\tilde{E}_{k'}}{2} - \tilde{E}_{k''} - \tilde{\omega}_q z_{k''}\right)^2 + \left(\frac{\tilde{E}_k-\tilde{E}_{k'}}{2}\right)^2} \right) |\phi_{k''}\rangle\langle\phi_{k''}|$$

$$= \sum_{k''\in I} \left(\sum_{\nu\in\{+,-\}} \frac{1}{\frac{\tilde{E}_k+\tilde{E}_{k'}}{2} - \tilde{E}_{k''} - \nu\tilde{\omega}_q} \frac{1+\nu z_{k''}}{2} \right) |\phi_{k''}\rangle\langle\phi_{k''}| + O(\text{small}^4)$$

This applies, since we dropped afore in the Coulomb self-energy (2.49) already terms being small of fourth order. Transforming the summation over photon wave vectors **q** in (2.50) to an integral and passing to spherical coordinates one obtains eventually

$$\hat{\mathcal{M}}_{\perp}^{(e)} = mc^2 \sum_{k,k'\in I_+} \langle U_k|\, Y_{\perp}\, |U_{k'}\rangle\ \hat{c}_k^{\dagger}\hat{c}_{k'} \tag{2.52}$$

whereby in terms of $\tilde{q} \equiv \tilde{\omega}_q$ and taking into account Dirac's high energy cut off

$$Y_{\perp} = \frac{\alpha_{FS}}{\pi} \int_0^{\infty} d\tilde{q} \int \frac{d\Omega_n}{4\pi} \left(\delta_{b,b'} - n_b n_{b'}\right) \times \tag{2.53}$$

$$\times\ \alpha_b e^{-i\, q_a x_a} \sum_{k''\in I} \Theta_{k''}^{(q)} \left(\sum_{\nu\in\{+,-\}} \frac{\tilde{q}}{\frac{\tilde{E}_k+\tilde{E}_{k'}}{2} - \tilde{E}_{k''} - \nu\tilde{q}} \frac{1+\nu z_{k''}}{2} \right) |\phi_{k''}\rangle\ \langle\phi_{k''}|\, e^{i\, q_a x_a} \alpha_{b'}$$

Here, for small values of $\tilde{q}$ particular attention should be given to the terms with $\nu = +1$, as in this case the denominators in (2.53) possibly fall to *zero*.

A way out of this difficulty is to divide the domain of $\tilde{q}$-integration into a high-energy part and into a narrow low-energy part introducing a parameter $\tilde{Q} \ll 1$, so that $\tilde{q} > \tilde{Q}$ in the high-energy part and $\tilde{q} < \tilde{Q}$ in the low-energy part, concurrently assuming as well (for electrons) $\frac{\tilde{E}_k+\tilde{E}_{k'}}{2} - 1 - \tilde{\Phi} \ll \tilde{Q}$. So, the energy $\hbar c Q$ is in between the atomic binding energy of an electron (a few eV) and the rest energy mc^2 of the electron (half-million eV), a requirement obviously fulfilled choosing for instance $\hbar c Q = 10$ keV. In point of fact the precise value we give for $\tilde{Q} \equiv \lambda_C Q = \frac{\hbar c Q}{mc^2}$ is irrelevant, as we employ for the evaluation of the $\tilde{q}$-integral in (2.53) the method of asymptotic matching, often used in boundary-layer theory [17]. Eventually, that way a suitable approximation to the matrix element in (2.52) independent on the intermediate parameter $\tilde{Q}$ can be obtained, as we show in the ensuing.

To this end we decompose the integral (2.53) into two integrals

$$Y_{\perp} = Y_{\perp}^{(\text{low})} + Y_{\perp}^{(\text{high})} \tag{2.54}$$

Due to $\tilde{q} \ll 1$ the dipole-approximation $e^{\pm i\, q_a \mathsf{x}_a} \approx 1$ applies in the first integration interval, so that the first integral is within the intended accuracy given by

$$Y_{\perp}^{(\text{low})} = \frac{\alpha_{FS}}{\pi} \cdot \frac{2}{3} \alpha_b \sum_{k'' \in I} \sum_{\nu \in \{+,-\}} \int_0^{\tilde{Q}} d\tilde{q} \frac{\tilde{q}}{\frac{\tilde{E}_k + \tilde{E}_{k'}}{2} - \tilde{E}_{k''} - \nu \tilde{q}} \frac{1 + \nu z_{k''}}{2} |\phi_{k''}\rangle\langle\phi_{k''}| \alpha_b \tag{2.55}$$

In the second integration interval $\tilde{Q} \le \tilde{q} < \infty$ we get rid of the intermediate summation over labels $k'' \in I \equiv I_+ \cup I_-$ following the example set previously by the Coulomb self-energy (2.31). Making in (2.53) use of the spectral representation

$$\frac{\mathsf{1}_{4\times 4}}{\zeta \cdot \mathsf{1}_{4\times 4} - \mathsf{H}^{(D)}} = \sum_{k'' \in I} \frac{1}{\zeta - E_{k''}} |\phi_{k''}\rangle \langle \phi_{k''}| \tag{2.56}$$

and concurrently implementing Dirac's cut off in the guise of an operator introducing

$$\begin{aligned} \Theta^{(D)} &= \sum_{k'' \in I} \Theta_{k''}^{(q)} |\phi_{k''}\rangle \langle \phi_{k''}| \\ &= \Theta_H \left(\tilde{\Omega}_{\max} - \tilde{q} - \sqrt{1 + \tilde{\Pi}_a \tilde{\Pi}_a} \right) \end{aligned} \tag{2.57}$$

one obtains for the second integral

$$\begin{aligned} Y_{\perp}^{(\text{high})} &= \frac{\alpha_{FS}}{\pi} \int_{\tilde{Q}}^{\infty} d\tilde{q} \int \frac{d\Omega_n}{4\pi} \left(\delta_{b,b'} - n_b n_{b'} \right) e^{-i q_a \mathsf{x}_a} \Theta^{(D)} e^{i q_a \mathsf{x}_a} \times \\ &\times \alpha_b e^{-i q_a \mathsf{x}_a} \frac{\tilde{q} \cdot \mathsf{1}_{4\times 4}}{\frac{\tilde{E}_k + \tilde{E}_{k'}}{2} \cdot \mathsf{1}_{4\times 4} - \tilde{\mathsf{H}}^{(D)} - \tilde{q} \Lambda^{(D)}} e^{i q_a \mathsf{x}_a} \alpha_{b'} \end{aligned} \tag{2.58}$$

According to (2.34) the effect of the operators $e^{\pm i q_a \mathsf{x}_a}$ in (2.58) is to replace everywhere $\tilde{\Pi}_a$ by $\tilde{\Pi}_a + \tilde{q} n_a$. In view of $\Omega_{\max} \ggg 1$ this engenders for the q-shifted cut off operator

$$\begin{aligned} e^{-i q_a \mathsf{x}_a} \Theta^{(D)} e^{i q_a \mathsf{x}_a} &= \Theta_H \left(\tilde{\Omega}_{\max} - \tilde{q} - \sqrt{1 + \left(\tilde{\Pi}_a + \tilde{q} n_a \right) \left(\tilde{\Pi}_a + \tilde{q} n_a \right)} \right) \\ &= \Theta_H \left(\tilde{\Omega}_{\max} - 2\tilde{q} - n_b \tilde{\Pi}_b \right) \end{aligned} \tag{2.59}$$

Executing the q-shift as well in the other terms in (2.58) then the searched for transversal self-energy of electrons in (2.52) assumes the guise

$$\hat{\mathcal{M}}_{\perp}^{(e)} = mc^2 \sum_{k,k' \in I_+} \langle U_k | \left(Y_{\perp}^{(\text{low})} + Y_{\perp,+}^{(\text{high})} + Y_{\perp,-}^{(\text{high})} \right) | U_{k'} \rangle \; \hat{c}_k^{\dagger} \hat{c}_{k'} \tag{2.60}$$

whereby

$$
\begin{aligned}
Y_{\perp,\pm}^{(\text{high})} &= \frac{\alpha_{FS}}{\pi} \int_{\tilde{Q}}^{\infty} d\tilde{q} \int \frac{d\Omega_n}{4\pi} \left(\delta_{b,b'} - n_b n_{b'}\right) \times \\
&\times \alpha_b \frac{\tilde{q} \cdot \mathbb{1}_{4\times 4}}{\frac{\tilde{E}_k + \tilde{E}_{k'}}{2} \cdot \mathbb{1}_{4\times 4} - \tilde{\mathsf{H}}^{(D)} - \tilde{q} n_b \alpha_b - \tilde{q} \Lambda_q^{(D)}} \alpha_{b'} \times \\
&\times \frac{1}{2}\left(\Theta_H\left(\tilde{\Omega}_{\max} - 2\tilde{q} - n_b \tilde{\Pi}_b\right) \pm \Theta_H\left(\tilde{\Omega}_{\max} - 2\tilde{q} + n_b \tilde{\Pi}_b\right)\right)
\end{aligned} \tag{2.61}
$$

The method to evaluate the matrix elements in (2.60) is the same as employed before for the matrix elements in (2.32), namely to restrict to the low-energy sector of QED, where the electron is not relativistic in the states $|U_k\rangle$ and $|U_{k'}\rangle$. With this proviso the application of the operator $\Pi_a \equiv \mathsf{p}_a - e\mathsf{A}_a^{(\text{ext})}$ on those states produces a small result, so that we can expand the integrand in (2.58) in the quantity $\alpha_a \tilde{\Pi}_a$, again retaining first and second order terms $(\alpha_a \tilde{\Pi}_a)^2$, but neglecting the higher-order terms.

2.5.1 Evaluation of $Y_{\perp,-}^{(high)}$

The domain of integration for the $\tilde{q}$-integral defining the term $Y_{\perp,-}^{(\text{high})}$ is actually a thin shell around the cut off $\Omega_{\max}$, so that effectively

$$
\begin{aligned}
&\Theta_H\left(\tilde{\Omega}_{\max} - 2\tilde{q} - n_a \tilde{\Pi}_a\right) - \Theta_H\left(\tilde{\Omega}_{\max} - 2\tilde{q} + n_a \tilde{\Pi}_a\right) \\
&= -2 n_a \tilde{\Pi}_a \, \delta\left(\tilde{\Omega}_{\max} - 2\tilde{q}\right)
\end{aligned} \tag{2.62}
$$

For large values $\tilde{q}$ there holds $\Lambda_q^{(D)} - \Lambda_q^{(0)} = O\left(\frac{1}{\tilde{q}^2}\right)$. With $\Lambda_q^{(0)}\Lambda_q^{(0)} = \mathbb{1}_{4\times 4}$ and insertion of (2.39) eventually

$$
\begin{aligned}
&\frac{\tilde{q} \cdot \mathbb{1}_{4\times 4}}{\frac{\tilde{E}_k + \tilde{E}_{k'}}{2} \mathbb{1}_{4\times 4} - \tilde{\mathsf{H}}^{(D)} - \tilde{q} n_b \alpha_b - \tilde{q} \Lambda_q^{(D)}} \\
&\equiv \frac{\mathbb{1}_{4\times 4}}{\frac{1}{\tilde{q}}\left(\frac{\tilde{E}_k + \tilde{E}_{k'}}{2} - \tilde{\Phi}\right)\hat{\mathbb{1}}_{4\times 4} - \frac{1}{\tilde{q}}\left(\sqrt{1+\tilde{q}^2} + \tilde{q}\right)\Lambda_q^{(0)} - \left(\frac{\alpha_a \tilde{\Pi}_a}{\tilde{q}} + \Lambda_q^{(D)} - \Lambda_q^{(0)}\right)} \\
&= -\frac{1}{2} n_{a'} \alpha_{a'} + O\left(\frac{1}{\tilde{q}}\right)
\end{aligned} \tag{2.63}
$$

Combining the expansions (2.62) and (2.63) in the integrand of (2.61) the $\tilde{q}$-integral determining $Y_{\perp,-}^{(\text{high})}$ is readily evaluated

$$Y_{\perp,-}^{(\text{high})} = \frac{\alpha_{FS}}{\pi}\frac{1}{2}\int \frac{d\Omega_n}{4\pi}\left(-n_a\tilde{\Pi}_a\right)\left(\delta_{b,b'} - n_b n_{b'}\right)\alpha_b\left(-\frac{1}{2}n_{a'}\alpha_{a'}\right)\alpha_{b'} \tag{2.64}$$

With help of the identities (summation convention)

$$\left(\delta_{b,b'} - n_b n_{b'}\right)\alpha_b\alpha_{a'}\alpha_{b'} = -2n_{a'}n_b\alpha_b \tag{2.65}$$

$$\int \frac{d\Omega_n}{4\pi}\left(n_a\tilde{\Pi}_a\right)\Bigg(\underbrace{n_{a'}n_{a'}}_{=1}\Bigg)n_b\alpha_b = \frac{1}{3}\delta_{a,b}\alpha_b\tilde{\Pi}_a$$

then finally

$$Y_{\perp,-}^{(\text{high})} = -\frac{\alpha_{FS}}{\pi}\cdot\frac{1}{6}\alpha_b\tilde{\Pi}_b \tag{2.66}$$

Notably, this result is a direct consequence of Dirac's gauge invariant cutoff.

2.5.2 Evaluation of $Y_{\perp,+}^{(high)}$

The evaluation of $Y_{\perp,+}^{(\text{high})}$ is considerably more elaborate and therefore we transfer most of the details of the calculational methods involved to the Appendices B and C. As such we settle in what follows just for a summary of the main results.

As a starting point we first rewrite the integral determining $Y_{\perp,+}^{(\text{high})}$ in the guise

$$Y_{\perp,+}^{(\text{high})} = \frac{\alpha_{FS}}{\pi}\int_{\tilde{Q}}^{\frac{\tilde{\Omega}_{\max}}{2}} d\tilde{q}\int \frac{d\Omega_n}{4\pi}\left(\delta_{b,b'} - n_b n_{b'}\right)\alpha_b \mathsf{K}_q \alpha_{b'} \tag{2.67}$$

whereby in terms of the definitions

$$\begin{aligned} w &\equiv \sqrt{1+\tilde{q}^2} \\ \varepsilon_\Phi &\equiv \frac{\tilde{E}_k + \tilde{E}_{k'}}{2} - \tilde{\Phi} \end{aligned} \tag{2.68}$$

the kernel in (2.67) is

$$\mathsf{K}_q = \frac{\mathsf{1}_{4\times 4}}{\frac{\varepsilon_\Phi}{\tilde{q}}\mathsf{1}_{4\times 4} - \left(1 + \frac{w}{\tilde{q}}\right)\Lambda_q^{(0)} - \left(\frac{\alpha_a\tilde{\Pi}_a}{\tilde{q}} + \Lambda_q^{(D)} - \Lambda_q^{(0)}\right)} \tag{2.69}$$

Note that ε_Φ is not a small term, because it comprises the rest energy of the electron. However the terms $\frac{\alpha_a\tilde{\Pi}_a}{\tilde{q}} + \Lambda_q^{(D)} - \Lambda_q^{(0)}$ we may consider as being small. Introducing

$$\mathsf{G}_q^{-1} \equiv \frac{\varepsilon_\Phi}{\tilde{q}}\hat{\mathsf{1}}_{4\times 4} - \left(1 + \frac{w}{\tilde{q}}\right)\Lambda_q^{(0)} \tag{2.70}$$

there follows with the expansion (2.37) and neglecting third- and higher order terms

$$\begin{aligned}
\mathsf{K}_q &= \frac{1_{4\times 4}}{\mathsf{G}_q^{-1} - \left(\frac{\alpha_a \tilde{\Pi}_a}{\tilde{q}} + \Lambda_q^{(D)} - \Lambda_q^{(0)}\right)} \\
&= \frac{1_{4\times 4}}{1_{4\times 4} - \mathsf{G}_q\left(\frac{\alpha_a \tilde{\Pi}_a}{\tilde{q}} + \Lambda_q^{(D)} - \Lambda_q^{(0)}\right)}\mathsf{G}_q \\
&= \mathsf{G}_q + \mathsf{G}_q\left(\frac{\alpha_a \tilde{\Pi}_a}{\tilde{q}} + \Lambda_q^{(1)} + \Lambda_q^{(2)}\right)\mathsf{G}_q + \mathsf{G}_q\left(\frac{\alpha_a \tilde{\Pi}_a}{\tilde{q}} + \Lambda_q^{(1)}\right)\mathsf{G}_q\left(\frac{\alpha_a \tilde{\Pi}_a}{\tilde{q}} + \Lambda_q^{(1)}\right)\mathsf{G}_q + \cdots
\end{aligned} \tag{2.71}$$

In terms of the projectors

$$\begin{aligned}
\mathsf{P}_{q,\pm}^{(0)} &= \frac{1}{2}\left(1_{4\times 4} \pm \Lambda_q^{(0)}\right) \\
\Lambda_q^{(0)} &= \mathsf{P}_{q,+}^{(0)} - \mathsf{P}_{q,-}^{(0)} = \frac{1}{w}\left(\beta + \tilde{q} n_c \alpha_c\right)
\end{aligned} \tag{2.72}$$

then (2.70) assumes the guise of a spectral representation

$$\mathsf{G}_q^{-1} = \frac{\varepsilon_\Phi - (\tilde{q} + w)}{\tilde{q}}\mathsf{P}_{q,+}^{(0)} + \frac{\varepsilon_\Phi + (\tilde{q} + w)}{\tilde{q}}\mathsf{P}_{q,-}^{(0)}, \tag{2.73}$$

implying

$$\begin{aligned}
\mathsf{G}_q &= \left(\mathsf{G}_q^{-1}\right)^{-1} \\
&= \frac{\tilde{q}}{\varepsilon_\Phi - (\tilde{q} + w)}\mathsf{P}_{q,+}^{(0)} + \frac{\tilde{q}}{\varepsilon_\Phi + (\tilde{q} + w)}\mathsf{P}_{q,-}^{(0)} \\
&\equiv \sum_{s\epsilon\{+,-\}} \tilde{q} \cdot g_{q,s} \mathsf{P}_{q,s}^{(0)}
\end{aligned} \tag{2.74}$$

Notably, here the coefficients

$$g_{q,s} = \frac{1}{\varepsilon_\Phi - s \cdot (\tilde{q} + w)} \tag{2.75}$$

depend via ε_Φ on the external electrostatic potential $\tilde{\Phi} = \frac{e\Phi^{(\mathrm{ext})}}{mc^2}$, so that we have

$$\begin{aligned}
\left[\tilde{\Pi}_a, g_{q,s}\right] &= \left[\tilde{\Pi}_a, \frac{1}{\varepsilon_\Phi - s\,(\tilde{q} + w)}\right] \\
&= -\frac{1}{\varepsilon_\Phi - s\,(\tilde{q} + w)}\left[\tilde{\Pi}_a, \varepsilon_\Phi - s\,(\tilde{q} + w)\right]\frac{1}{\varepsilon_\Phi - s\,(\tilde{q} + w)} \\
&= g_{q,s}^2\, i\tilde{\mathscr{E}}_a
\end{aligned} \tag{2.76}$$

Our primary interest in the ensuing are the *specific* matrix elements of $Y_{\perp,+}^{(\text{high})}$, for whom now a representation in the guise of a perturbation series is sought

$$\langle\phi_k|\, Y_{\perp,+}^{(\text{high})}\, |\phi_{k'}\rangle = \frac{\alpha_{FS}}{\pi} \int_{\tilde{Q}}^{\frac{\tilde{\Omega}_{\max}}{2}} d\tilde{q}\, \left(\mathsf{Z}_q^{(I)} + \mathsf{Z}_q^{(II)} + \mathsf{Z}_q^{(III)} + \cdots\right) \tag{2.77}$$

Here, in view of (2.67) and the expansion (2.71) it is defined

$$\begin{aligned}
\mathsf{Z}_q^{(I)} &= \langle\phi_k| \int \frac{d\Omega_n}{4\pi} \left(\delta_{b,b'} - n_b n_{b'}\right) \alpha_b \mathsf{G}_q \alpha_{b'} |\phi_{k'}\rangle \\
\mathsf{Z}_q^{(II)} &= \langle\phi_k| \int \frac{d\Omega_n}{4\pi} \left(\delta_{b,b'} - n_b n_{b'}\right) \alpha_b \mathsf{G}_q \left(\frac{\alpha_a \tilde{\Pi}_a}{\tilde{q}} + \Lambda_q^{(1)} + \Lambda_q^{(2)}\right) \mathsf{G}_q \alpha_{b'} |\phi_{k'}\rangle \\
\mathsf{Z}_q^{(III)} &= \langle\phi_k| \int \frac{d\Omega_n}{4\pi} \left(\delta_{b,b'} - n_b n_{b'}\right) \alpha_b \mathsf{G}_q \left(\frac{\alpha_a \tilde{\Pi}_a}{\tilde{q}} + \Lambda_q^{(1)}\right) \mathsf{G}_q \left(\frac{\alpha_a \tilde{\Pi}_a}{\tilde{q}} + \Lambda_q^{(1)}\right) \mathsf{G}_q \alpha_{b'} |\phi_{k'}\rangle
\end{aligned} \tag{2.78}$$

The evaluation of these specific matrix elements is lengthy and somewhat tedious. As regards the first term in the expansion (2.77), its evaluation indeed exemplifies most of the technical problems encountered insisting on gauge invariance in every step of the calculations. So we present the determination of $\mathsf{Z}_q^{(I)}$ in some detail here, but desist giving full information on the calculational details of the other terms, as the evaluation of $\mathsf{Z}_q^{(II)}$ and $\mathsf{Z}_q^{(III)}$ is patterned completely along the lines of the calculation of $\mathsf{Z}_q^{(I)}$.

Evaluation of $\mathsf{Z}_q^{(I)}$

Upon insertion of (2.74) into (2.78) then

$$\begin{aligned}
\mathsf{Z}_q^{(I)} &= \langle\phi_k| \int \frac{d\Omega_n}{4\pi} \left(\delta_{b,b'} - n_b n_{b'}\right) \alpha_b \left(\sum_{s\in\{+,-\}} \tilde{q}\cdot g_{q,s} \mathsf{P}_{q,s}^{(0)}\right) \alpha_{b'} |\phi_{k'}\rangle \\
&= \langle\phi_k| \int \frac{d\Omega_n}{4\pi} \left(\delta_{b,b'} - n_b n_{b'}\right) \alpha_b \begin{pmatrix} \sum_{s\in\{+,-\}} \tilde{q}\cdot g_{q,s} \frac{1}{2}\left(1_{4\times 4} + \frac{s}{w}\beta\right) \\ + \text{odd under } \mathbf{n} \to -\mathbf{n} \end{pmatrix} \alpha_{b'} |\phi_{k'}\rangle \\
&= \langle\phi_k| \int \frac{d\Omega_n}{4\pi} \sum_{s\in\{+,-\}} \tilde{q}\cdot g_{q,s} \frac{1}{2} \left(\left(\delta_{b,b'} - n_b n_{b'}\right) \alpha_b \alpha_{b'} + \frac{s}{w} \left(\delta_{b,b'} - n_b n_{b'}\right) \alpha_b \beta \alpha_{b'}\right) |\phi_{k'}\rangle \\
&= \langle\phi_k| \sum_{s\in\{+,-\}} \tilde{q}\cdot g_{q,s} \left(1_{4\times 4} - s\cdot \frac{1}{w}\beta\right) |\phi_{k'}\rangle
\end{aligned} \tag{2.79}$$

Introducing

$$g_{q,s}^{(k)} = \frac{1}{\tilde{E}_k - \tilde{\Phi} - s\cdot(\tilde{q} + w)} \tag{2.80}$$

there holds neglecting fourth order corrections,

$$g_{q,s} = \frac{1}{2}\left(g_{q,s}^{(k)} + g_{q,s}^{(k')}\right) \tag{2.81}$$

thus leading within the intended accuracy to the equivalent representation

$$\mathsf{Z}_q^{(I)} = \langle\phi_k| \sum_{s\in\{+,-\}} \frac{\tilde{q}}{2}\left(g_{q,s}^{(k)}\left(1_{4\times4} - s\cdot\frac{1}{w}\beta\right) + \left(1_{4\times4} - s\cdot\frac{1}{w}\beta\right)g_{q,s}^{(k')}\right)|\phi_{k'}\rangle \tag{2.82}$$

Introducing an integral representation for the term in (2.80) enables now in point of fact a gauge invariant expansion of (2.82) in small quantities

$$g_{q,s}^{(k')} = i\int_0^\infty dt\; e^{-it\left(\tilde{E}_{k'} - \tilde{\Phi}(\mathsf{x}) - s(\tilde{q}+\tilde{w}) - i0^+\right)} \tag{2.83}$$

With $\tilde{E}_{k'}$ being an eigenvalue of the scaled Dirac Hamiltonian $\tilde{\mathsf{H}}^{(D)} = \beta + \alpha_b\tilde{\Pi}_b + \tilde{\Phi}$ corresponding to the eigenstate $|\phi_{k'}\rangle$, there holds

$$e^{-it\tilde{E}_{k'}}\;|\phi_{k'}\rangle = e^{-it\tilde{\mathsf{H}}^{(D)}}\;|\phi_{k'}\rangle \tag{2.84}$$

Using this we see that applying $g_{q,s}^{(k')}$ on the specific state $|\phi_{k'}\rangle$ is in fact equivalent to applying the *operator*

$$\Gamma_{q,s} = i\int_0^\infty dt\; e^{-it\left(-s(\tilde{q}+\tilde{w})-i0^+\right)}e^{-it\beta}e^{it\left(\beta+\tilde{\Phi}(\mathsf{x})\right)}e^{-it\left(\beta+\alpha_b\tilde{\Pi}_b+\tilde{\Phi}(\mathsf{x})\right)} \tag{2.85}$$

Indeed

$$\begin{aligned}
g_{q,s}^{(k')}\;|\phi_{k'}\rangle &= i\int_0^\infty dt\; e^{-it\left(\tilde{E}_{k'}-\tilde{\Phi}(\mathsf{x})-s(\tilde{q}+\tilde{w})-i0^+\right)}\;|\phi_{k'}\rangle \\
&= i\int_0^\infty dt\; e^{-it\left(-s(\tilde{q}+\tilde{w})-i0^+\right)}e^{it\tilde{\Phi}(\mathsf{x})}e^{-it\tilde{E}_{k'}}\;|\phi_{k'}\rangle \\
&= i\int_0^\infty dt\; e^{-it\left(-s(\tilde{q}+\tilde{w})-i0^+\right)}e^{it\tilde{\Phi}(\mathsf{x})}e^{-it\tilde{\mathsf{H}}^{(D)}}\;|\phi_{k'}\rangle \\
&= i\int_0^\infty dt\; e^{-it\left(-s(\tilde{q}+\tilde{w})-i0^+\right)}e^{-it\beta}e^{it\left(\beta+\tilde{\Phi}(\mathsf{x})\right)}e^{-it\left(\beta+\tilde{\Phi}(\mathsf{x})+\alpha_b\tilde{\Pi}_b\right)}\;|\phi_{k'}\rangle \\
&= \Gamma_{q,s}\;|\phi_{k'}\rangle\,,
\end{aligned} \tag{2.86}$$

and correspondingly

$$\langle\phi_k|\,g_{q,s}^{(k)} = \langle\phi_k|\,\Gamma_{q,s}^{\dagger} \tag{2.87}$$

So then, the matrix element (2.82) assumes the equivalent guise

$$\mathsf{Z}_q^{(I)} = \langle \phi_k | \sum_{s \in \{+,-\}} \frac{\tilde{q}}{2} \cdot \left(\Gamma_{q,s}^{\dagger} \left(1_{4\times4} - s \cdot \frac{1}{\tilde{w}} \beta \right) + \left(1_{4\times4} - s \cdot \frac{1}{\tilde{w}} \beta \right) \Gamma_{q,s} \right) |\phi_{k'}\rangle \tag{2.88}$$

Substantial progress comes now using standard perturbation theory, with the gradient term $\alpha_b \tilde{\Pi}_b$ in (2.85) identified as a perturbation. So we obtain, retaining the quadratic terms but neglecting the third order terms,

$$\begin{aligned} & e^{it\left(\beta+\tilde{\Phi}(\mathbf{x})\right)} e^{-it\left(\beta+\tilde{\Phi}(\mathbf{x})+\alpha_b\tilde{\Pi}_b\right)} \qquad (2.89) \\ &= T_t \exp\left(-i \int_0^t dt' e^{it'\left(\beta+\tilde{\Phi}(\mathbf{x})\right)} \alpha_b \tilde{\Pi}_b e^{-it'\left(\beta+\tilde{\Phi}(\mathbf{x})\right)} \right) \\ &= \begin{pmatrix} 1_{4\times4} - i \int_0^t dt' e^{it'\left(\beta+\tilde{\Phi}(\mathbf{x})\right)} \alpha_b \tilde{\Pi}_b e^{-it'\left(\beta+\tilde{\Phi}(\mathbf{x})\right)} + \\ +(-i)^2 \int_0^t dt' e^{it'\left(\beta+\tilde{\Phi}(\mathbf{x})\right)} \alpha_{b'} \tilde{\Pi}_{b'} e^{-it'\left(\beta+\tilde{\Phi}(\mathbf{x})\right)} \int_0^{t'} dt'' e^{it''\left(\beta+\tilde{\Phi}(\mathbf{x})\right)} \alpha_{b''} \tilde{\Pi}_{b''} e^{-it''\left(\beta+\tilde{\Phi}(\mathbf{x})\right)} + \\ + \cdots \end{pmatrix} \end{aligned}$$

With the identity

$$\begin{aligned} e^{it\left(\beta+\tilde{\Phi}(\mathbf{x})\right)} \alpha_b \tilde{\Pi}_b e^{-it\left(\beta+\tilde{\Phi}(\mathbf{x})\right)} &= \left(e^{it\beta} \alpha_b e^{-it\beta} \right) \left(e^{it\tilde{\Phi}(\mathbf{x})} \tilde{\Pi}_b e^{-it\tilde{\Phi}(\mathbf{x})} \right) \\ &= \left(e^{2it\beta} \alpha_b \right) \left(\tilde{\Pi}_b + t \tilde{\mathscr{E}}_b \right) \end{aligned} \tag{2.90}$$

there emerges combining (2.85) with (2.89) eventually the perturbation series

$$\Gamma_{q,s} = \Gamma_{q,s}^{(0)} + \Gamma_{q,s}^{(1)} + \Gamma_{q,s}^{(2)} + \cdots \tag{2.91}$$

whereby

$$\begin{aligned} \Gamma_{q,s}^{(0)} &= i \int_0^\infty dt\, e^{-it\left(-s(\tilde{q}+w)-i0^+\right)} e^{-it\beta} \qquad (2.92) \\ \Gamma_{q,s}^{(1)} &= i \int_0^\infty dt\, e^{-it\left(-s(\tilde{q}+w)-i0^+\right)} e^{-it\beta} (-i) \int_0^t dt' e^{2it'\beta} \left(\tilde{\Pi}_b + t' \tilde{\mathscr{E}}_b \right) \alpha_b \\ \Gamma_{q,s}^{(2)} &= i \int_0^\infty dt\, e^{-it\left(-s(\tilde{q}+w)-i0^+\right)} e^{-it\beta} (-i)^2 \int_0^t dt' e^{2it'\beta} \left(\tilde{\Pi}_{b'} + t' \tilde{\mathscr{E}}_{b'} \right) \alpha_{b'} \int_0^{t'} dt'' e^{2it''\beta} \left(\tilde{\Pi}_{b''} + t'' \tilde{\mathscr{E}}_{b''} \right) \alpha_{b''} \end{aligned}$$

The calculations here are elementary and give for the terms in the expansion (2.91) the following gauge invariant results

$$\begin{aligned} \Gamma_{q,s}^{(0)} &= -\frac{1}{2\tilde{q}} \left(s \cdot 1_{4\times4} + \frac{1}{\tilde{q}+w} \beta \right) \qquad (2.93) \\ \Gamma_{q,s}^{(1)} &= -\frac{1}{2\tilde{q}\,(\tilde{q}+w)} \left(\alpha_b \tilde{\Pi}_b - \frac{1}{2} \left(1 - \frac{w}{\tilde{q}} \right) i \tilde{\mathscr{E}}_b \alpha_b \beta + \frac{1}{2\tilde{q}} s \cdot i \tilde{\mathscr{E}}_b \alpha_b \right) \\ \Gamma_{q,s}^{(2)} &= \begin{pmatrix} -\frac{1}{2} \left(\frac{1}{1-(\tilde{q}+w)^2} + \frac{1}{(\beta-s(\tilde{q}+w))^2} \right) \tilde{\Pi}_{b'} \tilde{\Pi}_{b''} \beta \\ + \left(\frac{1}{4} \frac{1}{1-(\tilde{q}+w)^2} 1_{4\times4} + \frac{1}{4} \frac{1}{(\beta-s(\tilde{q}+w))^2} + \frac{1}{2} \frac{1}{(\beta-s(\tilde{q}+w))^3} \beta \right) \tilde{\Pi}_{b'} i \tilde{\mathscr{E}}_{b''} \\ + \left(-\frac{1}{4} \frac{1}{1-(\tilde{q}+w)^2} 1_{4\times4} - \frac{1}{4} \frac{1}{(\beta+s(\tilde{q}+w))^2} + \frac{1}{2} \frac{1}{(\beta-s(\tilde{q}+w))^3} \beta \right) i \tilde{\mathscr{E}}_{b'} \tilde{\Pi}_{b''} \\ + \left(-\frac{1}{8} \frac{1}{1-(\tilde{q}+w)^2} - \frac{1}{8} \frac{1}{(\beta+s(\tilde{q}+w))^2} + \frac{1}{4} \frac{1}{(\beta-s(\tilde{q}+w))^3} \beta + \frac{1}{(\beta-s(\tilde{q}+w))^4} \right) \tilde{\mathscr{E}}_{b'} \tilde{\mathscr{E}}_{b''} \beta \end{pmatrix} \alpha_{b'} \alpha_{b''} \end{aligned}$$

Inserting the expansion (2.91) into (2.88), straightforward algebra leads finally to

$$\mathsf{Z}_q^{(I)} = \langle\phi_k| \begin{pmatrix} \left(\frac{1}{w} - \frac{1}{\tilde{q}+w}\right)\beta \\ -\frac{1}{\tilde{q}+w}\alpha_b\tilde{\Pi}_b + \left(\frac{1}{2}\frac{1}{w} + \frac{1}{\tilde{q}+w} - \frac{1}{\tilde{q}}\right) i\tilde{\mathscr{E}}_b\alpha_b\beta \\ +\left(\frac{1}{\tilde{q}+w} - \frac{1}{2w}\right)\left(\tilde{\Pi}_b\tilde{\Pi}_b - \tilde{\mathscr{B}}_c\sigma_c\right)\beta \\ +\left(\frac{1}{2\tilde{q}} - \frac{1}{\tilde{q}+w}\right)\left(\frac{1}{2}\left[\tilde{\Pi}_b, i\tilde{\mathscr{E}}_b\right]1_{4\times4} + \tilde{\mathscr{E}}_{b'}\tilde{\Pi}_{b''}\varepsilon_{b'b''c}\sigma_c\right) \\ +O\left(\tilde{\mathscr{E}}_b\tilde{\mathscr{E}}_b\beta\right) \end{pmatrix} |\phi_{k'}\rangle \tag{2.94}$$

Actually, the term $\Gamma_{q,s}^{(2)}$ in the expansion (2.91) is required exclusively for the evaluation of $\mathsf{Z}_q^{(I)}$, but isn't needed anymore for the other terms in (2.77), as we already agreed to neglect the third order terms.

Summary of Findings for $\mathsf{Z}_q^{(I)} + \mathsf{Z}_q^{(II)} + \mathsf{Z}_q^{(III)}$

It is expedient to decompose the terms in (2.78) so that

$$\begin{aligned} \mathsf{Z}_q^{(II)} &= \mathsf{Z}_q^{(II,a)} + \mathsf{Z}_q^{(II,b)} + \mathsf{Z}_q^{(II,c)} \\ \mathsf{Z}_q^{(III)} &= \mathsf{Z}_q^{(III,a)} + \mathsf{Z}_q^{(III,b)} + \mathsf{Z}_q^{(III,c)} \end{aligned}$$

whereby

$$\begin{aligned} \mathsf{Z}_q^{(II,a)} &= \langle\phi_k| \int \frac{d\Omega_n}{4\pi}\left(\delta_{b,b'} - n_b n_{b'}\right)\alpha_b \mathsf{G}_q \frac{\alpha_a\tilde{\Pi}_a}{\tilde{q}}\mathsf{G}_q\alpha_{b'}\,|\phi_{k'}\rangle \\ \mathsf{Z}_q^{(II,b)} &= \langle\phi_k| \int \frac{d\Omega_n}{4\pi}\left(\delta_{b,b'} - n_b n_{b'}\right)\alpha_b \mathsf{G}_q \Lambda_q^{(1)}\mathsf{G}_q\alpha_{b'}\,|\phi_{k'}\rangle \\ \mathsf{Z}_q^{(II,c)} &= \langle\phi_k| \int \frac{d\Omega_n}{4\pi}\left(\delta_{b,b'} - n_b n_{b'}\right)\alpha_b \mathsf{G}_q \Lambda_q^{(2)}\mathsf{G}_q\alpha_{b'}\,|\phi_{k'}\rangle \\ \mathsf{Z}_q^{(III,a)} &= \langle\phi_k| \frac{1}{\tilde{q}^2}\int \frac{d\Omega_n}{4\pi}\left(\delta_{b,b'} - n_b n_{b'}\right)\alpha_b \mathsf{G}_q\left(\alpha_a\tilde{\Pi}_a\right)\mathsf{G}_q\left(\alpha_{a'}\tilde{\Pi}_{a'}\right)\mathsf{G}_q\alpha_{b'}\,|\phi_{k'}\rangle \\ \mathsf{Z}_q^{(III,b)} &= \langle\phi_k| \frac{1}{\tilde{q}}\int \frac{d\Omega_n}{4\pi}\left(\delta_{b,b'} - n_b n_{b'}\right)\alpha_b \mathsf{G}_q \begin{pmatrix} \left(\alpha_a\tilde{\Pi}_a\right)\mathsf{G}_q\Lambda_q^{(1)} \\ +\Lambda_q^{(1)}\mathsf{G}_q\left(\alpha_a\tilde{\Pi}_a\right) \end{pmatrix} \mathsf{G}_q\alpha_{b'}\,|\phi_{k'}\rangle \\ \mathsf{Z}_q^{(III,c)} &= \langle\phi_k| \int \frac{d\Omega_n}{4\pi}\left(\delta_{b,b'} - n_b n_{b'}\right)\alpha_b \mathsf{G}_q \Lambda_q^{(1)}\mathsf{G}_q\Lambda_q^{(1)}\mathsf{G}_q\alpha_{b'}\,|\phi_{k'}\rangle \end{aligned} \tag{2.95}$$

Actually the evaluation of these six matrix elements proceeds to a great extend analogously to the evaluation of $\mathsf{Z}_q^{(I)}$. That is to say, after insertion of the spectral representation (2.74) in place of G_q, first the angular averages are calculated, whereby it proves advisable using the identities

$$\begin{aligned} \mathsf{P}^{(0)}_{q,s}\alpha_a &= \alpha_a \mathsf{P}^{(0)}_{q,-s} + s \cdot \frac{\tilde{q}}{w} n_a \mathsf{1}_{4\times4} \\ \mathsf{P}^{(0)}_{q,s}\mathsf{P}^{(0)}_{q,s'} &= \delta_{s,s'}\mathsf{P}^{(0)}_{q,s} \end{aligned} \tag{2.96}$$

and also making repeatedly use of (2.65). So then, minding $w \equiv \sqrt{1+\tilde{q}^2}$, it is found[4]

$$\begin{aligned} W_{a;s,s'} &\equiv \int \frac{d\Omega_n}{4\pi} \left(\delta_{b,b'} - n_b n_{b'}\right) \alpha_b \mathsf{P}^{(0)}_{q,s}\alpha_a \mathsf{P}^{(0)}_{q,s'}\alpha_{b'} \\ &= -\frac{1}{3}\alpha_a \left(\delta_{-s,s'}\left(\mathsf{1}_{4\times4} + s\frac{1}{w}\beta\right) + ss'\frac{\tilde{q}^2}{w^2}\mathsf{1}_{4\times4}\right) \end{aligned} \tag{2.97}$$

and

$$\begin{aligned} W_{a,a';s,s',s''} &\equiv \int \frac{d\Omega_n}{4\pi} \left(\delta_{b,b'} - n_b n_{b'}\right) \alpha_b \mathsf{P}^{(0)}_{q,s}\alpha_a \mathsf{P}^{(0)}_{q,s''}\alpha_{a'} \mathsf{P}^{(0)}_{q,s'}\alpha_{b'} \\ &= \delta_{a,a'} \begin{pmatrix} \delta_{-s,s''}\delta_{-s',s''}\left(\mathsf{1}_{4\times4} + s''\cdot\frac{1}{w}\beta\right) \\ -\frac{1}{3}\frac{\tilde{q}^2}{w^2}ss'\left(\delta_{-s,s''} + \delta_{-s',s''}\right)\mathsf{1}_{4\times4} \\ +\frac{1}{3}\frac{\tilde{q}^2}{w^2}ss'\left(\mathsf{1}_{4\times4} - s''\cdot\frac{1}{\tilde{w}}\beta\right) \end{pmatrix} - \frac{1}{3}\delta_{-s,s''}\delta_{-s',s''} i\varepsilon_{aa'c}\sigma_c\left(\mathsf{1}_{4\times4} + s''\cdot\frac{1}{\tilde{w}}\beta\right) \end{aligned} \tag{2.98}$$

In the sequel we apply the equivalences (2.86) and (2.87) and thereupon insert the expansion (2.91) with an eye toward neglecting systematically the third order terms.

By way of example, this strategy applied to the matrix element $\mathsf{Z}^{(III,a)}_q$ gives within the intended accuracy the equivalent expression

$$\mathsf{Z}^{(III,a)}_q = \langle\phi_k|\,\tilde{q}\sum_{s,s',s''}\Gamma^{(0)}_{q,s}\left(\tilde{\Pi}_a g_{q,s''}\tilde{\Pi}_{a'}\right) W_{a,a';s,s',s''}\Gamma^{(0)}_{q,s'}\,|\phi_{k'}\rangle \tag{2.99}$$

The term $g_{q,s''}$ here positioned in between the operators $\tilde{\Pi}_a$ at first sight prevents a transit to the operator $\Gamma_{q,s''}$. But with the identity

$$\tilde{\Pi}_a g_{q,s}\tilde{\Pi}_a = \frac{1}{2}\left(\tilde{\Pi}_a\tilde{\Pi}_a g_{q,s} + g_{q,s}\tilde{\Pi}_a\tilde{\Pi}_a + \left[\tilde{\Pi}_a, \left[g_{q,s}, \tilde{\Pi}_a\right]\right]\right) \tag{2.100}$$

and minding the commutator (2.76) that's not an issue any more.

Eventually, by this means the following results are obtained

[4] Details of the derivation are presented in the Appendix C.

$$
\mathsf{Z}_q^{(III,a)} = \langle\phi_k| \begin{pmatrix} \left(\frac{1}{6}\frac{1}{w} + \frac{1}{3}\frac{\tilde{q}}{w^2} - \frac{1}{\tilde{q}+w} + \frac{1}{3}\frac{1}{w^3}\right)\tilde{\Pi}_a\tilde{\Pi}_a\beta \\ +\left(\frac{1}{6}\frac{1}{w} - \frac{1}{3}\frac{1}{\tilde{q}+w}\right)\tilde{\mathcal{B}}_c\sigma_c\beta \\ +\left(\frac{1}{2}\frac{1}{\tilde{q}+w} - \frac{1}{4}\frac{1}{\tilde{q}} + \frac{1}{3}\left(\frac{1}{w} - \frac{\tilde{q}}{w^2}\right) - \frac{1}{6}\frac{1}{w^3}\right)\left[\tilde{\Pi}_a, i\tilde{\mathcal{E}}_a\right]1_{4\times4} \\ +\left(\frac{1}{6}\frac{1}{\tilde{q}} - \frac{1}{3}\frac{1}{\tilde{q}+w}\right)\tilde{\mathcal{E}}_a\tilde{\Pi}_{a'}\varepsilon_{aa'c}\sigma_c \\ +O\left(\tilde{\mathcal{E}}_a\tilde{\mathcal{E}}_a\beta\right) \end{pmatrix} |\phi_{k'}\rangle
$$

$$
\mathsf{Z}_q^{(III,b)} = \langle\phi_k| \begin{pmatrix} \left(\frac{4}{3}\frac{1}{\tilde{q}+w} - \frac{1}{3}\frac{\tilde{q}}{w^2} - \frac{1}{3}\frac{1}{w} - \frac{1}{w^3} + \frac{2}{3}\frac{\tilde{q}}{w^4}\right)\tilde{\Pi}_a\tilde{\Pi}_a\beta \\ +\frac{2}{3}\left(\frac{1}{\tilde{q}+w} + \frac{1}{2}\frac{\tilde{q}}{w^2} - \frac{1}{w}\right)\tilde{\mathcal{B}}_c\sigma_c\beta \\ +\left(\frac{1}{3}\left(\frac{\tilde{q}}{w^2} - \frac{2}{\tilde{q}+w}\right) + \frac{1}{6}\frac{1}{w^3} - \frac{1}{12}\frac{\tilde{q}}{w^4}\right)\left[\tilde{\Pi}_a, i\tilde{\mathcal{E}}_a\right]1_{4\times4} \\ +\left(\frac{1}{3}\frac{1}{\tilde{q}+w} - \frac{1}{4}\frac{1}{w} + \frac{1}{12}\frac{q}{w^2} + \frac{1}{12}\frac{1}{w^3}\right)2\tilde{\mathcal{E}}_a\tilde{\Pi}_{a'}\varepsilon_{aa'c}\sigma_c \\ +O\left(\tilde{\mathcal{E}}_a\tilde{\mathcal{E}}_a\beta\right) \end{pmatrix} |\phi_{k'}\rangle
$$

$$
\mathsf{Z}_q^{(III,c)} = \langle\phi_k| \begin{pmatrix} \left(-\frac{2}{3}\frac{1}{\tilde{q}+w} + \frac{2}{3}\frac{1}{w} - \frac{1}{3}\frac{\tilde{q}}{w^2} + \frac{1}{6}\frac{1}{w^3} - \frac{1}{3}\frac{\tilde{q}}{w^4} - \frac{1}{6}\frac{1}{w^5}\right)\tilde{\Pi}_a\tilde{\Pi}_a\beta \\ +\left(-\frac{1}{3}\frac{1}{\tilde{q}+w} + \frac{1}{2}\frac{1}{w} - \frac{1}{3}\frac{\tilde{q}}{w^2} - \frac{1}{6}\frac{1}{w^3}\right)\tilde{\mathcal{B}}_c\sigma_c\beta \\ +\left(\frac{1}{3}\frac{1}{\tilde{q}+w} - \frac{1}{3}\frac{1}{w} + \frac{1}{6}\frac{\tilde{q}}{w^2} - \frac{1}{12}\frac{1}{w^3} + \frac{1}{12}\frac{1}{w^5} - \frac{1}{12}\frac{\tilde{q}}{w^6}\right)\left[\tilde{\Pi}_a, i\tilde{\mathcal{E}}_a\right]1_{4\times4} \\ +\left(-\frac{1}{6}\frac{1}{\tilde{q}+w} + \frac{1}{4}\frac{1}{w} - \frac{1}{6}\frac{\tilde{q}}{w^2} - \frac{1}{12}\frac{1}{w^3} + \frac{1}{12}\frac{\tilde{q}}{w^4}\right)2\tilde{\mathcal{E}}_a\tilde{\Pi}_{a'}\varepsilon_{aa'c}\sigma_c \\ +O\left(\tilde{\mathcal{E}}_a\tilde{\mathcal{E}}_a\beta\right) \end{pmatrix} |\phi_{k'}\rangle
$$

and all the same

$$
\mathsf{Z}_q^{(II,a)} = \langle\phi_k| \begin{pmatrix} \frac{1}{3}\left(\frac{1}{\tilde{q}+w} - \frac{\tilde{q}}{w^2}\right)\alpha_a\tilde{\Pi}_a \\ +\frac{1}{6}\left(\frac{2}{\tilde{q}+w} - \frac{1}{w}\right)i\tilde{\mathcal{E}}_a\alpha_a\beta \\ +\frac{1}{3}\left(\frac{1}{\tilde{q}} - \frac{2}{\tilde{q}+w} + \frac{\tilde{q}}{w^2} - \frac{1}{w}\right)\left(\frac{1}{2}\left[\tilde{\Pi}_a, i\tilde{\mathcal{E}}_a\right]1_{4\times4} + \tilde{\mathcal{E}}_a\tilde{\Pi}_b\varepsilon_{abc}\sigma_c\right) \end{pmatrix} |\phi_{k'}\rangle
$$

$$
\mathsf{Z}_q^{(II,b)} = \langle\phi_k| \begin{pmatrix} \frac{1}{3}\left(\frac{1}{w^3} + \frac{\tilde{q}}{w^2} - \frac{1}{w}\right)\alpha_a\tilde{\Pi}_a \\ +\frac{1}{3}\left(\frac{1}{w} - \frac{\tilde{q}}{w^2}\right)i\tilde{\mathcal{E}}_a\alpha_a\beta \\ +O\left(\tilde{\mathcal{E}}_a\tilde{\mathcal{E}}_a\beta\right) \end{pmatrix} |\phi_{k'}\rangle
$$

$$
\mathsf{Z}_q^{(II,c)} = \langle\phi_k| \begin{pmatrix} \frac{1}{3}\left(\frac{\tilde{q}}{w^2} - \frac{1}{w} + \frac{2}{w^3} - \frac{\tilde{q}}{w^4} - \frac{1}{w^5}\right)\tilde{\Pi}_a\tilde{\Pi}_a\beta \\ +\left(\frac{1}{6}\left(\frac{1}{w} - \frac{\tilde{q}}{w^2}\right) - \frac{1}{12}\frac{1}{w^3} + \frac{1}{12}\frac{\tilde{q}}{w^4} - \frac{1}{12}\frac{1}{w^5} + \frac{1}{12}\frac{\tilde{q}}{w^6}\right)\left[\tilde{\Pi}_a, i\tilde{\mathcal{E}}_a\right]1_{4\times4} \\ +\frac{1}{12}\left(\frac{\tilde{q}}{w^2} - \frac{1}{w} + \frac{1}{w^3} - \frac{\tilde{q}}{w^4}\right)2\tilde{\mathcal{E}}_a\tilde{\Pi}_{a'}\varepsilon_{aa'c}\sigma_c \\ +O\left(\tilde{\mathcal{E}}_a\tilde{\mathcal{E}}_a\beta\right) \end{pmatrix} |\phi_{k'}\rangle
$$

Making use of the equivalence relations (2.48) the terms $i\tilde{\mathscr{E}}_a\alpha_a\beta$ can be eliminated and there follows

$$\langle\phi_k|\,Y^{(\text{high})}_{\perp,+}\,|\phi_{k'}\rangle = \frac{\alpha_{FS}}{\pi}\int_{\tilde{Q}}^{\frac{\tilde{\Omega}_{\max}}{2}} d\tilde{q}\,\left(\mathsf{Z}_q^{(I)}+\mathsf{Z}_q^{(II)}+\mathsf{Z}_q^{(III)}\right) \tag{2.101}$$

$$= \frac{\alpha_{FS}}{\pi}\int_{\tilde{q}_c}^{\frac{\tilde{\Omega}_{\max}}{2}} d\tilde{q}\,\left(\mathsf{Z}_q^{(I)}+\mathsf{Z}_q^{(II,a)}+\mathsf{Z}_q^{(II,b)}+\mathsf{Z}_q^{(II,c)}+\mathsf{Z}_q^{(III,a)}+\mathsf{Z}_q^{(III,b)}+\mathsf{Z}_q^{(III,c)}\right)$$

$$= \frac{\alpha_{FS}}{\pi}\langle\phi_k|\int_{\tilde{Q}}^{\frac{\tilde{\Omega}_{\max}}{2}} d\tilde{q}\begin{pmatrix} \left(\frac{1}{w}-\frac{1}{\tilde{q}+w}\right)\beta \\ +\left(-\frac{2}{3}\frac{1}{\tilde{q}+w}-\frac{1}{3}\frac{1}{w}+\frac{1}{3}\frac{1}{w^3}\right)\alpha_a\tilde{\Pi}_a \\ +\left(\frac{1}{3}\left(\frac{2}{\tilde{q}+w}-\frac{1}{w}\right)+\frac{1}{6}\frac{1}{w^3}-\frac{1}{2}\frac{1}{w^5}\right)\tilde{\Pi}_a\tilde{\Pi}_a\beta \\ +\left(\frac{1}{2}\frac{1}{w}-\frac{1}{\tilde{q}+w}-\frac{1}{6}\frac{1}{w^3}\right)\tilde{\mathscr{B}}_c\sigma_c\beta \\ +\left(\frac{1}{6}\frac{1}{w}-\frac{1}{3}\frac{1}{\tilde{q}+w}+\frac{1}{6}\frac{1}{w^3}\right)\tilde{\mathscr{E}}_a\tilde{\Pi}_b\varepsilon_{abc}\sigma_c \\ +\left(\frac{1}{3}\frac{1}{w}-\frac{1}{3}\frac{1}{\tilde{q}}-\frac{1}{6}\frac{1}{w^3}\right)\left[\tilde{\Pi}_a, i\tilde{\mathscr{E}}_a\right]\mathbb{1}_{4\times 4} \\ +O\left(\tilde{\mathscr{E}}_a\tilde{\mathscr{E}}_a\beta\right) \end{pmatrix}|\phi_{k'}\rangle$$

With $\tilde{Q} \ll 1$ and $\tilde{\Omega}_{\max} \ggg 1$ the calculation of the integrals gives

$$\int_0^{\frac{\tilde{\Omega}_{\max}}{2}} d\tilde{q}\left(\frac{1}{w}-\frac{1}{\tilde{q}+w}\right) = \frac{1}{2}\ln\left(\tilde{\Omega}_{\max}\right)-\frac{1}{4} \tag{2.102}$$

$$\int_0^{\frac{\tilde{\Omega}_{\max}}{2}} d\tilde{q}\left(-\frac{2}{3}\frac{1}{\tilde{q}+w}-\frac{1}{3}\frac{1}{w}+\frac{1}{3}\frac{1}{w^3}\right) = \frac{1}{6}-\frac{2}{3}\ln\left(\tilde{\Omega}_{\max}\right)$$

The coefficient of the second order derivative term $\tilde{\Pi}_a\tilde{\Pi}_a\beta$ is in fact identical zero

$$\int_0^{\infty} d\tilde{q}\left(\frac{1}{3}\left(\frac{2}{\tilde{q}+w}-\frac{1}{w}\right)+\frac{1}{6}\frac{1}{w^3}-\frac{1}{2}\frac{1}{w^5}\right) \equiv 0 \tag{2.103}$$

Further there holds

$$\int_0^{\infty} d\tilde{q}\left(\frac{1}{2}\frac{1}{\tilde{w}}-\frac{1}{\tilde{q}+\tilde{w}}-\frac{1}{6}\frac{1}{\tilde{w}^3}\right) = -\frac{5}{12} \tag{2.104}$$

$$\int_0^{\frac{\tilde{\Omega}_{\max}}{2}} d\tilde{q}\left(\frac{1}{6}\frac{1}{\tilde{w}}-\frac{1}{3}\frac{1}{\tilde{q}+\tilde{w}}+\frac{1}{6}\frac{1}{\tilde{w}^3}\right) = \frac{1}{12}$$

But regarding the coefficient of the term $\left[\tilde{\Pi}_a, i\tilde{\mathscr{E}}_a\right]$ there enters a dependence on the lower cutoff $\tilde{Q}$, as here we cannot take zero as a lower integration limit

$$\int_{\tilde{Q}}^{\infty} d\tilde{q} \left(\frac{1}{3}\frac{1}{\tilde{w}} - \frac{1}{3}\frac{1}{\tilde{q}} - \frac{1}{6}\frac{1}{\tilde{w}^3} \right) = -\frac{1}{6} - \frac{1}{3}\ln\left(\frac{1}{2\tilde{Q}}\right) \tag{2.105}$$

Now that we have evaluated the integrals it remains to collect the terms. Specializing to *electrons* the searched for specific matrix elements are given by

$$\langle U_k | Y_{\perp,+}^{(\text{high})} \left| U_{k'} \right\rangle = \frac{\alpha_{FS}}{\pi} \langle U_k | \begin{pmatrix} \left(\frac{1}{2}\ln\left(\tilde{\Omega}_{\max}\right) - \frac{1}{4}\right)\beta \\ +\left(\frac{1}{6} - \frac{2}{3}\ln\left(\tilde{\Omega}_{\max}\right)\right)\alpha_a \tilde{\Pi}_a - \frac{5}{12}\tilde{\mathscr{B}}_c \sigma_c \beta \\ +\left(-\frac{1}{6} + \frac{1}{3}\ln\left(2\tilde{Q}\right)\right)\left[\tilde{\Pi}_a, i\tilde{\mathscr{E}}_a\right] 1_{4\times 4} + \frac{1}{12}\tilde{\mathscr{E}}_a \tilde{\Pi}_b \varepsilon_{abc}\sigma_c \end{pmatrix} \left| U_{k'} \right\rangle \tag{2.106}$$

For zero electric field $\mathscr{E}_a \equiv 0$ the integrands enlisted in (2.101) turn out (after elementary algebraic manipulations) all being identical to the ones obtained afore by Dupont-Roc and Cohen-Tannoudji in their inspiring work [2], but the feature (2.103) and so the *disappearance* of the "even" gradient term $\Pi_a \Pi_a$ in point of fact escapes notice in their final result.

2.5.3 Evaluation of $Y_{\perp}^{(low)}$

An expansion of the rational function in the integrand of (2.55) with a factor ν enables to re-write in the denominator $\nu \tilde{E}_{k''} = z_{k''} \tilde{E}_{k''} = \left| \tilde{E}_{k''} \right|$, observing the numerator and hence the integrand vanishes for $\nu = -z_{k''}$ anyway. So then

$$\begin{aligned} Y_{\perp}^{(\text{low})} &= \frac{\alpha_{FS}}{\pi} \cdot \frac{2}{3}\alpha_b \sum_{k'' \in I} \sum_{\nu \in \{+,-\}} \int_0^{\tilde{Q}} d\tilde{q} \frac{\tilde{q}}{\nu \frac{\tilde{E}_k + \tilde{E}_{k'}}{2} - \left|\tilde{E}_{k''}\right| - \tilde{q}} \frac{\nu + z_{k''}}{2} \left|\phi_{k''}\right\rangle \left\langle \phi_{k''}\right| \alpha_b \qquad (2.107) \\ &= \frac{\alpha_{FS}}{\pi}\frac{2}{3}\alpha_b \sum_{k''} \begin{pmatrix} -\tilde{Q} z_{k''} - \frac{1-z_{k''}}{2}\left(\left|\tilde{E}_{k''}\right| + \frac{\tilde{E}_k + \tilde{E}_{k'}}{2}\right) \ln\left(1 + \frac{\tilde{Q}}{\left|\tilde{E}_{k''}\right| + \frac{\tilde{E}_k + \tilde{E}_{k'}}{2}}\right) \\ + \frac{1+z_{k''}}{2}\left(\left|\tilde{E}_{k''}\right| - \frac{\tilde{E}_k + \tilde{E}_{k'}}{2}\right) \ln\left(\frac{\tilde{Q} + \left|\tilde{E}_{k''}\right| - \frac{\tilde{E}_k + \tilde{E}_{k'}}{2}}{\left|\left|\tilde{E}_{k''}\right| - \frac{\tilde{E}_k + \tilde{E}_{k'}}{2}\right|}\right) \end{pmatrix} \left|\phi_{k''}\right\rangle \left\langle \phi_{k''}\right| \alpha_b \end{aligned}$$

For electrons $\tilde{E}_k > 0$, $\tilde{E}_{k'} > 0$ and thus, expanding the logarithmic terms proportional $\frac{1-z_{k''}}{2}$ to first order in $\tilde{Q} \ll 1$ and minding thereafter

$$\frac{1 + z_{k''}}{2} \left|\phi_{k''}\right\rangle \left\langle \phi_{k''}\right| = \left|U_{k''}\right\rangle \left\langle U_{k''}\right| \tag{2.108}$$

we get

$$Y_{\perp}^{(\mathrm{low})} = \frac{\alpha_{FS}}{\pi}\frac{2}{3}\alpha_b \sum_{k''\in I_+}\left(-\tilde{Q} + \left(\tilde{E}_{k''} - \frac{\tilde{E}_k+\tilde{E}_{k'}}{2}\right)\ln\left(\frac{\tilde{Q}+\tilde{E}_{k''}-\frac{\tilde{E}_k+\tilde{E}_{k'}}{2}}{\left|\tilde{E}_{k''}-\frac{\tilde{E}_k+\tilde{E}_{k'}}{2}\right|}\right)\right)|U_{k''}\rangle\langle U_{k''}|\alpha_b$$

Here a simplification is permitted observing the fermion modes labeled by k'' can contribute for low lying modes $K \in \{k, k'\}$ significantly only provided

$$\left|\tilde{E}_{k''} - \tilde{E}_K\right| \ll \tilde{Q} \ll 1 \tag{2.109}$$

Under this premise the searched for matrix element of $Y_{\perp}^{(\mathrm{low})}$ is

$$\begin{aligned}&\langle U_k|\, Y_{\perp}^{(\mathrm{low})}\,|U_{k'}\rangle \\ &= \frac{\alpha_{FS}}{\pi}\cdot\frac{2}{3}\sum_{k''\in I_+}\langle U_k|\,\alpha_b\,|U_{k''}\rangle\langle U_{k''}|\,\alpha_b\,|U_{k'}\rangle\left(\tilde{E}_{k''} - \frac{\tilde{E}_k+\tilde{E}_{k'}}{2}\right)\ln\left(\frac{\tilde{Q}}{\left|\tilde{E}_{k''}-\frac{\tilde{E}_k+\tilde{E}_{k'}}{2}\right|}\right)\end{aligned} \tag{2.110}$$

Because the logarithm here varies only slowly as a function of $\tilde{E}_{k''}$ it is expedient to replace in the argument of the logarithm the dependence on $\tilde{E}_{k''}$ by some average, see for instance [16, 18]

$$\begin{aligned}\tilde{M}_{k,k'} &= \sqrt{\tilde{M}_k\tilde{M}_{k'}} \\ \tilde{M}_k &= \frac{\sum_{k''\in I_+}|\langle U_k|\,\alpha_b\,|U_{k''}\rangle|^2\left(\tilde{E}_{k''}-\tilde{E}_k\right)\ln\left|\tilde{E}_{k''}-\tilde{E}_k\right|}{\sum_{k''\in I_+}|\langle U_k|\,\alpha_b\,|U_{k''}\rangle|^2\left(\tilde{E}_{k''}-\tilde{E}_k\right)}\end{aligned} \tag{2.111}$$

the idea of introducing such an average being the summation over fermion mode labels k'' in (2.110) should enable by now an evaluation in closed form

$$\begin{aligned}&\langle U_k|\, Y_{\perp}^{(\mathrm{low})}\,|U_{k'}\rangle \\ &= \frac{\alpha_{FS}}{\pi}\cdot\frac{2}{3}\sum_{k''\in I_+}\langle U_k|\,\alpha_b\,|U_{k''}\rangle\,\langle U_{k''}|\,\alpha_b\,|U_{k'}\rangle\left(\tilde{E}_{k''} - \frac{\tilde{E}_k+\tilde{E}_{k'}}{2}\right)\ln\left(\frac{\tilde{Q}}{\tilde{M}_{k,k'}}\right)\end{aligned} \tag{2.112}$$

To this end we use the identity $\Lambda^{(D)}\,|U_{k''}\rangle = z_{k''}\,|U_{k''}\rangle$ where for electrons $z_{k''} = 1$ and thus re-write first of all

$$\langle U_k|\,\alpha_b\,|U_{k''}\rangle \equiv \langle U_k|\,\frac{1}{2}\left(\Lambda^{(D)}\alpha_b + \alpha_b\Lambda^{(D)}\right)|U_{k''}\rangle \tag{2.113}$$

Next, because only the low lying modes contribute significantly one can expand the operator $\Lambda^{(D)} \equiv \Lambda_{q=0}^{(D)}$ along the lines (2.37), and keeping only the zeroth order and first order terms we have

$$\Lambda^{(D)} = \beta + \alpha_a \tilde{\Pi}_a - \frac{1}{2} i \tilde{\mathscr{E}}_a \alpha_a \beta + \cdots \tag{2.114}$$

Inserting this into (2.113) gives with help of the algebra of Dirac matrices eventually

$$\langle U_k | \alpha_b | U_{k''} \rangle \equiv \langle U_k | \tilde{\Pi}_b | U_{k''} \rangle + \cdots \tag{2.115}$$

By this means the matrix element (2.112) assumes the guise

$$\begin{aligned}
&\langle U_k | Y_\perp^{(\text{low})} | U_{k'} \rangle \\
&= \frac{\alpha_{FS}}{\pi} \cdot \frac{1}{3} \sum_{k'' \in I_+} \begin{pmatrix} \langle U_k | \tilde{\Pi}_b \left(\tilde{E}_{k''} - \tilde{E}_k \right) | U_{k''} \rangle \langle U_{k''} | \tilde{\Pi}_b | U_{k'} \rangle \\ + \langle U_k | \tilde{\Pi}_b | U_{k''} \rangle \langle U_{k''} | \left(\tilde{E}_{k''} - \tilde{E}_{k'} \right) \tilde{\Pi}_b | U_{k'} \rangle \end{pmatrix} \ln \left(\frac{\tilde{Q}}{\tilde{M}_{k,k'}} \right) \\
&= \frac{\alpha_{FS}}{\pi} \cdot \frac{1}{3} \begin{pmatrix} \langle U_k | \left[\tilde{\Pi}_b, \tilde{\text{H}}^{(D)} \right] \sum_{k'' \in I_+} | U_{k''} \rangle \langle U_{k''} | \tilde{\Pi}_b | U_{k'} \rangle \\ - \langle U_k | \tilde{\Pi}_b \sum_{k'' \in I_+} | U_{k''} \rangle \langle U_{k''} | \left[\tilde{\Pi}_b, \tilde{\text{H}}^{(D)} \right] | U_{k'} \rangle \end{pmatrix} \ln \left(\frac{\tilde{Q}}{\tilde{M}_{k,k'}} \right) \\
&\approx \frac{\alpha_{FS}}{\pi} \cdot \frac{1}{3} \begin{pmatrix} \langle U_k | \left[\tilde{\Pi}_b, \tilde{\text{H}}^{(D)} \right] \left(\sum_{k'' \in I_+} | U_{k''} \rangle \langle U_{k''} | + \sum_{k'' \in I_-} | V_{k''} \rangle \langle V_{k''} | \right) \tilde{\Pi}_b | U_{k'} \rangle \\ - \langle U_k | \tilde{\Pi}_b \left(\sum_{k'' \in I_+} | U_{k''} \rangle \langle U_{k''} | + \sum_{k'' \in I_-} | V_{k''} \rangle \langle V_{k''} | \right) \left[\tilde{\Pi}_b, \tilde{\text{H}}^{(D)} \right] | U_{k'} \rangle \end{pmatrix} \ln \left(\frac{\tilde{Q}}{\tilde{M}_{k,k'}} \right)
\end{aligned} \tag{2.116}$$

Here the transition to the last line is permitted in view of a tiny contribution $\langle U_k | \tilde{\Pi}_b | V_{k''} \rangle \approx 0$ for the low lying eigenstates of $\tilde{\text{H}}^{(D)} = \beta + \alpha_a \tilde{\Pi}_a + \tilde{\Phi}$.

So, with $\left[\tilde{\Pi}_b, \tilde{\Phi} \right] = i \tilde{\mathscr{E}}_b$ and minding as well the completeness relation

$$\sum_{k'' \in I_+} | U_{k''} \rangle \langle U_{k''} | + \sum_{k'' \in I_-} | V_{k''} \rangle \langle V_{k''} | \equiv 1_{4 \times 4} \tag{2.117}$$

there follows within the intended accuracy to order 1 in α_{FS} the result first obtained by Bethe [18]

$$\langle U_k | Y_\perp^{(\text{low})} | U_{k'} \rangle = \frac{\alpha_{FS}}{\pi} \langle U_k | \left(-\frac{1}{3} \right) \left[\tilde{\Pi}_b, i \tilde{\mathscr{E}}_b \right] \ln \left(\frac{\tilde{Q}}{\tilde{M}_{k,k'}} \right) | U_{k'} \rangle \tag{2.118}$$

2.6 Renormalization of Mass and the Energy Change of Electrons

Collecting the results obtained afore in (2.49), (2.66), (2.106), and (2.118), it so happens that the coefficients of the gradient terms $\alpha_a \tilde{\Pi}_a$ neutralize! So we obtain

$$\hat{\mathcal{M}}_C^{(e)} + \hat{\mathcal{M}}_\perp^{(e)} = mc^2 \sum_{k,k' \in I_+} \langle U_k | \left(Y_C + Y_{\perp,-}^{(\text{high})} + Y_{\perp,+}^{(\text{high})} + Y_\perp^{(\text{low})} \right) | U_{k'} \rangle \, \hat{c}_k^\dagger \hat{c}_{k'} \tag{2.119}$$

$$= mc^2 \sum_{k,k' \in I_+} \langle U_k | \begin{pmatrix} \left(\frac{\delta m}{m} + \frac{\alpha_{FS}}{\pi} \left(\left(1 + \frac{1}{2}\right) \ln\left(\tilde{\Omega}_{max}\right) - \frac{1}{4} \right) \right) \beta \\ + \frac{\alpha_{FS}}{\pi} \left(\frac{1}{6} - \frac{5}{12} \right) \frac{1}{4} \tilde{\mathcal{B}}_c \sigma_c \beta \\ + \frac{\alpha_{FS}}{\pi} \left(\frac{1}{12} - \frac{1}{3} \right) \tilde{\mathcal{E}}_a \tilde{\Pi}_b \varepsilon_{abc} \sigma_c \\ - \frac{\alpha_{FS}}{\pi} \left(\frac{1}{9} + \frac{1}{6} + \frac{1}{3} \ln(\frac{1}{2\tilde{Q}}) + \frac{1}{3} \ln(\frac{\tilde{Q}}{\tilde{M}_{k,k'}}) \right) \left[\tilde{\Pi}_a, i \tilde{\mathcal{E}}_a \right] 1_{4\times 4} \end{pmatrix} | U_{k'} \rangle \, \hat{c}_k^\dagger \hat{c}_{k'}$$

Now choosing

$$\frac{\delta m}{m} \stackrel{!}{=} -\frac{\alpha_{FS}}{\pi} \left(\frac{3}{2} \ln \tilde{\Omega}_{max} - \frac{1}{4} \right) \tag{2.120}$$

eliminates the coefficient of the matrix β in accord with the prescription of mass renormalization. In point of fact (2.120) concurs with Dirac's derivation in [7, p. 110], even though his calculational method and his choice of gauge differs completely from ours.

What's more, (2.120) concurs as well with what has been stated by Dupont-Roc and Cohen-Tannoudji in [2, Eq. (56)], the only distinction being the appearance of a ln 2 constant that can be attributed to a factor of two difference in the definition of the cutoff. But yet, the absence of an "odd" gradient term $\alpha_b \tilde{\Pi}_b$ in our result (2.119) for the self-energy is in sharp contrast to what has been found by Dupont-Roc and Cohen-Tannoudji in [2], the reason being these authors implemented at high energy merely a sharp cutoff *after* having q-shifted the integrand and thus missed out the neutralizing contribution of $Y_{\perp,-}^{(\text{high})}$ given by (2.66). It is the combination $Y_C + Y_{\perp,-}^{(\text{high})} + Y_{\perp,+}^{(\text{high})}$ that in effect restores with consideration of (2.48) the manifest relativistic invariance of the renormalized Dirac Hamiltonian.

Another important feature of (2.119) is the drop out of the intermediate parameter $\tilde{Q}$ in accord with the procedure of matching [17]. Adding to (2.119) now as well the contribution of vacuum polarization, which according to (2.30) is given in scaled units by $\hat{\mathcal{M}}_{VP}^{(e)} = mc^2 \frac{\alpha_{FS}}{\pi} \frac{1}{15} \left[\tilde{\Pi}_a, i \tilde{\mathcal{E}}_a \right] 1_{4\times 4}$, then the self-energy of electrons to order 1 in the fine structure constant $\alpha_{FS} = \frac{e^2}{4\pi \varepsilon_0 \hbar c} \approx \frac{1}{137}$ eventually emerges as

$$\hat{\mathcal{M}}^{(e)} = \hat{\mathcal{M}}_C^{(e)} + \hat{\mathcal{M}}_\perp^{(e)} + \hat{\mathcal{M}}_{VP}^{(e)} \tag{2.121}$$
$$= mc^2 \frac{\alpha_{FS}}{\pi} \sum_{k,k' \in I_+} \langle U_k| \begin{pmatrix} -\frac{1}{4}\tilde{\mathscr{B}}_c \sigma_c \beta - \frac{1}{4}\tilde{\mathscr{E}}_a \tilde{\Pi}_b \varepsilon_{abc} \sigma_c \\ -\frac{1}{3}\left(\frac{19}{30} + \ln(\frac{1}{2\tilde{M}_{k,k'}})\right)\left[\tilde{\Pi}_a, i\tilde{\mathscr{E}}_a\right] 1_{4\times 4} \end{pmatrix} |U_{k'}\rangle \; \hat{c}_k^\dagger \hat{c}_{k'}$$

The term with the magnetic induction field $\tilde{\mathscr{B}}_c$ here represents the electron magnetic moment anomaly of Schwinger [4] and the terms with the electrostatic field $\tilde{\mathscr{E}}_a$ comprises the well known Lamb shift terms including a correction to the spin-orbit interaction *and* the effects of vacuum polarization [13, 19].

Bringing the units in one obtains in view of (2.130), also identifying the Bohr magneton $\mu_B = \frac{e\hbar}{2m}$ and the Compton length $\lambda_C = \frac{\hbar}{mc}$, now for the energy change of electrons in the presence of external electromagnetic (static) background fields

$$\hat{\mathcal{M}}^{(e)} = \hat{\mathcal{M}}_C^{(e)} + \hat{\mathcal{M}}_\perp^{(e)} + \hat{\mathcal{M}}_{VP}^{(e)} \tag{2.122}$$
$$= \frac{\alpha_{FS}}{\pi} \sum_{k,k'} \langle U_k| \begin{pmatrix} -\frac{1}{2} \cdot \mu_B B_c^{(\text{ext})} \sigma_c \beta \\ -\frac{1}{4}\lambda_C \cdot e\left(-\nabla_a \Phi^{(\text{ext})}\right) \frac{1}{mc} \Pi_b \varepsilon_{abc} \sigma_c \\ -\frac{1}{3}\left(\frac{19}{30} + \ln(\frac{1}{2\tilde{M}_{k,k'}})\right) e\lambda_C^2 \left(-\nabla^2 \Phi^{(\text{ext})}\right) 1_{4\times 4} \end{pmatrix} |U_{k'}\rangle \; \hat{c}_k^\dagger \hat{c}_{k'}$$

For the bound state $k = k'$ of the electron corresponding to $2s$ in hydrogen a value $-\ln \tilde{M}_{k,k} = 7.6876$ has been authorized by Heitler [9, p. 346], implying the Bethe logarithm provides the main contribution.

The corresponding energy change for positrons we desist writing here, as it can be obtained from (2.122) replacing $U_k \to V_k$ and $\hat{c}_k^\dagger \hat{c}_{k'} \to \hat{b}_{k'}^\dagger \hat{b}_k$ and of course $e \to -e$ and $\sigma_c \to i\sigma_y \sigma_c \, i\sigma_y$.

In closing we emphasize, the obtained result (2.122) agrees to order 1 in the fine structure constant α_{FS} in every detail with what can be obtained with S-matrix based methods [20, 21]. A derivation that deals with all circumbendibus of S-matrix theory and renormalization theory in a meticulously precise manner can be found in [16]. Our derivation of the self-energy terms (2.122) resembles—as regards the spirit—Dirac's derivation [7, Sect. 29]. Like Dirac we avoid the in- and out-states of the S-matrix formalism. But there are conspicuous differences, other than our use of the Coulomb gauge and Dirac's use of the radiation gauge instead. With our Wegner flow equation based approach a unitary equivalent Hamiltonian to the QED-Hamiltonian is constructed in the guise of a perturbation series respecting conservation of particle number, for details see [1], while Dirac uses perturbation theory to solve for the one-electron problem the Heisenberg equations of motion for the fermion and photon field operators [7, Sect. 18].

Appendix A

Calculation of $\varrho^{(pol)}(\mathbf{r})$ and $j_b^{(pol)}(\mathbf{r})$

Instead of calculating the vacuum polarization charge density $\varrho^{(\mathrm{pol})}(\mathbf{r})$ and the vacuum polarization current density $j_b^{(\mathrm{pol})}(\mathbf{r})$ separately, it was Schwinger's idea to derive both quantities from a functional [5], so that the *total* variation of that functional determines $\varrho^{(\mathrm{pol})}(\mathbf{r})$ and $j_b^{(\mathrm{pol})}(\mathbf{r})$ from the requirement, the variation δW of that functional around the *prescribed* external potentials $\Phi^{(\mathrm{ext})}(\mathbf{r})$ and $\mathsf{A}^{(\mathrm{ext})}(\mathbf{r})$ being

$$\delta W \stackrel{!}{=} \int d^3r' \left(\varrho^{(\mathrm{pol})}(\mathbf{r}')\,\delta\Phi(\mathbf{r}') - j_b^{(\mathrm{pol})}(\mathbf{r}')\,\delta\mathsf{A}_b(\mathbf{r}')\right) \tag{2.123}$$

On physical grounds such a functional $W\left[\Phi(\cdot),\mathsf{A}(\cdot)\right]$ undoubtedly exists, because (2.123) obviously relates to the familiar Hamiltonian for the interaction of a polarizable medium with (fixed) external electromagnetic sources [22].

As a concurrent variation of $\mathsf{A}(\cdot)$ by $\delta\mathsf{A}(\cdot)$ and of $\Phi(\cdot)$ by $\delta\Phi(\cdot)$ implies a corresponding variation of the Dirac Hamiltonian $\mathsf{H}^{(D)}$ being

$$\delta\mathsf{H}^{(D)} = e\,\delta\Phi(\mathsf{x})\,\mathbb{1}_{4\times4} - e\,c\,\alpha_b\delta\mathsf{A}_b(\mathsf{x}) \tag{2.124}$$

the variation (2.123) of the sought functional may be rewritten by merging now (2.17) and (2.18) in compliance with (2.19) into the compact guise

$$\begin{aligned}\delta W &= -\frac{e}{2}\int d^3r \sum_{k\in I}\sum_{\mu,\nu}\phi_\mu^\star(\mathbf{r};k)\left(\delta_{\nu,\mu}\delta\Phi(\mathsf{x}) - (c\alpha_b)_{\mu,\nu}\,\delta\mathsf{A}_b(\mathsf{x})\right)\Lambda_{\nu,\mu}^{(D)}\phi_\mu(\mathbf{r};k)\\ &= -\frac{1}{2}\mathrm{tr}\left(\delta\mathsf{H}^{(D)}\,\Lambda^{(D)}\right)\end{aligned} \tag{2.125}$$

What we are looking for in the ensuing is therefore a suitable representation of that functional enabling a *gauge invariant* expansion in the coupling constant e.

To this end we write

$$z_k = \frac{E_k}{|E_k|} = \frac{1}{\pi}\int_{-\infty}^{\infty} d\omega \frac{E_k}{\omega^2 + E_k^2} = \int_{-\infty}^{\infty}\frac{d\omega}{\pi}\frac{1}{-i\omega + E_k} \tag{2.126}$$

and so obtain for the energy-sign operator $\Lambda^{(D)}$ a suitable integral representation

$$\Lambda^{(D)} = \sum_k z_k\,|\phi_k\rangle\,\langle\phi_k| = \int_{-\infty}^{\infty}\frac{d\omega}{\pi}\frac{\mathbb{1}_{4\times4}}{-i\omega\mathbb{1}_{4\times4} + \mathsf{H}^{(D)}} \tag{2.127}$$

Inspired by a remark of R.P. Feynman in [23], ibid. "Seventeenth Lecture," let us rewrite (2.127) by insertion of unity in the guise $\mathbb{1}_{4\times4} = \beta\beta$. This way

$$\begin{aligned}\Lambda^{(D)} &= \int_{-\infty}^{\infty}\frac{d\omega}{\pi}\frac{\mathsf{1}_{4\times4}}{\beta\left(-i\omega\mathsf{1}_{4\times4}+\mathsf{H}^{(D)}\right)}\beta\\ &= \int_{-\infty}^{\infty}\frac{d\omega}{\pi}\frac{\mathsf{1}_{4\times4}}{\beta\left(-i\omega\mathsf{1}_{4\times4}+\beta mc^2+c\alpha_b\Pi_b+e\Phi\left(\mathsf{x}\right)\right)}\beta\\ &= \int_{-\infty}^{\infty}\frac{d\tilde{\omega}}{\pi}\frac{\mathsf{1}_{4\times4}}{\mathsf{1}_{4\times4}-\gamma_\mu\tilde{\Pi}_\mu}\gamma_0 \end{aligned}\tag{2.128}$$

where for convenience we introduced a common (pseudo-) four vector notation

$$\begin{aligned}\gamma_\mu\tilde{\Pi}_\mu &\equiv \gamma_0\tilde{\Pi}_0-\gamma_b\tilde{\Pi}_b\\ \gamma_\mu\tilde{p}_\mu &\equiv \gamma_0\tilde{p}_0-\gamma_b\tilde{\mathsf{p}}_b\end{aligned}\tag{2.129}$$

and also agreed on the definition of *scaled* variables

$$\begin{aligned}\tilde{\Pi}_0 &= \frac{\Pi_0}{mc^2}=\tilde{\mathsf{p}}_0-\tilde{\mathsf{A}}_0\left(\mathsf{x}\right)\\ \tilde{\Pi}_b &= \frac{\Pi_b}{mc}=\tilde{\mathsf{p}}_b-\tilde{\mathsf{A}}_b\left(\mathsf{x}\right)\ \text{ for b} \in \{\text{x,y,z}\}\\ \tilde{\omega} &= \frac{\omega}{mc^2}\\ \tilde{\mathsf{p}}_0 &= i\tilde{\omega}\\ \tilde{\mathsf{p}}_b &= \frac{\mathsf{p}_b}{mc}\\ \tilde{\mathsf{A}}_0\left(\mathsf{x}\right) &= \tilde{\Phi}\left(\mathsf{x}\right)=\frac{e\Phi^{(\text{ext})}\left(\mathsf{x}\right)}{mc^2}\\ \tilde{\mathsf{A}}_b\left(\mathsf{x}\right) &= \frac{e\mathsf{A}_b^{(\text{ext})}\left(\mathsf{x}\right)}{mc}\\ \tilde{W} &= \frac{W}{mc^2}\end{aligned}\tag{2.130}$$

Actually, because $\gamma_b \equiv \beta\alpha_b$ indeed is anti-Hermitean, but $\gamma_0 \equiv \beta$ is Hermitean, so then the set γ_μ cannot transform like the components of a proper four vector, and not against expectation, the operator $\gamma_\mu\tilde{\Pi}_\mu$ is *not* Hermitean.

Upon insertion of (2.128) into (2.125), thereafter using the cyclic invariance of the trace and rewriting

$$\gamma_0\,\delta\mathsf{H}^{(D)} = -mc^2\delta\left(\gamma_\mu\tilde{\Pi}_\mu\right)\tag{2.131}$$

then

$$\delta\tilde{W} = \int_{-\infty}^{\infty}\frac{d\tilde{\omega}}{2\pi}\,\text{tr}\left(\delta\left(\gamma_\mu\tilde{\Pi}_\mu\right)\frac{\mathsf{1}_{4\times4}}{\mathsf{1}_{4\times4}-\gamma_\mu\tilde{\Pi}_\mu}\right)\tag{2.132}$$

As is well known, the identity

$$\gamma_5 \gamma_\mu \gamma_5 = -\gamma_\mu \quad \text{for } \mu \in \{0, x, y, z\} \tag{2.133}$$

together with the cyclic invariance of the trace implies the vanishing of the trace of any *odd* number of γ_μ matrices. So, expanding the fraction under the trace in (2.132) it follows

$$\begin{aligned}
\delta \tilde{W} &= \int_{-\infty}^{\infty} \frac{d\tilde{\omega}}{2\pi} \operatorname{tr}\left(\delta \left(\gamma_\mu \tilde{\Pi}_\mu \right) \frac{\mathsf{1}_{4\times4} + \gamma_\mu \tilde{\Pi}_\mu}{\mathsf{1}_{4\times4} - \left(\gamma_\mu \tilde{\Pi}_\mu \right)^2} \right) \\
&= \int_{-\infty}^{\infty} \frac{d\tilde{\omega}}{2\pi} \operatorname{tr}\left(\delta \left(\gamma_\mu \tilde{\Pi}_\mu \right) \left(\gamma_\mu \tilde{\Pi}_\mu \right) \frac{\mathsf{1}_{4\times4}}{\mathsf{1}_{4\times4} - \left(\gamma_\mu \tilde{\Pi}_\mu \right)^2} \right)
\end{aligned}$$

Observing

$$\delta \left(\left(\gamma_\mu \tilde{\Pi}_\mu \right)^2 \right) = \delta \left(\gamma_\mu \tilde{\Pi}_\mu \right) \left(\gamma_\mu \tilde{\Pi}_\mu \right) + \left(\gamma_\mu \tilde{\Pi}_\mu \right) \delta \left(\gamma_\mu \tilde{\Pi}_\mu \right) \tag{2.134}$$

and employing once more the cyclic invariance of the trace, thereafter

$$\delta \tilde{W} = \int_{-\infty}^{\infty} \frac{d\tilde{\omega}}{2\pi} \operatorname{tr}\left(\frac{1}{2} \delta \left(\left(\gamma_\mu \tilde{\Pi}_\mu \right)^2 \right) \frac{\mathsf{1}_{4\times4}}{\mathsf{1}_{4\times4} - \left(\gamma_\mu \tilde{\Pi}_\mu \right)^2} \right) \tag{2.135}$$

Because we seek after a functional $\tilde{W}$ providing the variation $\delta \tilde{W}$ stated in (2.135), it is expedient to introduce an integration over a dummy variable ξ serving the purpose to represent δW now as a *total* differential

$$\begin{aligned}
\delta \tilde{W} &= \frac{1}{2} \int_{-\infty}^{\infty} \frac{d\tilde{\omega}}{2\pi} \operatorname{tr}\left(\delta \left(\left(\gamma_\mu \tilde{\Pi}_\mu \right)^2 \right) (-1) \int_0^{\infty} d\xi \frac{\partial}{\partial \xi} \frac{\mathsf{1}_{4\times4}}{(\xi + 1)\, \mathsf{1}_{4\times4} - \left(\gamma_\mu \tilde{\Pi}_\mu \right)^2} \right) \\
&= \frac{1}{2} \int_{-\infty}^{\infty} \frac{d\tilde{\omega}}{2\pi} \operatorname{tr}\left(\delta \left(\left(\gamma_\mu \tilde{\Pi}_\mu \right)^2 \right) \int_0^{\infty} d\xi \frac{\mathsf{1}_{4\times4}}{\left((\xi + 1)\, \mathsf{1}_{4\times4} - \left(\gamma_\mu \tilde{\Pi}_\mu \right)^2 \right)^2} \right) \\
&= \frac{1}{2} \int_{-\infty}^{\infty} \frac{d\tilde{\omega}}{2\pi} \operatorname{tr}\left(\delta \int_0^{\infty} d\xi \, \frac{\mathsf{1}_{4\times4}}{(\xi + 1)\, \mathsf{1}_{4\times4} - \left(\gamma_\mu \tilde{\Pi}_\mu \right)^2} \right)
\end{aligned} \tag{2.136}$$

As the functional $W\left[\Phi(\cdot), \mathsf{A}(\cdot)\right]$ obtainable from (2.136) is determined except for a constant, the subtraction of a suitable field-independent term ensuring $W\left[0, \mathsf{0}\right] = 0$

is permitted. By this means the sought functional W, with its variation δW given by (2.136), is readily identified in the guise

$$W\left[\Phi(\cdot), \mathsf{A}(\cdot)\right] \tag{2.137}$$
$$= \frac{1}{2}mc^2 \int_0^\infty d\xi \int_{-\infty}^{\infty} \frac{d\tilde{\omega}}{2\pi} \operatorname{tr}\left(\frac{1_{4\times4}}{(\xi+1)\,1_{4\times4} - \left(\gamma_\mu \tilde{\Pi}_\mu\right)^2} - \frac{1_{4\times4}}{(\xi+1) - \tilde{p}_\mu \tilde{p}_\mu} \right)$$

throughout assuming that altering the order of integrations was allowed.

In order to find the response functions in (2.24) connecting the polarization charge density and the polarization current density to the external potentials, we now evaluate the functional (2.137) with an expansion in powers of the coupling constant e. To this end let us line up the terms $\left(\gamma_\mu \tilde{\Pi}_\mu\right)^2$ in the denominator of (2.137) according to the individual terms being of zero order, respectively being of first and second order in the coupling constant e. That way

$$\begin{aligned}
\left(\gamma_\mu \tilde{\Pi}_\mu\right)^2 &= \left(\gamma_0 \tilde{\Pi}_0 - \gamma_b \tilde{\Pi}_b\right)^2 \\
&= \tilde{\Pi}_0 \tilde{\Pi}_0 + \left(\gamma_b \tilde{\Pi}_b\right)^2 - \gamma_0 \gamma_b \left[\tilde{\Pi}_0, \tilde{\Pi}_b\right] \\
&= \tilde{\Pi}_0 \tilde{\Pi}_0 - \left(\alpha_b \tilde{\Pi}_b\right)^2 - \alpha_b \left[-\tilde{\mathsf{A}}_0, \tilde{\mathsf{p}}_b\right] \\
&= \tilde{\Pi}_0 \tilde{\Pi}_0 - \tilde{\Pi}_b \tilde{\Pi}_b + \tilde{\mathscr{B}}_c \sigma_c - i\alpha_b \tilde{\mathscr{E}}_b \\
&= \underbrace{\tilde{p}_\mu \tilde{p}_\mu}_{0^{th}\text{-order}} - \underbrace{\left(\tilde{p}_\mu \tilde{\mathsf{A}}_\mu + \tilde{\mathsf{A}}_\mu \tilde{p}_\mu + i\alpha_b \tilde{\mathscr{E}}_b - \tilde{\mathscr{B}}_c \sigma_c\right)}_{1^{st}\text{-order}} + \underbrace{\tilde{\mathsf{A}}_\mu \tilde{\mathsf{A}}_\mu}_{2^{nd}\text{-order}} \\
&= \tilde{p}_\mu \tilde{p}_\mu + \mathsf{K}_1 + \mathsf{K}_2
\end{aligned} \tag{2.138}$$

whereby $\tilde{p}_\mu \tilde{p}_\mu \equiv -\tilde{\omega}^2 - \tilde{\mathsf{p}}_b \tilde{\mathsf{p}}_b$ and

$$\begin{aligned}
\tilde{\mathscr{E}}_b &= -i\left[\tilde{\mathsf{p}}_b\,, \tilde{\mathsf{A}}_0\right] = \frac{\hbar}{mc}\frac{e}{mc^2}\left(-\nabla_b \Phi^{(\text{ext})}\right) = \frac{\lambda_C \cdot e\mathscr{E}_b^{(\text{ext})}}{mc^2} \\
\tilde{\mathscr{B}}_c &= i\,\varepsilon_{bb'c}\left[\tilde{\mathsf{p}}_b\,, \tilde{\mathsf{A}}_{b'}\right] = \frac{\hbar}{mc}\frac{e}{mc}\mathscr{B}_c^{(\text{ext})} = \frac{\lambda_C \cdot e\,c\,\mathscr{B}_c^{(\text{ext})}}{mc^2}
\end{aligned} \tag{2.139}$$

the contributions of the external electromagnetic fields (scaled units).

Next, a straightforward expansion of the integrand in (2.137) with regard to the small quantities,

$$\begin{aligned}
\mathsf{K}_1 &= -\tilde{p}_\mu \tilde{\mathsf{A}}_\mu(\mathsf{x}) - \tilde{\mathsf{A}}_\mu(\mathsf{x})\,\tilde{p}_\mu - i\alpha_b \tilde{\mathscr{E}}_b(\mathsf{x}) + \tilde{\mathscr{B}}_c(\mathsf{x})\,\sigma_c \\
\mathsf{K}_2 &= \tilde{\mathsf{A}}_\mu(\mathsf{x})\,\tilde{\mathsf{A}}_\mu(\mathsf{x})
\end{aligned} \tag{2.140}$$

leads within the intended accuracy to the representation

$$W\left[\Phi\left(\cdot\right),\mathsf{A}\left(\cdot\right)\right] = \frac{1}{2}mc^2\int_0^\infty d\xi\int_{-\infty}^{\infty}\frac{d\tilde{\omega}}{2\pi}\,\mathrm{tr}\begin{pmatrix}\frac{1}{\xi+1+\tilde{\omega}^2+\tilde{\mathsf{p}}_b\tilde{\mathsf{p}}_b}\circ\left(\mathsf{K}_1+\mathsf{K}_2\right)\circ\frac{1}{\xi+1+\tilde{\omega}^2+\tilde{\mathsf{p}}_b\tilde{\mathsf{p}}_b}\\ +\frac{1}{\xi+1+\tilde{\omega}^2+\tilde{\mathsf{p}}_b\tilde{\mathsf{p}}_b}\circ\mathsf{K}_1\circ\frac{1}{\xi+1+\tilde{\omega}^2+\tilde{\mathsf{p}}_b\tilde{\mathsf{p}}_b}\circ\mathsf{K}_1\circ\frac{1}{\xi+1+\tilde{\omega}^2+\tilde{\mathsf{p}}_b\tilde{\mathsf{p}}_b}\\ +o\left(e^3\right)\end{pmatrix} \tag{2.141}$$

As is well known the trace operation can be evaluated in any complete basis system. For our purpose it is expedient to evaluate the trace operation in (2.141) in a complete basis system spanned by the states $|\mathbf{k},\zeta\rangle \equiv |\mathbf{k}\rangle\otimes|\zeta\rangle$, whereby these basis states are simultaneous eigenstates of the momentum operator p_b and of β

$$\begin{aligned}\mathsf{p}_b\left|\mathbf{k}\right\rangle &= \hbar k_b\left|\mathbf{k}\right\rangle\\ \beta\left|\zeta\right\rangle &= \left(\delta_{\zeta,1}+\delta_{\zeta,2}-\delta_{\zeta,3}-\delta_{\zeta,4}\right)\left|\zeta\right\rangle\end{aligned} \tag{2.142}$$

with $\zeta\in\{1,2,3,4\}$ here enumerating all eigenstates of β. Mind the orthonormality and completeness relations for these basis states

$$\left\langle\mathbf{k},\zeta\mid\mathbf{k}',\zeta'\right\rangle = \left\langle\mathbf{k}\mid\mathbf{k}'\right\rangle\left\langle\zeta\mid\zeta'\right\rangle = \delta_{\mathbf{k},\mathbf{k}'}\delta_{\zeta,\zeta'} \tag{2.143}$$

$$\sum_{\mathbf{k},\zeta}\left|\mathbf{k},\zeta\right\rangle\left\langle\mathbf{k},\zeta\right| = \sum_{\mathbf{k}}\left|\mathbf{k}\right\rangle\left\langle\mathbf{k}\right|\otimes\sum_{\zeta}\left|\zeta\right\rangle\left\langle\zeta\right| = \mathsf{1}\otimes\mathsf{1}_{4\times4} \tag{2.144}$$

In the evaluation of the trace with the basis functions $|\mathbf{k},\zeta\rangle$ a *single* term K_1 gives no contribution to the functional (2.141). Upon insertion of the completeness relation (2.144) and writing $\tilde{\mathbf{k}} = \lambda_C\cdot\mathbf{k}$, we get eventually for the leading order contribution of the wanted expansion of W in the coupling e

$$\begin{aligned}W\left[\Phi\left(\cdot\right),\mathsf{A}\left(\cdot\right)\right] &= \frac{1}{2}mc^2\int_0^\infty d\xi\int_{-\infty}^{\infty}\frac{d\tilde{\omega}}{2\pi}\sum_{\mathbf{k}',\mathbf{k}''}\frac{1}{\left(\xi+1+\tilde{\omega}^2+\left|\tilde{\mathbf{k}}'\right|^2\right)^2}\times\\ &\times\sum_{\zeta',\zeta''}\begin{pmatrix}\left\langle\mathbf{k}',\zeta'\right|\tilde{\mathsf{A}}_\mu\left(\mathsf{x}\right)\left|\mathbf{k}'',\zeta''\right\rangle\left\langle\mathbf{k}'',\zeta''\right|\tilde{\mathsf{A}}_\mu\left(\mathsf{x}\right)\left|\mathbf{k}',\zeta'\right\rangle\\ +\left\langle\mathbf{k}',\zeta'\right|\mathsf{K}_1\left|\mathbf{k}'',\zeta''\right\rangle\frac{1}{\xi+1+\tilde{\omega}^2+\left|\tilde{\mathbf{k}}''\right|^2}\left\langle\mathbf{k}'',\zeta''\right|\mathsf{K}_1\left|\mathbf{k}',\zeta'\right\rangle\end{pmatrix}\end{aligned} \tag{2.145}$$

Regarding the matrix elements in (2.145) we have

$$\left\langle\mathbf{k}'',\zeta''\right|\tilde{\mathsf{A}}_\mu\left(\mathsf{x}\right)\left|\mathbf{k}',\zeta'\right\rangle = \tilde{\mathsf{A}}_{F,\mu}(\mathbf{k}''-\mathbf{k}')\delta_{\zeta',\zeta''}, \tag{2.146}$$

with

$$\tilde{\mathsf{A}}_{F,\mu}(\mathbf{k}''-\mathbf{k}') = \frac{1}{V_P}\int_{V_P}d^3re^{-i(\mathbf{k}''-\mathbf{k}')\cdot\mathbf{r}}\tilde{\mathsf{A}}_\mu\left(\mathbf{r}\right) \tag{2.147}$$

the Fourier coefficient of the external four-potential.

Evaluation of the trace operation in (2.145) with regard to the indices ζ', ζ'' gives

$$\begin{aligned}
&\sum_{\zeta',\zeta''} \left\langle \mathbf{k}', \zeta' \right| \tilde{\mathsf{A}}_\mu(\mathsf{x}) \left| \mathbf{k}'', \zeta'' \right\rangle \left\langle \mathbf{k}'', \zeta'' \right| \tilde{\mathsf{A}}_\mu(\mathsf{x}) \left| \mathbf{k}', \zeta' \right\rangle \\
&= \sum_{\zeta',\zeta''} \delta_{\zeta',\zeta''} \delta_{\zeta'',\zeta'} \tilde{\mathsf{A}}_{F,\mu}(\mathbf{k}' - \mathbf{k}'') \tilde{\mathsf{A}}_{F,\mu}(\mathbf{k}'' - \mathbf{k}') \\
&= 4\Big(\tilde{\Phi}_F(\mathbf{k}' - \mathbf{k}'') \tilde{\Phi}_F(\mathbf{k}'' - \mathbf{k}') - \tilde{\mathsf{A}}_{F,b}(\mathbf{k}' - \mathbf{k}'') \tilde{\mathsf{A}}_{F,b}(\mathbf{k}'' - \mathbf{k}') \Big)
\end{aligned} \tag{2.148}$$

and in a similar way

$$\begin{aligned}
&\sum_{\zeta',\zeta''} \left\langle \mathbf{k}', \zeta' \right| \mathsf{K}_1 \left| \mathbf{k}'', \zeta'' \right\rangle \left\langle \mathbf{k}'', \zeta'' \right| \mathsf{K}_1 \left| \mathbf{k}', \zeta' \right\rangle \\
&= 4 \begin{pmatrix} \left(\tilde{k}'_a + \tilde{k}''_a\right)\left(\tilde{k}''_b + \tilde{k}'_b\right) \tilde{\mathsf{A}}_{F,a}\left(\mathbf{k}' - \mathbf{k}''\right) \tilde{\mathsf{A}}_{F,b}\left(\mathbf{k}'' - \mathbf{k}'\right) \\
-4\tilde{\omega}^2 \tilde{\Phi}\left(\mathbf{k}' - \mathbf{k}''\right) \tilde{\Phi}\left(\mathbf{k}'' - \mathbf{k}'\right) \\
+\tilde{\mathscr{B}}_{F,c}\left(\mathbf{k}' - \mathbf{k}''\right) \tilde{\mathscr{B}}_{F,c}\left(\mathbf{k}'' - \mathbf{k}'\right) - \tilde{\mathscr{E}}_{F,c}\left(\mathbf{k}' - \mathbf{k}''\right) \tilde{\mathscr{E}}_{F,c}\left(\mathbf{k}'' - \mathbf{k}'\right) \end{pmatrix}
\end{aligned} \tag{2.149}$$

Insertion of the partial traces (2.148), (2.149) into (2.145) and thereafter making an index transformation $\left(\mathbf{k}', \mathbf{k}''\right) \to \left(\mathbf{k}'', \mathbf{k}'\right)$ enables to rewrite the integrand as a derivative with respect to the variable ξ, so that

$$\begin{aligned}
W\left[\Phi(\cdot), \mathsf{A}(\cdot)\right] &= \frac{1}{2} mc^2 \int_{-\infty}^{\infty} \frac{d\tilde{\omega}}{2\pi} \sum_{\mathbf{k}',\mathbf{k}''} \int_0^{\infty} d\xi \left(-2\frac{d}{d\xi}\right) \times \\
&\times \begin{pmatrix} \left(\frac{1}{\xi+1+\tilde{\omega}^2+\left|\tilde{\mathbf{k}}'\right|^2} + \frac{1}{\xi+1+\tilde{\omega}^2+\left|\tilde{\mathbf{k}}''\right|^2} \right) \begin{pmatrix} \tilde{\Phi}_F\left(\mathbf{k}' - \mathbf{k}''\right) \tilde{\Phi}_F\left(\mathbf{k}'' - \mathbf{k}'\right) \\ -\tilde{\mathsf{A}}_{F,b}\left(\mathbf{k}' - \mathbf{k}''\right) \tilde{\mathsf{A}}_{F,b}\left(\mathbf{k}'' - \mathbf{k}'\right) \end{pmatrix} \\
+ \frac{1}{\xi+1+\tilde{\omega}^2+\left|\tilde{\mathbf{k}}'\right|^2} \cdot \frac{1}{\xi+1+\tilde{\omega}^2+\left|\tilde{\mathbf{k}}''\right|^2} \begin{pmatrix} \left(\tilde{k}'_a + \tilde{k}''_a\right)\left(\tilde{k}''_b + \tilde{k}'_b\right) \times \\ \times \tilde{\mathsf{A}}_{F,a}\left(\mathbf{k}' - \mathbf{k}''\right) \tilde{\mathsf{A}}_{F,b}\left(\mathbf{k}'' - \mathbf{k}'\right) \\ -4\tilde{\omega}^2 \tilde{\Phi}\left(\mathbf{k}' - \mathbf{k}''\right) \tilde{\Phi}\left(\mathbf{k}'' - \mathbf{k}'\right) \\ +\tilde{\mathscr{B}}_{F,c}\left(\mathbf{k}' - \mathbf{k}''\right) \tilde{\mathscr{B}}_{F,c}\left(\mathbf{k}'' - \mathbf{k}'\right) \\ -\tilde{\mathscr{E}}_{F,c}\left(\mathbf{k}' - \mathbf{k}''\right) \tilde{\mathscr{E}}_{F,c}\left(\mathbf{k}'' - \mathbf{k}'\right) \end{pmatrix} \end{pmatrix}
\end{aligned} \tag{2.150}$$

and finally

$$W\left[\Phi(\cdot),\mathsf{A}(\cdot)\right]=mc^2\int_{-\infty}^{\infty}\frac{d\tilde{\omega}}{2\pi}\sum_{\mathbf{k}',\mathbf{k}''}\times \tag{2.151}$$

$$\times\left(\begin{array}{c}\left(\frac{1}{1+\tilde{\omega}^2+\left|\tilde{\mathbf{k}}'\right|^2}+\frac{1}{1+\tilde{\omega}^2+\left|\tilde{\mathbf{k}}''\right|^2}\right)\left(\begin{array}{c}\tilde{\Phi}_F\left(\mathbf{k}'-\mathbf{k}''\right)\tilde{\Phi}_F\left(\mathbf{k}''-\mathbf{k}'\right)\\-\tilde{\mathsf{A}}_{F,b}\left(\mathbf{k}'-\mathbf{k}''\right)\tilde{\mathsf{A}}_{F,b}\left(\mathbf{k}''-\mathbf{k}'\right)\end{array}\right)\\ +\frac{1}{1+\tilde{\omega}^2+\left|\tilde{\mathbf{k}}'\right|^2}\cdot\frac{1}{1+\tilde{\omega}^2+\left|\tilde{\mathbf{k}}''\right|^2}\cdot\left(\begin{array}{c}\left(\tilde{k}'_a+\tilde{k}''_a\right)\left(\tilde{k}''_b+\tilde{k}'_b\right)\times\\ \times\tilde{\mathsf{A}}_{F,a}\left(\mathbf{k}'-\mathbf{k}''\right)\tilde{\mathsf{A}}_{F,b}\left(\mathbf{k}''-\mathbf{k}'\right)\\ \\ -4\tilde{\omega}^2\tilde{\Phi}\left(\mathbf{k}'-\mathbf{k}''\right)\tilde{\Phi}\left(\mathbf{k}''-\mathbf{k}'\right)\\+\tilde{\mathscr{B}}_{F,c}\left(\mathbf{k}'-\mathbf{k}''\right)\tilde{\mathscr{B}}_{F,c}\left(\mathbf{k}''-\mathbf{k}'\right)\\-\tilde{\mathscr{E}}_{F,c}\left(\mathbf{k}'-\mathbf{k}''\right)\tilde{\mathscr{E}}_{F,c}\left(\mathbf{k}''-\mathbf{k}'\right)\end{array}\right)\end{array}\right)$$

To evaluate (2.151) it is expedient to substitute new summation variables

$$k'_a=k_a+\frac{q_a}{2}\ ,\ k''_a=k_a-\frac{q_a}{2} \tag{2.152}$$

After some reordering of terms and thereafter replacing the summation over wave numbers $\mathbf{k}$ by an integral according to $\frac{1}{V_P}\sum_{\mathbf{k}}f(\mathbf{k})=\int\frac{d^3k}{(2\pi)^3}f(\mathbf{k})$ the result is

$$W\left[\Phi(\cdot),\mathsf{A}(\cdot)\right]=mc^2\sum_{\mathbf{q}}\int_{-\infty}^{\infty}\frac{d\tilde{\omega}}{2\pi}V_P\left(\frac{mc}{\hbar}\right)^3\int\frac{d^3\tilde{k}}{(2\pi)^3}\times \tag{2.153}$$

$$\times\left(\begin{array}{c}+\left(\begin{array}{c}\frac{1}{1+\tilde{\omega}^2+\left|\tilde{\mathbf{k}}+\frac{1}{2}\tilde{\mathbf{q}}\right|^2}+\frac{1}{1+\tilde{\omega}^2+\left|\tilde{\mathbf{k}}-\frac{1}{2}\tilde{\mathbf{q}}\right|^2}\\-\frac{4\tilde{\omega}^2}{\left(1+\tilde{\omega}^2+\left|\tilde{\mathbf{k}}+\frac{1}{2}\tilde{\mathbf{q}}\right|^2\right)\left(1+\tilde{\omega}^2+\left|\tilde{\mathbf{k}}-\frac{1}{2}\tilde{\mathbf{q}}\right|^2\right)}\end{array}\right)\tilde{\Phi}_F(\mathbf{q})\tilde{\Phi}_F(-\mathbf{q})\\ \\ -\left(\begin{array}{c}\delta_{a,b}\left(\frac{1}{1+\tilde{\omega}^2+\left|\tilde{\mathbf{k}}+\frac{1}{2}\tilde{\mathbf{q}}\right|^2}+\frac{1}{1+\tilde{\omega}^2+\left|\tilde{\mathbf{k}}-\frac{1}{2}\tilde{\mathbf{q}}\right|^2}\right)\\-\frac{4\tilde{k}_a\tilde{k}_b}{\left(1+\tilde{\omega}^2+\left|\tilde{\mathbf{k}}+\frac{1}{2}\tilde{\mathbf{q}}\right|^2\right)\left(1+\tilde{\omega}^2+\left|\tilde{\mathbf{k}}-\frac{1}{2}\tilde{\mathbf{q}}\right|^2\right)}\end{array}\right)\tilde{\mathsf{A}}_{F,a}(\mathbf{q})\tilde{\mathsf{A}}_{F,b}(-\mathbf{q})\\ \\ +\frac{1}{\left(1+\tilde{\omega}^2+\left|\tilde{\mathbf{k}}+\frac{1}{2}\tilde{\mathbf{q}}\right|^2\right)\left(1+\tilde{\omega}^2+\left|\tilde{\mathbf{k}}-\frac{1}{2}\tilde{\mathbf{q}}\right|^2\right)}\left(\tilde{\mathscr{B}}_{F,c}(\mathbf{q})\tilde{\mathscr{B}}_{F,c}(-\mathbf{q})-\tilde{\mathscr{E}}_{F,c}(\mathbf{q})\tilde{\mathscr{E}}_{F,c}(-\mathbf{q})\right)\end{array}\right),$$

whereby $\tilde{k}_a=\frac{\hbar}{mc}k_a$ and $\tilde{q}_a=\frac{\hbar}{mc}q_a$. As far as concerns the integrations with respect to the variables $\tilde{\omega}$ and $\tilde{\mathbf{k}}$ the summation index $\mathbf{q}$ can be considered being constant, so we can choose an orthogonal decomposition $\mathbf{k}=\mathbf{k}_\perp+\mathbf{k}_\parallel$, with $\mathbf{k}_\parallel$ being aligned in the direction of $\mathbf{q}$ and $\mathbf{k}_\perp\cdot\mathbf{q}=0$. Because of the Coulomb gauge constraint $\mathbf{q}\cdot\mathsf{A}_F^{(\text{ext})}(\mathbf{q})=0$ then $\mathbf{k}\cdot\mathsf{A}_F(\mathbf{q})=\mathbf{k}_\perp\cdot\mathsf{A}_F(\mathbf{q})$, of course. Thus the terms $a\neq b$ appearing in the integrand (2.153) give no contribution to the integral. That way

$$
W[\Phi(\cdot), \mathsf{A}(\cdot)] = mc^2 V_P \left(\frac{mc}{\hbar}\right)^3 \sum_{\mathbf{q}} \int_{-\infty}^{\infty} \frac{d\tilde{\omega}}{2\pi} \int_{-\infty}^{\infty} \frac{d\tilde{k}_{\perp,1}}{2\pi} \int_{-\infty}^{\infty} \frac{d\tilde{k}_{\perp,2}}{2\pi} \int_{-\infty}^{\infty} \frac{d\tilde{k}_{\parallel}}{2\pi} \times \tag{2.154}
$$

$$
\times \left(\begin{array}{l} + \left(\begin{array}{l} \frac{1}{1+\tilde{\omega}^2+\tilde{k}_{\perp,1}^2+\tilde{k}_{\perp,2}^2+\left|\tilde{k}_{\parallel}+\frac{1}{2}\tilde{q}\right|^2} + \frac{1}{1+\tilde{\omega}^2+\tilde{k}_{\perp,1}^2+\tilde{k}_{\perp,2}^2+\left|\tilde{k}_{\parallel}-\frac{1}{2}\tilde{q}\right|^2} \\ - \frac{4\tilde{\omega}^2}{\left(1+\tilde{\omega}^2+\tilde{k}_{\perp,1}^2+\tilde{k}_{\perp,2}^2+\left|\tilde{k}_{\parallel}+\frac{1}{2}\tilde{q}\right|^2\right)\left(1+\tilde{\omega}^2+\tilde{k}_{\perp,1}^2+\tilde{k}_{\perp,2}^2+\left|\tilde{k}_{\parallel}-\frac{1}{2}\tilde{q}\right|^2\right)} \end{array} \right) \tilde{\Phi}_F(\mathbf{q})\,\tilde{\Phi}_F(-\mathbf{q}) \\ - \sum_{j=1,2} \left(\begin{array}{l} \frac{1}{1+\tilde{\omega}^2+\tilde{k}_{\perp,1}^2+\tilde{k}_{\perp,2}^2+\left|\tilde{k}_{\parallel}+\frac{1}{2}\tilde{q}\right|^2} + \frac{1}{1+\tilde{\omega}^2+\tilde{k}_{\perp,1}^2+\tilde{k}_{\perp,2}^2+\left|\tilde{k}_{\parallel}-\frac{1}{2}\tilde{q}\right|^2} \\ - \frac{4\tilde{k}_{\perp,j}^2}{\left(1+\tilde{\omega}^2+\tilde{k}_{\perp,1}^2+\tilde{k}_{\perp,2}^2+\left|\tilde{k}_{\parallel}+\frac{1}{2}\tilde{q}\right|^2\right)\left(1+\tilde{\omega}^2+\tilde{k}_{\perp,1}^2+\tilde{k}_{\perp,2}^2+\left|\tilde{k}_{\parallel}-\frac{1}{2}\tilde{q}\right|^2\right)} \end{array} \right) \tilde{\mathsf{A}}_{F,j}(\mathbf{q})\,\tilde{\mathsf{A}}_{F,j}(-\mathbf{q}) \\ + \frac{1}{\left(1+\tilde{\omega}^2+\tilde{k}_{\perp,1}^2+\tilde{k}_{\perp,2}^2+\left|\tilde{k}_{\parallel}+\frac{1}{2}\tilde{q}\right|^2\right)\left(1+\tilde{\omega}^2+\tilde{k}_{\perp,1}^2+\tilde{k}_{\perp,2}^2+\left|\tilde{k}_{\parallel}-\frac{1}{2}\tilde{q}\right|^2\right)} \left(\tilde{\mathscr{B}}_{F,c}(\mathbf{q})\,\tilde{\mathscr{B}}_{F,c}(-\mathbf{q}) - \tilde{\mathscr{E}}_{F,c}(\mathbf{q})\,\tilde{\mathscr{E}}_{F,c}(-\mathbf{q})\right) \end{array} \right)
$$

Renaming now the integration variables $\left(\tilde{\omega}, \tilde{k}_{\perp,j}\right) \to \left(\tilde{k}_{\perp,j}, \tilde{\omega}\right)$ for $j = 1, 2$ in the terms with a factor $\tilde{\mathsf{A}}_{F,j}(\mathbf{q})\,\tilde{\mathsf{A}}_{F,j}(-\mathbf{q})$ the result is

$$
W[\Phi(\cdot), \mathsf{A}(\cdot)] = mc^2 V_P \left(\frac{mc}{\hbar}\right)^3 \sum_{\mathbf{q}} \int_{-\infty}^{\infty} \frac{d\tilde{k}_{\perp,1}}{2\pi} \int_{-\infty}^{\infty} \frac{d\tilde{k}_{\perp,2}}{2\pi} \int_{-\infty}^{\infty} \frac{d\tilde{k}_{\parallel}}{2\pi} \times \tag{2.155}
$$

$$
\times \int_{-\infty}^{\infty} \frac{d\tilde{\omega}}{2\pi} \left(\begin{array}{l} + \left(\frac{1}{\tilde{\omega}^2+\tilde{E}_{k,+}^2} + \frac{1}{\tilde{\omega}^2+\tilde{E}_{k,-}^2} - \frac{4\tilde{\omega}^2}{\left(\tilde{\omega}^2+\tilde{E}_{k,+}^2\right)\left(\tilde{\omega}^2+\tilde{E}_{k,-}^2\right)} \right) \left(\tilde{\Phi}_F(\mathbf{q})\,\tilde{\Phi}_F(-\mathbf{q}) - \sum_{j=1,2} \tilde{\mathsf{A}}_{F,j}(\mathbf{q})\,\tilde{\mathsf{A}}_{F,j}(-\mathbf{q}) \right) \\ + \frac{1}{\left(\tilde{\omega}^2+\tilde{E}_{k,+}^2\right)\left(\tilde{\omega}^2+\tilde{E}_{k,-}^2\right)} \left(\tilde{\mathscr{B}}_{F,c}(\mathbf{q})\,\tilde{\mathscr{B}}_{F,c}(-\mathbf{q}) - \tilde{\mathscr{E}}_{F,c}(\mathbf{q})\,\tilde{\mathscr{E}}_{F,c}(-\mathbf{q})\right) \end{array} \right),
$$

whereby we abbreviated

$$
\tilde{E}_{k,\pm} = \sqrt{1 + \tilde{k}_{\perp,1}^2 + \tilde{k}_{\perp,2}^2 + \left(\tilde{k}_{\parallel} \pm \frac{1}{2}\tilde{q}\right)^2} \tag{2.156}
$$

Here, the integrals with regard to the variable $\tilde{\omega}$ are given by

$$
\int_{-\infty}^{\infty} \frac{d\tilde{\omega}}{\pi} \frac{1}{\tilde{\omega}^2 + \tilde{E}_{k,\pm}^2} = \frac{1}{\tilde{E}_{k,\pm}} \tag{2.157}
$$

$$
\int_{-\infty}^{\infty} \frac{d\tilde{\omega}}{\pi} \frac{\tilde{\omega}^2}{\left(\tilde{\omega}^2 + \tilde{E}_{k,+}^2\right)\left(\tilde{\omega}^2 + \tilde{E}_{k,-}^2\right)} = \frac{1}{\tilde{E}_{k,+} + \tilde{E}_{k,-}}
$$

$$
\int_{-\infty}^{\infty} \frac{d\tilde{\omega}}{\pi} \frac{1}{\left(\tilde{\omega}^2 + \tilde{E}_{k,+}^2\right)\left(\tilde{\omega}^2 + \tilde{E}_{k,-}^2\right)} = \frac{1}{\tilde{E}_{k,+} \cdot \tilde{E}_{k,-} \cdot \left(\tilde{E}_{k,+} + \tilde{E}_{k,-}\right)}
$$

The Fourier coefficients of electromagnetic fields being connected to those of the potentials by

$$
\mathscr{B}_{F,c}(\mathbf{q}) = \left(-i\mathbf{q} \wedge \mathsf{A}_F(\mathbf{q})\right)_c \tag{2.158}
$$

$$
\mathscr{E}_{F,c}(\mathbf{q}) = iq_c \Phi_F(\mathbf{q})
$$

we have

$$\tilde{\mathscr{B}}_{F,c}(\mathbf{q})\tilde{\mathscr{B}}_{F,c}(-\mathbf{q}) - \tilde{\mathscr{E}}_{F,c}(\mathbf{q})\tilde{\mathscr{E}}_{F,c}(-\mathbf{q}) \equiv -\tilde{\mathbf{q}}^2\left(\tilde{\Phi}_F(\mathbf{q})\tilde{\Phi}_F(-\mathbf{q}) - \tilde{\mathsf{A}}_{F,c}(\mathbf{q})\tilde{\mathsf{A}}_{F,c}(-\mathbf{q})\right) \tag{2.159}$$

After insertion of the integrals (2.157) thereupon the searched for functional assumes the guise

$$W[\Phi(\cdot),\mathsf{A}(\cdot)] = mc^2 V_P\left(\frac{mc}{\hbar}\right)^3 \sum_{\mathbf{q}} \int_{-\infty}^{\infty}\frac{d^2\tilde{k}_\perp}{(2\pi)^2}\int_{-\infty}^{\infty}\frac{d\tilde{k}_\parallel}{2\pi}\times \tag{2.160}$$
$$\times\frac{1}{2}\left(\frac{1}{\tilde{E}_{k,+}}+\frac{1}{\tilde{E}_{k,-}}-\frac{4}{\tilde{E}_{k,+}+\tilde{E}_{k,-}}-\frac{\tilde{\mathbf{q}}^2}{\tilde{E}_{k,+}\cdot\tilde{E}_{k,-}\cdot\left(\tilde{E}_{k,+}+\tilde{E}_{k,-}\right)}\right)\left(\tilde{\Phi}_F(\mathbf{q})\tilde{\Phi}_F(-\mathbf{q}) - \tilde{\mathsf{A}}_{F,c}(\mathbf{q})\tilde{\mathsf{A}}_{F,c}(-\mathbf{q})\right)$$

A simplification of the integrand is enabled observing

$$\frac{1}{\tilde{E}_{k,+}}+\frac{1}{\tilde{E}_{k,-}}-\frac{4}{\tilde{E}_{k,+}+\tilde{E}_{k,-}}-\frac{\tilde{\mathbf{q}}^2}{\tilde{E}_{k,+}\cdot\tilde{E}_{k,-}\cdot\left(\tilde{E}_{k,+}+\tilde{E}_{k,-}\right)} \tag{2.161}$$
$$= -\frac{\tilde{\mathbf{q}}^2}{\left(\tilde{E}_{k,+}+\tilde{E}_{k,-}\right)\tilde{E}_{k,+}\cdot\tilde{E}_{k,-}}\left(1-\frac{4\tilde{k}_\parallel^2}{\left(\tilde{E}_{k,+}+\tilde{E}_{k,-}\right)^2}\right)$$

Upon insertion of (2.161), thereafter passing to cylindrical variables and bringing in the physical units too, eventually the expression (2.160) for W becomes

$$W[\Phi(\cdot),\mathsf{A}(\cdot)] = -V_P\sum_{\mathbf{q}} I(\mathbf{q})\frac{1}{2}\left(\Phi_F^{(\mathrm{ext})}(\mathbf{q})\Phi_F^{(\mathrm{ext})}(-\mathbf{q}) - c^2\mathsf{A}_{F,b}^{(\mathrm{ext})}(\mathbf{q})\mathsf{A}_{F,b}^{(\mathrm{ext})}(-\mathbf{q})\right)+\cdots, \tag{2.162}$$

whereby

$$I(\mathbf{q}) = \epsilon_0\mathbf{q}^2\frac{\alpha_{FS}}{\pi}\int_0^\infty d\tilde{k}_\perp\tilde{k}_\perp\int_{-\infty}^{\infty}d\tilde{k}_\parallel\frac{\Theta_H\left(\tilde{\Omega}_{max}-\tilde{E}_{k,-}-\tilde{E}_{k,+}\right)}{\left(\tilde{E}_{k,+}+\tilde{E}_{k,-}\right)\cdot\tilde{E}_{k,+}\cdot\tilde{E}_{k,-}}\left(1-\frac{4\tilde{k}_\parallel^2}{\left(\tilde{E}_{k,+}+\tilde{E}_{k,-}\right)^2}\right). \tag{2.163}$$

and $\alpha_{FS} = \frac{e^2}{4\pi\epsilon_0\hbar c} \approx \frac{1}{137}$ the fine structure constant.

In (2.163), conforming with Dirac's cut-off procedure [7], the step function in the integral limits the *sum* of the (scaled) kinetic energies $\tilde{E}_{k,\pm}$ by the cutoff parameter $\tilde{\Omega}_{max}$, thus suggesting to introduce new integration variables W and K

$$\begin{aligned} W &= \tilde{E}_{k,+}+\tilde{E}_{k,-} \\ &= \sqrt{1+\tilde{k}_\perp^2+\tilde{k}_\parallel^2+\frac{\tilde{q}^2}{4}+\tilde{q}\tilde{k}_\parallel}+\sqrt{1+\tilde{k}_\perp^2+\tilde{k}_\parallel^2+\frac{\tilde{q}^2}{4}-\tilde{q}\tilde{k}_\parallel} \\ K &= \tilde{k}_\parallel \end{aligned} \tag{2.164}$$

With help of the Jacobian

$$\frac{\partial(W,K)}{\partial\left(\tilde{k}_\perp,\tilde{k}_z\right)} = \det\begin{pmatrix} \frac{\partial W}{\partial\tilde{k}_\perp}, & \frac{\partial K}{\partial\tilde{k}_\perp} \\ \frac{\partial W}{\partial\tilde{k}_\parallel}, & \frac{\partial K}{\partial\tilde{k}_\parallel}\end{pmatrix} = \det\begin{pmatrix} \frac{\partial W}{\partial\tilde{k}_\perp}, & 0 \\ \frac{\partial W}{\partial\tilde{k}_\parallel}, & 1\end{pmatrix} = \frac{\partial W}{\partial\tilde{k}_\perp} = \tilde{k}_\perp\left(\frac{1}{\tilde{E}_{k,+}}+\frac{1}{\tilde{E}_{k,-}}\right)$$

then

$$dWdK = d\tilde{k}_\perp d\tilde{k}_\parallel \frac{\partial(W,K)}{\partial\left(\tilde{k}_\perp,\tilde{k}_\parallel\right)} = d\tilde{k}_\perp\tilde{k}_\perp\, d\tilde{k}_\parallel\, \frac{W}{\tilde{E}_{k,+}\cdot\tilde{E}_{k,-}}, \tag{2.165}$$

and in further consequence

$$\frac{1}{\varepsilon_0}\frac{I(\mathbf{q})}{\mathbf{q}^2} = \frac{\alpha_{FS}}{\pi}\int_{W_{min}}^{W_{max}} dW\frac{2}{W^2}\int_0^{K_{max}(W)} dK\left(1-\frac{4K^2}{W^2}\right) \tag{2.166}$$

The limits of integration here being

$$\begin{aligned} \sqrt{4+\tilde{q}^2} \equiv W_{min} \le W \le W_{max} \equiv \tilde{\Omega}_{max} \\ 0 \le K \le K_{max}(W) \equiv \frac{W}{2}\cdot\sqrt{\frac{W^2-4-\tilde{q}^2}{W^2-\tilde{q}^2}} \end{aligned} \tag{2.167}$$

This is, because the peak value for the variable $K \equiv \tilde{k}_\parallel$ at a *fixed* value $\tilde{E}_{k,+}+\tilde{E}_{k,-} = W$ can be attained only for $\tilde{k}_\perp \equiv 0$.

The integral with respect to K in (2.166) is elementary and so there is only one integration left

$$\frac{1}{\varepsilon_0}\frac{I(\mathbf{q})}{\mathbf{q}^2} = \frac{\alpha_{FS}}{\pi}\int_{W_{min}}^{W_{max}}\frac{dW}{W}\sqrt{\frac{W^2-W_{min}^2}{W^2+4-W_{min}^2}}\left(1-\frac{1}{3}\frac{W^2-W_{min}^2}{W^2+4-W_{min}^2}\right) \tag{2.168}$$

This integral can be solved exactly, as one readily confirms with help of the substitution

$$y = \sqrt{\frac{W^2-W_{min}^2}{W^2+4-W_{min}^2}} = \sqrt{1-\frac{4}{W^2+4-W_{min}^2}}$$

That way eventually

$$\begin{aligned} \frac{1}{\varepsilon_0}\frac{I(\mathbf{q})}{\mathbf{q}^2} &= \frac{\alpha_{FS}}{\pi}\int_0^{y_{max}} dy\frac{1-\frac{1}{3}y^2}{1+\frac{\tilde{q}^2}{4}(1-y^2)}\cdot\frac{y^2}{1-y^2} \\ &= \frac{\alpha_{FS}}{\pi}\left(\frac{1}{3}\int_0^{y_{max}} dy\left(\frac{2}{1-y^2}+y^2-2\right)-\frac{\tilde{q}^2}{4}\int_0^{y_{max}} dy\frac{\left(1-\frac{1}{3}y^2\right)y^2}{1+\frac{\tilde{q}^2}{4}(1-y^2)}\right) \end{aligned} \tag{2.169}$$

with the upper limit of integration given by

$$y_{max} = \sqrt{\frac{W_{max}^2 - W_{min}^2}{W_{max}^2 + 4 - W_{min}^2}} \tag{2.170}$$

Restricting the spatial variation of the external field to wavelengths long compared to the Compton length $\lambda_C = \frac{\hbar}{mc}$ and taking reasonably for the cutoff parameter $\tilde{\Omega}_{max}$ a very large value, i.e.,

$$\tilde{\Omega}_{max} \equiv \frac{\hbar \Omega_{max}}{mc^2} \gg 1 \tag{2.171}$$
$$\frac{\hbar |\mathbf{q}|}{mc} = \tilde{q} \ll \tilde{\Omega}_c$$

the expansion of (2.170) with regard to $\tilde{q}$ reveals the upper integration limit y_{max} being independent on $\tilde{q}$. With neglect of irrelevant terms $O(\tilde{\Omega}_{max}^{-2})$ finally

$$\frac{1}{\varepsilon_0} \frac{I(\mathbf{q})}{\mathbf{q}^2} = \frac{\alpha_{FS}}{\pi} \left(\frac{2}{3} \ln \tilde{\Omega}_{max} - \frac{5}{9} - \frac{\tilde{q}^2}{4} \int_0^1 dy \frac{y^2 \left(1 - \frac{1}{3} y^2\right)}{1 + \frac{\tilde{q}^2}{4} \left(1 - y^2\right)} \right) \tag{2.172}$$
$$= \frac{\alpha_{FS}}{\pi} \left(\frac{2}{3} \ln \tilde{\Omega}_{max} - \frac{5}{9} - \frac{\tilde{q}^2}{15} + \frac{\tilde{q}^4}{140} + \cdots . \right)$$

Proceeding from (2.162) the variation of the functional W leads in view of the symmetry $I(\mathbf{q}) = I(-\mathbf{q})$ simply to

$$\delta W = -V_P \sum_{\mathbf{q}} I(\mathbf{q}) \left(\Phi_F^{(\text{ext})}(\mathbf{q}) \, \delta\Phi_F(-\mathbf{q}) - c^2 \mathsf{A}_{F,b}^{(\text{ext})}(\mathbf{q}) \, \delta\mathsf{A}_{F,b}(-\mathbf{q}) \right), \tag{2.173}$$

whereby (conforming with the definition of a Fourier coefficient) there holds

$$\delta\Phi_F(\mathbf{q}) = \frac{1}{V_P} \int_{V_P} d^3 r e^{-i\mathbf{q}\cdot\mathbf{r}} \delta\Phi(\mathbf{r})$$
$$\delta\mathsf{A}_{F,b}(\mathbf{q}) = \frac{1}{V_P} \int_{V_P} d^3 r e^{-i\mathbf{q}\cdot\mathbf{r}} \delta\mathsf{A}_b(\mathbf{r}) \tag{2.174}$$

Next, upon insertion of (2.174) into (2.173) and by comparison with (2.123)

$$\delta W = -\int_{V_P} d^3 r \sum_{\mathbf{q}} e^{+i\mathbf{q}\cdot\mathbf{r}} I(\mathbf{q}) \left(\Phi_F^{(\text{ext})}(\mathbf{q}) \, \delta\Phi(\mathbf{r}) - c^2 \mathsf{A}_{F,b}^{(\text{ext})}(\mathbf{q}) \, \delta\mathsf{A}_b(\mathbf{r}) \right) \tag{2.175}$$
$$\stackrel{!}{=} \int_{V_P} d^3 r \left(\varrho^{(\text{pol})}(\mathbf{r}) \, \delta\Phi(\mathbf{r}) - j_b^{(\text{pol})}(\mathbf{r}) \, \delta\mathsf{A}_b(\mathbf{r}) \right)$$

From this the searched for polarization charge density and polarization current densities are readily identified in the form of Fourier series

$$\varrho^{(\text{pol})}(\mathbf{r}) = \sum_{\mathbf{q}} e^{i\mathbf{q}\cdot\mathbf{r}} \varrho_F^{(\text{pol})}(\mathbf{q}) = -\sum_{\mathbf{q}} e^{i\mathbf{q}\cdot\mathbf{r}} I(\mathbf{q})\, \Phi_F^{(\text{ext})}(\mathbf{q}) \tag{2.176}$$
$$j_b^{(\text{pol})}(\mathbf{r}) = \sum_{\mathbf{q}} e^{i\mathbf{q}\cdot\mathbf{r}} j_{F,b}^{(\text{pol})}(\mathbf{q}) = -\sum_{\mathbf{q}} e^{i\mathbf{q}\cdot\mathbf{r}} I(\mathbf{q})\, c^2 \mathsf{A}_{F,b}^{(\text{ext})}(\mathbf{q})$$

with the respective Fourier coefficients (all) given, in terms of this *only one* integral $I(\mathbf{q})$ in (2.172), finally as

$$\varrho_F^{(\text{pol})}(\mathbf{q}) = -I(\mathbf{q})\, \Phi_F^{(\text{ext})}(\mathbf{q}) \tag{2.177}$$
$$j_{F,b}^{(\text{pol})}(\mathbf{q}) = -I(\mathbf{q})\, c^2 \mathsf{A}_{F,b}^{(\text{ext})}(\mathbf{q})$$

Appendix B

Abstract In the ensuing a convenient integral representation for Dirac's energy-sign operator $\Lambda^{(D)}$ is introduced and with that for the unitary transformed operator $\Lambda_q^{(D)} \equiv e^{-i\mathbf{q}\cdot\mathbf{r}} \Lambda^{(D)} e^{i\mathbf{q}\cdot\mathbf{r}}$ a gauge invariant expansion

$$\Lambda_q^{(D)} = \Lambda_q^{(0)} + \Lambda_q^{(1)} + \Lambda_q^{(2)} + \cdots$$

is derived, that not only applies for arbitrary wave numbers q, but is as well accurate to second order in quantities rated as small in the low-energy sector of Quantum Electrodynamics.

Keywords: Dirac energy-sign operator · Integral representation · Canonical transformation · Spectral representation · Gauge invariant perturbation series

Dirac's Energy-Sign Operator and a Proof of its Gauge Invariance

The energy-sign operator $\Lambda^{(D)}$ has been introduced by P.A.M. Dirac in [24, Eq. (5)]. Using (his) bra- and ket-notation

$$\Lambda^{(D)} = \sum_k z_k \, |\phi_k\rangle \langle \phi_k| \tag{2.178}$$

$$z_k = \frac{E_k}{|E_k|} \tag{2.179}$$

Here E_k denotes an eigenvalue of the Dirac Hamiltonian $\mathsf{H}^{(D)} \equiv mc^2 \tilde{\mathsf{H}}^{(D)}$, with k a multi-index to label the associated (possibly degenerate) eigenstates $|\phi_k\rangle$ of $\mathsf{H}^{(D)}$

forming a complete system of orthonormal basis states. In the presence of (weak) static external electromagnetic fields in the background, as represented by the vector potential $A_b^{(\text{ext})}(\mathsf{x})$ and the electrostatic potential $\Phi^{(\text{ext})}(\mathsf{x})$, there holds

$$\begin{aligned}
\tilde{\mathsf{H}}^{(D)} &= \beta + \alpha_b \tilde{\Pi}_b + \tilde{\Phi}^{(\text{ext})}(\mathsf{x})\, \mathsf{1}_{4\times 4} \qquad (2.180)\\
\tilde{\Pi}_b &= \frac{1}{mc}\left(\mathsf{p}_b - eA_b^{(\text{ext})}(\mathsf{x})\right)\\
\tilde{\Phi}(\mathsf{x}) &= \frac{e}{mc^2}\Phi^{(\text{ext})}(\mathsf{x})\\
\left[\mathsf{p}_b, \mathsf{x}_a\right] &= \frac{\hbar}{i}\delta_{b,a}\cdot \mathsf{1}\\
\tilde{\mathsf{H}}^{(D)}\left|\phi_k\right\rangle &= \tilde{E}_k\left|\phi_k\right\rangle\\
\sum_k \left|\phi_k\right\rangle\left\langle\phi_k\right| &= \mathsf{1}_{4\times 4} \qquad (2.181)\\
\left\langle\phi_k, \phi_{k'}\right\rangle &= \delta_{k,k'}
\end{aligned}$$

With help of the completeness relation (2.181) one obtains for $\Lambda^{(D)}$ now instead of (2.178) as a common alternative [2]

$$\Lambda^{(D)} = \frac{\tilde{\mathsf{H}}^{(D)}}{\sqrt{\tilde{\mathsf{H}}^{(D)}\tilde{\mathsf{H}}^{(D)}}} \qquad (2.182)$$

Prima facie a problem with the representation (2.182) being, even though the kinetic term $\alpha_b\tilde{\Pi}_b$ in the Dirac Hamiltonian (2.180) is indeed manifestly gauge invariant, now a shift of the electrostatic potential by a constant, $\tilde{\Phi}^{(\text{ext})} \to \tilde{\Phi}^{(\text{ext})} + \tilde{\Phi}_0$, changes the Dirac Hamiltonian and its eigenvalues accordingly, that is $\tilde{\mathsf{H}}^{(D)} \to \tilde{\mathsf{H}}^{(D)} + \tilde{\Phi}_0\, \mathsf{1}_{4\times 4}$ and $\tilde{E}_k \to \tilde{E}_k + \tilde{\Phi}_0$, thus rising the question of $\Lambda^{(D)}$, when given by (2.182), being gauge invariant, even though $\tilde{\mathsf{H}}^{(D)}$ isn't gauge invariant.

A simple proof of the gauge invariance of the energy-sign operator relies on the observation that $\Lambda^{(D)}$ is connected to Dirac's β matrix by a specific unitary transformation [25, 26]. Based on the Hamiltonian flow equation approach as detailed in our article [26], there holds

$$\begin{aligned}
\Lambda^{(D)} &= \lim_{s\to\infty} \mathsf{Z}(s) \qquad (2.183)\\
\frac{1}{2}\frac{d}{ds}\mathsf{Z}(s) &= \tilde{\mathsf{H}}^{(D)} - \mathsf{Z}(s)\circ\tilde{\mathsf{H}}^{(D)}\circ\mathsf{Z}(s)\\
\mathsf{Z}(s)\circ\mathsf{Z}(s) &= \mathsf{1}_{4\times 4}\\
\mathsf{Z}(0) &= \beta
\end{aligned}$$

Shifting the external potential by a constant $\tilde{\Phi}_0$ then changes here the (scaled) Dirac Hamiltonian according to $\tilde{\mathsf{H}}^{(D)} \to \tilde{\mathsf{H}}^{(D)} + \tilde{\Phi}_0$, but obviously such a shift doesn't change the differential equation for $\mathsf{Z}(s)$ at all, because $\left[\mathsf{Z}(s), \tilde{\Phi}_0\right] \equiv 0$.

Consequently, $\mathsf{Z}(s)$ is indeed gauge invariant and so the limiting value $\Lambda^{(D)}$ is according to (2.183) gauge invariant as well.

Integral Representation of $\Lambda_q^{(D)}$

In the ensuing we agree on integrals with bounds of integration given by $\pm\infty$ should be evaluated as

$$\int_{-\infty}^{\infty} d\tilde{\omega}\,\mathcal{R}(\tilde{\omega}) = \lim_{L\to\infty} \int_{-L}^{L} d\tilde{\omega}\,\mathcal{R}(\tilde{\omega}) \tag{2.184}$$

So, with help of the identity

$$z_k = \frac{\tilde{E}_k}{\left|\tilde{E}_k\right|} = \frac{1}{\pi} \int_{-\infty}^{\infty} d\tilde{\omega} \frac{1}{i\tilde{\omega} + \tilde{E}_k} \tag{2.185}$$

a convenient integral representation for the energy-sign operator (2.178) is obtained in the guise

$$\Lambda^{(D)} = \int_{-\infty}^{\infty} \frac{d\tilde{\omega}}{\pi} \frac{\mathsf{1}_{4\times4}}{i\tilde{\omega}\mathsf{1}_{4\times4} + \tilde{\mathsf{H}}^{(D)}} \tag{2.186}$$

Noticing the operator $e^{-i\,q_a \mathsf{x}_a}$ being a translation operator for p_b, i.e.,

$$\begin{aligned} e^{-i\,q_a\mathsf{x}_a}\,\tilde{\Pi}_b\,e^{i\,q_a\mathsf{x}_a} &= \tilde{\Pi}_b + \tilde{q}n_b \\ e^{-i\,q_a\mathsf{x}_a}\tilde{\mathsf{H}}^{(D)}\,e^{i\,q_a\mathsf{x}_a} &= \tilde{\mathsf{H}}^{(D)} + \tilde{q}n_b\alpha_b \\ \tilde{q} &= \frac{\hbar}{mc}\,|\mathbf{q}| \end{aligned} \tag{2.187}$$

one readily finds

$$\begin{aligned} \Lambda_q^{(D)} &\equiv e^{-i\,q_a\mathsf{x}_a}\,\Lambda^{(D)}\,e^{i\,q_a\mathsf{x}_a} \\ &= e^{-i\mathbf{q}\cdot\mathbf{r}} \int_{-\infty}^{\infty} \frac{d\tilde{\omega}}{\pi} \frac{\mathsf{1}_{4\times4}}{i\tilde{\omega}\mathsf{1}_{4\times4} + \tilde{\mathsf{H}}^{(D)}} e^{i\mathbf{q}\cdot\mathbf{r}} \\ &= \int_{-\infty}^{\infty} \frac{d\tilde{\omega}}{\pi} \frac{\mathsf{1}_{4\times4}}{i\tilde{\omega}\mathsf{1}_{4\times4} + \tilde{\mathsf{H}}^{(D)} + \tilde{q}n_b\alpha_b} \end{aligned} \tag{2.188}$$

Introducing a Hermitean operator Ω with eigenvalues $\tilde{\omega}$ and associated eigenstates $|\tilde{\omega}\rangle$ forming a complete system, with Ω operating on a Hilbert space being entirely *disconnected* from the physical Hilbert space spanned by the eigenstates $|\phi_k\rangle$ of $\tilde{\mathsf{H}}^{(D)}$, then one has a vanishing commutator $\left[\tilde{\mathsf{H}}^{(D)} + \tilde{q}n_b\alpha_b, \Omega\right] \equiv 0$ and there holds as a spectral representation

$$\frac{1}{i\Omega+\tilde{\mathsf{H}}^{(D)}+\tilde{q}n_b\alpha_b}=\int_{-\infty}^{\infty}d\tilde{\omega}\frac{\mathsf{1}_{4\times 4}}{i\tilde{\omega}\mathsf{1}_{4\times 4}+\tilde{\mathsf{H}}^{(D)}+\tilde{q}n_b\alpha_b}\left|\tilde{\omega}\right\rangle\left\langle\tilde{\omega}\right| \tag{2.189}$$

Consider next a Hermitean operator Z *conjugate* to Ω, so that

$$[\Omega, Z]=-i\cdot\mathsf{1} \tag{2.190}$$

Then it is true as well that $\left[\tilde{\mathsf{H}}^{(D)}+\tilde{q}n_b\alpha_b, Z\right]\equiv 0$. For now denoting the eigenvalues of Z as $\tilde{\zeta}$ and the associated eigenstates as $\left|\tilde{\zeta}\right\rangle$, there holds

$$\begin{aligned}\int_{-\infty}^{\infty}d\tilde{\omega}\left|\tilde{\omega}\right\rangle\left\langle\tilde{\omega}\right|=\mathsf{1}=\int_{-\infty}^{\infty}d\tilde{\zeta}\left|\tilde{\zeta}\right\rangle\left\langle\tilde{\zeta}\right|\\ \left\langle\tilde{\omega}|\tilde{\zeta}\right\rangle=\frac{e^{-i\tilde{\omega}\tilde{\zeta}}}{\sqrt{2\pi}}\end{aligned} \tag{2.191}$$

Such operators Ω and Z indeed exist, for example thinking of Ω being the momentum operator and Z the position operator of a fictitious particle moving in one dimension.

It follows from (2.189) and from what has been said, by now the representation

$$\begin{aligned}\Lambda_q^{(D)}&=2\int_{-\infty}^{\infty}d\tilde{\omega}\frac{1}{i\tilde{\omega}+\tilde{\mathsf{H}}^{(D)}+\tilde{q}n_b\alpha_b}\left\langle\tilde{\zeta}=0\left|\tilde{\omega}\right\rangle\left\langle\tilde{\omega}\right|\tilde{\zeta}=0\right\rangle\\ &=2\cdot\left\langle\tilde{\zeta}=0\right|\frac{1}{i\Omega+\tilde{\mathsf{H}}^{(D)}+\tilde{q}n_b\alpha_b}\left|\tilde{\zeta}=0\right\rangle\end{aligned} \tag{2.192}$$

The advantage of a reformulation of an integral like (2.188) as a specific matrix element of the type indicated in (2.192) is that it enables canonical transformations of the integrand that leave the expectation value (2.192) unaltered. This calculational trick we learned from a notice (in a different context though) of Julian Schwinger, see [27, Eq. 4–14.37].

To this end consider an operator $T=e^{-\tilde{\Phi}(\mathbf{x})Z}$. Using the eigenvalue property $Z\left|\tilde{\zeta}\right\rangle=\tilde{\zeta}\left|\tilde{\zeta}\right\rangle$ we readily obtain $T\left|\tilde{\zeta}=0\right\rangle=1\cdot\left|\tilde{\zeta}=0\right\rangle$ and therefore

$$\begin{aligned}\Lambda_q^{(D)}&=2\cdot\left\langle\tilde{\zeta}=0\right|T^{-1}T\frac{1}{\left(i\Omega+\tilde{\mathsf{H}}^{(D)}+\tilde{q}n_b\alpha_b\right)}T^{-1}T\left|\tilde{\zeta}=0\right\rangle\\ &=2\cdot\left\langle\tilde{\zeta}=0\right|T\frac{1}{\left(i\Omega+\tilde{\mathsf{H}}^{(D)}+\tilde{q}n_b\alpha_b\right)}T^{-1}\left|\tilde{\zeta}=0\right\rangle\\ &=2\cdot\left\langle\tilde{\zeta}=0\right|\frac{1}{T\left(i\Omega+\tilde{\mathsf{H}}^{(D)}+\tilde{q}n_b\alpha_b\right)T^{-1}}\left|\tilde{\zeta}=0\right\rangle\end{aligned} \tag{2.193}$$

Noticing that T acts as a shift operator for Ω then

$$T\Omega T^{-1} = \Omega - i\,\tilde{\Phi} \tag{2.194}$$

Likewise

$$\begin{aligned} T\Pi_b T^{-1} &= \Pi_b - Z\left[\tilde{\Phi}\,(\mathsf{x})\,,\,\tilde{\Pi}_b\right] \\ &= \tilde{\Pi}_b + i\tilde{\mathscr{E}}_b\,(\mathsf{x})\,Z \end{aligned} \tag{2.195}$$

where

$$\tilde{\mathscr{E}}_b = -i\left[\tilde{\Pi}_b, \tilde{\Phi}\right] = -i\left[\tilde{\mathsf{p}}_b, \tilde{\Phi}\right] = \frac{\hbar}{mc}\frac{e}{mc^2}\left(-\nabla_b\Phi^{(ext)}\right) = \frac{\lambda_C \cdot eE_b^{(ext)}}{mc^2} \tag{2.196}$$

denotes the electrostatic field (scaled units).

In effect, the potential term $\tilde{\Phi}$ in the denominator of (2.193) cancels and for $\Lambda_q^{(D)}$ the following manifestly gauge invariant representation is found

$$\Lambda_q^{(D)} = 2\cdot\left\langle\tilde{\zeta}=0\right|\frac{1}{i\Omega+\beta+\tilde{q}n_b\alpha_b+\alpha_b\left(\tilde{\Pi}_b+i\tilde{\mathscr{E}}_b Z\right)}\left|\tilde{\zeta}=0\right\rangle \tag{2.197}$$

Gauge Invariant Perturbation Series for $\Lambda_q^{(D)}$

The fact that the application of the operator $\tilde{\Pi}_b + i\tilde{\mathscr{E}}_b Z$ on states appendant to the low-energy sector of Quantum Electrodynamics produces a small contribution only enables to evaluate (2.197) perturbatively regarding $\tilde{\Pi}_b + i\tilde{\mathscr{E}}_b Z$ being small of first order

$$\Lambda_q^{(D)} = \Lambda_q^{(0)} + \Lambda_q^{(1)} + \Lambda_q^{(2)} + \cdots \tag{2.198}$$

The zero order term $\Lambda_q^{(0)}$ is readily identified from (2.197) being given by

$$\begin{aligned} \Lambda_q^{(0)} &= 2\cdot\left\langle\tilde{\zeta}=0\right|\frac{1}{i\Omega+\beta+\tilde{q}n_b\alpha_b}\left|\tilde{\zeta}=0\right\rangle \\ &= 2\cdot\int_{-\infty}^{\infty}d\tilde{\omega}\frac{1}{i\tilde{\omega}+\beta+\tilde{q}n_b\alpha_b}\underbrace{\left\langle\tilde{\zeta}=0|\tilde{\omega}\right\rangle\left\langle\tilde{\omega}|\tilde{\zeta}=0\right\rangle}_{=\frac{1}{2\pi}} \\ &= \int_{-\infty}^{\infty}\frac{d\tilde{\omega}}{\pi}\frac{i\tilde{\omega}1_{4\times4}-(\beta+\tilde{q}n_b\alpha_b)}{\left(i\tilde{\omega}1_{4\times4}+\beta+\tilde{q}n_b\alpha_b\right)\left(i\tilde{\omega}1_{4\times4}-(\beta+\tilde{q}n_b\alpha_b)\right)} \end{aligned} \tag{2.199}$$

Observing $(\beta+\tilde{q}n_b\alpha_b)^2 = \left(1+\tilde{q}^2\right)1_{4\times4}$ by this means

$$
\begin{aligned}
\Lambda_q^{(0)} &= \int_{-\infty}^{\infty}\frac{d\tilde{\omega}}{\pi}\left(\frac{i\tilde{\omega}}{\left(i\tilde{\omega}-\sqrt{1+\tilde{q}^2}\right)\left(i\tilde{\omega}+\sqrt{1+\tilde{q}^2}\right)}1_{4\times4}-\frac{1}{\left(i\tilde{\omega}-\sqrt{1+\tilde{q}^2}\right)\left(i\tilde{\omega}+\sqrt{1+\tilde{q}^2}\right)}(\beta+\tilde{q}n_b\alpha_b)\right)\\
&= \int_{-\infty}^{\infty}\frac{d\tilde{\omega}}{\pi}\left(\frac{1}{i\tilde{\omega}-\sqrt{1+\tilde{q}^2}}\frac{1}{2}\left(1_{4\times4}-\frac{\beta+\tilde{q}n_b\alpha_b}{\sqrt{1+\tilde{q}^2}}\right)+\frac{1}{i\tilde{\omega}+\sqrt{1+\tilde{q}^2}}\frac{1}{2}\left(1_{4\times4}+\frac{\beta+\tilde{q}n_b\alpha_b}{\sqrt{1+\tilde{q}^2}}\right)\right)\\
&= \mathsf{P}_+^{(0)}-\mathsf{P}_-^{(0)}
\end{aligned}
\tag{2.200}
$$

whereby

$$
\mathsf{P}_\pm^{(0)}=\frac{1}{2}\left(1_{4\times4}\pm\frac{\beta+\tilde{q}n_b\alpha_b}{\sqrt{1+\tilde{q}^2}}\right)=\left(\mathsf{P}_\pm^{(0)}\right)^\dagger \tag{2.201}
$$

and

$$
\begin{aligned}
\mathsf{P}_+^{(0)}+\mathsf{P}_-^{(0)} &= 1_{4\times4}\\
\mathsf{P}_+^{(0)}\circ\mathsf{P}_-^{(0)} &= 0_{4\times4}=\mathsf{P}_-^{(0)}\circ\mathsf{P}_+^{(0)}\\
\mathsf{P}_\pm^{(0)}\mathsf{P}_\pm^{(0)} &= \mathsf{P}_\pm^{(0)}
\end{aligned}
\tag{2.202}
$$

Clearly, the matrices $\mathsf{P}_\pm^{(0)}$ are projection operators and there follows from (2.200)

$$
\begin{aligned}
\Lambda_q^{(0)} &= \frac{\beta+\tilde{q}n_b\alpha_b}{\sqrt{1+\tilde{q}^2}}\\
\Lambda_q^{(0)}\circ\Lambda_q^{(0)} &= 1_{4\times4}
\end{aligned}
\tag{2.203}
$$

To obtain the higher order terms $\Lambda_q^{(j)}$ for $j=1,2,\ldots$ in the perturbation series (2.198) it is expedient to rewrite (2.197) in the guise

$$
\begin{aligned}
\Lambda_q^{(D)} &= 2\cdot\left\langle\tilde{\zeta}=0\right|\frac{1_{4\times4}}{\mathsf{M}^{-1}+\alpha_a\left(\tilde{\Pi}_a+i\tilde{\mathscr{E}}_aZ\right)}\left|\tilde{\zeta}=0\right\rangle\\
&= 2\cdot\left\langle\tilde{\zeta}=0\right|\frac{1_{4\times4}}{1_{4\times4}+\mathsf{M}\alpha_a\left(\tilde{\Pi}_a+i\tilde{\mathscr{E}}_aZ\right)}\mathsf{M}\left|\tilde{\zeta}=0\right\rangle
\end{aligned}
\tag{2.204}
$$

whereby

$$
\begin{aligned}
\mathsf{M}^{-1} &\equiv i\Omega+\beta+\tilde{q}n_a\alpha_a=i\Omega+\sqrt{1+\tilde{q}^2}\Lambda_q^{(0)}\\
&= \left(i\Omega+\sqrt{1+\tilde{q}^2}\right)\mathsf{P}_{q,+}^{(0)}+\left(i\Omega-\sqrt{1+\tilde{q}^2}\right)\mathsf{P}_{q,-}^{(0)}
\end{aligned}
\tag{2.205}
$$

The second line in (2.205) is a spectral representation.

Therefore $\mathsf{M} = (\mathsf{M}^{-1})^{-1}$ is given by

$$\mathsf{M} = \frac{1}{i\Omega + \sqrt{1+\tilde{q}^2}} \mathsf{P}^{(0)}_{q,+} + \frac{1}{i\Omega - \sqrt{1+\tilde{q}^2}} \mathsf{P}^{(0)}_{q,-} \equiv M_+ \mathsf{P}^{(0)}_{q,+} + M_- \mathsf{P}^{(0)}_{q,-}$$
$$M_\pm = \frac{1}{i\Omega \pm \sqrt{1+\tilde{q}^2}} \tag{2.206}$$

Upon expansion with respect to the small term $\tilde{\Pi}_a + i\tilde{\mathscr{E}}_a Z$ we readily identify the terms $\Lambda_q^{(j)}$ for $j = 0, 1, 2, \ldots$ of the perturbation series (2.198) being

$$\Lambda_q^{(0)} = 2\left\langle \tilde{\zeta} = 0 \right| \mathsf{M} \left| \tilde{\zeta} = 0 \right\rangle \tag{2.207}$$
$$\Lambda_q^{(1)} = 2\left\langle \tilde{\zeta} = 0 \right| - \mathsf{M}\alpha_a \left(\tilde{\Pi}_a + i\tilde{\mathscr{E}}_a Z\right) \mathsf{M} \left| \tilde{\zeta} = 0 \right\rangle$$
$$\Lambda_q^{(2)} = 2\left\langle \tilde{\zeta} = 0 \right| \mathsf{M}\alpha_a \left(\tilde{\Pi}_a + i\tilde{\mathscr{E}}_a Z\right) \mathsf{M}\alpha_{a'} \left(\tilde{\Pi}_{a'} + i\tilde{\mathscr{E}}_{a'} Z\right) \mathsf{M} \left| \tilde{\zeta} = 0 \right\rangle$$

To evaluate these matrix elements we reorder all terms so that the operator Z is positioned far left or else far right, all along meticulously carrying about the commutators $\left[Z, M_\pm^n\right]$ for $n = 1, 2, \ldots$ With help of the rule $\left[\mathsf{U}, \frac{1}{\mathsf{V}}\right] = -\frac{1}{\mathsf{V}}\left[\mathsf{U}, \mathsf{V}\right]\frac{1}{\mathsf{V}}$ and taking into account (2.190) one readily confirms

$$\left[Z, M_\pm^n\right] = -n\, M_\pm^{n+1} \text{ for n = 1,2,3,...} \tag{2.208}$$

Let us agree on the summation convention applies not only for indices $a, a', a'' \in \{x, y, z\}$ but as well for indices $s, s', s'' \in \{+, -\}$. Upon insertion of $\mathsf{M} = M_s \mathsf{P}^{(0)}_{q,s}$ into (2.207) there follows, reordering the Z operators by taking into account the commutation rule (2.208) and minding the eigenvalue property $Z \left| \tilde{\zeta} = 0 \right\rangle = 0$ or else $\left\langle \tilde{\zeta} = 0 \right| Z = 0$, that de facto all terms generated that way, with an operator Z positioned to the far right (or else to the far left), eventually disappear, thus leaving over

$$\Lambda_q^{(0)} = 2\left\langle \tilde{\zeta} = 0 \right| M_s \left| \tilde{\zeta} = 0 \right\rangle \mathsf{P}^{(0)}_{q,s} \tag{2.209}$$
$$\Lambda_q^{(1)} = -2\left\langle \tilde{\zeta} = 0 \right| M_s(\tilde{\Pi}_a + i\tilde{\mathscr{E}}_a Z) M_{s'} \left| \tilde{\zeta} = 0 \right\rangle \mathsf{P}^{(0)}_{q,s}\alpha_a \mathsf{P}^{(0)}_{q,s'}$$
$$= -2\left\langle \tilde{\zeta} = 0 \right| \left(\tilde{\Pi}_a M_s M_{s'} + i\tilde{\mathscr{E}}_a M_s^2 M_{s'}\right) \left| \tilde{\zeta} = 0 \right\rangle \mathsf{P}^{(0)}_{q,s}\alpha_a \mathsf{P}^{(0)}_{q,s'}$$
$$\Lambda_q^{(2)} = 2\left\langle \tilde{\zeta} = 0 \right| M_s(\tilde{\Pi}_a + i\tilde{\mathscr{E}}_a Z) M_{s''}(\tilde{\Pi}_{a'} + i\tilde{\mathscr{E}}_{a'} Z) M_{s'} \left| \tilde{\zeta} = 0 \right\rangle \mathsf{P}^{(0)}_{q,s}\alpha_a \mathsf{P}^{(0)}_{q,s''}\alpha_{a'} \mathsf{P}^{(0)}_{q,s'}$$
$$= 2\left\langle \tilde{\zeta} = 0 \right| \begin{pmatrix} \tilde{\Pi}_a \tilde{\Pi}_{a'} M_s M_{s''} M_{s'} \\ +i\tilde{\mathcal{E}}_a \tilde{\Pi}_{a'} M_s^2 M_{s''} M_{s'} \\ -\tilde{\Pi}_a i\tilde{\mathcal{E}}_{a'} M_s M_{s''} M_{s'}^2 \\ +\tilde{\mathcal{E}}_a \tilde{\mathcal{E}}_{a'} M_s^2 M_{s''} M_{s'}^2 \end{pmatrix} \left| \tilde{\zeta} = 0 \right\rangle \mathsf{P}^{(0)}_{q,s}\alpha_a \mathsf{P}^{(0)}_{q,s''}\alpha_{a'} \mathsf{P}^{(0)}_{q,s'}$$

The next thing is to evaluate the matrix elements $\left\langle M_{\pm}^{n}\right\rangle_0 \equiv 2\left\langle \tilde{\zeta}=0\right| M_{\pm}^{n} \left|\tilde{\zeta}=0\right\rangle$ and $\left\langle M_{+}^{n_1} M_{-}^{n_2}\right\rangle_0 \equiv 2\left\langle \tilde{\zeta}=0\right| M_{+}^{n_1} M_{-}^{n_2} \left|\tilde{\zeta}=0\right\rangle$.

Proceeding like in (2.199) then

$$\begin{aligned}
\left\langle M_{\pm}^{n}\right\rangle_0 &= 2\left\langle \tilde{\zeta}=0\right| \frac{1}{\left(i\Omega \pm \sqrt{1+\tilde{q}^2}\right)^n} \left|\tilde{\zeta}=0\right\rangle \\
&= 2\left\langle \tilde{\zeta}=0\right| \left(\int_{-\infty}^{\infty} d\tilde{\omega} \frac{1}{\left(i\tilde{\omega} \pm \sqrt{1+\tilde{q}^2}\right)^n} \left|\tilde{\omega}\right\rangle \left\langle\tilde{\omega}\right| \right) \left|\tilde{\zeta}=0\right\rangle \\
&= 2\int_{-\infty}^{\infty} d\tilde{\omega} \frac{1}{\left(i\tilde{\omega} \pm \sqrt{1+\tilde{q}^2}\right)^n} \underbrace{\left\langle \tilde{\zeta}=0|\tilde{\omega}\right\rangle \left\langle \tilde{\omega}|\tilde{\zeta}=0\right\rangle}_{=\frac{1}{2\pi}} \\
&= \int_{-\infty}^{\infty} \frac{d\tilde{\omega}}{\pi} \frac{1}{\left(i\tilde{\omega} \pm \sqrt{1+\tilde{q}^2}\right)^n} \\
&= \pm\,\delta_{n,1},
\end{aligned} \tag{2.210}$$

and similarly

$$\begin{aligned}
\left\langle M_{+}^{n_1} M_{-}^{n_2}\right\rangle_0 &= 2\left\langle \tilde{\zeta}=0\right| \frac{1}{\left(i\Omega + \sqrt{1+\tilde{q}^2}\right)^{n_1}} \frac{1}{\left(i\Omega - \sqrt{1+\tilde{q}^2}\right)^{n_2}} \left|\tilde{\zeta}=0\right\rangle \\
&= \int_{-\infty}^{\infty} \frac{d\tilde{\omega}}{\pi} \frac{1}{\left(i\tilde{\omega} + \sqrt{1+\tilde{q}^2}\right)^{n_1} \left(i\tilde{\omega} - \sqrt{1+\tilde{q}^2}\right)^{n_2}}
\end{aligned}$$

For $n_1, n_2 \in \{1, 2, 3, 4, 5\}$ and abbreviating $w \equiv \sqrt{1+\tilde{q}^2}$ it is found

$$\left\langle M_{+}^{n_1} M_{-}^{n_2}\right\rangle_0 = \begin{pmatrix}
-\frac{1}{w}, & +\frac{1}{2w^2}, & -\frac{1}{4w^3}, & +\frac{1}{8w^4}, & -\frac{1}{16w^5}, & \dots \\
-\frac{1}{2w^2}, & +\frac{1}{2w^3}, & -\frac{3}{8w^4}, & +\frac{1}{4w^5}, & -\frac{5}{32w^6}, & \dots \\
-\frac{1}{4w^3}, & \frac{3}{8w^4}, & -\frac{3}{8w^5}, & +\frac{5}{16w^6}, & -\frac{15}{64w^7}, & \dots \\
-\frac{1}{8w^4}, & +\frac{1}{4w^5}, & -\frac{5}{16w^6}, & +\frac{5}{16w^7}, & -\frac{35}{128w^8}, & \dots \\
-\frac{1}{16w^5}, & +\frac{5}{32w^6}, & -\frac{15}{64w^7}, & +\frac{35}{128w^8}, & -\frac{35}{128w^9}, & \dots \\
\dots & \dots & \dots & \dots & \dots &
\end{pmatrix}_{n_1,n_2} \tag{2.211}$$

Insertion of these matrix elements into (2.209) gives for the zeroth order term in agreement with (2.200)

$$\Lambda_q^{(0)} = s \cdot \mathsf{P}_{q,s}^{(0)} \tag{2.212}$$

Taking into account (2.210) and (2.211) the first order term is given by

$$\begin{aligned}\Lambda_q^{(1)} &= -\left(\tilde{\Pi}_a \left\langle M_s M_{-s}\right\rangle_0 + i\tilde{\mathcal{E}}_a \left\langle M_s^2 M_{-s}\right\rangle_0\right) \mathsf{P}_{q,s}^{(0)} \alpha_a \mathsf{P}_{q,-s}^{(0)} \\ &= \left(\frac{1}{w}\tilde{\Pi}_a + \frac{s}{2w^2} i\tilde{\mathcal{E}}_a\right) \mathsf{P}_{q,s}^{(0)} \alpha_a \mathsf{P}_{q,-s}^{(0)}\end{aligned} \tag{2.213}$$

The evaluation of the second order term is more elaborate. At first

$$\Lambda_q^{(2)} = \begin{pmatrix} \tilde{\Pi}_a \tilde{\Pi}_{a'} \left\langle M_s M_{s''} M_{s'}\right\rangle_0 \\ + i\tilde{\mathcal{E}}_a \tilde{\Pi}_{a'} \left\langle M_s^2 M_{s''} M_{s'}\right\rangle_0 \\ - \tilde{\Pi}_a i\tilde{\mathcal{E}}_{a'} \left\langle M_s M_{s''} M_{s'}^2\right\rangle_0 \\ + \tilde{\mathcal{E}}_a \tilde{\mathcal{E}}_{a'} \left\langle M_s^2 M_{s''} M_{s'}^2\right\rangle_0 \end{pmatrix} \mathsf{P}_{q,s}^{(0)} \alpha_a \mathsf{P}_{q,s''}^{(0)} \alpha_{a'} \mathsf{P}_{q,s'}^{(0)} \tag{2.214}$$

Progress comes observing

$$\begin{aligned}\mathsf{P}_{q,s}^{(0)} \alpha_a &= \alpha_a \mathsf{P}_{q,-s}^{(0)} + s\frac{\tilde{q}}{w} n_a \mathbb{1}_{4\times 4} \\ \alpha_a \mathsf{P}_{q,s}^{(0)} &= \mathsf{P}_{q,-s}^{(0)} \alpha_a + s\frac{\tilde{q}}{w} n_a \mathbb{1}_{4\times 4}\end{aligned} \tag{2.215}$$

which leads with help of the multiplication rules (2.202) for projection operators to

$$\begin{aligned}&\mathsf{P}_{q,s}^{(0)} \alpha_a \mathsf{P}_{q,s''}^{(0)} \alpha_{a'} \mathsf{P}_{q,s'}^{(0)} \\ &= \left(\alpha_a \mathsf{P}_{q,-s}^{(0)} + s\frac{\tilde{q}}{w} n_a\right) \mathsf{P}_{q,s''}^{(0)} \left(\mathsf{P}_{q,-s'}^{(0)} \alpha_{a'} + s'\frac{\tilde{q}}{w} n_{a'}\right) \\ &= \delta_{-s,s''}\delta_{-s',s''} \alpha_a \mathsf{P}_{q,s''}^{(0)} \alpha_{a'} + \delta_{-s,s''} s' \frac{\tilde{q}}{w} n_{a'} \alpha_a \mathsf{P}_{q,s''}^{(0)} + \delta_{-s',s''} s \frac{\tilde{q}}{w} n_a \mathsf{P}_{q,s''}^{(0)} \alpha_{a'} + s's\frac{\tilde{q}^2}{w^2} n_a n_{a'} \mathsf{P}_{q,s''}^{(0)}\end{aligned} \tag{2.216}$$

By this means it follows from (2.214) and inserting the integrals (2.211) eventually

$$\Lambda_q^{(2)} = \begin{pmatrix} \left(\frac{1}{2w^2}\tilde{\Pi}_a\tilde{\Pi}_{a'} + \frac{1}{8w^4}\tilde{\mathcal{E}}_a\tilde{\mathcal{E}}_{a'}\right)\alpha_a\Lambda_q^{(0)}\alpha_{a'} + \frac{1}{4w^3}\begin{pmatrix}\left[\tilde{\Pi}_a, i\tilde{\mathcal{E}}_a\right]\mathbb{1}_{4\times 4} \\ + \left(\tilde{\mathcal{E}}_a\tilde{\Pi}_{a'} + \tilde{\Pi}_{a'}\tilde{\mathcal{E}}_a\right)\varepsilon_{aa'c}\sigma_c\end{pmatrix} \\ + \left(-\frac{1}{w^2}\tilde{\Pi}_a\tilde{\Pi}_{a'} + \frac{1}{4w^4}\tilde{\mathcal{E}}_a\tilde{\mathcal{E}}_{a'}\right)\frac{\tilde{q}}{w} n_{a'}\alpha_a + \frac{3}{4w^3}\frac{\tilde{q}}{w} n_{a'}\, i\tilde{\mathcal{E}}_a\tilde{\Pi}_{a'}\alpha_a\Lambda_q^{(0)} \\ + \left(-\frac{1}{w^2}\tilde{\Pi}_{a'}\tilde{\Pi}_a + \frac{1}{4w^4}\tilde{\mathcal{E}}_a\tilde{\mathcal{E}}_{a'}\right)\frac{\tilde{q}}{w} n_{a'}\alpha_a - \frac{3}{4w^3}\frac{\tilde{q}}{w} n_{a'}\tilde{\Pi}_{a'} i\tilde{\mathcal{E}}_a\Lambda_q^{(0)}\alpha_a \\ + \left(\frac{3}{2w^2}\tilde{\Pi}_a\tilde{\Pi}_{a'} - \frac{5}{8w^4}\tilde{\mathcal{E}}_a\tilde{\mathcal{E}}_{a'}\right)\frac{\tilde{q}}{w^2}^2 n_a n_{a'}\Lambda_q^{(0)} + \frac{1}{2w^3}\frac{\tilde{q}}{w^2}^2 n_a n_{a'}\left[\tilde{\Pi}_{a'}, i\tilde{\mathcal{E}}_a\right]\mathbb{1}_{4\times 4} \end{pmatrix} \tag{2.217}$$

Compatibility Relations

Because $\Lambda_q^{(D)} = e^{-i\,q_a x_a}\Lambda^{(D)}e^{i\,q_a x_a}$ and $\Lambda^{(D)}\Lambda^{(D)} = 1_{4\times4}$, of course there holds as well $\Lambda_q^{(D)}\Lambda_q^{(D)} = 1_{4\times4}$. This property engenders the coefficients $\Lambda_q^{(j)}$ for $j = 0, 1, 2, \ldots$ of the perturbation expansion (2.198) should obey to the following compatibility relations

$$\Lambda_q^{(0)}\Lambda_q^{(0)} = 1_{4\times4} \tag{2.218}$$

$$\Lambda_q^{(0)}\Lambda_q^{(1)} + \Lambda_q^{(1)}\Lambda_q^{(0)} = 0_{4\times4} \tag{2.219}$$

$$\Lambda_q^{(0)}\Lambda_q^{(2)} + \Lambda_q^{(2)}\Lambda_q^{(0)} + \Lambda_q^{(1)}\Lambda_q^{(1)} = 0_{4\times4} \tag{2.220}$$

$$\ldots \quad \ldots$$

To check (2.219) we use $\Lambda_q^{(0)}\mathsf{P}_{q,s}^{(0)} = s\mathsf{P}_{q,s}^{(0)} = \mathsf{P}_{q,s}^{(0)}\Lambda_q^{(0)}$. Upon insertion of (2.213) then

$$\begin{aligned}
&\Lambda_q^{(0)}\Lambda_q^{(1)} + \Lambda_q^{(1)}\Lambda_q^{(0)} \\
&= \Lambda_q^{(0)}\left(\frac{1}{w}\tilde{\Pi}_a + \frac{s}{2w^2}i\tilde{\mathcal{E}}_a\right)\mathsf{P}_{q,s}^{(0)}\alpha_a\mathsf{P}_{q,-s}^{(0)} + \left(\frac{1}{w}\tilde{\Pi}_a + \frac{s}{2w^2}i\tilde{\mathcal{E}}_a\right)\mathsf{P}_{q,s}^{(0)}\alpha_a\mathsf{P}_{q,-s}^{(0)}\Lambda_q^{(0)} \\
&= \left(\frac{1}{w}\tilde{\Pi}_a + \frac{s}{2w^2}i\tilde{\mathcal{E}}_a\right)\left(\Lambda_q^{(0)}\mathsf{P}_{q,s}^{(0)}\alpha_a\mathsf{P}_{q,-s}^{(0)} + \mathsf{P}_s^{(0)}\alpha_a\mathsf{P}_{q,-s}^{(0)}\Lambda_q^{(0)}\right) \\
&= \left(\frac{1}{w}\tilde{\Pi}_a + \frac{s}{2w^2}i\tilde{\mathcal{E}}_a\right)s\left(\underbrace{\mathsf{P}_{q,s}^{(0)}\alpha_a\mathsf{P}_{q,-s}^{(0)} - \mathsf{P}_{q,s}^{(0)}\alpha_a\mathsf{P}_{q,-s}^{(0)}}_{=0_{4\times4}}\right) = 0_{4\times4}
\end{aligned}$$

The proof of (2.220) proceeds in a similar way, but is more elaborate.

One find inserting (2.213)

$$\begin{aligned}
\Lambda_q^{(1)}\Lambda_q^{(1)} &= \left(\frac{1}{w}\tilde{\Pi}_a + \frac{s}{2w^2}i\tilde{\mathcal{E}}_a\right)\left(\frac{1}{w}\tilde{\Pi}_{a'} + \frac{s'}{2w^2}i\tilde{\mathcal{E}}_{a'}\right)\mathsf{P}_{q,s}^{(0)}\alpha_a\underbrace{\mathsf{P}_{q,-s}^{(0)}\mathsf{P}_{q,s'}^{(0)}}_{=\delta_{0,s+s'}\mathsf{P}_{q,-s}^{(0)}}\alpha_{a'}\mathsf{P}_{q,-s'}^{(0)} \\
&= \left(\frac{1}{w}\tilde{\Pi}_a + \frac{s}{2w^2}i\tilde{\mathcal{E}}_a\right)\left(\frac{1}{w}\tilde{\Pi}_{a'} - \frac{s}{2w^2}i\tilde{\mathcal{E}}_{a'}\right)\mathsf{P}_{q,s}^{(0)}\alpha_a\mathsf{P}_{q,-s}^{(0)}\alpha_{a'}\mathsf{P}_{q,s}^{(0)}
\end{aligned} \tag{2.221}$$

Correspondingly it follows inserting (2.214) and taking into account (2.210)

$$\Lambda_q^{(0)}\Lambda_q^{(2)} + \Lambda_q^{(2)}\Lambda_q^{(0)} = \begin{pmatrix} \tilde{\Pi}_a\tilde{\Pi}_{a'}\left\langle M_s M_{s''} M_{s'}\right\rangle_0 \\ +i\tilde{\mathcal{E}}_a\tilde{\Pi}_{a'}\left\langle M_s^2 M_{s''} M_{s'}\right\rangle_0 \\ -\tilde{\Pi}_a i\tilde{\mathcal{E}}_{a'}\left\langle M_s M_{s''} M_{s'}^2\right\rangle_0 \\ +\tilde{\mathcal{E}}_a\tilde{\mathcal{E}}_{a'}\left\langle M_s^2 M_{s''} M_{s'}^2\right\rangle_0 \end{pmatrix}(s+s')\,\mathsf{P}_{q,s}^{(0)}\alpha_a\mathsf{P}_{q,s''}^{(0)}\alpha_{a'}\mathsf{P}_{q,s'}^{(0)}$$

$$
= \begin{pmatrix} \tilde{\Pi}_a \tilde{\Pi}_{a'} \langle M_s M_{s''} M_s \rangle_0 \\ +i\tilde{\mathcal{E}}_a \tilde{\Pi}_{a'} \langle M_s^2 M_{s''} M_s \rangle_0 \\ -\tilde{\Pi}_a i\tilde{\mathcal{E}}_{a'} \langle M_s M_{s''} M_s^2 \rangle_0 \\ +\tilde{\mathcal{E}}_a \tilde{\mathcal{E}}_{a'} \langle M_s^2 M_{s''} M_s^2 \rangle_0 \end{pmatrix} 2s \mathsf{P}^{(0)}_{q,s} \alpha_a \mathsf{P}^{(0)}_{q,s''} \alpha_{a'} \mathsf{P}^{(0)}_{q,s}
$$

$$
= \begin{pmatrix} \tilde{\Pi}_a \tilde{\Pi}_{a'} \langle M_s^2 M_{-s} \rangle_0 \\ +i\tilde{\mathcal{E}}_a \tilde{\Pi}_{a'} \langle M_s^3 M_{-s} \rangle_0 \\ -\tilde{\Pi}_a i\tilde{\mathcal{E}}_{a'} \langle M_s^3 M_{-s} \rangle_0 \\ +\tilde{\mathcal{E}}_a \tilde{\mathcal{E}}_{a'} \langle M_s^4 M_{-s} \rangle_0 \end{pmatrix} 2s \mathsf{P}^{(0)}_{q,s} \alpha_a \mathsf{P}^{(0)}_{q,-s} \alpha_{a'} \mathsf{P}^{(0)}_{q,s} \tag{2.222}
$$

For the matrix elements we have according to (2.211)

$$
\begin{aligned}
\langle M_s^2 M_{-s} \rangle_0 &= -\frac{s}{2\tilde{w}^2} \\
\langle M_s^3 M_{-s} \rangle_0 &= -\frac{1}{4\tilde{w}^3} \\
\langle M_s^4 M_{-s} \rangle_0 &= -\frac{s}{8\tilde{w}^4}
\end{aligned} \tag{2.223}
$$

and therefore

$$
\begin{pmatrix} \tilde{\Pi}_a \tilde{\Pi}_{a'} \langle M_s^2 M_{-s} \rangle_0 \\ +i\tilde{\mathcal{E}}_a \tilde{\Pi}_{a'} \langle M_s^3 M_{-s} \rangle_0 \\ -\tilde{\Pi}_a i\tilde{\mathcal{E}}_{a'} \langle M_s^3 M_{-s} \rangle_0 \\ +\tilde{\mathcal{E}}_a \tilde{\mathcal{E}}_{a'} \langle M_s^4 M_{-s} \rangle_0 \end{pmatrix} \cdot 2s = -\left(\frac{1}{w}\tilde{\Pi}_a + \frac{s}{2w^2} i\tilde{\mathcal{E}}_a\right)\left(\frac{1}{w}\tilde{\Pi}_{a'} - \frac{s}{2w^2} i\tilde{\mathcal{E}}_{a'}\right) \tag{2.224}
$$

Thus adding (2.221) to (2.222) and taking into account the factorization (2.224) confirms $\Lambda_q^{(0)}\Lambda_q^{(2)} + \Lambda_q^{(2)}\Lambda_q^{(0)} + \Lambda_q^{(1)}\Lambda_q^{(1)} = 0_{4\times 4}$.

Appendix C

Abstract In the ensuing two kinds of angular averages (please mind the summation convention) are evaluated for $s, s', s'' \in \{+, -\}$ and $a, a', b, b', c \in \{x, y, z\}$

$$
\begin{aligned}
W_{a;ss'} &= \int \frac{d\Omega_n}{4\pi} \left(\delta_{b,b'} - n_b n_{b'}\right) \alpha_b \mathsf{P}^{(0)}_{q,s} \alpha_a \mathsf{P}^{(0)}_{q,s'} \alpha_{b'} \\
W_{a,a';s,s',s''} &= \int \frac{d\Omega_n}{4\pi} \left(\delta_{b,b'} - n_b n_{b'}\right) \alpha_b \mathsf{P}^{(0)}_{q,s} \alpha_a \mathsf{P}^{(0)}_{q,s''} \alpha_{a'} \mathsf{P}^{(0)}_{q,s'} \alpha_{b'} \\
\mathsf{P}^{(0)}_{q,s} &= \frac{1}{2}\left(1_{4\times 4} + s\frac{\beta + \tilde{q} n_c \alpha_c}{\sqrt{1+\tilde{q}^2}}\right)
\end{aligned}
$$

Keywords: Angular average · Products of Dirac matrices.

Angular Average $W_{a;ss'}$

The task here is to evaluate the (transversal) angular average of the following linear combination of 4 × 4-matrix products (mind the summation convention)

$$W_{a;ss'} \equiv \int \frac{d\Omega_n}{4\pi} \left(\delta_{b,b'} - n_b n_{b'}\right) \alpha_b \mathsf{P}^{(0)}_{q,s} \alpha_a \mathsf{P}^{(0)}_{q,s'} \alpha_{b'} \tag{2.225}$$

To this end we use the identity (2.215) and obtain

$$\begin{aligned} W_{a;ss'} &= \int \frac{d\Omega_n}{4\pi} \left(\delta_{b,b'} - n_b n_{b'}\right) \alpha_b \left(\alpha_a \mathsf{P}^{(0)}_{q,-s} + s\frac{\tilde{q}}{\tilde{w}} n_a 1_{4\times4}\right) \mathsf{P}^{(0)}_{q,s'} \alpha_{b'} \\ &= \int \frac{d\Omega_n}{4\pi} \left(\delta_{b,b'} - n_b n_{b'}\right) \alpha_b \left(\alpha_a \delta_{-s,s'} + s\frac{\tilde{q}}{\tilde{w}} n_a 1_{4\times4}\right) \mathsf{P}^{(0)}_{q,s'} \alpha_{b'} \\ &= \int \frac{d\Omega_n}{4\pi} \left(\delta_{b,b'} - n_b n_{b'}\right) \alpha_b \left(\alpha_a \delta_{-s,s'} + s\frac{\tilde{q}}{\tilde{w}} n_a 1_{4\times4}\right) \frac{1}{2}\left(1_{4\times4} + s'\frac{1}{\tilde{w}}\beta + s'\frac{\tilde{q}}{\tilde{w}} n_c \alpha_c\right) \alpha_{b'} \\ &= \int \frac{d\Omega_n}{4\pi} \left(\delta_{b,b'} - n_b n_{b'}\right) \alpha_b \begin{pmatrix} \alpha_a \delta_{-s,s'} \frac{1}{2}\left(1_{4\times4} + s'\frac{1}{\tilde{w}}\beta\right) + \frac{1}{2} ss' \frac{\tilde{q}^2}{\tilde{w}^2} n_a n_c \alpha_c \\ +\text{odd under } \mathbf{n} \to -\mathbf{n} \end{pmatrix} \alpha_{b'} \end{aligned} \tag{2.226}$$

With $\beta\alpha_{b'} = -\alpha_{b'}\beta$ next

$$W_{a;ss'} = \begin{pmatrix} \int \frac{d\Omega_n}{4\pi} \left(\delta_{b,b'} - n_b n_{b'}\right) \alpha_b \alpha_a \alpha_{b'} \delta_{-s,s'} \frac{1}{2}\left(1_{4\times4} - s'\frac{1}{\tilde{w}}\beta\right) \\ +\frac{1}{2} ss' \frac{\tilde{q}^2}{\tilde{w}^2} \int \frac{d\Omega_n}{4\pi} \left(\delta_{b,b'} - n_b n_{b'}\right) \alpha_b \alpha_c \alpha_{b'} n_a n_c \end{pmatrix} \tag{2.227}$$

Eventually one finds with help of the identity

$$\left(\delta_{b,b'} - n_b n_{b'}\right) \alpha_b \alpha_c \alpha_{b'} = -2 n_c n_b \alpha_b \tag{2.228}$$

the result

$$\begin{aligned} W_{a;ss'} &= \begin{pmatrix} \int \frac{d\Omega_n}{4\pi} \left(-2 n_a n_b \alpha_b\right) \delta_{-s,s'} \frac{1}{2}\left(1_{4\times4} - s'\frac{1}{\tilde{w}}\beta\right) \\ +\frac{1}{2} ss' \frac{\tilde{q}^2}{\tilde{w}^2} \int \frac{d\Omega_n}{4\pi} \left(-2 n_c n_b \alpha_b\right) n_a n_c \end{pmatrix} \\ &= -\left(\underbrace{\int \frac{d\Omega_n}{4\pi} n_a n_b}_{\frac{1}{3}\delta_{a,b}}\right) \alpha_b \begin{pmatrix} \delta_{-s,s'} \left(1_{4\times4} - s'\frac{1}{\tilde{w}}\beta\right) \\ +ss' \frac{\tilde{q}^2}{\tilde{w}^2} \underbrace{n_c n_c}_{=1} 1_{4\times4} \end{pmatrix} \\ &= -\frac{1}{3}\alpha_a \left(\delta_{-s,s'} \left(1_{4\times4} + s\frac{1}{\tilde{w}}\beta\right) + ss' \frac{\tilde{q}^2}{\tilde{w}^2} 1_{4\times4}\right) \end{aligned} \tag{2.229}$$

Note the symmetry

$$\left(W_{a;ss'}\right)^\dagger = W_{a;s's} \tag{2.230}$$

Angular Average $W_{aa';ss's''}$

The task here is to evaluate the (transversal) angular average of the following linear combination of 4 × 4-matrix products (mind the summation convention)

$$W_{a,a';s,s'',s'} \equiv \int \frac{d\Omega_n}{4\pi} \left(\delta_{b,b'} - n_b n_{b'}\right) \alpha_b \mathsf{P}^{(0)}_{q,s} \alpha_a \mathsf{P}^{(0)}_{q,s''} \alpha_{a'} \mathsf{P}^{(0)}_{q,s'} \alpha_{b'} \tag{2.231}$$

To this end we use the identity (2.216) and obtain

$$W_{a,a';s,s'',s'} = \int \frac{d\Omega_n}{4\pi} \left(\delta_{b,b'} - n_b n_{b'}\right) \alpha_b \begin{pmatrix} \delta_{-s,s''} \delta_{-s',s''} \alpha_a \mathsf{P}^{(0)}_{q,s''} \alpha_{a'} \\ + \frac{\tilde{q}}{\tilde{w}} \left(\delta_{-s,s''} s' n_{a'} \alpha_a \mathsf{P}^{(0)}_{q,s''} + \delta_{-s',s''} s n_a \mathsf{P}^{(0)}_{q,s''} \alpha_{a'}\right) \\ + \frac{\tilde{q}^2}{\tilde{w}^2} s s' n_a n_{a'} \mathsf{P}^{(0)}_{q,s''} \end{pmatrix} \alpha_{b'} \tag{2.232}$$

Insertion of (2.201) into (2.232) gives bearing in mind all terms being odd under $\mathbf{n} \to -\mathbf{n}$ give no contribution to the angular average

$$W_{a,a';s,s'',s'} = \frac{1}{2} \begin{pmatrix} \delta_{-s,s''} \delta_{-s',s''} \left(\int \frac{d\Omega_n}{4\pi} \left(\delta_{b,b'} - n_b n_{b'}\right) \alpha_b \alpha_a \alpha_{a'} \alpha_{b'}\right) \left(1_{4\times 4} + s'' \cdot \frac{1}{\tilde{w}} \beta\right) \\ + \frac{\tilde{q}^2}{\tilde{w}^2} \begin{pmatrix} \delta_{-s,s''} s' s'' \left(\int \frac{d\Omega_n}{4\pi} n_{a'} n_c \left(\delta_{b,b'} - n_b n_{b'}\right) \alpha_b \alpha_a \alpha_c \alpha_{b'}\right) \\ + \delta_{-s',s''} s s'' \left(\int \frac{d\Omega_n}{4\pi} n_a n_c \left(\delta_{b,b'} - n_b n_{b'}\right) \alpha_b \alpha_c \alpha_{a'} \alpha_{b'}\right) \end{pmatrix} \\ + \frac{\tilde{q}^2}{\tilde{w}^2} s s' \left(\int \frac{d\Omega_n}{4\pi} n_a n_{a'} \left(\delta_{b,b'} - n_b n_{b'}\right) \alpha_b \alpha_{b'}\right) \left(1_{4\times 4} - s'' \cdot \frac{1}{\tilde{w}} \beta\right) \end{pmatrix} \tag{2.233}$$

With the algebra of Dirac matrices

$$\begin{aligned} \beta \alpha_b + \alpha_b \beta &= 0 \\ \alpha_a \alpha_b &= \delta_{a,b} 1_{4\times 4} + i\, \varepsilon_{abc} \sigma_c \end{aligned} \tag{2.234}$$

and taking into account $n_c n_c = 1$ and $\varepsilon_{abc} n_a n_b = 0$ one readily confirms the matrix identities

$$\begin{aligned} \left(\delta_{b,b'} - n_b n_{b'}\right) \alpha_b \alpha_{b'} &= 2 \cdot 1_{4\times 4} \\ \left(\delta_{b,b'} - n_b n_{b'}\right) \alpha_b \beta \alpha_{b'} &= -2\beta \\ \left(\delta_{b,b'} - n_b n_{b'}\right) \alpha_b \sigma_c \alpha_{b'} &= -2 n_c n_b \sigma_b \\ \left(\delta_{b,b'} - n_b n_{b'}\right) \alpha_b \sigma_c \beta \alpha_{b'} &= +2 n_c n_b \sigma_b \beta \\ \left(\delta_{b,b'} - n_b n_{b'}\right) \alpha_b \alpha_c \alpha_{b'} &= -2 n_c n_b \alpha_b \end{aligned} \tag{2.235}$$

and out of this the associated angular averages

$$\begin{aligned} \int \frac{d\Omega_n}{4\pi} \left(\delta_{b,b'} - n_b n_{b'}\right) \alpha_b \alpha_{b'} &= 2 \cdot 1_{4\times 4} \\ \int \frac{d\Omega_n}{4\pi} n_a n_{a'} \left(\delta_{b,b'} - n_b n_{b'}\right) \alpha_b \alpha_{b'} &= \frac{2}{3} \delta_{a,a'} \cdot 1_{4\times 4} \\ \int \frac{d\Omega_n}{4\pi} \left(\delta_{b,b'} - n_b n_{b'}\right) \alpha_b \alpha_a \alpha_{a'} \alpha_{b'} &= 2\delta_{a,a'} \cdot 1_{4\times 4} - \frac{2}{3} i \varepsilon_{aa'c} \sigma_c \end{aligned} \tag{2.236}$$

$$\int \frac{d\Omega_n}{4\pi} n_{a'} n_c \left(\delta_{b,b'} - n_b n_{b'}\right) \alpha_b \alpha_a \alpha_c \alpha_{b'} = \frac{2}{3} \delta_{a,a'} \cdot 1_{4\times4}$$

Eventually one obtains inserting the angular averages (2.236) into (2.233) the result

$$W_{a,a';s,s'',s'} = \delta_{a,a'} \begin{pmatrix} \delta_{-s,s''} \delta_{-s',s''} \left(1_{4\times4} + s'' \cdot \frac{1}{\tilde{w}} \beta\right) \\ -\frac{1}{3} \frac{\tilde{q}^2}{\tilde{w}^2} ss' \left(\delta_{-s,s''} + \delta_{-s',s''}\right) 1_{4\times4} \\ +\frac{1}{3} \frac{\tilde{q}^2}{\tilde{w}^2} ss' \left(1_{4\times4} - s'' \cdot \frac{1}{\tilde{w}} \beta\right) \end{pmatrix} - \frac{1}{3} \delta_{-s,s''} \delta_{-s',s''} i \varepsilon_{aa'c} \sigma_c \left(1_{4\times4} + s'' \cdot \frac{1}{\tilde{w}} \beta\right) \tag{2.237}$$

References

1. Schopohl N, Cetin NS. Advances in Quantum Science I - Theoretical and Methodological Approaches, Chapter 1: From the Relativistic Field Theory of Quantum Electrodynamics to a Hamiltonian for Electrons and Positrons as Point Particles with Attributes Mass, Charge and Spin. Springer Nature; 2026.
2. Roc JD, Tannoudji CC. Tendances Actuelles En Physique Atomique, Les Houches, Ecoles d'Ete de Physique, Session 38, volume 1, chapter "Effective Hamiltonian approach to g-2 relativistic calculation", pp. 157–180. North-Holland Physics Publishing Amsterdam. Oxford. New York. Tokyo; 1982.
3. Schopohl N, Cetin NS. Advances in Quantum Science I – Theoretical and Methodological Approaches, Chapter 3: The Nonrelativistic Hamiltonian of Electrons and Positrons: A Derivation from Quantum Electrodynamics. Springer Nature; 2026.
4. Schwinger J. On quantum-electrodynamics and the magnetic moment of the electron. Phys Rev. 1948;73(4):416.
5. Schwinger J. On gauge invariance and vacuum polarization. Phys Rev. 1950;82:664–79.
6. Lamb WE, Retherford RC. Fine structure of the hydrogen atom by a microwave method. Phys Rev. 1947;72:241.
7. Dirac PAM. Lectures on quantum field theory. Academic Press London; 1966.
8. Pauli W. Pauli lectures on physics 6. Selected topics in field quantization. Cambridge MIT; 1973.
9. Heitler W. The quantum theory of radiation. 3rd ed. New York: Dover Publications Inc.; 1984.
10. Zee A. Quantum field theory in a Nutshell, 2nd ed. Princeton University Press; 2010.
11. Heisenberg W. Bemerkungen zur Diracschen Theorie des Positrons. Z Physik. 1934;90:209–31.
12. Ma ST. Vacuum polarization in the positron theory. Lond Edinb Dublin Philos Mag J Sci. 1949;40(310):1112–28.
13. French JB, Weisskopf VF. The electromagnetic shift of energy levels. Phys Rev. 1949;75(8):1240–8.
14. Uehling EA. Polarization effects in the positron theory. Phys Rev. 1935;48:55.
15. Serber R. Linear modifications in the Maxwell field equations. Phys Rev. 1935;48:49–54.
16. Itzykson C, Bernard Zuber J. Quantum field theory. McGRAW HILL International Editions; 1985.
17. Bender CM, Orszag SA. Advanced mathematical methods for scientists and engineers. McGraw-Hill International Edition; 1978.
18. Bethe HA, Salpeter EE. Quantum mechanics of one- and two-electron atoms. A Plenum/Rosetta edition; 1977.
19. Kroll NM, Lamb WE Jr. On the self-energy of a bound electron. Phys Rev. 1949;75:388.
20. Landau LD, Lifschitz JM. Quantenelektrodynamik. Verlag Harri Deutsch; 2009.
21. Schweber SS. An introduction to relativistic quantum field theory. Harper International Student Reprint; 1964.

22. Kirzhnitz DA. Chapter 2: General properties of electromagnetic response functions, volume 24 of Modern Problems in Condensed Matter Sciences. Elsevier Science Publishers B.V. North Holland Physics; 1989.The dielectric function of condensed systems.
23. Feynman RP. Quantum electrodynamics. Advanced book classics. Addison Wesley; 1961.
24. Dirac PAM. Discussion of the infinite distribution of electrons in the theory of the positron. Proc Cambr Phil Soc. 1934;30:150.
25. Eriksen E. Foldy-Wouthuysen-Tranformation. Exact solution with generalization to the two-particle problem. Phys Rev. 1958;111(1011).
26. Schopohl N, Cetin NS. Advances in Quantum Science I - Theoretical and Methodological Approaches, Chapter 4: A Flow Equation Approach Transforming the Dirac Hamiltonian into a Unitary Equivalent and Energy-Separating Hamiltonian. Springer Nature; 2026.
27. Schwinger J. Particles, sources, and fields, volume II. Perseus; 1998.

Chapter 3
Derivation of a Nonrelativistic Hamiltonian for Interacting Electrons and/or Positrons from Quantum Electrodynamics

Nils Schopohl and Nadine Suzan Cetin

Abstract A particle number conserving quantum field Hamiltonian $\hat{\mathcal{H}}_U$, previously obtained in [1] from a Hamiltonian flow equation, that we solved perturbatively to third order in the coupling e for the original relativistic field theory Hamiltonian of Quantum Electrodynamics in Coulomb gauge, is briefly summarized. Proceeding from there it is shown with help of the energy separating Eriksen transformation, how to derive in an entirely symmetric fashion from $\hat{\mathcal{H}}_U$ the effective nonrelativistic Hamiltonian for interacting electrons and/or positrons as individual particles with attributes mass, charge and spin.

Keywords Quantum electrodynamics · Wegner flow equation method · Unitary transformation · Fock-space · Quantum field · Dirac hamiltonian · Eriksen transformation · Spin-orbit and other spin-orbit interaction · Magnetic dipole-dipole interaction · Darwin contact interaction · Annihilation interaction · Positronium

3.1 Introduction

We transformed in [1, Sect. 6] the Hamiltonian of relativistic quantum electrodynamics in Coulomb gauge [2, 3] by means of the flow equation method of Wegner [4] to a unitary equivalent quantum field Hamiltonian in the guise of a perturbation series with the coupling e to electromagnetic fields as expansion parameter, whereby the limiting value $\hat{\mathcal{H}}_U$ of that flow by construction exhibits the

N. Schopohl (✉)
Center for Quantum Science, Eberhard Karls Universität Tübingen, Tübingen, Germany
e-mail: nils.schopohl@uni-tuebingen.de

N. Suzan Cetin
Charles University Prague, Mathematical Institute, Prague, Czech Republic
e-mail: nadine-suzan.cetin@matfyz.cuni.cz

T. Onishi (ed.), *Advances in Quantum Science I*,
https://doi.org/10.1007/978-981-95-9717-8_3

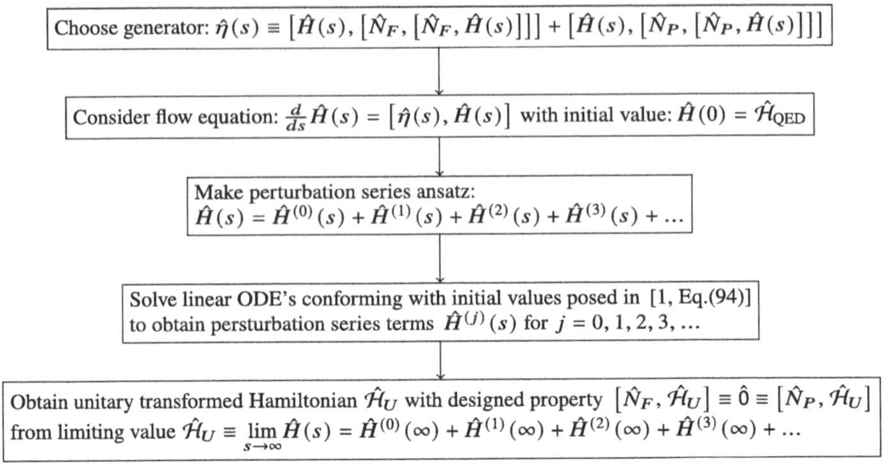

Fig. 3.1 Conceptual schema to explicate the Hamiltonian flow equation method employed in [1, Sect. 6] to find a fermion number and photon number conserving Hamiltonian $\hat{\mathcal{H}}_U$ being unitary equivalent to the fundamental field theory Hamiltonian $\hat{\mathcal{H}}_{\mathrm{QED}}$ of Quantum Electrodynamics

virtue of *conserving* particle number. A flow chart of the essential steps of Wegner's method is given in Fig. 3.1.

The result for $\hat{\mathcal{H}}_U$ as enlisted in Eq. (3.1), for details see [1, Sect. 7], consists in the main of three types of contributions: the renormalized one body Dirac Hamiltonian $\hat{\mathcal{H}}_D^{(\mathrm{ren})}$, the term $\hat{\mathcal{H}}_{\mathrm{low}}^{(\mathrm{rad})}$ describing the interaction of the Dirac fermions with quantized *long* wave-length electromagnetic radiation, and thirdly the variant effective two-body interactions that originated from the flow, for now denoted collectively as $\hat{\mathcal{U}}_{\mathrm{int}}$. With regard to the other terms also enlisted in [1, Sect. 7], namely the Compton interaction terms, the vacuum-energy and the self-energy of photons in strong external fields, we desist to discuss these contributions here. Instead our focus in what follows is on the low-energy sector of QED, where electrons and/or positrons may be considered as wave packets that move at nonrelativistic speed.

Altogether

$$\hat{\mathcal{H}}_U = \hat{\mathcal{H}}_D^{(\mathrm{ren})} + \hat{\mathcal{H}}_{\mathrm{low}}^{(\mathrm{rad})} + \hat{\mathcal{U}}_{\mathrm{int}} \tag{3.1}$$

$$\hat{\mathcal{H}}_D^{(\mathrm{ren})} = \hat{\mathcal{H}}_D + \hat{\mathcal{M}}^{(e)} + \hat{\mathcal{M}}^{(p)} \tag{3.2}$$

$$\hat{\mathcal{H}}_{\mathrm{low}}^{(\mathrm{rad})} = \hat{\mathcal{H}}_{\mathrm{low}}^{(\mathrm{e,rad})} + \hat{\mathcal{H}}_{\mathrm{low}}^{(\mathrm{p,rad})} \tag{3.3}$$

$$\hat{\mathcal{U}}_{\mathrm{int}} = \hat{\mathcal{U}}_C^{(e,e)} + \hat{\mathcal{U}}_C^{(p,p)} + \hat{\mathcal{U}}_C^{(e,p)} + \hat{\mathcal{U}}_\perp^{(e,e)} + \hat{\mathcal{U}}_\perp^{(p,p)} + \hat{\mathcal{U}}_\perp^{(e,p)}. \tag{3.4}$$

The renormalized one-particle Hamiltonian $\hat{\mathcal{H}}_D^{(\mathrm{ren})}$ comprising the terms enlisted in (3.2) we discuss in Sect. (3.3), the nonrelativistic matter-light interaction $\hat{\mathcal{H}}_{\mathrm{low}}^{(\mathrm{rad})}$ comprising the terms enlisted in (3.3) we discuss in Sect. (3.4). As a last point we treat in the subsequent Sect. (3.5) all the terms in (3.4) representing en bloc the effective two-particle interactions $\hat{\mathcal{U}}_{\text{int}}$ between electrons (e) and/or positrons (p).

For a clear understanding, how the one-particle Hamiltonian $\hat{\mathcal{H}}_D^{(\text{ren})}$ emerges eventually from adding to the field Hamiltonian $\hat{\mathcal{H}}_D$ the self-energy terms $\hat{\mathcal{M}}^{(e)}$ and $\hat{\mathcal{M}}^{(p)}$, we refer to our chapter [5], where we particularize, based on a careful evaluation of these self-energy terms with Dirac's cut-off method, how the terms δe and δm get annulled by the procedure of charge and mass renormalization, and how then the magnetic moment correction of Schwinger and what's more how the terms contributing to the Lamb shift are obtained by a consistent expansion in an accurate logical way without violating gauge invariance.

Our objective in what follows is first of all to disentangle the matter and antimatter degrees of freedom embodied within $\hat{\mathcal{H}}_U$ with help of the energy separating Eriksen transformation [6, 7], and subsequently to extract from $\hat{\mathcal{U}}_{\text{int}}$ the effective two-body interaction terms between electrons and/or positrons, when these particles are restricted to move at slow speed v compared to the speed of light c. By expanding in the small parameter $\bar{v} = v/c$ and aiming at an accuracy $O(\bar{v}^3)$ only, indeed all the well known relativistic correction terms adding to the Coulomb interaction between electrons (positrons) are obtained, namely the Darwin contact and the spin-orbit interaction term, moreover the orbit-orbit, spin-other-orbit and magnetic dipole-dipole interactions.

By transforming the obtained expansion for $\hat{\mathcal{H}}_U$ eventually back from second quantization to first quantization, the *classical* conception of a cloud consisting of interacting *point* particles with attributes mass, charge and spin emerges. Specializing to *electrons* only, our results are in complete accord with the results enlisted bei Itoh in [8], even though our method of derivation indeed differs substantially from his.

With regard to the *positrons* we obtain the corresponding *mirrored* Hamiltonian (this feature being as important as being trivial of course), and what's more interesting than this, we obtain the Hamiltonian for electrons interacting with positrons, comprising besides the expected relativistic corrections to the Coulomb attraction between electron and positron, now as well the so called *annihilation* term [9].

Specializing our result to the case of positronium (that is *one* electron is interacting with *one* positron), in deed full agreement can be attested with a two-body Hamiltonian deduced for positronium by other methods of derivation. To name but a few, there is the derivation of Feynman [10], the derivation of Landau as summarized in [11], and last not least the detailed elaboration of Schwinger [12].

3.2 Approximate Evaluation of Expectation Values in the Basis of Newton-Wigner Eigenstates

In [1] the Newton-Wigner Hamiltonian was introduced as a specific unitary transformation of the Dirac Hamiltonian $\mathsf{H}^{(D)}$ given by

$$\mathsf{H}^{(NW)} = \mathsf{T}\,\mathsf{H}^{(D)}\mathsf{T}^{\dagger} \qquad (3.5)$$
$$\mathsf{T} = \exp(-\Omega_u),$$

whereas T denotes the Eriksen transformation [6, 7]. In point of fact, in the low energy sector of QED the eigenfunctions $\phi_\nu^{(NW)}(\mathbf{r}, k)$ of $\mathsf{H}^{(NW)}$ change only slowly if regarded on the length scale of the Compton length $\lambda_C = \frac{h}{mc} \ll a_B$. This is because expanding the generator Ω_u of the Eriksen transformation to lowest order in the velocity parameter $\bar{v} = v/c$ it is found that $\mathsf{H}^{(NW)}$ coincides with the Schrödinger-Pauli Hamiltonian of nonrelativistic quantum mechanics [6, 7].

Moreover, a feature of the eigenfunctions $\phi_\nu^{(NW)}(\mathbf{r}, k)$ of $\mathsf{H}^{(NW)}$ is that they are concurrently eigenfunctions of the operator β with eigenvalue $z_k = \frac{E_k}{|E_k|} = \pm 1$, thus entailing the property of *energy separation* [6, 7].

This means there holds for all eigenfunction $\phi_\nu^{(NW)}(\mathbf{r}, k)$ corresponding to a *positive* eigenvalue, i.e. $z_k = +1$,

$$U_\nu^{(NW)}(\mathbf{r}, k) = \frac{1 + z_k}{2}\phi_\nu^{(NW)}(\mathbf{r}, k) = \sum_{\nu'} \frac{1}{2}\left(1_{4\times 4} + \beta\right)_{\nu,\nu'} \phi_{\nu'}^{(NW)}(\mathbf{r}, k), \qquad (3.6)$$

implying (in the Dirac presentation of matrices α_b and β) the components $\nu = 3, 4$ of $\phi_\nu^{(NW)}(\mathbf{r}, k)$ vanish identically for $k \in I_+$.

Likewise there holds for all eigenfunction $\phi_\nu^{(NW)}(\mathbf{r}, k)$ corresponding to a *negative* eigenvalue, i.e. $z_k = -1$,

$$V_\nu^{(NW)}(\mathbf{r}, k) = \frac{1 - z_k}{2}\phi_\nu^{(NW)}(\mathbf{r}, k) = \sum_{\nu'} \frac{1}{2}\left(1_{4\times 4} - \beta\right)_{\nu,\nu'} \phi_{\nu'}^{(NW)}(\mathbf{r}, k), \qquad (3.7)$$

implying (in the Dirac presentation of matrices α_b and β) the components $\nu = 1, 2$ of $\phi_\nu^{(NW)}(\mathbf{r}, k)$ vanish identically for $k \in I_-$.

So it appears natural to recompile the matrix elements of an operator G, originally to be evaluated in the basis of the Dirac eigenmodes $\phi_\mu(\mathbf{r}, k)$, now as a matrix elements of the unitary equivalent operator $\mathsf{TGT}^\dagger$ to be evaluated in the basis of the NW-eigenfunctions $\phi_\nu^{(NW)}(\mathbf{r}, k)$. The advantage of this approach being first of all, that an expansion in the parameter $\bar{v}$ is enabled, because the NW-eigenfunctions at low excitation energy are spatially slowly varying, and in the second place the advantage being the *repartition* of the matrix elements of $\mathsf{TGT}^\dagger$ in the NW-basis now as 2×2-block matrices, with the specific outcome of that repartition procedure depending on fermion mode labels $k, k' \in I_+$ or $k, k' \in I_-$, or rather $k \in I_+, k' \in I_-$ or $k \in I_-, k' \in I_+$, whereas $I_\pm = \{\, k \mid z_k = \pm 1\}$.

This way a complete separation of the matter and antimatter degrees of freedom in the unitary transformed quantum field Hamiltonian $\hat{\mathcal{H}}_U$ succeeds, as is exemplified in the ensuing.

As we restrict to the low energy sector of QED, all particles move at non-relativistic speed $\bar{v} = \frac{v}{c} \ll 1$. This enables to find a suitable approximation for the matrix elements of an operator G by first writing

$$\langle \phi_k | \mathsf{G} | \phi_{k'} \rangle = \left\langle \phi_k^{(NW)} \middle| e^{-\Omega_u} \mathsf{G} e^{\Omega_u} \middle| \phi_{k'}^{(NW)} \right\rangle, \tag{3.8}$$

and then expanding the generator Ω_u of the Eriksen transformation as a perturbation series

$$\Omega_u = \Omega_u^{(1)} + \Omega_u^{(3)} + \Omega_u^{(5)} + \ldots \tag{3.9}$$

The details of this expansion are given in our chapter [7, Eq. (170)].

In what follows we restrict to the first order term $\Omega_u^{(1)}$ in this expansion, as the contributions of the remainder in (3.9), when applied to states $\left|\phi_{k'}^{(NW)}\right\rangle$ in the low energy sector of QED, are at least small of order in $\bar{v}^3$. Thus

$$\Omega_u \approx \Omega_u^{(1)} \equiv \frac{1}{2}\tilde{\Pi}_b \alpha_b \beta \tag{3.10}$$

$$\tilde{\Pi}_b = \frac{1}{mc}\left(\mathsf{p}_b - e\mathsf{A}_b^{(\mathrm{ext})}(\mathsf{x})\right).$$

Notably, working with this first order approximation (3.10) one finds the matrix element (3.8) being already accurate to second order $\bar{v}^2$ with the expansion

$$e^{-\Omega_u}\mathsf{G}e^{\Omega_u} = \mathsf{G} - \left[\Omega_u^{(1)}, \mathsf{G}\right] + \frac{1}{2}\left[\Omega_u^{(1)}, \left[\Omega_u^{(1)}, \mathsf{G}\right]\right] + O(\bar{v}^3). \tag{3.11}$$

Specializing to $\mathsf{G} = e^{iq_a \mathsf{x}_a}$ or else to $\mathsf{G} = \alpha_b e^{iq_a \mathsf{x}_a}$ one finds, calculating the commutators as indicated in (3.11), as well identifying

$$\alpha_a \sigma_b + \sigma_b \alpha_a = 2\delta_{a,b}\,\gamma_5 \tag{3.12}$$
$$\gamma_5 = \begin{pmatrix} 0_{2\times 2}\,, 1_{2\times 2} \\ 1_{2\times 2}\,, 0_{2\times 2} \end{pmatrix}$$

and abbreviating $\tilde{q}_b \equiv n_b \tilde{q}$ with $n_b n_b = 1$ and $\tilde{q} \equiv \frac{\hbar}{mc}|\mathbf{q}|$, eventually the approximations

$$e^{-\Omega_u} e^{iq_a \mathsf{x}_a} e^{\Omega_u} \simeq e^{iq_a \mathsf{x}_a} \begin{pmatrix} \left(1 - \frac{\tilde{q}^2}{8}\right) 1_{4\times 4} + \\ + \frac{1}{2}\beta\alpha_b \tilde{q} n_b + \frac{i}{4}\varepsilon_{bb'b''}\,\tilde{q} n_b \tilde{\Pi}_{b'}\sigma_{b''} \end{pmatrix}, \tag{3.13}$$

respectively

$$e^{-\Omega_u}\alpha_b e^{iq_a x_a} e^{\Omega_u} \simeq e^{iq_a x_a}\begin{pmatrix} \left(\delta_{b,a'} - \frac{\tilde{q}^2}{8} n_b n_{a'}\right)\alpha_{a'} + \\ +\frac{1}{2}\left(\tilde{q}n_b + 2\tilde{\Pi}_b - i\varepsilon_{bb'b''}\,\tilde{q}n_{b'}\sigma_{b''}\right)\beta + \\ -\frac{1}{4}\left(\tilde{q}n_{a'}\tilde{\Pi}_b + \tilde{q}n_b\tilde{\Pi}_{a'} + \tilde{\Pi}_b\tilde{\Pi}_{a'} + \tilde{\Pi}_{a'}\tilde{\Pi}_b\right)\alpha_{a'} + \\ +\frac{i}{4}\varepsilon_{bb'b''}\,\tilde{q}n_{b'}\tilde{\Pi}_{b''}\gamma_5 \end{pmatrix}. \tag{3.14}$$

As follows immediately from (3.6), (3.7), the property of energy separation of the eigenstates $\left|U_k^{(NW)}\right\rangle$ and $\left|V_k^{(NW)}\right\rangle$ engenders for the operators α_b and γ_5 the vanishing of the matrix elements

$$\begin{aligned} \left\langle U_k^{(NW)}\right|\alpha_b\left|U_{k'}^{(NW)}\right\rangle &\equiv 0 \equiv \left\langle V_k^{(NW)}\right|\alpha_b\left|V_{k'}^{(NW)}\right\rangle \\ \left\langle U_k^{(NW)}\right|\gamma_5\left|U_{k'}^{(NW)}\right\rangle &\equiv 0 \equiv \left\langle V_k^{(NW)}\right|\gamma_5\left|V_{k'}^{(NW)}\right\rangle. \end{aligned} \tag{3.15}$$

With (3.15) and minding

$$\begin{aligned} \beta\left|U_k^{(NW)}\right\rangle &= \left|U_k^{(NW)}\right\rangle \\ \beta\left|V_k^{(NW)}\right\rangle &= -\left|V_k^{(NW)}\right\rangle, \end{aligned} \tag{3.16}$$

one readily confirms inserting the expansions (3.13) and (3.14)

$$\begin{aligned} \langle U_k|\, e^{iq_a x_a}\,|U_{k'}\rangle &= \left\langle U_k^{(NW)}\right| e^{-\Omega_u} e^{iq_a x_a} e^{\Omega_u}\left|U_{k'}^{(NW)}\right\rangle \\ &= \int d^3r \sum_{\nu,\nu'\in\{1,2\}} U_\nu^{\star(NW)}(\mathbf{r},k)\, e^{i\mathbf{q}\cdot\mathbf{r}}\left(\left(1-\frac{\tilde{q}^2}{8}\right)1_{4\times4} + \frac{1}{4}i\varepsilon_{bb'b''}\,\tilde{q}n_b\tilde{\mathsf{D}}_{b'}\sigma_{b''}\right)_{\nu,\nu'} U_{\nu'}^{(NW)}(\mathbf{r},k') \end{aligned} \tag{3.17}$$

and

$$\begin{aligned} \langle U_k|\,\alpha_b e^{iq_a x_a}\,|U_{k'}\rangle &= \left\langle U_k^{(NW)}\right| e^{-\Omega_u}\alpha_b e^{iq_a x_a} e^{\Omega_u}\left|U_{k'}^{(NW)}\right\rangle \\ &= \int d^3r \sum_{\nu,\nu'\in\{1,2\}} U_\nu^{\star(NW)}(\mathbf{r},k)\, e^{i\mathbf{q}\cdot\mathbf{r}}\left(\left(\frac{1}{2}\tilde{q}n_b + \tilde{\mathsf{D}}_b\right)1_{4\times4} - \frac{1}{2}i\varepsilon_{ba'a''}\,\tilde{q}n_{a'}\sigma_{a''}\right)_{\nu,\nu'} U_{\nu'}^{(NW)}(\mathbf{r},k'), \end{aligned} \tag{3.18}$$

whereby in the evaluation of the matrix elements the gauge invariant operator $\Pi_b = \mathsf{p}_b - e\mathsf{A}_b^{(\mathrm{ext})}(\mathsf{x})$ mutates in position space representation to

$$\mathsf{D}_b \equiv \frac{\hbar}{i}\frac{\partial}{\partial r_b} - e\mathsf{A}_b^{(\mathrm{ext})}(\mathbf{r}) \ \text{ and } \ \tilde{\mathsf{D}}_b = \frac{\mathsf{D}_b}{mc}. \tag{3.19}$$

In a similar way one finds

$$\langle V_k | e^{iq_a x_a} | V_{k'} \rangle = \left\langle V_k^{(NW)} \right| e^{-\Omega_u} e^{iq_a x_a} e^{\Omega_u} \left| V_{k'}^{(NW)} \right\rangle \tag{3.20}$$

$$= \int d^3r \sum_{\nu,\nu' \in \{3,4\}} V_\nu^{\star(NW)}(\mathbf{r},k)\, e^{i\mathbf{q}\cdot\mathbf{r}} \left((1 - \tfrac{\tilde{q}^2}{8}) 1_{4\times4} + \tfrac{1}{4} i\varepsilon_{bb'b''}\, \tilde{q} n_b \tilde{\mathsf{D}}_{b'} \sigma_{b''} \right)_{\nu,\nu'} V_{\nu'}^{(NW)}(\mathbf{r},k')$$

$$= \int d^3r \sum_{\nu,\nu' \in \{3,4\}} V_{\nu'}^{(NW)}(\mathbf{r},k') \left((1 - \tfrac{\tilde{q}^2}{8}) 1_{4\times4} + \tfrac{1}{4} i\varepsilon_{bb'b''}\, \tilde{q} n_b \tilde{\mathsf{D}}^\star_{b'} \sigma_{b''} \right)_{\nu,\nu'} e^{i\mathbf{q}\cdot\mathbf{r}}\, V_\nu^{\star(NW)}(\mathbf{r},k)$$

and likewise

$$\langle V_k | \alpha_b e^{iq_a x_a} | V_{k'} \rangle = \left\langle V_k^{(NW)} \right| e^{-\Omega_u} \alpha_b e^{iq_a x_a} e^{\Omega_u} \left| V_{k'}^{(NW)} \right\rangle \tag{3.21}$$

$$= -\int d^3r \sum_{\nu,\nu' \in \{3,4\}} V_{\nu'}^{(NW)}(\mathbf{r},k') \left((\tfrac{1}{2}\tilde{q} n_b + \tilde{\mathsf{D}}^\star_b) 1_{4\times4} - \tfrac{1}{2} i\varepsilon_{ba'a''}\, \tilde{q} n_{a'} \sigma_{a''} \right)_{\nu,\nu'} e^{i\mathbf{q}\cdot\mathbf{r}}\, V_\nu^{\star(NW)}(\mathbf{r},k)\,.$$

Here, the terms on the right-hand side of the last equal sign in (3.20) and (3.21) result from partial integration, that way effectively replacing $e \to -e$ in the gauge invariant derivative for the positrons.

As regards the matrix elements shaping the mutual electron-positron interactions one finds upon insertion of the expansion (3.13) and minding the energy separation property (3.6), (3.7) of the Newton-Wigner eigenstates

$$\langle V_k | e^{iq_a x_a} | U_{k'} \rangle = \left\langle V_k^{(NW)} \right| e^{-\Omega_u} e^{iq_a x_a} e^{\Omega_u} \left| U_{k'}^{(NW)} \right\rangle \tag{3.22}$$

$$= \int d^3r \sum_{\substack{\nu \in \{3,4\} \\ \nu' \in \{1,2\}}} V_\nu^{\star(NW)}(\mathbf{r},k)\, e^{i\mathbf{q}\cdot\mathbf{r}} \frac{\tilde{q}}{2} n_b\, (\beta\alpha_b)_{\nu,\nu'}\, U_{\nu'}^{(NW)}(\mathbf{r},k')\,.$$

Likewise, insertion of the expansion (3.14) leads eventually to

$$\langle V_k | \alpha_b e^{iq_a x_a} | U_{k'} \rangle = \left\langle V_k^{(NW)} \right| e^{-\Omega_u} \alpha_b e^{iq_a x_a} e^{\Omega_u} \left| U_{k'}^{(NW)} \right\rangle \tag{3.23}$$

$$= \int d^3r \sum_{\substack{\nu \in \{3,4\} \\ \nu' \in \{1,2\}}} V_\nu^{\star(NW)}(\mathbf{r},k)\, e^{i\mathbf{q}\cdot\mathbf{r}} \begin{pmatrix} \alpha_b \; + \\ -\frac{\tilde{q}^2}{8} n_b n_a \alpha_a \; + \\ -\frac{1}{4}\left(\tilde{q} n_a \tilde{\Pi}_b + \tilde{q} n_b \tilde{\Pi}_a + \tilde{\Pi}_b \tilde{\Pi}_a + \tilde{\Pi}_a \tilde{\Pi}_b \right) \alpha_a \; + \\ + \frac{i}{4} \varepsilon_{bb'b''}\, \tilde{q} n_{b'} \tilde{\Pi}_{b''} \gamma_5 \end{pmatrix}_{\nu,\nu'} U_{\nu'}^{(NW)}(\mathbf{r},k')\,.$$

3.3 The Renormalized Schrödinger-Pauli Hamiltonian

In order to establish the low energy representation of the *one* particle Hamiltonian in the perturbation expansion of the unitary transformed QED-Hamiltonian (3.1), it proofs expedient to rewrite the one-particle term $\hat{\mathcal{H}}_D$ in (3.2)–as introduced in [1, Eq. (9)] and rewritten in [1, Eq. (21)]–now in the guise

$$
\begin{aligned}
&\hat{\mathcal{H}}_D + \sum_{k \in I_-} |E_k| \,\hat{1}_4 \qquad (3.24)\\
&= \sum_{k \in I_+} |E_k| \,\hat{c}_k^\dagger \hat{c}_k + \sum_{k \in I_-} |E_k| \,\hat{b}_k^\dagger \hat{b}_k \\
&= \sum_{k,k' \in I_+} \langle U_k | \,\mathsf{H}^{(D)} \,| U_{k'} \rangle \,\hat{c}_k^\dagger \hat{c}_{k'} + \sum_{k,k' \in I_-} \langle V_k | - \mathsf{H}^{(D)} \,| V_{k'} \rangle \,\hat{b}_{k'}^\dagger \hat{b}_k \\
&= \sum_{k,k' \in I_+} \left\langle U_k^{(NW)} \middle| \mathsf{H}^{(NW)} \middle| U_{k'}^{(NW)} \right\rangle \hat{c}_k^\dagger \hat{c}_{k'} + \sum_{k,k' \in I_-} \left\langle V_k^{(NW)} \middle| - \mathsf{H}^{(NW)} \middle| V_{k'}^{(NW)} \right\rangle \hat{b}_{k'}^\dagger \hat{b}_k \\
&\equiv \hat{\mathcal{H}}^{(e)} + \hat{\mathcal{H}}^{(p)}.
\end{aligned}
$$

With the energy-separation property, see (3.6) and (3.7), the summation over the mode indices $k, k' \in I_\pm$ in (3.24) thus reduces straight to the respective linear combinations of Dirac fermion mode creation and destruction operators defining the field operators already introduced in [1, Eq. (69)], namely for electrons

$$
\begin{aligned}
\hat{\psi}_1(\mathbf{r}) &= \sum_{k \in I_+} U_1^{(NW)}(\mathbf{r}, k)\, \hat{c}_k \qquad (3.25)\\
\hat{\psi}_2(\mathbf{r}) &= \sum_{k \in I_+} U_2^{(NW)}(\mathbf{r}, k)\, \hat{c}_k,
\end{aligned}
$$

whereas for positrons

$$
\begin{aligned}
\hat{\chi}_1^\dagger(\mathbf{r}) &= \sum_{k \in I_-} V_3^{(NW)}(\mathbf{r}; k)\, \hat{b}_k^\dagger \qquad (3.26)\\
\hat{\chi}_2^\dagger(\mathbf{r}) &= \sum_{k \in I_-} V_4^{(NW)}(\mathbf{r}; k)\, \hat{b}_k^\dagger.
\end{aligned}
$$

In point of fact, these linear combinations of fermion mode creation and destruction operators satisfy along with their hermitean adjoints the familiar anti-commutation relations of *fermion field* operators. Notably these field operators, having just two components $s \in \{1, 2\}$, comprehensively describe the spin-$\frac{1}{2}$ fermion quantum fields of electrons and positrons as being truly *dientangled* entities.

So we obtain directly from (3.24)

$$\hat{\mathcal{H}}^{(e)} = \int d^3r \sum_{s,s'\in\{1,2\}} \hat{\psi}_s^{\dagger}(\mathbf{r}) \left(\mathsf{H}^{(NW)}\right)_{s,s'} \hat{\psi}_{s'}(\mathbf{r}) \tag{3.27}$$

$$\hat{\mathcal{H}}^{(p)} = \int d^3r \sum_{s,s'\in\{3,4\}} \left(\left(-\mathsf{H}^{(NW)}\right)_{s',s} \hat{\chi}_s^{\dagger}(\mathbf{r})\right) \hat{\chi}_{s'}(\mathbf{r}). \tag{3.28}$$

Note that due to (3.26) the order of indices s, s' in the positron Hamiltonian $\hat{\mathcal{H}}^{(p)}$ is reversed. Eventually an expansion up to the leading order relativistic corrections gives, see for instance [1, 6, 7]

$$\begin{aligned} \mathsf{H}^{(NW)} &\equiv e^{-\Omega_u}\mathsf{H}^{(D)}e^{\Omega_u} \\ &= \begin{cases} mc^2\beta + \frac{(\alpha_b\Pi_b)^2}{2m}\beta + e\Phi^{(\text{ext})} \\ -\frac{1}{2mc^2}\left(\frac{(\alpha_b\Pi_b)^2}{2m}\right)^2\beta - \frac{1}{8}\left[\frac{\alpha_b\Pi_b}{mc}, \left[\frac{\alpha_{b'}\Pi_{b'}}{mc}, e\Phi^{(ext)}\right]\right] + \dots \end{cases} \end{aligned} \tag{3.29}$$

Writing

$$\mathbf{E}^{(\text{ext})} = -\nabla\Phi^{(\text{ext})} \quad , \quad \mathbf{B}^{(\text{ext})} = \mathbf{rot}\,\mathsf{A}^{(\text{ext})}$$

and minding the multiplication rule

$$\sigma_b\sigma_{b'} = \delta_{b,b'}1_{4\times4} + i\varepsilon_{bb'b''}\sigma_{b''} \quad \text{for } b, b', b'' \in \{x, y, z\}$$

one finds with $\mu_B = \frac{e\hbar}{2m}$ the Bohr magneton and $\lambda_C = \frac{\hbar}{mc}$ the Compton wavelength

$$\frac{(\alpha_b\Pi_b)^2}{2m} = \frac{\Pi_b\Pi_b}{2m}1_{4\times4} - \mu_B B_a^{(\text{ext})}\sigma_a \tag{3.30}$$

$$\left[\frac{\alpha_b\Pi_b}{mc}, \left[\frac{\alpha_{b'}\Pi_{b'}}{mc}, e\Phi^{(\text{ext})}\right]\right] = e\lambda_C^2\left(-\nabla^2\Phi^{(\text{ext})}\right)1_{4\times4} + \frac{2e\lambda_C}{mc}\left(\mathbf{E}^{(\text{ext})} \wedge \Pi\right)_b\sigma_b. \tag{3.31}$$

The first and the second term in (3.31) are the Darwin contact and the spin-orbit interaction. Note the prefactors of these terms here being in a ratio 1 to 2.

So, the first three terms in (3.29) comprise the rest energy and the common Schrödinger-Pauli Hamiltonian of nonrelativistic quantum mechanics in the presence of external electrostatic and magnetostatic background fields, while the last two terms stated represent the energy scale of fine structure (of relevance to the spectrum of the hydrogen atom), in fact this fine structure originating from two distinct roots, namely the relativistic correction to the kinetic energy and the double commutator (3.31) comprising the Darwin contact and the spin-orbit interaction.

Insertion of the expansion (3.29) into the electron Hamiltonian (3.27) , minding the definition of the electron field operators (3.25) and taking into account the eigenvalue property $\beta\left|U_k^{(NW)}\right\rangle = \left|U_k^{(NW)}\right\rangle$, see (3.16), eventually leads within the intended accuracy $O(\bar{v}^3)$ to a second quantized Hamiltonian for the nonrelativistic electrons

$$\hat{\mathcal{H}}^{(e)} = \int d^3r \sum_{s,s'\in\{1,2\}} \hat{\psi}_s^\dagger(\mathbf{r}) \begin{pmatrix} \left(mc^2 + \frac{\mathsf{D}_b\mathsf{D}_b}{2m}\left(1 - \frac{1}{2mc^2}\frac{\mathsf{D}_b\mathsf{D}_b}{2m}\right) + e\Phi^{(\mathrm{ext})}\right)\delta_{s,s'} + \\ -\mu_B B_a^{(\mathrm{ext})}(\sigma_a^{(P)})_{s,s'} + \\ -\frac{e}{8}\lambda_C^2\left(-\nabla^2\Phi^{(\mathrm{ext})}\right)\delta_{s,s'} + \\ -\frac{e}{4}\lambda_C\frac{1}{mc}\epsilon_{abb'}E_a^{(\mathrm{ext})}\mathsf{D}_b\left(\sigma_{b'}^{(P)}\right)_{s,s'} \end{pmatrix} \hat{\psi}_{s'}(\mathbf{r}) \tag{3.32}$$

with D_a defined in (3.19) and $\sigma_a^{(P)}$ the 2 × 2-Pauli matrices.

An expedient partial integration in the positron Hamiltonian (3.28) replaces

$$\left(\mathsf{D}_a\mathsf{D}_b...\hat{\chi}_{s'}^\dagger(\mathbf{r})\right)\hat{\chi}_s(\mathbf{r}) \rightarrow \hat{\chi}_{s'}^\dagger(\mathbf{r})\left(...\mathsf{D}_b^\star\mathsf{D}_a^\star\right)\hat{\chi}_s(\mathbf{r})$$
$$\mathsf{D}_b^\star = -\left(\frac{\hbar}{i}\frac{\partial}{\partial r_b} + e\mathsf{A}_b^{(\mathrm{ext})}(\mathbf{r})\right).$$

Then insertion of the expansion (3.29) into the positron Hamiltonian (3.28) , minding the definition of the positron field operators (3.26) and taking into account the eigenvalue property $\beta\left|V_k^{(NW)}\right\rangle = -\left|V_k^{(NW)}\right\rangle$, see (3.16), eventually leads to a second quantized (mirrored) Hamiltonian for the nonrelativistic positrons in the guise

$$\hat{\mathcal{H}}^{(p)} = \int d^3r \sum_{s,s'\in\{1,2\}} \hat{\chi}_{s'}^\dagger(\mathbf{r}) \begin{pmatrix} \left(mc^2 + \frac{\mathsf{D}_b^\star\mathsf{D}_b^\star}{2m}\left(1 - \frac{1}{2mc^2}\frac{\mathsf{D}_b^\star\mathsf{D}_b^\star}{2m}\right) - e\Phi^{(\mathrm{ext})}\right)\delta_{s,s'} \\ +\mu_B B_a^{(\mathrm{ext})}(\sigma_a^{(P)})_{s,s'} \\ +\frac{e}{8}\lambda_C^2\left(-\nabla^2\Phi^{(\mathrm{ext})}\right)\delta_{s,s'} + \frac{e}{4}\lambda_C\frac{1}{mc}\epsilon_{abb'}E_a^{(\mathrm{ext})}\mathsf{D}_b^\star(\sigma_{b'}^{(P)})_{s,s'} \end{pmatrix} \hat{\chi}_s(\mathbf{r}). \tag{3.33}$$

Conforming with a special case of charge conjugation symmetry the positron Hamiltonian is connected to the electron Hamiltonian by replacing charge $e \rightarrow -e$, concurrently as well reversing the *order* of indices s, s' of the Pauli 2 × 2-matrices, whereby $\left(\sigma_a^{(P)}\right)_{s,s'} = \left(i\sigma_y^{(P)}\sigma_a^{(P)}i\sigma_y^{(P)}\right)_{s',s}$.

To account for the key effects of QED as regards the *electrons*, the associated self-energy terms $\hat{\mathcal{M}}^{(e)}$ need to be included, that we elaborated afore in our chapter [5, Eq. (177)]). Here as well a replacement $|U_{k'}\rangle \rightarrow \left|U_{k'}^{(\mathrm{NW})}\right\rangle$ is made in the matrix elements defining the self-energy $\hat{\mathcal{M}}^{(e)}$ in [5, Eq. (177)]), which approximation should be unobjectionable in front of a factor $\frac{\alpha_{FS}}{\pi}$ anyway, as we restrict the wave function of the particles to the low energy sector of QED, where their motion is governed by the Newton-Wigner Hamiltonian $\mathsf{H}^{(\mathrm{NW})}$ expanded so that it comprises as well the lowest order relativistic corrections, see (3.29). This way the *renormalized* low energy Hamiltonian of nonrelativistic electrons is eventually obtained as

$$\hat{\mathcal{H}}^{(e,\mathrm{ren})} = \hat{\mathcal{H}}^{(e)} + \hat{\mathcal{M}}^{(e)} \tag{3.34}$$
$$= \int d^3r \sum_{s,s'\in\{1,2\}} \hat{\psi}_s^\dagger(\mathbf{r}) \begin{pmatrix} (mc^2 + \frac{\mathsf{D}_b\mathsf{D}_b}{2m}\left(1 - \frac{1}{2mc^2}\frac{\mathsf{D}_b\mathsf{D}_b}{2m}\right) + e\Phi^{(\mathrm{ext})})\delta_{s,s'} \\ -\mu_B\left(1 + \frac{\alpha_{FS}}{2\pi}\right) B_a^{(\mathrm{ext})}(\sigma_a^{(P)})_{s,s'} \\ -\frac{1}{8}\left(1 + \frac{\alpha_{FS}}{\pi}\frac{8}{3}\left(\frac{19}{30} + \ln\frac{1}{2\tilde{M}_{k,k'}}\right)\right)e\lambda_C^2\left(-\nabla^2\Phi^{(\mathrm{ext})}\right)\delta_{s,s'} \\ -\frac{1}{4}\left(1 + \frac{\alpha_{FS}}{\pi}\right)e\lambda_C\frac{1}{mc}\epsilon_{abb'}E_a^{(\mathrm{ext})}\mathsf{D}_b\ (\sigma_{b'})_{s,s'} \end{pmatrix} \hat{\psi}_{s'}(\mathbf{r})\,.$$

Notably, the prefactors of the Darwin term $-\nabla^2\Phi^{(\mathrm{ext})}$ and the spin-orbit term $\epsilon_{abb'}E_a^{(\mathrm{ext})}\mathsf{D}_b\ (\sigma_{b'})_{s,s'}$ are in (3.34) not anymore in a ratio 1 to 2 like in Schrödinger-Pauli quantum mechanics, in point of fact this (tiny) discrepancy constituting a famous key result of Quantum Electrodynamics. The correction terms proportional to $\frac{\alpha_{FS}}{\pi}$ enlisted in (3.34) indeed engender, that the g-factor of the electron is enhanced so that $g - 2 = \frac{\alpha_{FS}}{\pi}$, a result first disclosed by Schwinger in [13] and rederived by him with other methods in [14]. Further these correction terms provoke–identifying now the external potential $\Phi^{(\mathrm{ext})}$ with the Coulomb field of the atomic nucleus–that the energy difference between $2p_{3/2}$ and $2p_{1/2}$ states in the hydrogen atom actually differ from the energy difference as deduced from the spectrum calculated with the bare, i.e. not renormalized Dirac equation. Along with this goes a lift of the degeneracy of the $2s_{1/2}$ and the $2p_{1/2}$ energy levels in the hydrogen atom too, as has been observed in the pioneering experiments of Lamb and Retherford. For an insightful review see [15] and also [16].

We desist writing here the corresponding renormalized Hamiltonian of positrons, as it can be obtained readily from (3.34) replacing charge $e \to -e$, $\mu_B \to -\mu_B$, $\sigma_a^{(P)} \to i\sigma_y^{(P)}\sigma_a^{(P)}i\sigma_y^{(P)}$ and in addition, of course, writing field operators $\hat{\chi}_s(\mathbf{r})$ instead of $\hat{\psi}_s(\mathbf{r})$.

Compliant with Schrödinger-Pauli quantum mechanics the derived renormalized Hamiltonian (3.34) thus governs in the presence of external electromagnetic background fields the dynamics of nonrelativistic electrons as particles with individual attributes mass, charge and spin.

3.4 The Nonrelativistic Matter-Light Interaction

Insertion of the expansions (3.18) and (3.21) into the term in $\hat{\mathcal{H}}_{\mathrm{low}}^{(\mathrm{rad})}$ describing the matter-light interaction as stated afore in [1], and minding the Coulomb gauge constraint

$$\sum_{b} q_b \left(\hat{\mathsf{A}}^{(-)}_{b,\mathbf{q}} + \hat{\mathsf{A}}^{(+)}_{b,\mathbf{q}} \right) \equiv 0, \tag{3.35}$$

one readily confirms the relevant operator describing emission and absorption of low energy photons by slowly moving electrons or else positrons being

$$\hat{\mathcal{H}}^{(\text{e,rad})}_{\text{low}} = -\frac{e}{m} \int d^3r \left(\begin{array}{c} \hat{\mathfrak{A}}_b(\mathbf{r}) \left(\sum\limits_{s\in\{1,2\}} \hat{\psi}^\dagger_s(\mathbf{r}) \left(\frac{\hbar}{i} \nabla_b - e\mathsf{A}^{(\text{ext})}_b(\mathbf{r}) \right) \hat{\psi}_s(\mathbf{r}) \right) \\ +\varepsilon_{a'ba''} \frac{\partial \hat{\mathfrak{A}}_b(\mathbf{r})}{\partial r_{a'}} \left(\sum\limits_{s,s'\in\{1,2\}} \hat{\psi}^\dagger_s(\mathbf{r}) \left(\frac{\hbar}{2} \sigma^{(P)}_{a''} \right)_{s,s'} \hat{\psi}_{s'}(\mathbf{r}) \right) \end{array} \right) \tag{3.36}$$

$$\hat{\mathcal{H}}^{(\text{p,rad})}_{\text{low}} = +\frac{e}{m} \int d^3r \left(\begin{array}{c} \hat{\mathfrak{A}}_b(\mathbf{r}) \left(\sum\limits_{s\in\{1,2\}} \hat{\chi}^\dagger_s(\mathbf{r}) \left(\frac{\hbar}{i} \nabla_b + e\mathsf{A}^{(\text{ext})}_b(\mathbf{r}) \right) \hat{\chi}_s(\mathbf{r}) \right) \\ -\varepsilon_{a'ba''} \frac{\partial \hat{\mathfrak{A}}_b(\mathbf{r})}{\partial r_{a'}} \left(\sum\limits_{s,s'\in\{1,2\}} \hat{\chi}^\dagger_{s'}(\mathbf{r}) \left(\frac{\hbar}{2} \sigma^{(P)}_{a''} \right)_{s,s'} \hat{\chi}_s(\mathbf{r}) \right) \end{array} \right), \tag{3.37}$$

whereby $\hat{\psi}_s(\mathbf{r})$ is the *two-component* fermion field operators for *electrons* (3.25) and $\hat{\chi}_s(\mathbf{r})$ is the *two-component* fermion field operators for *positrons* (3.26). The linear combination of Photon creation and destruction operators in (3.36) and (3.37), as given by

$$\hat{\mathfrak{A}}_b(\mathbf{r}) \equiv \frac{1}{\sqrt{V_P}} \sum_{q<q_B} \sum_{\nu\in\{I,II\}} \tilde{A}_b(\mathbf{q},\nu) \left(e^{i\mathbf{q}\cdot\mathbf{r}} \hat{a}_{\mathbf{q},\nu} + e^{-i\mathbf{q}\cdot\mathbf{r}} \hat{a}^\dagger_{\mathbf{q},\nu} \right), \tag{3.38}$$

comprises the *long wavelength* part of the electromagnetic field operator introduced in [1, Eq. (26)] whereby $q_B = \frac{2\pi}{a_B}$ with a_B the Bohr radius.

In terms of the familiar paramagnetic and diamagnetic current density operators for nonrelativistic electrons

$$\hat{j}^{(e,\text{para})}_a(\mathbf{r}) = \frac{e}{2m} \frac{\hbar}{i} \sum_{s\in\{1,2\}} \left(\hat{\psi}^\dagger_s(\mathbf{r}) \frac{\partial}{\partial r_a} \hat{\psi}_s(\mathbf{r}) - \left(\frac{\partial}{\partial r_a} \hat{\psi}^\dagger_s(\mathbf{r}) \right) \hat{\psi}_s(\mathbf{r}) \right) \tag{3.39}$$

$$\hat{j}^{(e,\text{dia})}_a(\mathbf{r}) = -\frac{e^2}{m} \left(\sum_{s\in\{1,2\}} \hat{\psi}^\dagger_s(\mathbf{r}) \hat{\psi}_s(\mathbf{r}) \right) A^{(\text{ext})}_a(\mathbf{r}), \tag{3.40}$$

and in like manner for nonrelativistic positrons

$$\hat{j}^{(p,\text{para})}_a(\mathbf{r}) = -\frac{e}{2m} \frac{\hbar}{i} \sum_{s\in\{1,2\}} \left(\hat{\chi}^\dagger_s(\mathbf{r}) \frac{\partial}{\partial r_a} \hat{\chi}_s(\mathbf{r}) - \left(\frac{\partial}{\partial r_a} \hat{\chi}^\dagger_s(\mathbf{r}) \right) \hat{\chi}_s(\mathbf{r}) \right)$$

$$\hat{j}^{(p,\text{dia})}_a(\mathbf{r}) = -\frac{e^2}{m} \left(\sum_{s\in\{1,2\}} \hat{\chi}^\dagger_s(\mathbf{r}) \hat{\chi}_s(\mathbf{r}) \right) A^{(\text{ext})}_a(\mathbf{r}).$$

Also let us introduce the (magnetization) spin current density operators

$$j_a^{(e,\text{spin})}(\mathbf{r}) = \frac{e}{m}\varepsilon_{aa'a''}\frac{\partial}{\partial r_{a'}}\hat{S}_{a''}^{(e)}(\mathbf{r}) \tag{3.41}$$

$$j_a^{(p,\text{spin})}(\mathbf{r}) = \frac{-e}{m}\varepsilon_{aa'a''}\frac{\partial}{\partial r_{a'}}\hat{S}_{a''}^{(p)}(\mathbf{r})$$

formed with the respective spin-density operators for electrons (e) and positrons (p)

$$\hat{S}_a^{(e)}(\mathbf{r}) = \sum_{s,s'\in\{1,2\}} \hat{\psi}_s^\dagger(\mathbf{r})\left(\frac{\hbar}{2}\sigma_a^{(P)}\right)_{s,s'}\hat{\psi}_{s'}(\mathbf{r}) \tag{3.42}$$

$$\hat{S}_a^{(p)}(\mathbf{r}) = \sum_{s,s'\in\{1,2\}} \hat{\chi}_{s'}^\dagger(\mathbf{r})\left(\sigma_y^{(P)}\frac{\hbar}{2}\sigma_a^{(P)}\sigma_y^{(P)}\right)_{s',s}\hat{\chi}_s(\mathbf{r})\,. \tag{3.43}$$

Then one finds after a partial integration of the derivatives $\frac{\partial\hat{\mathfrak{A}}_b(\mathbf{r})}{\partial r_{a'}}$ in (3.36), (3.37) eventually the familiar matter-light interaction operators describing emission and absorption processes of (low energy) photons with electrons (e) respectively positrons (p), indeed now in their standard guise customary in nonrelativistic many body physics, see for instance [17–20]

$$\hat{\mathcal{H}}_{int}^{(\text{e,rad})} = -\int d^3r\left(\hat{j}_b^{(\text{e,para})}(\mathbf{r}) + \hat{j}_b^{(\text{e,dia})}(\mathbf{r}) + j_b^{(\text{e,spin})}(\mathbf{r})\right)\hat{\mathfrak{A}}_b(\mathbf{r}) \tag{3.44}$$

$$\hat{\mathcal{H}}_{int}^{(\text{p,rad})} = -\int d^3r\left(\hat{j}_b^{(\text{p,para})}(\mathbf{r}) + \hat{j}_b^{(\text{p,dia})}(\mathbf{r}) + j_b^{(\text{p,spin})}(\mathbf{r})\right)\hat{\mathfrak{A}}_b(\mathbf{r})\,.$$

If one aims here at a better accuracy beyond order $O\left(\bar{v}^3\right)$, one needs to include in the expansion of the generator Ω_u of the Eriksen transformation in (3.10), besides the first order term $\Omega_u^{(1)}$ now as well the third order term $\Omega_u^{(3)}$, the latter now comprising also the effect of the external electrostatic field $\mathcal{E}_b^{\text{ext}} = -\nabla_b\Phi^{(\text{ext})}$. For details see in [7].

3.5 The Effective Electron-Electron Interactions

In what follows the effective two-particle interactions enlisted in (3.4) are evaluated with the restriction the electrons (or else positrons) all move at a nonrelativistic speed. In this case second order gradients applied to the NW-Eigenfunctions in front of a Dirac matrix α_b may be safely discarded, as for a low excitation energy of the particles their eigenfunctions display on the scale of the Compton wavelength λ_C only slow variations. So, upon insertion of the expansion of the matrix elements (3.17) occurring in the effective electron-electron interaction terms obtained previously in [1, Eq. (150)] it is found

$$\begin{aligned}
\hat{\mathcal{U}}_C^{(e,e)} &\equiv \frac{e^2}{2\varepsilon_0} \sum_{k,k',K,K' \in I_+} \int \frac{d^3q}{(2\pi)^3} \frac{1}{|\mathbf{q}|^2} \langle U_k | e^{-iq_a x_a} | U_{k'} \rangle \langle U_K | e^{iq_a x_a} | U_{K'} \rangle \hat{c}_k^\dagger \hat{c}_K^\dagger \hat{c}_{K'} \hat{c}_{k'} \\
&= \frac{e^2}{2\varepsilon_0} \int d^3r \int d^3r' \sum_{k,k',K,K' \in I_+} \hat{c}_k^\dagger \hat{c}_K^\dagger \hat{c}_{K'} \hat{c}_{k'} \int \frac{d^3q}{(2\pi)^3} \frac{e^{iq_a \cdot (r'_a - r_a)}}{|\mathbf{q}|^2} \times \\
&\times \left(\begin{array}{c}
\left(1 - \frac{\tilde{q}^2}{4}\right) \sum\limits_{\nu,\nu' \in \{1,2\}} U_\nu^{\star(NW)}(\mathbf{r},k) U_\nu^{(NW)}(\mathbf{r},k') U_{\nu'}^{\star(NW)}(\mathbf{r}',K) U_{\nu'}^{(NW)}(\mathbf{r}',K') + \\
- \sum\limits_{\nu,\nu',\nu'' \in \{1,2\}} \frac{i}{4} \varepsilon_{bb'b''} \tilde{q} n_b U_\nu^{\star(NW)}(\mathbf{r},k) (\sigma_{b''})_{\nu,\nu'} \tilde{\mathsf{D}}_{b'} U_{\nu'}^{(NW)}(\mathbf{r},k') \times U_{\nu''}^{\star(NW)}(\mathbf{r}',K) U_{\nu''}^{(NW)}(\mathbf{r}',K') + \\
+ \sum\limits_{\nu,\nu',\nu'' \in \{1,2\}} U_{\nu''}^{\star(NW)}(\mathbf{r},k) U_{\nu''}^{(NW)}(\mathbf{r},k') \times \frac{i}{4} \varepsilon_{aa'a''} \tilde{q} n_a U_\nu^{\star(NW)}(\mathbf{r}',K) (\sigma_{a''})_{\nu,\nu'} \tilde{\mathsf{D}}'_{a'} U_{\nu'}^{(NW)}(\mathbf{r}',K')
\end{array} \right).
\end{aligned} \tag{3.45}$$

As regards the photon mediated effective electron-electron interaction terms derived in [1, Eq. (153)], it is given by

$$\begin{aligned}
\hat{\mathcal{U}}_\perp^{(e,e)} &= mc^2 \left(\frac{e}{mc}\right)^2 \sum_{k,k' \in I_+} \sum_{K,K' \in I_+} \int \frac{d^3q}{(2\pi)^3} \frac{\hbar}{2\varepsilon_0 \omega(\mathbf{q})} (\delta_{b,b'} - n_b n_{b'}) \times \\
&\times \frac{(\tilde{E}_k - \tilde{E}_{k'} - \tilde{\omega}_q) + (\tilde{E}_{K'} - \tilde{E}_K - \tilde{\omega}_q)}{(\tilde{E}_k - \tilde{E}_{k'} - \tilde{\omega}_q)^2 + (\tilde{E}_{K'} - \tilde{E}_K - \tilde{\omega}_q)^2} \langle U_k | \alpha_b e^{iq_a x_a} | U_{k'} \rangle \langle U_K | \alpha_{b'} e^{-iq_a x_a} | U_{K'} \rangle \hat{c}_k^\dagger \hat{c}_K^\dagger \hat{c}_{K'} \hat{c}_{k'}.
\end{aligned} \tag{3.46}$$

A simplification is enabled here, because the excitation energies of the electrons are restricted to be markedly below their rest energy mc^2. Defining

$$X \equiv \tilde{E}_k - \tilde{E}_{k'} - \tilde{\omega}_q \text{ and } Y \equiv \tilde{E}_{K'} - \tilde{E}_K - \tilde{\omega}_q$$

there holds as an identity

$$\frac{X+Y}{X^2+Y^2} = 2\frac{X+Y}{(X+Y)^2 + (X-Y)^2}. \tag{3.47}$$

Minding $\tilde{\omega}_q \equiv \lambda_C \cdot |\mathbf{q}| = \frac{\hbar}{mc} |\mathbf{q}| \equiv |\tilde{\mathbf{q}}|$ and in view of only major photon wavenumbers $|\mathbf{q}| \geq \frac{2\pi}{a_B}$ being relevant in the integration over wave vectors $\mathbf{q}$, then

$$(X-Y)^2 \ll (X+Y)^2, \tag{3.48}$$

thus enabling the approximation

$$\frac{X+Y}{X^2+Y^2} \approx \frac{2}{X+Y}, \tag{3.49}$$

or else

$$\frac{(\tilde{E}_k - \tilde{E}_{k'} - \tilde{\omega}_q) + (\tilde{E}_{K'} - \tilde{E}_K - \tilde{\omega}_q)}{(\tilde{E}_k - \tilde{E}_{k'} - \tilde{\omega}_q)^2 + (\tilde{E}_{K'} - \tilde{E}_K - \tilde{\omega}_q)^2} \simeq \frac{2}{\tilde{E}_k - \tilde{E}_{k'} + \tilde{E}_{K'} - \tilde{E}_K - 2\tilde{\omega}_q} \simeq -\frac{1}{|\tilde{\mathbf{q}}|}. \tag{3.50}$$

Upon insertion of the expansions (3.18) into (3.46) and using the identities

$$\delta_{b,b'} - n_b n_{b'} = \left(\delta_{b,b''} - n_b n_{b''}\right)\left(\delta_{b'',b'} - n_{b''} n_{b'}\right) \quad (3.51)$$
$$\varepsilon_{ba'a''}\,\varepsilon_{bb'b''}\, n_{a'} n_{b'} (\sigma_{a''})_{\mu,\mu'} (\sigma_{b''})_{\nu,\nu'} = \left(\delta_{a'',b''} - n_{a''} n_{b''}\right)(\sigma_{a''})_{\mu,\mu'}\,(\sigma_{b''})_{\nu,\nu'}$$

one obtains within the range of validity of the approximation (3.50)

$$\hat{\mathcal{U}}_{\perp}^{(e,e)} = -\frac{e^2}{2\varepsilon_0}\sum_{k,k'}\sum_{K,K'} \hat{c}_k^\dagger \hat{c}_K^\dagger \hat{c}_{K'} \hat{c}_{k'} \int d^3r \int d^3r' \times \quad (3.52)$$
$$\times \left(\begin{array}{c} \sum\limits_{\nu,\nu'\in\{1,2\}} U_\nu^{\star(NW)}(\mathbf{r},k)\, U_{\nu'}^{\star(NW)}(\mathbf{r}',K)\, \frac{\mathsf{D}_b}{mc} \times \\ \times \int \frac{d^3q}{(2\pi)^3}\frac{1}{|\mathbf{q}|^2} e^{i\mathbf{q}\cdot(\mathbf{r}-\mathbf{r}')}\left(\delta_{b,b'} - n_b n_{b'}\right) \times \frac{\mathsf{D}'_{b'}}{mc} U_{\nu'}^{(NW)}(\mathbf{r}',K')\, U_\nu^{(NW)}(\mathbf{r},k') + \\ + \sum\limits_{\mu,\mu',\nu,\nu'\in\{1,2\}} U_\mu^{\star(NW)}(\mathbf{r},k)\,(\sigma_a)_{\mu,\mu'}\, U_{\mu'}^{(NW)}(\mathbf{r},k') \times \\ \times \left(\frac{\hbar}{2mc}\right)^2 \int \frac{d^3q}{(2\pi)^3} e^{i\mathbf{q}\cdot(\mathbf{r}-\mathbf{r}')}\left(\delta_{a,b} - n_a n_b\right) U_\nu^{\star(NW)}(\mathbf{r}',K')\,(\sigma_b)_{\nu,\nu'}\, U_{\nu'}^{(NW)}(\mathbf{r}',K) + \\ + \sum\limits_{\nu,\nu',\nu''\in\{1,2\}} U_\nu^{\star(NW)}(\mathbf{r},k) \int \frac{d^3q}{(2\pi)^3}\frac{e^{i\mathbf{q}\cdot(\mathbf{r}-\mathbf{r}')}}{|\mathbf{q}|^2}(iq_{b'})\frac{\hbar}{2mc}\varepsilon_{bb'b''}\,(\sigma_{b''})_{\nu,\nu'}\, U_{\nu'}^{(NW)}(\mathbf{r},k') \times \\ \times U_{\nu''}^{\star(NW)}(\mathbf{r}',K)\,\frac{\mathsf{D}'_b}{mc}\, U_{\nu''}^{(NW)}(\mathbf{r}',K') + \\ + \sum\limits_{\nu,\nu',\nu''\in\{1,2\}} U_{\nu''}^{\star(NW)}(\mathbf{r},k)\,\frac{\mathsf{D}_b}{mc}\, U_{\nu''}^{(NW)}(\mathbf{r},k') \times \\ \times U_\nu^{\star(NW)}(\mathbf{r}',K) \int \frac{d^3q}{(2\pi)^3}\frac{e^{i\mathbf{q}\cdot(\mathbf{r}-\mathbf{r}')}}{|\mathbf{q}|^2}(-iq_{b'})\frac{\hbar}{2mc}\varepsilon_{bb'b''}\,(\sigma_{b''})_{\nu,\nu'}\, U_{\nu'}^{(NW)}(\mathbf{r}',K') \end{array} \right).$$

Evaluation of the Fourier integrals in (3.45) gives

$$\int \frac{d^3q}{(2\pi)^3}\frac{e^{iq_a\cdot(r'_a-r_a)}}{|\mathbf{q}|^2} = \frac{1}{4\pi\,|\mathbf{r}'-\mathbf{r}|} \quad (3.53)$$
$$\int \frac{d^3q}{(2\pi)^3}\frac{e^{iq_a\cdot(r'_a-r_a)}}{|\mathbf{q}|^2} iq_b = \nabla'_b\frac{1}{4\pi\,|\mathbf{r}'-\mathbf{r}|} = -\frac{r'_b-r_b}{4\pi\,|\mathbf{r}'-\mathbf{r}|^3}$$
$$\int \frac{d^3q}{(2\pi)^3} e^{iq_a\cdot(r'_a-r_a)} = \delta^{(3)}\left(\mathbf{r}-\mathbf{r}'\right), \quad (3.54)$$

whereas the evaluation of the Fourier integrals in (3.52) is more elaborate. A very careful deduction is presented in [2, Complement A_I], where it is shown that

$$\int \frac{d^3q}{(2\pi)^3}\frac{1}{|\mathbf{q}|^2} e^{i\mathbf{q}\cdot(\mathbf{r}'-\mathbf{r})}\left(\delta_{a,b} - n_a n_b\right) = \frac{1}{8\pi}\left(\frac{\delta_{a,b}}{|\mathbf{r}'-\mathbf{r}|} + \frac{(r'_a-r_a)(r'_b-r_b)}{|\mathbf{r}'-\mathbf{r}|^3}\right) \quad (3.55)$$
$$\int \frac{d^3q}{(2\pi)^3} e^{i\mathbf{q}\cdot(\mathbf{r}'-\mathbf{r})}\left(\delta_{a,b} - n_a n_b\right) = \frac{1}{4\pi}\left(\frac{8}{3}\pi\,\delta^{(3)}(\mathbf{r}'-\mathbf{r})\,\delta_{a,b} + \frac{3(r'_a-r_a)(r'_b-r_b) - |\mathbf{r}'-\mathbf{r}|^2\delta_{a,b}}{|\mathbf{r}'-\mathbf{r}|^5}\right).$$

Upon insertion of these Fourier integrals into (3.45), (3.52) and keeping in (3.45) and (3.52) the order of the creation and annihilation operators in tact meticulously, then the summation over the fermion-mode indices k, k', K, K' leads at once to the afore introduced *two component* field operators (3.25). This way there results the searched for effective electron-electron interactions in the guise

$$\hat{\mathcal{U}}_C^{(e,e)} = \frac{e^2}{2\varepsilon_0}\int d^3r \int d^3r' \times \tag{3.56}$$
$$\times \left(\begin{array}{c} \sum_{s,s'\in\{+,-\}} \hat{\psi}_s^\dagger(\mathbf{r})\,\hat{\psi}_{s'}^\dagger(\mathbf{r}')\,\frac{1}{4\pi|\mathbf{r}'-\mathbf{r}|}\,\hat{\psi}_{s'}(\mathbf{r}')\,\hat{\psi}_s(\mathbf{r}) \\ -\frac{1}{4}\left(\frac{\hbar}{mc}\right)^2 \sum_{s,s'\in\{+,-\}} \hat{\psi}_s^\dagger(\mathbf{r})\,\hat{\psi}_{s'}^\dagger(\mathbf{r}')\,\delta^{(3)}(\mathbf{r}-\mathbf{r}')\,\hat{\psi}_{s'}(\mathbf{r}')\,\hat{\psi}_s(\mathbf{r}) \\ +\sum_{s,\bar{s},s'\in\{+,-\}} \hat{\psi}_s^\dagger(\mathbf{r})\,\hat{\psi}_{s'}^\dagger(\mathbf{r}')\,\varepsilon_{bb'b''}\frac{1}{mc}\left(-\frac{r_b-r_b'}{4\pi|\mathbf{r}-\mathbf{r}'|^3}\right)\hat{\psi}_{s'}(\mathbf{r}')\left(\frac{\hbar}{2}\sigma_{b''}^{(P)}\right)_{s,\bar{s}}\frac{\mathsf{D}_{b'}}{mc}\hat{\psi}_{\bar{s}}(\mathbf{r}), \end{array} \right)$$

and also

$$\hat{\mathcal{U}}_\perp^{(e,e)} = -\frac{e^2}{2\varepsilon_0}\int d^3r \int d^3r' \times \tag{3.57}$$
$$\times \left(\begin{array}{c} \sum_{s,s'\in\{+,-\}} \hat{\psi}_s^\dagger(\mathbf{r})\,\hat{\psi}_{s'}^\dagger(\mathbf{r}')\,\frac{\mathsf{D}_a}{mc}\frac{1}{8\pi}\left(\frac{\delta_{a,b}}{|\mathbf{r}-\mathbf{r}'|}+\frac{(r_a-r_a')(r_b-r_b')}{|\mathbf{r}-\mathbf{r}'|^3}\right)\frac{\mathsf{D}_b'}{mc}\,\hat{\psi}_{s'}(\mathbf{r}')\,\hat{\psi}_s(\mathbf{r})+ \\ +\left(\frac{1}{mc}\right)^2\frac{1}{4\pi}\left(\frac{8}{3}\pi\delta^{(3)}(\mathbf{r}-\mathbf{r}')\,\delta_{a,b}+\frac{3(r_a-r_a')(r_b-r_b')-|\mathbf{r}-\mathbf{r}'|^2\delta_{a,b}}{|\mathbf{r}-\mathbf{r}'|^5}\right)\times \\ \times\sum_{s,\bar{s},s',\bar{s}'\in\{+,-\}}\left(\frac{\hbar}{2}\sigma_b^{(P)}\right)_{s',\bar{s}'}\left(\frac{\hbar}{2}\sigma_a^{(P)}\right)_{s,\bar{s}}\hat{\psi}_s^\dagger(\mathbf{r})\,\hat{\psi}_{s'}^\dagger(\mathbf{r}')\,\hat{\psi}_{\bar{s}'}(\mathbf{r}')\,\hat{\psi}_{\bar{s}}(\mathbf{r})+ \\ +\sum_{s,\bar{s},s'\in\{+,-\}}\hat{\psi}_s^\dagger(\mathbf{r})\,\hat{\psi}_{s'}^\dagger(\mathbf{r}')\,\frac{1}{mc}\left(\frac{r_b-r_b'}{2\pi|\mathbf{r}-\mathbf{r}'|^3}\right)\varepsilon_{bb'b''}\,\frac{\mathsf{D}_{b'}'}{mc}\,\hat{\psi}_{s'}(\mathbf{r}')\left(\frac{\hbar}{2}\sigma_{b''}^{(P)}\right)_{s,\bar{s}}\hat{\psi}_{\bar{s}}(\mathbf{r}) \end{array} \right).$$

Actually, the third lines in (3.56) and in (3.58) arise consolidating the third and fourth lines in (3.45) and in (3.52) with the substitution $(\mathbf{q},\mathbf{r},\mathbf{r}') \to (-\mathbf{q},\mathbf{r}',\mathbf{r})$.

Finally, upon retranslation of (3.56) and (3.58) back from *second* quantization to *first* quantization, well known expressions for the effective electron-electron interactions emerge in standard guise as

$$\mathcal{U}_C^{(e,e)} = \frac{1}{2}\sum_{j\neq j'}^{N_e}\left(\begin{array}{c} \frac{e^2}{4\pi\varepsilon_0\left|\mathbf{r}^{(j)}-\mathbf{r}^{(j')}\right|}+ \\ -\frac{e^2}{4\pi\varepsilon_0}\left(\frac{\hbar}{mc}\right)^2\pi\,\delta^{(3)}\left(\mathbf{r}^{(j)}-\mathbf{r}^{(j')}\right)+ \\ -\left(\frac{1}{mc}\right)^2\left(\frac{\hbar}{2}\boldsymbol{\sigma}^{(j)}\right)\cdot\left(\frac{e^2}{4\pi\varepsilon_0}\frac{\mathbf{r}^{(j)}-\mathbf{r}^{(j')}}{\left|\mathbf{r}^{(j)}-\mathbf{r}^{(j')}\right|^3}\wedge\boldsymbol{\Pi}^{(j)}\right) \end{array} \right) \tag{3.58}$$

and

$$\mathcal{U}_{\perp}^{(e,e)} = \frac{1}{2}\sum_{j\neq j'}^{N_e}\left(\begin{array}{c} \frac{\Pi_a^{(j)}}{mc}\left(-\frac{e^2}{4\pi\varepsilon_0}\right)\frac{1}{2}\left(\frac{\delta_{a,b}}{\left|\mathbf{r}^{(j)}-\mathbf{r}^{(j')}\right|}+\frac{\left(r_a^{(j)}-r_a^{(j')}\right)\left(r_b^{(j)}-r_b^{(j')}\right)}{\left|\mathbf{r}^{(j)}-\mathbf{r}^{(j')}\right|^3}\right)\frac{\Pi_b^{(j')}}{mc}+ \\ +\left(-\frac{e^2}{4\pi\varepsilon_0}\right)\left(\frac{1}{mc}\right)^2\left(\frac{\hbar}{2}\sigma_a^{(j)}\right)\left(\begin{array}{c}\frac{8}{3}\pi\delta^{(3)}\left(\mathbf{r}^{(j)}-\mathbf{r}^{(j')}\right)\delta_{a,b}+ \\ +\frac{3\left(r_a^{(j)}-r_a^{(j')}\right)\left(r_b^{(j)}-r_b^{(j')}\right)-3\left|\mathbf{r}^{(j)}-\mathbf{r}^{(j')}\right|^2\delta_{a,b}}{\left|\mathbf{r}^{(j)}-\mathbf{r}^{(j')}\right|^5}\end{array}\right)\left(\frac{\hbar}{2}\sigma_b^{(j')}\right)+ \\ -\left(\frac{1}{mc}\right)^2\left(\frac{\hbar}{2}\boldsymbol{\sigma}^{(j)}\right)\cdot\left(\frac{e^2}{2\pi\varepsilon_0}\frac{\mathbf{r}^{(j')}-\mathbf{r}^{(j)}}{\left|\mathbf{r}^{(j')}-\mathbf{r}^{(j)}\right|^3}\wedge\boldsymbol{\Pi}^{(j')}\right)\end{array}\right). \tag{3.59}$$

Noteably, all the interaction terms in (3.58) and in (3.59) comprise *gauge invariant* derivatives only: $\Pi_b^{(j)} = \mathsf{p}_b^{(j)} - e\mathsf{A}_b\left(\mathbf{r}^{(j)}\right)$.

The leading order term in (3.58) is the mutual *Coulomb* repulsion between electrons. Thereto adds the (two-body) *Darwin* contact potential term in the second line and in the third line of (3.58) one recognizes the *spin-orbit* interaction of an electron (j) moving in the Coulomb electric field created by all other electrons $(j' \neq j)$.

In the photon mediated effective interaction terms (3.59), the first line represents the *orbit-orbit* interaction, the second line is the *magnetic dipole-dipole* interaction between the electrons and the third line in (3.59) is the *other-spin-orbit* interaction.

As a matter of fact, (3.58) and (3.59) *together* comprise the effective electron-electron interactions of the *Breit-Pauli* Hamiltonian well known in the relativistic many-electron theory, as utilized, e.g. in the field of computational atomic structure see [21, chapter 7].

With regard to the contact potentials occuring in (3.58) and in (3.59), these are in view of the magnitude of their prefactors, after evaluation of the integrals, in effect just a small correction to the predominant Coulomb interaction. Notably, comparing (3.58) with the result obtained by Bethe and Salpeter (BS) in [22, Eq. (42.1)], a discrepancy exists. Instead of the two-body *Darwin* contact potential term occuring in the second line of (3.58), there superseds now in BS [22, Eq. (42.1)] a non-hermitian interaction term H_4. Notwithstanding the above, our results (3.58) and (3.59) agree completely with the results obtained by Itoh [8, Eqs. (2a)–(2g)], who (to the best of our knowledge) first called attention to that deficiency and explained as well the underlying cause.

Incidentally, the authors of reference [21] claim in the introduction to their chapter 7, that in order to include relativistic effects in a *rigorous* way the Dirac equation needs to be solved for a many-electron system. As a matter of fact, there exists *no* relativistic *many-electron* Dirac equation! There exists instead the not really covariant approach introduced by G. Breit [23] to obtain the leading order relativistic corrections to the mutual Coulomb interaction between electrons, by bringing with a physical argumentation the current-current interactions between moving electrons in as well. A lucent summary of the genesis of the Breit Hamiltonian and a discussion of its deficiencies can be found in [22, Sect. 38]. As regards applications in quantum chemistry these are reviewed in [24].

In closing let us reiterate, that a rigorous justification of the Breit-Pauli Hamiltonian for a plurality of electrons in any case requires a derivation from quantum electrodynamics.

3.6 The Effective Electron-Positron Interactions

In analogy to the derivation of the effective *electron-electron* interactions obtained afore in (3.58) and (3.59), we derive now the effective interactions between *electrons* and *positrons*. Starting point is the effective electron-positron interaction obtained previously in [1, Eq. (152)], respectively in [1, Eq. (155)]. Minding $\tilde{E}_k > 0$ for $k \in I_+$ but $\tilde{E}_K < 0$ for $k \in I_-$, we have

$$\hat{\mathcal{U}}_C^{(e,p)} = \frac{e^2}{\varepsilon_0} \sum_{k,k' \in I_+} \sum_{K,K' \in I_-} \int \frac{d^3q}{(2\pi)^3} \frac{1}{|\mathbf{q}|^2} \begin{pmatrix} -\langle U_k | e^{-iq_a \mathrm{x}_a} | U_{k'} \rangle \langle V_K | e^{iq_a \mathrm{x}_a} | V_{K'} \rangle + \\ + \langle U_k | e^{-iq_a \mathrm{x}_a} | V_{K'} \rangle \langle V_K | e^{iq_a \mathrm{x}_a} | U_{k'} \rangle \end{pmatrix} \hat{c}_k^\dagger \hat{c}_{k'} \hat{b}_{K'}^\dagger \hat{b}_K , \tag{3.60}$$

and what's more

$$\begin{aligned} \hat{\mathcal{U}}_\perp^{(e,p)} = \frac{e^2}{m} \sum_{k,k' \in I_+} \sum_{K,K' \in I_-} \int \frac{d^3q}{(2\pi)^3} \frac{\hbar}{2\varepsilon_0 \omega(\mathbf{q})} \left(\delta_{b.b'} - \frac{q_b q_{b'}}{|\mathbf{q}|^2} \right) \hat{c}_k^\dagger \hat{c}_{k'} \hat{b}_{K'}^\dagger \hat{b}_K \times \\ \times \left(\begin{array}{l} \left(\begin{array}{l} -\frac{(\tilde{E}_k - \tilde{E}_{k'} - \tilde{\omega}_q) + (|\tilde{E}_K| - |\tilde{E}_{K'}| - \tilde{\omega}_q)}{(\tilde{E}_k - \tilde{E}_{k'} - \tilde{\omega}_q)^2 + (|\tilde{E}_K| - |\tilde{E}_{K'}| - \tilde{\omega}_q)^2} + \\ + \frac{(|\tilde{E}_K| - |\tilde{E}_{K'}| + \tilde{\omega}_q) + (\tilde{E}_k - \tilde{E}_{k'} + \tilde{\omega}_q)}{(|\tilde{E}_K| - |\tilde{E}_{K'}| + \tilde{\omega}_q)^2 + (\tilde{E}_k - \tilde{E}_{k'} + \tilde{\omega}_q)^2} \end{array} \right) \langle U_k | \alpha_b e^{iq_a \mathrm{x}_a} | U_{k'} \rangle \langle V_K | \alpha_{b'} e^{-iq_a \mathrm{x}_a} | V_{K'} \rangle + \\ + \left(\begin{array}{l} \frac{(\tilde{E}_k + |\tilde{E}_{K'}| - \tilde{\omega}_q) + (|\tilde{E}_K| + \tilde{E}_{k'} - \tilde{\omega}_q)}{(\tilde{E}_k + |\tilde{E}_{K'}| - \tilde{\omega}_q)^2 + (|\tilde{E}_K| + \tilde{E}_{k'} - \tilde{\omega}_q)^2} + \\ - \frac{(|\tilde{E}_K| + \tilde{E}_{k'} + \tilde{\omega}_q) + (\tilde{E}_k + |\tilde{E}_{K'}| + \tilde{\omega}_q)}{(|\tilde{E}_K| + \tilde{E}_{k'} + \tilde{\omega}_q)^2 + (\tilde{E}_k + |\tilde{E}_{K'}| + \tilde{\omega}_q)^2} \end{array} \right) \langle U_k | \alpha_b e^{iq_a \mathrm{x}_a} | V_{K'} \rangle \langle V_K | \alpha_{b'} e^{-iq_a \mathrm{x}_a} | U_{k'} \rangle \end{array} \right) . \end{aligned} \tag{3.61}$$

As electrons and positrons are here assumed being (nearly) nonrelativistic, their excitation energy differences, $E_k - E_{k'}$ and $|E_K| - |E_{K'}|$, are certainly restricted to values much smaller than their rest energy mc^2, and so one finds, proceeding as before in (3.49), the valid approximation

$$-\frac{(\tilde{E}_k - \tilde{E}_{k'} - \tilde{\omega}_q) + (|\tilde{E}_K| - |\tilde{E}_{K'}| - \tilde{\omega}_q)}{(\tilde{E}_k - \tilde{E}_{k'} - \tilde{\omega}_q)^2 + (|\tilde{E}_K| - |\tilde{E}_{K'}| - \tilde{\omega}_q)^2} + \frac{(|\tilde{E}_K| - |\tilde{E}_{K'}| + \tilde{\omega}_q) + (\tilde{E}_k - \tilde{E}_{k'} + \tilde{\omega}_q)}{(|\tilde{E}_K| - |\tilde{E}_{K'}| + \tilde{\omega}_q)^2 + (\tilde{E}_k - \tilde{E}_{k'} + \tilde{\omega}_q)^2} \simeq \frac{2}{|\tilde{\mathbf{q}}|} . \tag{3.62}$$

Furthermore, because the matrix elements occuring in the second line of (3.61) differ for (nearly) nonrelativistic particle states noticeably from zero only for photon wavenumbers $\mathbf{q}$ obeying to the constraint $|\mathbf{q}| \ll \frac{mc}{\hbar}$, one can approximate, nota bene in front of a factor $\langle U_k | \alpha_b e^{iq_a \mathrm{x}_a} | V_{K'} \rangle \langle V_K | \alpha_{b'} e^{-iq_a \mathrm{x}_a} | U_{k'} \rangle$, then as well

$$\begin{aligned} &\frac{\left(\tilde{E}_k+\left|\tilde{E}_{K'}\right|-\tilde{\omega}_q\right)+\left(\left|\tilde{E}_K\right|+\tilde{E}_{k'}-\tilde{\omega}_q\right)}{\left(\tilde{E}_k+\left|\tilde{E}_{K'}\right|-\tilde{\omega}_q\right)^2+\left(\left|\tilde{E}_K\right|+\tilde{E}_{k'}-\tilde{\omega}_q\right)^2} - \frac{\left(\left|\tilde{E}_K\right|+\tilde{E}_{k'}+\tilde{\omega}_q\right)+\left(\tilde{E}_k+\left|\tilde{E}_{K'}\right|+\tilde{\omega}_q\right)}{\left(\left|\tilde{E}_K\right|+\tilde{E}_{k'}+\tilde{\omega}_q\right)^2+\left(\tilde{E}_k+\left|\tilde{E}_{K'}\right|+\tilde{\omega}_q\right)^2} \\ &\simeq \frac{2}{\tilde{E}_k+\left|\tilde{E}_{K'}\right|+\left|\tilde{E}_K\right|+\tilde{E}_{k'}-2\tilde{\omega}_q} - \frac{2}{\tilde{E}_k+\left|\tilde{E}_{K'}\right|+\left|\tilde{E}_K\right|+\tilde{E}_{k'}+2\tilde{\omega}_q} \simeq \frac{1}{2-|\tilde{\mathbf{q}}|} - \frac{1}{2+|\tilde{\mathbf{q}}|} \simeq \frac{|\tilde{\mathbf{q}}|}{2} \end{aligned} \quad (3.63)$$

This applies, because only small wave vectors $\tilde{\mathbf{q}} \equiv \frac{\hbar}{mc}\mathbf{q} \ll 1$ contribute most to (3.61). The pertaining photon mediated interaction term $\hat{\mathcal{U}}_{\perp}^{(e,p)}$ in (3.61) thus assumes with the approximations (3.62) and (3.63) eventually the guise

$$\begin{aligned} \hat{\mathcal{U}}_{\perp}^{(e,p)} &= mc^2 \cdot \alpha_{FS} \cdot 4\pi \sum_{k,k' \in I_+} \sum_{K,K' \in I_-} \int \frac{d^3\tilde{q}}{(2\pi)^3} \frac{1}{2\,|\tilde{\mathbf{q}}|} \left(\delta_{b,b'} - n_b n_{b'}\right) \times \\ &\times \begin{pmatrix} \frac{2}{|\tilde{\mathbf{q}}|} \langle U_k | \alpha_b e^{iq_a \mathrm{x}_a} | U_{k'} \rangle \langle V_K | \alpha_{b'} e^{-iq_a \mathrm{x}_a} | V_{K'} \rangle + \\ + \frac{|\tilde{\mathbf{q}}|}{2} \langle U_k | \alpha_b e^{iq_a \mathrm{x}_a} | V_{K'} \rangle \langle V_K | \alpha_{b'} e^{-iq_a \mathrm{x}_a} | U_{k'} \rangle \end{pmatrix} \hat{c}_k^\dagger \hat{c}_{k'} \hat{b}_{K'}^\dagger \hat{b}_K . \end{aligned} \quad (3.64)$$

Up next we approximate the matrix elements that shape the electron-positron interaction terms $\hat{\mathcal{U}}_C^{(e,p)}$ in (3.60) and the photon mediated electron-positron interaction terms $\hat{\mathcal{U}}_{\perp}^{(e,p)}$ in (3.64). Insertion of the expansions (3.17), (3.20) and (3.22) into (3.60) leads to

$$\begin{aligned} \hat{\mathcal{U}}_C^{(e,p)} &= mc^2 \cdot \alpha_{FS} \int d^3r \int d^3r' \sum_{k,k' \in I_+} \sum_{K,K' \in I_-} \hat{c}_k^\dagger \hat{c}_{k'} \hat{b}_{K'}^\dagger \hat{b}_K \cdot 4\pi \int \frac{d^3\tilde{q}}{(2\pi)^3} \frac{e^{iq_a \cdot (r'_a - r_a)}}{|\tilde{\mathbf{q}}|^2} \times \\ &\times \begin{pmatrix} -\left(1 - \frac{|\tilde{\mathbf{q}}|^2}{4}\right) \sum_{\substack{\nu \in \{1,2\} \\ \nu' \in \{3,4\}}} U_\nu^{\star(NW)}(\mathbf{r},k) U_\nu^{(NW)}(\mathbf{r},k') V_{\nu'}^{(NW)}(\mathbf{r}',K') V_{\nu'}^{\star(NW)}(\mathbf{r}',K) + \\ - \sum_{\substack{\nu,\nu' \in \{1,2\} \\ \nu'' \in \{3,4\}}} U_\nu^{\star(NW)}(\mathbf{r},k) \left(-i\,|\tilde{\mathbf{q}}|\, n_b \epsilon_{bb'b''} \tilde{\mathsf{D}}_{b'} \sigma_{b''}\right)_{\nu,\nu'} U_{\nu'}^{(NW)}(\mathbf{r},k') V_{\nu''}^{(NW)}(\mathbf{r}',K') V_{\nu''}^{\star(NW)}(\mathbf{r}',K) + \\ - \sum_{\substack{\nu,\nu' \in \{3,4\} \\ \nu'' \in \{1,2\}}} U_{\nu''}^{\star(NW)}(\mathbf{r},k) U_{\nu''}^{(NW)}(\mathbf{r},k') V_{\nu'}^{(NW)}(\mathbf{r}',K') \left(-i\,|\tilde{\mathbf{q}}|\, n_b \epsilon_{bb'b''} \left(-\tilde{\mathsf{D}}'^{\star}_{b'}\right) \sigma_{b''}\right)_{\nu,\nu'} V_\nu^{\star(NW)}(\mathbf{r}',K) + \\ + \sum_{\substack{\mu,\mu' \in \{1,2\} \\ \nu,\nu' \in \{3,4\}}} \frac{|\tilde{\mathbf{q}}|^2}{4} n_b n_{b'} U_\mu^{\star(NW)}(\mathbf{r},k) U_{\mu'}^{(NW)}(\mathbf{r}',k') V_\nu^{(NW)}(\mathbf{r},K') V_{\nu'}^{\star(NW)}(\mathbf{r}',K) (\alpha_b)_{\mu,\nu} (\alpha_{b'})_{\nu',\mu'} \end{pmatrix} \end{aligned} \quad (3.65)$$

and insertion of the expansions (3.18), (3.21) and (3.23) into (3.64) leads to

$$
\begin{aligned}
\hat{\mathcal{U}}_{\perp}^{(e,p)} = mc^2 \cdot \alpha_{FS} \int d^3r \int d^3r' \sum_{k,k' \in I_+} \sum_{K,K' \in I_-} \hat{c}_k^{\dagger} \hat{c}_{k'} \hat{b}_{K'}^{\dagger} \hat{b}_K \cdot 4\pi \int \frac{d^3\tilde{q}}{(2\pi)^3} \frac{e^{iqa\cdot(r'_a - ra)}}{2\,|\tilde{\mathbf{q}}|} \left(\delta_{b.b'} - n_b n_{b'}\right) \times \\
\times \left(
\begin{array}{l}
\frac{2}{|\tilde{\mathbf{q}}|} \sum\limits_{\substack{\nu \in \{1,2\} \\ \nu' \in \{3,4\}}} U_{\nu}^{\star(NW)}(\mathbf{r},k) \tilde{\mathsf{D}}_b U_{\nu}^{(NW)}(\mathbf{r},k') V_{\nu'}^{(NW)}(\mathbf{r}',K') \left(-\tilde{\mathsf{D}}'^{\star}_{b'}\right) V_{\nu'}^{\star(NW)}(\mathbf{r}',K) + \\
+ \frac{2}{|\tilde{\mathbf{q}}|} \sum\limits_{\substack{\mu \in \{1,2\} \\ \nu,\nu' \in \{3,4\}}} U_{\mu}^{\star(NW)}(\mathbf{r},k) \tilde{\mathsf{D}}_b U_{\mu}^{(NW)}(\mathbf{r},k') V_{\nu'}^{(NW)}(\mathbf{r}',K') (-i \tfrac{|\tilde{\mathbf{q}}|}{2} n_{b'}) \epsilon_{bb'b''} (\sigma_{b''})_{\nu,\nu'} V_{\nu}^{\star(NW)}(\mathbf{r}',K) + \\
+ \frac{2}{|\tilde{\mathbf{q}}|} \sum\limits_{\substack{\mu,\mu' \in \{1,2\} \\ \nu \in \{3,4\}}} U_{\mu}^{\star(NW)}(\mathbf{r},k) (-i \tfrac{|\tilde{\mathbf{q}}|}{2} n_{b'}) \epsilon_{bb'b''} (\sigma_{b''})_{\mu,\mu'} U_{\mu'}^{(NW)}(\mathbf{r},k') V_{\nu}^{(NW)}(\mathbf{r}',K') \left(-\tilde{\mathsf{D}}_b^{\star}\right) V_{\nu}^{\star(NW)}(\mathbf{r}',K) + \\
+ \frac{|\tilde{\mathbf{q}}|}{2} \sum\limits_{\substack{\mu,\mu' \in \{1,2\} \\ \nu,\nu' \in \{3,4\}}} U_{\mu}^{\star(NW)}(\mathbf{r},k) U_{\mu'}^{(NW)}(\mathbf{r}',k') V_{\nu}^{(NW)}(\mathbf{r},K') V_{\nu'}^{\star(NW)}(\mathbf{r}',K) (\alpha_b)_{\mu,\nu} (\alpha_{b'})_{\nu',\mu'}
\end{array}
\right).
\end{aligned}
\tag{3.66}
$$

For consistency, in view of an already small factor $\frac{|\tilde{\mathbf{q}}|}{2}$ in the last line of (3.66), the expressions resulting from the insertion of the expansion enlisted in (3.23) have been truncated keeping just the leading order term $(\alpha_b)_{\mu,\nu}(\alpha_{b'})_{\nu',\mu'}$.

Recasting next the interaction term (3.65) it follows from what has been said

$$
\begin{aligned}
\hat{\mathcal{U}}_{C}^{(e,p)} = mc^2 \cdot \alpha_{FS} \int d^3r \int d^3r' 4\pi \int \frac{d^3\tilde{q}}{(2\pi)^3} \frac{e^{iqa\cdot(r'_a - ra)}}{|\tilde{\mathbf{q}}|^2} \times \\
\times \left(
\begin{array}{l}
-\left(1 - \frac{|\tilde{\mathbf{q}}|^2}{4}\right) \sum\limits_{\nu \in \{1,2\}} \left(\sum\limits_{k \in I_+} U_{\nu}^{\star(NW)}(\mathbf{r},k)\, \hat{c}_k^{\dagger} \right) \left(\sum\limits_{k' \in I_+} U_{\nu}^{(NW)}(\mathbf{r},k')\, \hat{c}_{k'} \right) \times \\
\times \sum\limits_{\nu' \in \{3,4\}} \left(\sum\limits_{K' \in I_-} V_{\nu'}^{(NW)}(\mathbf{r}',K')\, \hat{b}_{K'}^{\dagger} \right) \left(\sum\limits_{K \in I_-} V_{\nu'}^{\star(NW)}(\mathbf{r}',K)\, \hat{b}_K \right) + \\
- \sum\limits_{\nu,\nu' \in \{1,2\}} \left(\sum\limits_{k \in I_+} U_{\nu}^{\star(NW)}(\mathbf{r},k)\, \hat{c}_k^{\dagger} \right) \left(-i\, |\tilde{\mathbf{q}}|\, n_b \epsilon_{bb'b''} \tilde{\mathsf{D}}_{b'} \sigma_{b''} \right)_{\nu,\nu'} \left(\sum\limits_{k' \in I_+} U_{\nu'}^{(NW)}(\mathbf{r},k')\, \hat{c}_{k'} \right) \times \\
\times \sum\limits_{\nu'' \in \{3,4\}} \left(\sum\limits_{K' \in I_-} V_{\nu''}^{(NW)}(\mathbf{r}',K')\, \hat{b}_{K'}^{\dagger} \right) \left(\sum\limits_{K \in I_-} V_{\nu''}^{\star(NW)}(\mathbf{r}',K)\, \hat{b}_K \right) + \\
- \sum\limits_{\nu'' \in \{1,2\}} \left(\sum\limits_{k \in I_+} U_{\nu''}^{\star(NW)}(\mathbf{r},k)\, \hat{c}_k^{\dagger} \right) \left(\sum\limits_{k' \in I_+} U_{\nu''}^{(NW)}(\mathbf{r},k')\, \hat{c}_{k'} \right) \times \\
\times \sum\limits_{\nu,\nu' \in \{3,4\}} \left(\sum\limits_{K' \in I_-} V_{\nu'}^{(NW)}(\mathbf{r}',K')\, \hat{b}_{K'}^{\dagger} \right) \left(-i\, |\tilde{\mathbf{q}}|\, n_b \epsilon_{bb'b''} \left(-\tilde{\mathsf{D}}'^{\star}_{b'}\right) \sigma_{b''} \right)_{\nu,\nu'} \left(\sum\limits_{K \in I_-} V_{\nu}^{\star(NW)}(\mathbf{r}',K)\, \hat{b}_K \right) + \\
+ \frac{|\tilde{\mathbf{q}}|^2}{4} n_b n_{b'} \sum\limits_{\mu,\mu' \in \{1,2\}} \sum\limits_{\nu,\nu' \in \{3,4\}} (\alpha_b)_{\mu,\nu} (\alpha_{b'})_{\nu',\mu'} \times \\
\times \left(\sum\limits_{k \in I_+} U_{\mu}^{\star(NW)}(\mathbf{r},k) \hat{c}_k^{\dagger} \right) \left(\sum\limits_{k' \in I_+} U_{\mu'}^{(NW)}(\mathbf{r}',k') \hat{c}_{k'} \right) \left(\sum\limits_{K' \in I_-} V_{\nu}^{(NW)}(\mathbf{r},K') \hat{b}_{K'}^{\dagger} \right) \left(\sum\limits_{K \in I_-} V_{\nu'}^{\star(NW)}(\mathbf{r}',K) \hat{b}_K \right) \times
\end{array}
\right)
\end{aligned}
\tag{3.67}
$$

With the two-component electron and positron field operators (3.25) and (3.26) one can rewrite the interaction term (3.67) by replacing there the 4×4-Dirac matrices σ_b and α_b now by 2×2-Pauli matrices $\sigma_b^{(P)}$, this way obtaining

$$\hat{\mathcal{U}}_C^{(e,p)} = mc^2 \cdot \alpha_{FS} \int d^3r \int d^3r' \, 4\pi \int \frac{d^3\tilde{q}}{(2\pi)^3} \frac{e^{iq_a\cdot(r'_a - r_a)}}{|\tilde{\mathbf{q}}|^2} \times$$
$$\times \left(\begin{array}{c} -\left(1-\frac{|\tilde{\mathbf{q}}|^2}{4}\right) \sum\limits_{s,\tilde{s}\in\{+,-\}} \hat{\psi}_s^\dagger(\mathbf{r})\hat{\psi}_s(\mathbf{r})\hat{\chi}_{\tilde{s}}^\dagger(\mathbf{r}')\hat{\chi}_{\tilde{s}}(\mathbf{r}') + \\ - \sum\limits_{s,s'\in\{+,-\}} \hat{\psi}_s^\dagger(\mathbf{r})(-i\,|\tilde{\mathbf{q}}|\, n_b \epsilon_{bb'b''} \tilde{\mathsf{D}}_{b'} \sigma_{b''}^{(P)})_{s,s'} \hat{\psi}_{s'}(\mathbf{r}) \sum\limits_{\tilde{s}\in\{+,-\}} \hat{\chi}_{\tilde{s}}^\dagger(\mathbf{r}')\hat{\chi}_{\tilde{s}}(\mathbf{r}') + \\ - \sum\limits_{s\in\{+,-\}} \hat{\psi}_s^\dagger(\mathbf{r})\hat{\psi}_s(\mathbf{r}) \sum\limits_{\tilde{s},\tilde{s}'\in\{+,-\}} \hat{\chi}_{\tilde{s}'}^\dagger(\mathbf{r}') \left(-i\,|\tilde{\mathbf{q}}|\, n_b \epsilon_{bb'b''}\left(-\tilde{\mathsf{D}}'^{\star}_{b'}\right)\sigma_{b''}^{(P)}\right)_{\tilde{s},\tilde{s}'} \hat{\chi}_{\tilde{s}}(\mathbf{r}') + \\ + \frac{|\tilde{\mathbf{q}}|^2}{4} n_b n_{b'} \sum\limits_{\substack{s,\tilde{s}\in\{+,-\}\\ s',\tilde{s}'\in\{+,-\}}} (\sigma_b^{(P)})_{s,\tilde{s}} (\sigma_{b'}^{(P)})_{\tilde{s}',s'} \hat{\psi}_s^\dagger(\mathbf{r})\hat{\psi}_{s'}(\mathbf{r}')\hat{\chi}_{\tilde{s}}^\dagger(\mathbf{r})\,\hat{\chi}_{\tilde{s}'}(\mathbf{r}') \end{array} \right) \tag{3.68}$$

In like manner rewriting the interaction term (3.66) gives

$$\hat{\mathcal{U}}_\perp^{(e,p)} = mc^2 \cdot \alpha_{FS} \int d^3r \int d^3r' \, 4\pi \int \frac{d^3\tilde{q}}{(2\pi)^3} \frac{e^{iq_a\cdot(r'_a - r_a)}}{2\,|\tilde{\mathbf{q}}|} (\delta_{b,b'} - n_b n_{b'}) \times$$
$$\times \left(\begin{array}{c} \frac{2}{|\tilde{\mathbf{q}}|} \sum\limits_{s,\tilde{s}\in\{+,-\}} \hat{\psi}_s^\dagger(\mathbf{r})\tilde{\mathsf{D}}_b\hat{\psi}_s(\mathbf{r})\; \hat{\chi}_{\tilde{s}}^\dagger(\mathbf{r}')\left(-\tilde{\mathsf{D}}^{\star}_{b'}\right)'\hat{\chi}_{\tilde{s}}(\mathbf{r}') + \\ + \frac{2}{|\tilde{\mathbf{q}}|} \sum\limits_{s\in\{+,-\}} \hat{\psi}_s^\dagger(\mathbf{r})\tilde{\mathsf{D}}_b\hat{\psi}_s(\mathbf{r}) \sum\limits_{\tilde{s},\tilde{s}'\in\{+,-\}} \hat{\chi}_{\tilde{s}'}^\dagger(\mathbf{r}')(-i\frac{|\tilde{\mathbf{q}}|}{2} n_{b'})\epsilon_{bb'b''}(\sigma_{b''}^{(P)})_{\tilde{s}',\tilde{s}}\hat{\chi}_{\tilde{s}'}(\mathbf{r}') + \\ + \frac{2}{|\tilde{\mathbf{q}}|} \sum\limits_{s,s'\in\{+,-\}} \hat{\psi}_s^\dagger(\mathbf{r})(-i\frac{|\tilde{\mathbf{q}}|}{2} n_{b'})\epsilon_{bb'b''}(\sigma_{b''}^{(P)})_{s,s'}\hat{\psi}_{s'}(\mathbf{r}) \sum\limits_{\tilde{s}\in\{+,-\}} \hat{\chi}_{\tilde{s}}^\dagger(\mathbf{r}')\left(-\tilde{\mathsf{D}}^{\star}_{b}\right)'\hat{\chi}_{\tilde{s}}(\mathbf{r}') + \\ + \frac{|\tilde{\mathbf{q}}|}{2} \sum\limits_{\substack{s,\tilde{s}\in\{+,-\}\\ s',\tilde{s}'\in\{+,-\}}} (\sigma_b^{(P)})_{s,\tilde{s}} (\sigma_{b'}^{(P)})_{\tilde{s}',s'} \hat{\psi}_s^\dagger(\mathbf{r})\hat{\psi}_{s'}(\mathbf{r}')\hat{\chi}_{\tilde{s}}^\dagger(\mathbf{r})\,\hat{\chi}_{\tilde{s}'}(\mathbf{r}') \end{array} \right) \tag{3.69}$$

Upon insertion of the Fourier integrals (3.53) and (3.55) and bringing the units in, the effective electron-positron interactions (3.68) and (3.69) eventually assume the guise

$$
\hat{\mathcal{U}}_C^{(e,p)} = \frac{e^2}{\varepsilon_0} \int d^3r \int d^3r' \times
$$
$$
\times \left(\begin{array}{c}
\sum\limits_{s,\tilde{s}\in\{+,-\}} \hat{\psi}_s^\dagger(\mathbf{r})\hat{\psi}_s(\mathbf{r}) \left(-\frac{1}{4\pi|\mathbf{r}'-\mathbf{r}|}\right) \hat{\chi}_{\tilde{s}}^\dagger(\mathbf{r}')\hat{\chi}_{\tilde{s}}(\mathbf{r}') + \\
+ \sum\limits_{s,\tilde{s}\in\{+,-\}} \hat{\psi}_s^\dagger(\mathbf{r})\hat{\psi}_s(\mathbf{r})(\frac{\hbar}{2mc})^2 \delta^{(3)}(\mathbf{r}-\mathbf{r}')\, \hat{\chi}_{\tilde{s}}^\dagger(\mathbf{r}')\hat{\chi}_{\tilde{s}}(\mathbf{r}') + \\
+\frac{1}{2mc} \sum\limits_{s,s'\in\{+,-\}} \hat{\psi}_s^\dagger(\mathbf{r}) \frac{r_b-r'_b}{4\pi|\mathbf{r}'-\mathbf{r}|^3} \epsilon_{bb'b''} \frac{\mathsf{D}_{b'}}{mc} (\frac{\hbar}{2}\sigma_{b''}^{(P)})_{s,s'} \hat{\psi}_{s'}(\mathbf{r}) \sum\limits_{\tilde{s}\in\{+,-\}} \hat{\chi}_{\tilde{s}}^\dagger(\mathbf{r}')\hat{\chi}_{\tilde{s}}(\mathbf{r}') + \\
+\frac{1}{2mc} \sum\limits_{s\in\{+,-\}} \hat{\psi}_s^\dagger(\mathbf{r})\hat{\psi}_s(\mathbf{r}) \sum\limits_{\tilde{s},\tilde{s}'\in\{+,-\}} \hat{\chi}_{\tilde{s}'}^\dagger(\mathbf{r}')(\frac{r_b-r'_b}{4\pi|\mathbf{r}'-\mathbf{r}|^3} \epsilon_{bb'b''} \frac{\left(-\mathsf{D}_{b'}^\star\right)'}{mc} (\frac{\hbar}{2}\sigma_{b''}^{(P)})_{\tilde{s},\tilde{s}'}) \hat{\chi}_{\tilde{s}}(\mathbf{r}') + \\
+(\frac{\hbar}{2mc})^2 \int \frac{d^3q}{(2\pi)^3} e^{iq_a\cdot(r'_a-r_a)} n_b n_{b'} \sum\limits_{\substack{s,\tilde{s}\in\{+,-\}\\ s',\tilde{s}'\in\{+,-\}}} (\sigma_b^{(P)})_{s,\tilde{s}} (\sigma_{b'}^{(P)})_{\tilde{s}',s'} \hat{\psi}_s^\dagger(\mathbf{r})\hat{\psi}_{s'}(\mathbf{r}')\hat{\chi}_{\tilde{s}}^\dagger(\mathbf{r})\; \hat{\chi}_{\tilde{s}'}(\mathbf{r}')
\end{array} \right) \tag{3.70}
$$

and also

$$
\hat{\mathcal{U}}_\perp^{(e,p)} = \frac{e^2}{\varepsilon_0} \int d^3r \int d^3r' \times
$$
$$
\times \left(\begin{array}{c}
\sum\limits_{s,\tilde{s}\in\{+,-\}} \hat{\psi}_s^\dagger(\mathbf{r}) \frac{\mathsf{D}_b}{mc} \hat{\psi}_s(\mathbf{r}) \frac{1}{8\pi} \left(\frac{\delta_{b,b'}}{|\mathbf{r}-\mathbf{r}'|} + \frac{(r_b-r'_b)(r_{b'}-r'_{b'})}{|\mathbf{r}-\mathbf{r}'|^3} \right) \hat{\chi}_{\tilde{s}}^\dagger(\mathbf{r}') \frac{\left(-\mathsf{D}_{b'}^\star\right)'}{mc} \hat{\chi}_{\tilde{s}}(\mathbf{r}') + \\
+ \sum\limits_{s,\tilde{s},\tilde{s}'\in\{+,-\}} \hat{\psi}_s^\dagger(\mathbf{r}) \frac{\mathsf{D}_b}{mc} \hat{\psi}_s(\mathbf{r}) \hat{\chi}_{\tilde{s}}^\dagger(\mathbf{r}')(-\frac{r'_{b'}-r_{b'}}{4\pi|\mathbf{r}'-\mathbf{r}|^3}) \frac{1}{mc} \epsilon_{bb'b''} (\frac{\hbar}{2}\sigma_{b''}^{(P)})_{\tilde{s}',\tilde{s}} \hat{\chi}_{\tilde{s}'}(\mathbf{r}') + \\
+ \sum\limits_{s,s',\tilde{s}\in\{+,-\}} \hat{\psi}_s^\dagger(\mathbf{r})(\frac{r_{b'}-r'_{b'}}{4\pi|\mathbf{r}'-\mathbf{r}|^3}) \frac{1}{mc} \epsilon_{bb'b''} (\frac{\hbar}{2}\sigma_{b''}^{(P)})_{s,s'} \hat{\psi}_{s'}(\mathbf{r}) \hat{\chi}_{\tilde{s}}^\dagger(\mathbf{r}') \frac{\left(-\mathsf{D}_b^\star\right)'}{mc} \hat{\chi}_{\tilde{s}}(\mathbf{r}') + \\
- \hat{\psi}_s^\dagger(\mathbf{r})(\frac{\hbar}{2}\sigma_b^{(P)})_{s,s'} \hat{\psi}_{s'}(\mathbf{r})(\frac{1}{mc})^2 \left(\begin{array}{c} \frac{2}{3}\delta^{(3)}(\mathbf{r}'-\mathbf{r})\,\delta_{b,b'} + \\ + \frac{3(r'_b-r_b)(r'_{b'}-r_{b'})-|\mathbf{r}'-\mathbf{r}|^2\delta_{b,b'}}{4\pi|\mathbf{r}'-\mathbf{r}|^5} \end{array} \right) \hat{\chi}_{\tilde{s}}^\dagger(\mathbf{r}')(\frac{\hbar}{2}\sigma_{b'}^{(P)})_{\tilde{s}',\tilde{s}} \hat{\chi}_{\tilde{s}'}(\mathbf{r}') + \\
+(\frac{\hbar}{2mc})^2 \int \frac{d^3q}{(2\pi)^3} e^{i\mathbf{q}\cdot(\mathbf{r}'-\mathbf{r})} \left(\delta_{b,b'} - n_b n_{b'}\right) \sum\limits_{\substack{s,\tilde{s}\in\{+,-\}\\ s',\tilde{s}'\in\{+,-\}}} \left(\sigma_b^{(P)}\right)_{s,\tilde{s}} \left(\sigma_{b'}^{(P)}\right)_{\tilde{s}',s'} \hat{\psi}_s^\dagger(\mathbf{r})\hat{\psi}_{s'}(\mathbf{r}')\hat{\chi}_{\tilde{s}}^\dagger(\mathbf{r})\hat{\chi}_{\tilde{s}'}(\mathbf{r}')
\end{array} \right) \tag{3.71}
$$

Taking a closer look at (3.70), apparently the first line there describes (as expected of course) the Coulomb *attraction* between electrons and positrons. The term in the second line we identify as a repulsive (two-body) Darwin contact. The other terms in (3.70) represent leading order relativistic corrections, as there are in the third and fourth lines the spin-orbit interaction of electrons moving in the (longitudinal) electric field created by the positrons, and, vice versa, the spin-orbit interaction of positrons moving in the (longitudinal) electric field created by the electrons.

Likewise, taking a closer look at the photon mediated interaction terms in (3.71), the top four lines conform to the analogous effective electron-electron interaction terms listed in (3.58), namely the *orbit-orbit* interaction in the first line, the *other-spin-orbit* interaction in the second and third lines, and the *magnetic dipole-dipole* interaction between electrons and positrons in the fourth line.

Yet the interaction terms occuring in the fifth line of (3.70) and in the fifth line of (3.71), respectively, have no counterpart and represent a new type of particle-particle interaction, commonly encountered in covariant S-matrix based perturbation theory as the *annihilation* interaction, see for instance [10, p.133], [11, Eq. (83,20)], [25, Sect. 6-1-3], and last not least [12, Chapter 5–5].

Addition of the interactions (3.70) and (3.71) decomposes as

$$\hat{\mathcal{U}}_C^{(e,p)} + \hat{\mathcal{U}}_\perp^{(e,p)} = \hat{\mathcal{U}}_{\text{Breit}}^{(e,p)} + \hat{\mathcal{U}}_{an}^{(e,p)}. \tag{3.72}$$

Here $\hat{\mathcal{U}}_{\text{Breit}}^{(e,p)}$ represents the well-known two-body interaction terms for electrons and positrons, formed in analogy to the electron-electron interaction terms (3.56) and (3.58), comprising besides the Coulomb interaction the leading order relativistic corrections

$$\hat{\mathcal{U}}_{\text{Breit}}^{(e,p)} = \frac{e^2}{\varepsilon_0} \int d^3r \int d^3r' \times$$
$$\times \left(\begin{array}{c} \sum\limits_{s,\tilde{s}\in\{+,-\}} \hat{\psi}_s^\dagger(\mathbf{r})\hat{\psi}_s(\mathbf{r}) \left(-\frac{1}{4\pi|\mathbf{r}'-\mathbf{r}|} \right) \hat{\chi}_{\tilde{s}}^\dagger(\mathbf{r}')\hat{\chi}_{\tilde{s}}(\mathbf{r}') + \\ + \sum\limits_{s,\tilde{s}\in\{+,-\}} \hat{\psi}_s^\dagger(\mathbf{r})\hat{\psi}_s(\mathbf{r}) (\frac{\hbar}{2mc})^2 \delta^{(3)}(\mathbf{r}-\mathbf{r}')\, \hat{\chi}_{\tilde{s}}^\dagger(\mathbf{r}')\hat{\chi}_{\tilde{s}}(\mathbf{r}') + \\ + \frac{1}{2mc} \sum\limits_{s,s'\in\{+,-\}} \hat{\psi}_s^\dagger(\mathbf{r}) \frac{r_b - r'_b}{4\pi|\mathbf{r}'-\mathbf{r}|^3} \epsilon_{bb'b''} \frac{\mathsf{D}_{b'}}{mc} (\frac{\hbar}{2}\sigma_{b''}^{(P)})_{s,s'} \hat{\psi}_{s'}(\mathbf{r}) \sum\limits_{\tilde{s}\in\{+,-\}} \hat{\chi}_{\tilde{s}}^\dagger(\mathbf{r}')\hat{\chi}_{\tilde{s}}(\mathbf{r}') + \\ + \frac{1}{2mc} \sum\limits_{s\in\{+,-\}} \hat{\psi}_s^\dagger(\mathbf{r})\hat{\psi}_s(\mathbf{r}) \sum\limits_{\tilde{s},\tilde{s}'\in\{+,-\}} \hat{\chi}_{\tilde{s}'}^\dagger(\mathbf{r}') (\frac{r_b - r'_b}{4\pi|\mathbf{r}'-\mathbf{r}|^3} \epsilon_{bb'b''} \frac{\left(-\mathsf{D}_{b'}^\star\right)'}{mc} (\frac{\hbar}{2}\sigma_{b''}^{(P)})_{\tilde{s},\tilde{s}'}) \hat{\chi}_{\tilde{s}}(\mathbf{r}') + \\ \sum\limits_{s,\tilde{s}\in\{+,-\}} \hat{\psi}_s^\dagger(\mathbf{r}) \frac{\mathsf{D}_b}{mc} \hat{\psi}_s(\mathbf{r}) \frac{1}{8\pi} \left(\frac{\delta_{b,b'}}{|\mathbf{r}-\mathbf{r}'|} + \frac{(r_b - r'_b)(r_{b'} - r'_{b'})}{|\mathbf{r}-\mathbf{r}'|^3} \right) \hat{\chi}_{\tilde{s}}^\dagger(\mathbf{r}') \frac{\left(-\mathsf{D}_{b'}^\star\right)'}{mc} \hat{\chi}_{\tilde{s}}(\mathbf{r}') + \\ + \frac{1}{mc} \sum\limits_{s,\tilde{s},\tilde{s}'\in\{+,-\}} \hat{\psi}_s^\dagger(\mathbf{r}) \frac{\mathsf{D}_b}{mc} \hat{\psi}_s(\mathbf{r}) \hat{\chi}_{\tilde{s}}^\dagger(\mathbf{r}') (-\frac{r'_{b'} - r_{b'}}{4\pi|\mathbf{r}'-\mathbf{r}|^3}) \epsilon_{bb'b''} (\frac{\hbar}{2}\sigma_{b''}^{(P)})_{\tilde{s}',\tilde{s}} \hat{\chi}_{\tilde{s}'}(\mathbf{r}') + \\ + \frac{1}{mc} \sum\limits_{s,s',\tilde{s}\in\{+,-\}} \hat{\psi}_s^\dagger(\mathbf{r}) (\frac{r_{b'} - r'_{b'}}{4\pi|\mathbf{r}'-\mathbf{r}|^3}) \epsilon_{bb'b''} (\frac{\hbar}{2}\sigma_{b''}^{(P)})_{s,s'} \hat{\psi}_{s'}(\mathbf{r}) \hat{\chi}_{\tilde{s}}^\dagger(\mathbf{r}') \frac{\left(-\mathsf{D}_b^\star\right)'}{mc} \hat{\chi}_{\tilde{s}}(\mathbf{r}') + \\ - \hat{\psi}_s^\dagger(\mathbf{r}) (\frac{\hbar}{2}\sigma_b^{(P)})_{s,s'} \hat{\psi}_{s'}(\mathbf{r}) (\frac{1}{mc})^2 \left(\begin{array}{c} \frac{2}{3}\delta^{(3)}(\mathbf{r}'-\mathbf{r})\, \delta_{b,b'} + \\ + \frac{3(r'_b - r_b)(r'_{b'} - r_{b'}) - |\mathbf{r}'-\mathbf{r}|^2 \delta_{b,b'}}{4\pi|\mathbf{r}'-\mathbf{r}|^5} \end{array} \right) \hat{\chi}_{\tilde{s}}^\dagger(\mathbf{r}') (\frac{\hbar}{2}\sigma_{b'}^{(P)})_{\tilde{s}',\tilde{s}} \hat{\chi}_{\tilde{s}'}(\mathbf{r}') + \end{array} \right) \tag{3.73}$$

Reading off from (3.73) the *first quantized* version of these particle-particle interactions and considering just *one* electron and *one* positron, full agreement with [11, Eq. (83,15)] is found, specializing there to $m_1 = m_2 = m$ and $e_1 = -e_2 = e$.

The term $\hat{\mathcal{U}}_{an}^{(e,p)}$ in (3.72) represents the *sum* of the annihilation interactions encountered in the fifth lines in (3.70) and in (3.71). Insertion of the integral (3.54) leads then to

$$\hat{\mathcal{U}}_{an}^{(e,p)} = \frac{e^2}{\varepsilon_0} \int d^3r \int d^3r' \left(\frac{\hbar}{2mc}\right)^2 \delta^{(3)}\left(\mathbf{r} - \mathbf{r}'\right) \times$$
$$\times \sum_{\substack{s,\tilde{s}\in\{+,-\} \\ s',\tilde{s}'\in\{+,-\}}} (\sigma_b^{(P)})_{s,\tilde{s}} (\sigma_{b'}^{(P)})_{\tilde{s}',s'} \hat{\psi}_s^{\dagger}(\mathbf{r}) \hat{\psi}_{s'}(\mathbf{r}') \hat{\chi}_{\tilde{s}}^{\dagger}(\mathbf{r})\ \hat{\chi}_{\tilde{s}'}(\mathbf{r}'). \tag{3.74}$$

Apparently, electron and positron field operators here appear with regard to their coupling to spin in an unsuitably *entangled* guise, thus preventing an intuitively accessible physical interpretation.

Fortunately, such an entanglement as displayed in (3.74) can be removed making use of an identity between Pauli matrices[1]

$$\left(\sigma_b^{(P)}\right)_{s,\tilde{s}} \left(\sigma_b^{(P)}\right)_{\tilde{s}',s'} = \frac{3}{2}\delta_{s,s'}\delta_{\tilde{s}',\tilde{s}} + \frac{1}{2}\left(\sigma_b^{(P)}\right)_{s,s'} \left(\sigma_y^{(P)}\sigma_b^{(P)}\sigma_y^{(P)}\right)_{\tilde{s},\tilde{s}'}. \tag{3.75}$$

As a result the annihilation term (3.74) assumes with help of the identity (3.75) the equivalent guise of a contact interaction between electrons and positrons, incorporating not only their particle densities, but as well their spin densities,[2] thus resolving the mentioned difficulty of physical interpretation

$$\hat{\mathcal{U}}_{an}^{(e,p)} = \frac{e^2}{\varepsilon_0} \int d^3r \int d^3r' \times$$
$$\times \left(\frac{\hbar}{2mc}\right)^2 \left(\begin{array}{l} \frac{3}{2} \sum\limits_{s\in\{+,-\}} \hat{\psi}_s^{\dagger}(\mathbf{r})\hat{\psi}_s(\mathbf{r})\ \delta^3(\mathbf{r}-\mathbf{r}') \sum\limits_{\tilde{s}\in\{+,-\}} \hat{\chi}_{\tilde{s}}^{\dagger}(\mathbf{r}')\hat{\chi}_{\tilde{s}}(\mathbf{r}') + \\ +\frac{1}{2} \sum\limits_{s,s'\in\{+,-\}} \hat{\psi}_s^{\dagger}(\mathbf{r})(\sigma_b^{(P)})_{s,s'}\hat{\psi}_{s'}(\mathbf{r})\ \delta^3(\mathbf{r}-\mathbf{r}') \sum\limits_{\tilde{s},\tilde{s}'\in\{+,-\}} \hat{\chi}_{\tilde{s}}^{\dagger}(\mathbf{r}')(\sigma_y^{(P)}\sigma_b^{(P)}\sigma_y^{(P)})_{\tilde{s},\tilde{s}'}\hat{\chi}_{\tilde{s}'}(\mathbf{r}') \end{array} \right) \tag{3.76}$$

Retranslating (3.72) back from second quantization to first quantization and restricting to the special case of positronium, i.e. setting $N_e = 1 = N_p$, then after some elementary reshaping the following effective electron-positron interaction is obtained eventually

$$\mathcal{U}_C^{(e,p)} + \mathcal{U}_{\perp}^{(e,p)} = -\frac{e^2}{4\pi\epsilon_0} \frac{1}{\left|\mathbf{r}^{(e)} - \mathbf{r}^{(p)}\right|} 1_{2\times 2} \otimes 1_{2\times 2} \ +$$

[1] A proof is given in the appendix.

[2] See the definitions in Eqs. (3.42) and (3.43).

$$
+\ \frac{1}{\epsilon_0 c^2}\left(\frac{e\hbar}{2m}\right)^2 \left(\begin{array}{c} \delta^{(3)}\left(\mathbf{r}^{(e)}-\mathbf{r}^{(p)}\right)\mathsf{1}\ + \\ +\left(\boldsymbol{\sigma}^{(e)}+2\boldsymbol{\sigma}^{(p)}\right)\cdot\left(\frac{\mathbf{r}^{(e)}-\mathbf{r}^{(p)}}{4\pi\left|\mathbf{r}^{(p)}-\mathbf{r}^{(e)}\right|^3}\wedge\frac{\boldsymbol{\Pi}^{(e)}}{\hbar}\right)\ + \\ -\left(\boldsymbol{\sigma}^{(p)}+2\boldsymbol{\sigma}^{(e)}\right)\cdot\left(\frac{\mathbf{r}^{(e)}-\mathbf{r}^{(p)}}{4\pi\left|\mathbf{r}^{(e)}-\mathbf{r}^{(p)}\right|^3}\wedge\frac{\boldsymbol{\Pi}^{(p)}}{\hbar}\right)\ + \\ +\frac{1}{2\pi}\left(\frac{\delta_{b,b'}}{\left|\mathbf{r}^{(e)}-\mathbf{r}^{(p)}\right|}+\frac{(r_b^{(e)}-r_b^{(p)})(r_{b'}^{(e)}-r_{b'}^{(p)})}{\left|\mathbf{r}^{(e)}-\mathbf{r}^{(p)}\right|^3}\right)\frac{\Pi_b^{(e)}\Pi_{b'}^{(p)}}{\hbar^2}\mathsf{1} \\ +\left(\begin{array}{c} \frac{2}{3}\delta^{(3)}\left(\mathbf{r}^{(e)}-\mathbf{r}^{(p)}\right)\delta_{b,b'} \\ +\frac{3(r_b^{(e)}-r_b^{(p)})(r_{b'}^{(e)}-r_{b'}^{(p)})-\left|\mathbf{r}^{(e)}-\mathbf{r}^{(p)}\right|^2\delta_{b,b'}}{4\pi\left|\mathbf{r}^{(e)}-\mathbf{r}^{(p)}\right|^5} \end{array} \right)\sigma_b^{(e)}\sigma_{b'}^{(p)} \\ +\delta^3(\mathbf{r}^{(e)}-\mathbf{r}^{(p)})\left(\frac{3}{2}\cdot\mathsf{1}+\frac{1}{2}\sigma_b^{(e)}\sigma_b^{(p)}\right) \end{array} \right) \tag{3.77}
$$

Here the energy scale of the leading order relativistic correction terms adding to the dominant Coulomb interaction is set in terms of Bohr's magneton $\mu_B = \frac{e\hbar}{2m}$.

Just for clearness, in the two-particle subspace there holds for the electron (e) and the positron (p)

$$
\begin{aligned}
\sigma_b^{(p)} &= \mathsf{1}_{2\times 2}\otimes(\sigma_y^{(P)}\sigma_b^{(P)}\sigma_y^{(P)}) \\
\sigma_b^{(e)} &= \sigma_b^{(P)}\otimes\mathsf{1}_{2\times 2} \\
\mathsf{1} &= \mathsf{1}_{2\times 2}\otimes\mathsf{1}_{2\times 2},
\end{aligned} \tag{3.78}
$$

whereby the respective gauge invariant derivative operators read

$$
\begin{aligned}
\Pi_b^{(e)} &= \mathsf{p}_b^{(e)} - e\mathsf{A}_b^{(ext)}(\mathsf{x}^{(e)}) \\
\Pi_b^{(p)} &= \mathsf{p}_b^{(p)} + e\mathsf{A}_b^{(ext)}(\mathsf{x}^{(p)}).
\end{aligned} \tag{3.79}
$$

To compare with the result as stated in [11, Eq. (84,1)] for the special case of *zero* external field, it is expedient to restrict to the center of mass system. In this case

$$
\begin{aligned}
\mathsf{A}_b^{(ext)} &= 0 \\
\mathsf{R}_b &= \frac{1}{2}(r_b^{(e)}+r_b^{(p)}) = 0 \\
\mathsf{P}_b &= \mathsf{p}_b^{(e)}+\mathsf{p}_b^{(p)} = 0,
\end{aligned} \tag{3.80}
$$

and in terms of relative co-ordinates now

$$
\begin{aligned}
\mathsf{p}_b^{(e)} &= -\mathsf{p}_b^{(p)} \equiv \mathsf{p}_b \\
r_b^{(e)} - r_b^{(p)} &\equiv r_b.
\end{aligned} \tag{3.81}
$$

In this case (3.77) assumes the guise

$$\mathcal{U}_C^{(e,p)} + \mathcal{U}_\perp^{(e,p)} = -\frac{e^2}{4\pi\epsilon_0}\frac{1}{|\mathbf{r}|}1 + \frac{1}{4\pi\epsilon_0 c^2}\left(\frac{e\hbar}{2m}\right)^2 \begin{pmatrix} 4\pi\delta^{(3)}(\mathbf{r})1 - 2\left(\frac{\delta_{b,b'}}{|\mathbf{r}|} + \frac{r_b r_{b'}}{|\mathbf{r}|^3}\right)\frac{\mathsf{p}_b\mathsf{p}_{b'}}{\hbar^2}1 + \\ + 3\cdot\frac{1}{|\mathbf{r}|^3}\left(\mathbf{r}\wedge\frac{\mathsf{p}}{\hbar}\right)_b\left(\sigma_b^{(e)} + \sigma_b^{(p)}\right) + \\ + \left(4\pi\delta^{(3)}(\mathbf{r})\frac{2}{3}\delta_{b,b'} + \frac{3r_b r_{b'} - |\mathbf{r}|^2\delta_{b,b'}}{|\mathbf{r}|^5}\right)\sigma_b^{(e)}\sigma_{b'}^{(p)} \\ +4\pi\delta^3(\mathbf{r})\left(\frac{3}{2}\cdot 1 + \frac{1}{2}\sigma_b^{(e)}\sigma_b^{(p)}\right) \end{pmatrix}.$$

Introducing total spin

$$S_b = \frac{\hbar}{2}\left(\sigma_b^{(e)} + \sigma_b^{(p)}\right) \tag{3.82}$$

(mind the summation convention!) there holds[3]

$$S_b S_b = \frac{3}{2}\cdot 1 + \frac{1}{2}\sigma_b^{(e)}\sigma_b^{(p)} \tag{3.83}$$
$$\left(3\, r_b r_{b'} - |\mathbf{r}|^2\,\delta_{b,b'}\right) S_b S_{b'} = \frac{\hbar^2}{2}\left(3\, r_b r_{b'} - |\mathbf{r}|^2\,\delta_{b,b'}\right)\sigma_b^{(e)}\sigma_{b'}^{(p)}.$$

With help of these identities there follows eventually

$$\mathcal{U}_C^{(e,p)} + \mathcal{U}_\perp^{(e,p)} = -\frac{e^2}{4\pi\epsilon_0}\frac{1}{|\mathbf{r}|}1 + \frac{1}{4\pi\epsilon_0 c^2}\left(\frac{e\hbar}{2m}\right)^2 \begin{pmatrix} 4\pi\delta^{(3)}(\mathbf{r})\cdot 1 - 2\left(\frac{\delta_{b,b'}}{|\mathbf{r}|} + \frac{r_b r_{b'}}{|\mathbf{r}|^3}\right)\frac{\mathsf{p}_b\mathsf{p}_{b'}}{\hbar^2}1 + \\ + 6\cdot\frac{1}{|\mathbf{r}|^3}\left(\mathbf{r}\wedge\frac{\mathsf{p}}{\hbar}\right)_b\frac{1}{\hbar}S_b + \\ + 2\cdot\frac{3r_b r_{b'} - |\mathbf{r}|^2\delta_{b,b'}}{|\mathbf{r}|^5}\frac{1}{\hbar^2}S_b S_{b'} + \\ + 4\pi\delta^{(3)}(\mathbf{r})\left(\frac{7}{3}S_b S_b - 2\cdot 1\right) \end{pmatrix}. \tag{3.84}$$

Aside from our use of SI-units, there is full agreement with the well known interaction terms liable for the fine structure of *positronium*, see for instance [11, Eq. (84,1)].

[3] A proof is given in the appendix.

3.7 Concluding Remarks

The problem of obtaining the relativistic corrections to the nonrelativistic Hamiltonian of Schrödinger-Pauli quantum mechanics actually dates back to the early work of Breit [23, 26]. But as discussed comprehensively by Bethe and Salpeter [22], the approach of Breit needs modifications as it cannot be used to evaluate relativistic corrections to the interaction potential between electrons consistently.

Certainly a most influential work in (more) modern times is the fully relativistic perturbative S-matrix formalism of R.P. Feynman, directly aiming at the evaluation of all kind of expectation values for observables [27]. As regards for instance our obtained "correction" terms to the Coulomb interaction in (3.58) and (3.59), the corresponding Feynman diagrams, actually representing the scattering amplitudes for these interaction processes in momentum space representation, (albeit involving just *two* particles) can be found being concisely discussed in Feynman's lectures, see [10, lecture 26]. Equivalent to Feynman's diagrammatic method is the Schwinger-Dyson Green's function perturbation theory, as presented in great detail for instance in [11, 28, 29].

As regards the problem of obtaining relativistic corrections to the *bound* states of fermions, a Green's function based integral equation method is due to Bethe and Salpeter [30]. Notwithstanding their approach in principle being fully relativistic, it applies again just to *two* interacting particles in their center of mass frame, see for instance [25].

Related to this is the method of Schwinger [12], who identifies the relativistic corrections for *two* interacting particles from an equation of motion for a two-particle propagator, and in particular disseminates as well a thorough discussion of the spectrum of positronium.

The idea, to derive the effective nonrelativistic Hamiltonian for a system of electrons directly from quantum electrodynamics in Coulomb gauge, has actually been published in 1965 already by Itoh [8], who employed a perturbative procedure. Restricting to the field theory context of QED he aims at the elimination of the photon quantum fields, albeit in an unsystematic way, as regards his treatment of processes with electron-positron pairs coupling to external fields, and also his neglect of terms not conserving particle number, actually just speaking of them as being small.

Partly along the lines of Itoh [8], Bialynicki-Birula suggested a more systematic perturbation theory [3], based on a step-by-step sequence of unitary transformations of the QED-Hamiltonian in Coulomb gauge, aiming at the elimination of the interaction of the electron-positron pair current with the electromagnetic vector potential quantum field of photons. He derives the effective interaction terms afore-mentioned in (3.56), but his derivation lacks all the other effective two-particle interactions enlisted in our result (3.58). This is because in Birulas's approach the so called "small" components of the Dirac modes $\phi_\mu(\mathbf{r}, k)$ for either energy sign $z_k = \pm 1$ have been (tacitly) neglected from the outset.

We believe our derivation of the nonrelativistic particle number conserving Hamiltonian for interacting electrons and/or positrons is more systematic and more complete than all the ones given afore in the mentioned works [3, 8, 10–12, 22, 23, 26]. The reason being that the conditions under which there emerges an intuitively accessible particle picture for electrons and positrons have now been clarified by our approach with the following three steps:

- The first objective was the elimination of all processes regarding emission or else absorption of high-energy photons by the Dirac quantum fields. This is achieved by solving Hamiltonian flow equations, so that the limiting value $\hat{\mathcal{H}}_U$ of that flow is, besides being unitary equivalent to the relativistic field theory Hamiltonian $\hat{\mathcal{H}}_{QED}$ of quantum electrodynamics, as well conserving the number of occupied Dirac modes in the Fock space of QED (fermion number), see [1].
- Secondly, introducing the energy-separating Eriksen transformation [6, 7] to obtain in lieu of a *four*-component Dirac-fermion quantum field, now indeed in a completely *symmetric* and systematic fashion both, a *two*-component quantum field for electrons, and *separately* a *two*-component quantum field for positrons. Upon an expansion of the generator of the Eriksen transformation in the small parameter v/c, that way one succeeds to derive a nonrelativistic particle number conserving field theory Hamiltonian for *any plurality* of individually interacting electrons and/or positrons.
- Thirdly, upon retranslation of the field theory Hamiltonian obtained in step 2 back to *first* quantization there emerges a familiar classical interpretation, in as much the obtained first quantized Hamiltonian now governs the motion *and* the interactions of a *fixed* number N of point particles moving at nonrelativistic speed, whereas every particle is located at its individual (instantaneous) position $\mathbf{r}^{(j)}$ for $j \in \{1, 2, ..., N\}$ and carries besides (instantaneous) momentum $\mathbf{p}^{(j)}$ at this level of description the attributes mass, charge and also spin.

Appendix

It is asserted that[4]

$$\sum_{b\in\{x,y,z\}} \left(\sigma_b^{(P)}\right)_{s,\tilde{s}} \left(\sigma_b^{(P)}\right)_{\tilde{s}',s'} = \frac{3}{2}\delta_{s,s'}\delta_{\tilde{s}',\tilde{s}} + \frac{1}{2}\sum_{b\in\{x,y,z\}} \left(\sigma_b^{(P)}\right)_{s,s'} \left(\sigma_y^{(P)}\sigma_b^{(P)}\sigma_y^{(P)}\right)_{\tilde{s},\tilde{s}'} \tag{A.1}$$

The proof relies on the fact that any 2×2-matrix M enables a representation as a linear combination of the three Pauli matrices and unity

$$\sigma_x^{(P)} = \begin{pmatrix} 0, 1 \\ 1, 0 \end{pmatrix}, \quad \sigma_y^{(P)} = \begin{pmatrix} 0, -i \\ i, 0 \end{pmatrix}, \quad \sigma_z^{(P)} = \begin{pmatrix} 1, 0 \\ 0, -1 \end{pmatrix} \tag{A.2}$$

[4] In the interest of clarity the summation convention is suspended in the ensuing.

$$\sigma_0^{(P)} \equiv 1_{2\times 2} = \begin{pmatrix} 1, 0 \\ 0, 1 \end{pmatrix}$$

Therefore

$$\mathsf{M} = \sum_{A' \in \{0,x,y,z\}} c_{A'}^{(M)} \sigma_{A'}^{(P)} \tag{A.3}$$

Calculating the trace for A, $A' \in \{0, x, y, z\}$ one readily confirms

$$\frac{1}{2}\mathrm{tr}\left(\sigma_A^{(P)}\sigma_{A'}^{(P)}\right) = \delta_{A,A'} \tag{A.4}$$

Given a matrix M one thus determines the coefficients in the expansion (A.3) as

$$c_A^{(M)} = \frac{1}{2}\mathrm{tr}\left(\sigma_A^{(P)}\mathsf{M}\right) = \frac{1}{2}\sum_{j,k\in\{1,2\}}\left(\sigma_A^{(P)}\right)_{j,k} M_{k,j} \tag{A.5}$$

Insertion of (A.5) into (A.3) gives the representation

$$\mathsf{M} = \sum_{A'\in\{0,x,y,z\}}\left(\frac{1}{2}\sum_{j',k'\in\{1,2\}}\left(\sigma_{A'}^{(P)}\right)_{j',k'} M_{k',j'}\right)\sigma_{A'}^{(P)} \tag{A.6}$$

So there holds for any 2×2-matrix M with elements $M_{j,k}$

$$\mathsf{M}_{j,k} = \sum_{A'\in\{0,x,y,z\}}\left(\frac{1}{2}\sum_{j',k'\in\{1,2\}}\left(\sigma_{A'}^{(P)}\right)_{j',k'} M_{k',j'}\right)\left(\sigma_{A'}^{(P)}\right)_{j,k} \tag{A.7}$$

For the special 2×2-matrix N with matrix elements $N_{j,k} = \delta_{j,j''}\delta_{k,k''}$ it follows in particular $N_{k',j'} = \delta_{k',j''}\delta_{j',k''}$. This way one finds from (A.7) some kind of completeness relation in the guise

$$\begin{aligned}\delta_{j,j''}\delta_{k,k''} &= \sum_{A'\in\{0,x,y,z\}}\left(\frac{1}{2}\sum_{j',k'\in\{1,2\}}\left(\sigma_{A'}^{(P)}\right)_{j',k'} \delta_{k',j''}\delta_{j',k''}\right)\left(\sigma_{A'}^{(P)}\right)_{j,k} \\ &= \frac{1}{2}\delta_{k'',j''}\delta_{j,k} + \frac{1}{2}\sum_{a\in\{x,y,z\}}\left(\sigma_a^{(P)}\right)_{k'',j''}\left(\sigma_a^{(P)}\right)_{j,k}\end{aligned} \tag{A.8}$$

Next consider

$$\begin{aligned}&\sum_{b\in\{x,y,z\}}\left(\sigma_b^{(P)}\right)_{j,j'}\left(\sigma_b^{(P)}\right)_{k',k} \\ =\quad &\sum_{j'',k''\in\{1,2\}}\delta_{j,j''}\delta_{k,k''}\sum_{b\in\{x,y,z\}}\left(\sigma_b^{(P)}\right)_{j'',j'}\left(\sigma_b^{(P)}\right)_{k',k''}\end{aligned}$$

insert$(A.8)$

$$= \sum_{j'',k''\in\{1,2\}} \left(\frac{1}{2}\delta_{k'',j''}\delta_{j,k} + \frac{1}{2}\sum_{a\in\{x,y,z\}} \left(\sigma_a^{(P)}\right)_{k'',j''} \left(\sigma_a^{(P)}\right)_{j,k} \right) \sum_{b\in\{x,y,z\}} \left(\sigma_b^{(P)}\right)_{j'',j'} \left(\sigma_b^{(P)}\right)_{k',k''}$$

$$= \left(\begin{array}{c} \frac{1}{2}\sum_b \left(\sum_{j''\in\{1,2\}} \left(\sigma_b^{(P)}\right)_{k',j''} \left(\sigma_b^{(P)}\right)_{j'',j'} \right) \delta_{j,k} \\ +\frac{1}{2}\sum_{a,b} \left(\sum_{j'',k''\in\{1,2\}} \left(\sigma_b^{(P)}\right)_{k',k''} \left(\sigma_a^{(P)}\right)_{k'',j''} \left(\sigma_b^{(P)}\right)_{j'',j'} \right) \left(\sigma_a^{(P)}\right)_{j,k} \end{array} \right)$$

$$= \frac{1}{2}\left(\sum_b \sigma_b^{(P)}\sigma_b^{(P)}\right)_{k',j'} \cdot \delta_{j,k} + \frac{1}{2}\sum_a \left(\sum_b \sigma_b^{(P)}\sigma_a^{(P)}\sigma_b^{(P)}\right)_{k',j'} \left(\sigma_a^{(P)}\right)_{j,k}$$

Making use of well known relations obeyed by the Pauli matrices here the last line can be simplified

$$\sum_b \sigma_b^{(P)}\sigma_b^{(P)} = 3\cdot 1_{2\times 2} \tag{A.9}$$

$$\begin{aligned} \sum_b \sigma_b^{(P)}\sigma_a^{(P)}\sigma_b^{(P)} &= \sum_b \left(\sigma_b^{(P)}\sigma_a^{(P)} + \sigma_a^{(P)}\sigma_b^{(P)} - \sigma_a^{(P)}\sigma_b^{(P)}\right)\sigma_b^{(P)} \\ &= \sum_b \left(2\delta_{a,b}1_{2\times 2} - \sigma_a^{(P)}\sigma_b^{(P)}\right)\sigma_b^{(P)} = (2-3)\,\sigma_a^{(P)} \\ &= -\sigma_a^{(P)} \end{aligned}$$

So there follows from what has been said

$$\sum_{b\in\{x,y,z\}} \left(\sigma_b^{(P)}\right)_{j,j'} \left(\sigma_b^{(P)}\right)_{k',k} = \frac{3}{2}\delta_{j,k}\delta_{k',j'} - \frac{1}{2}\sum_{a\in\{x,y,z\}} \left(\sigma_a^{(P)}\right)_{j,k} \left(\sigma_a^{(P)}\right)_{k',j'} \tag{A.10}$$

Taking into account the identity

$$\left(\sigma_a^{(P)}\right)_{k',j'} = \left(-\sigma_y^{(P)}\sigma_a^{(P)}\sigma_y^{(P)}\right)_{j',k'} \tag{A.11}$$

then eventually

$$\sum_{b\in\{x,y,z\}} \left(\sigma_b^{(P)}\right)_{j,j'} \left(\sigma_b^{(P)}\right)_{k',k} = \frac{3}{2}\delta_{j,k}\delta_{k',j'} + \frac{1}{2}\sum_{a\in\{x,y,z\}} \left(\sigma_a^{(P)}\right)_{j,k} \left(\sigma_y^{(P)}\sigma_a^{(P)}\sigma_y^{(P)}\right)_{j',k'} \tag{A.12}$$

With the identification $\{j, j', k', k\} \equiv \{s, \tilde{s}, \tilde{s}', s'\}$ this becomes indentical to the assertion (A.1), which was to be proven.

Now, with the definitions introduced afore in (3.78) and in (3.82) we have

$$\begin{aligned} S_b S_{b'} &= \frac{\hbar^2}{4}\left(\sigma_b^{(e)} + \sigma_b^{(p)}\right)\left(\sigma_{b'}^{(e)} + \sigma_{b'}^{(p)}\right) \\ &= \frac{\hbar^2}{4}\left(\sigma_b^{(P)} \otimes 1_{2\times 2} + 1_{2\times 2} \otimes \sigma_y^{(P)}\sigma_b^{(P)}\sigma_y^{(P)}\right)\left(\sigma_{b'}^{(P)} \otimes 1_{2\times 2} + 1_{2\times 2} \otimes \sigma_y^{(P)}\sigma_{b'}^{(P)}\sigma_y^{(P)}\right) \end{aligned} \tag{A.13}$$

Using the multiplication rule

$$(A_I \otimes B_I)(A_{II} \otimes B_{II}) = (A_I A_{II}) \otimes (B_I B_{II}) \tag{A.14}$$

one finds with $\sigma_y^{(P)}\sigma_y^{(P)} = 1_{2\times 2}$ instead of (A.13) next

$$S_b S_{b'} = \frac{\hbar^2}{4}\begin{pmatrix} \sigma_b^{(P)}\sigma_{b'}^{(P)} \otimes 1_{2\times 2} + 1_{2\times 2} \otimes \sigma_y^{(P)}\sigma_b^{(P)}\sigma_{b'}^{(P)}\sigma_y^{(P)} + \\ + \sigma_b^{(P)} \otimes \sigma_y^{(P)}\sigma_{b'}^{(P)}\sigma_y^{(P)} + \sigma_{b'}^{(P)} \otimes \sigma_y^{(P)}\sigma_b^{(P)}\sigma_y^{(P)} \end{pmatrix} \tag{A.15}$$

Observing $\sum_{b,b'} r_b \sigma_b^{(P)} r_{b'} \sigma_{b'}^{(P)} = |\mathbf{r}|^2 \cdot 1_{2\times 2}$ and defining $1 \equiv 1_{2\times 2} \otimes 1_{2\times 2}$ leads to

$$\sum_{b,b'} (r_b S_b)(r_{b'} S_{b'}) = \frac{\hbar^2}{2}\left(|\mathbf{r}|^2 \cdot 1 \; + \; \sum_{b,b'} r_b r_{b'} \sigma_b^{(e)}\sigma_{b'}^{(p)}\right) \tag{A.16}$$

Likewise, specializing in (A.15) to $b = b'$ and taking into account (A.9) one finds

$$\sum_b S_b S_b = \frac{\hbar^2}{2}\left(3 \cdot 1 + \sum_b \sigma_b^{(e)}\sigma_b^{(p)}\right) \tag{A.17}$$

Combining (A.16) and (A.17) there results

$$\sum_{b,b'} \left(3\, r_b r_{b'} - |\mathbf{r}|^2\, \delta_{b,b'}\right) S_b S_{b'} = \frac{\hbar^2}{2}\sum_{b,b'} \left(3\, r_b r_{b'} - |\mathbf{r}|^2\, \delta_{b,b'}\right) \sigma_b^{(e)}\sigma_{b'}^{(p)}, \tag{A.18}$$

thus settling the proof of the identities (3.83).

References

1. Schopohl N, Cetin NS. Advances in quantum science I - theoretical and methodological approaches, chapter 1: From the relativistic field theory of quantum electrodynamics to a Hamiltonian for electrons and positrons as point particles with attributes mass. Cham: Springer Nature; 2026.

2. Cohen-Tannoudji C, Dupon-Roc J, Grynberg G. Photons and atoms: introduction to quantum electrodynamics. Weinheim: Wiley-VCH; 2004.
3. Bialynicki-Birula I. The Hamiltonian of quantum electrodynamics. In: Barut AO, editor. Quantum electrodynamics and quantum optics. New York: Plenum Press; 1984.
4. Wegner F. Flow equations and normal ordering: a survey. J Phys A: Math Gen. 2006;39(30):8221–30.
5. Schopohl N, Cetin NS. Advances in quantum science I - theoretical and methodological approaches, chapter 2: Obtaining Schwinger's magnetic moment anomaly and the Lamb shift terms from a particle number conserving Hamiltonian being unitary equivalent to the relativistic Hamiltonian of quantum electrodynamics. Cham: Springer Nature; 2026.
6. Eriksen E. Foldy-Wouthuysen-transformation. Exact solution with generalization to the two-particle problem. Phys Rev. 1958;111:1011.
7. Schopohl N, Cetin NS. Advances in quantum science I - theoretical and methodological approaches, chapter 4: A flow equation approach transforming the Dirac Hamiltonian into a unitary equivalent and energy-separating Hamiltonian. Cham: Springer Nature; 2026.
8. Itoh T. Derivation of nonrelativistic Hamiltonian for electrons from quantum electrodynamics. Rev Mod Phys. 1965;37(1):159–65.
9. Landau LD. On the electron-positron interaction. In: Collected papers of L. D. Landau. Oxford: Pergamon Press Ltd.; 1965. p. 532–9.
10. Feynman RP. Quantum electrodynamics. New York: Addison Wesley; 1961.
11. Landau LD, Lifschitz JM. Quantenelektrodynamik. Frankfurt: Verlag Harri Deutsch; 2009.
12. Schwinger J. Particles, sources and fields, vol. 3. Reading: Perseus; 1998.
13. Schwinger J. On quantum-electrodynamics and the magnetic moment of the electron. Phys Rev. 1948;73(4):416–7.
14. Schwinger J. On gauge invariance and vacuum polarization. Phys Rev. 1950;82:664–79.
15. Lamb WE. Fine structure of the hydrogen atom [Internet]. Nobel Prize Outreach AB; 1955 [cited 2026 Jun 3]. Available from: https://www.nobelprize.org/prizes/physics/1955/lamb/lecture.
16. Milonni PW. The quantum vacuum: an introduction to quantum electrodynamics. San Diego: Academic Press Inc; 1994.
17. Fetter AL, Walecka JD. Quantum theory of many particle systems. New York: McGraw-Hill; 1971.
18. Abrikosov AA, Gorkov LP, Dzyaloshinski IE. Methods of quantum field theory in statistical physics. New York: Dover; 1964.
19. Keldysh LV. The dielectric function of condensed systems. Chapter 1: Introduction. In: Modern problems in condensed matter sciences, volume 24. Amsterdam: Elsevier Science Publishers B.V. North Holland Physics; 1989.
20. Dolgov OV, Maksimov EG. The dielectric function of crystalline systems. In: Modern problems in condensed matter sciences, volume 24. Amsterdam: Elsevier Science Publishers B.V. North Holland Physics; 1989.
21. Froese-Fischer C, Brage T, Jönsson P. Computational atomic structure: an MCHF approach. Bristol: Institute of Physics Publishing; 1997.
22. Bethe HA, Salpeter EE. Quantum mechanics of one- and two-electron atoms. New York: Plenum/Rosetta; 1977.
23. Breit G. Dirac's equation and the spin-spin interactions of two electrons. Phys Rev. 1932;39:616–24.
24. Kutzelnigg W. Solved and unsolved problems in relativistic quantum chemistry. Chem Phys. 2012;395:16–34.
25. Itzykson C, Zuber JB. Quantum field theory. New York: McGraw-Hill; 1985.
26. Breit G. The effect of retardation on the interaction of two electrons. Phys Rev. 1929;34:553–73.
27. Feynman RP. Space-time approach to quantum electrodynamics. Phys Rev. 1949;76(6):769–89.
28. Bjorken JD, Drell SD. Relativistic quantum fields. New York: McGraw-Hill; 1965.
29. Schweber SS. An introduction to relativistic quantum field theory. New York: Harper & Row; 1964.
30. Salpeter EE, Bethe HA. A relativistic wave equation for bound-state problems. Phys Rev. 1951;84(6):1232–42.

Chapter 4
A Flow Equation Approach Transforming the Dirac Hamiltonian into a Unitary Equivalent and Energy-Separating Hamiltonian

Nils Schopohl and Nadine Suzan Cetin

Abstract The Dirac Hamiltonian $\mathsf{H}^{(D)} = mc^2\beta + c\alpha_b\Pi_b + e\Phi(\mathsf{x})$ is transformed with a purpose-built flow equation method, so that the result of that transformation is not only unitary equivalent to $\mathsf{H}^{(D)}$ *and* granted to strive toward a limiting value $\mathsf{H}^{(NW)}$ commuting with the Dirac β-matrix, but is in point of fact *energy-separating*. The latter property entails there holds as an identity $\frac{\mathsf{H}^{(NW)}}{\sqrt{\mathsf{H}^{(NW)}\mathsf{H}^{(NW)}}} = \beta$, a property which *unambiguously* singles out $\mathsf{H}^{(NW)}$ among all Hamiltonians just being unitary equivalent to $\mathsf{H}^{(D)}$ and commuting with β. The equivalence of the proposed flow equation method with the unitary transformation of Eriksen is established on the basis of an *exact* solution of a *reverse* flow equation, the limiting value of that reverse flow representing a unitary transformation that maps the β-matrix back into the energy-sign operator $\frac{\mathsf{H}^{(D)}}{\sqrt{\mathsf{H}^{(D)}\mathsf{H}^{(D)}}}$ associated with $\mathsf{H}^{(D)}$. In contradistinction to $\mathsf{H}^{(D)}$, it's unitary equivalent $\mathsf{H}^{(NW)}$ generates the motion of electrons and positrons in the presence of external fields as entirely *separated* wave packets devoid of any "Zitterbewegung." Upon expansion of $\mathsf{H}^{(NW)}$ in the Sommerfeld parameter $\kappa = \frac{v}{c}$ our results agree with results obtained recently by Silenko with a different method, based on a correction scheme developed for the original step-by-step Foldy–Wouthuysen transformation method, the latter long-since known for not generating a unitary equivalent Hamiltonian being as well energy-separating.

Keywords Eriksen transformation · Brockett flow equation · Dirac energy-sign operator · Newton–Wigner Hamiltonian · Magnus expansion

---

N. Schopohl (✉)
Center for Quantum Science, Eberhard Karls Universität Tübingen, Tübingen, Germany
e-mail: nils.schopohl@uni-tuebingen.de

N. S. Cetin
Mathematical Institute, Charles University Prague, Prague, Czech Republic
e-mail: nadine-suzan.cetin@matfyz.cuni.cz

T. Onishi (ed.), *Advances in Quantum Science I*,
https://doi.org/10.1007/978-981-95-9717-8_4

4.1 Introduction

Based on insight into the fundamental meaning of locality obtained by Newton and Wigner (NW) in 1949 for (freely moving) relativistic fermions [1], Foldy and Wouthuysen [2] established in 1950 with their unitary transformation of the free Dirac Hamiltonian an interpretation of concepts like position or velocity or spin or orbital angular momentum, void of the well-known paradoxical properties of operators in the Dirac representation (for instance, the velocity operator having eigenvalues $\pm c$) and thus agreeable to the physical intuition ascribed to a moving single particle [3]. For this reason we refer in the ensuing to the result of such a unitary transformation of the Dirac Hamiltonian $\mathsf{H}^{(D)}$ as the NW-Hamiltonian $\mathsf{H}^{(NW)}$ also if (weak) external electromagnetic fields are taken into account.

For a relativistic charged fermion, if minimally coupled solely to a magnetostatic vector potential, a generalization of the unitary Foldy–Wouthuysen transformation in closed form has been discovered by Case [4] and as a general proposition, along a different line of reasoning, by Eriksen [5], whose unitary transformation indeed applies as well if both, electrostatic and/or magnetostatic potentials, are taken into account.

In the ensuing a purpose-built flow equation approach is introduced resulting in a limiting value $\mathsf{H}^{(NW)}$ equivalent to a unitary transformation of the Dirac Hamiltonian for a charged spin-$\frac{1}{2}$ fermion with coupling to electromagnetic fields that turns out to be equivalent to the *energy-separating* result obtained with the Eriksen transformation. The advantage of our flow equation approach being that it provides in a lucid manner the expansion of $\mathsf{H}^{(NW)}$ in powers of the Sommerfeld parameter $\kappa = \frac{v}{c}$ to arbitrary order, the results of this new expansion method being in full agreement with corresponding results obtained by the (arduous) expansion of Eriksen's transformation.

4.2 The Flow Equation Method

The so-called Hamiltonian flow equation method, originally developed by Wegner for problems in nonrelativistic many body physics [6, 7], pertains to a continuous unitary transformation of a given Hamiltonian H via

$$\begin{aligned} \mathsf{H}(s) &= \mathsf{U}(s)\,\mathsf{H}\mathsf{U}^{\dagger}(s) \\ \frac{d}{ds}\mathsf{U}(s) &= \eta(s)\,\mathsf{U}(s) \\ \mathsf{U}(0) &= 1_{4\times 4} \\ \eta^{\dagger}(s) &= -\eta(s)\,. \end{aligned} \tag{4.1}$$

Given an anti-symmetric generator $\eta(s)$ for the flow, the unitary transformation of an Hermitian operator H is determined solving an initial value problem:

$$\begin{aligned}\frac{d}{ds}\mathsf{H}(s) &= \left[\eta(s), \mathsf{H}(s)\right] \\ \mathsf{H}(0) &= \mathsf{H}.\end{aligned} \tag{4.2}$$

Of course, the crux of the method is the choice of the generator $\eta(s)$, with the essential point being that $\eta(s)$ controls the properties of the limiting value $\mathsf{H}(\infty)$. A comprehensive review discussing various generators with a multitude of applications of the Wegner flow equation method in nonrelativistic many body physics has been given by Kehrein [8]. Let us choose a most simple generator that depends on a *constant* Hermitian operator Γ and consider a positive semi-definite functional as introduced by Brockett [9, 10]:

$$\Phi(s) \equiv \frac{1}{2}\mathrm{tr}\left((\mathsf{H}(s) - \Gamma)^2\right) \geq 0. \tag{4.3}$$

Here $\mathsf{H}(s)$ is the solution of a flow with generator $\eta(s)$, specified in terms of Γ as

$$\begin{aligned}\eta(s) &= \left[\Gamma, \mathsf{H}(s)\right] \\ \frac{d}{ds}\mathsf{H}(s) &= \left[\eta(s), \mathsf{H}(s)\right] \\ \mathsf{H}(0) &= \mathsf{H}.\end{aligned} \tag{4.4}$$

Exploiting the cyclic invariance of the trace one readily finds after a sequence of elementary adjustments

$$\begin{aligned}\frac{d}{ds}\Phi(s) &= \mathrm{tr}\left[(\mathsf{H}(s) - \Gamma)\frac{d}{ds}\mathsf{H}(s)\right] \\ &= \mathrm{tr}\left[(\mathsf{H}(s) - \Gamma)\left[\eta(s), \mathsf{H}(s)\right]\right] \\ &= \mathrm{tr}\left[(\mathsf{H}(s) - \Gamma)\left[\left[\Gamma, \mathsf{H}(s)\right], \mathsf{H}(s)\right]\right] \\ &= -\mathrm{tr}\left[\left[\mathsf{H}(s), \Gamma\right]\left[\Gamma, \mathsf{H}(s)\right]\right] \\ &= -\mathrm{tr}\left[\eta^{\dagger}(s)\,\eta(s)\right] \\ &\leq 0.\end{aligned} \tag{4.5}$$

Because $\Phi(s) \geq 0$ for all $s \geq 0$ the only possible conclusion being that

$$\lim_{s\to\infty}\frac{d}{ds}\Phi(s) = 0, \tag{4.6}$$

i.e.:

$$\lim_{s\to\infty}\eta(s) = \left[\Gamma, \mathsf{H}(\infty)\right] = 0_{4\times 4}. \tag{4.7}$$

So, a double-bracket flow of the type (4.4) indeed strives to a limiting value $\mathsf{H}(\infty)$ that commutes with the given Hermitian operator Γ.

A while ago Bylev and Pirner (BP) [11] suggested a flow equation approach to obtain a unitary transformation of the Hamiltonian $\mathsf{H}^{(D)}$ for a Dirac particle moving in an external *static* electromagnetic potential, choosing as a generator

$$\eta(s) = \left[\beta, \mathsf{H}(s)\right] \tag{4.8}$$

and choosing as initial data at s=0

$$\mathsf{H}(0) = \mathsf{H}^{(D)}. \tag{4.9}$$

The afore-propounded argument in the special case $\Gamma = \beta$ at once reveals the limiting value $\mathsf{H}(\infty)$ of that flow being *even*, i.e.:

$$0_{4\times 4} = \eta(\infty) = \beta\mathsf{H}(\infty) - \mathsf{H}(\infty)\beta. \tag{4.10}$$

Actually, the result $\mathsf{H}(\infty)$ being an even operator was obtained in [11] merely for a special case of *perturbation theory*, akin (but not identical) with the outcome of the *perturbative* procedure of consecutive step-by-step canonical transformations aiming at eliminating the odd terms in the Dirac Hamiltonian due to Foldy and Wouthuysen (FW) [2]. Comparing the results obtained with the BP-method with the ones obtained by the step-by-step FW-method expanding in the small Sommerfeld parameter $\kappa = \frac{v}{c}$ a discrepancy arises beyond order κ^4 in that expansion. Mind because of the ambiguity in the definition of the sequence of operators with such a step-by-step method, a block-diagonal operator resulting from *several* successive unitary 4×4-transformations is only guaranteed to give results being equivalent up to a 2×2-block-diagonal unitary transformation, as earlier on was already emphasized by Pursey [12] and by Eriksen and Kolsrud [13]. However, as has been concisely discussed in an elucidating article by Costella and McKellar [3], the crucial finding of Foldy and Wouthuysen is not so much concerned with the (certainly useful) perturbative step-by-step elimination of the odd terms in a Dirac Hamiltonian, but is in the main concerned with the unraveling of a problem of interpretation with the four-component amplitude $\Psi_\mu^{(D)}(\mathbf{r}, t)$ that solves the Dirac equation.

4.3 A Problem of Interpretation with the Four-Component Dirac Amplitude

In relativistic quantum mechanics, a fermion with attributes mass m, spin $\pm\frac{\hbar}{2}$, and charge $q_e = -|e|$, say moving in the presence of electrostatic and magnetostatic fields, is described in terms of a four-component amplitude $\Psi_\nu^{(D)}(\mathbf{r}, t)$ solving the Dirac equation[1]:

[1] As of now the summation convention always applies, if not suspended explicitly.

$$i\hbar\partial_t \Psi_\nu^{(D)}(\mathbf{r},t) = \mathsf{H}_{\nu,\nu'}^{(D)}\Psi_{\nu'}^{(D)}(\mathbf{r},t) \qquad \nu,\nu' \in \{1,2,3,4\}. \tag{4.11}$$

Here

$$\mathsf{H}^{(D)} = mc^2\beta + c\alpha_b\Pi_b + e\Phi(\mathsf{x})\,1_{4\times 4} \tag{4.12}$$

denotes the Hamiltonian of the relativistic fermion with minimal coupling to electromagnetic fields in Dirac–Pauli representation, with the well-known 4×4 matrices β and α_b anti-commuting and being of square equal to unity, see, for instance, [14–17]. The magnetic induction field $\mathsf{B} = \mathbf{rot}\,\mathsf{A}$ is encoded in the gauge invariant derivative operator composed of the Cartesian components of the conjugate momentum and position *operators* p_b and x_a of fundamental quantum mechanics:

$$\Pi_b = \mathsf{p}_b - e\mathsf{A}_b(\mathsf{x}) \qquad \left[\mathsf{p}_b, \mathsf{x}_a\right] = \frac{\hbar}{i}\delta_{a,b}\hat{1}. \tag{4.13}$$

According to the superposition principle, the Dirac amplitude $\Psi_\nu^{(D)}(\mathbf{r},t)$ may be represented as a linear combination of a *complete* system of orthonormal *four-component* eigenfunctions $U_\nu^{(D)}(\mathbf{r},k)$ and $V_\nu^{(D)}\left(\mathbf{r},\tilde{k}\right)$ of $\mathsf{H}^{(D)}$, entailing both the positive and the negative-energy eigenvalues $E_k > 0$ and $E_{\tilde{k}} < 0$

$$\Psi_\nu^{(D)}(\mathbf{r},t) = \sum_{k\in I_+} U_\nu^{(D)}(\mathbf{r},k)\,c_k e^{iE_k t} + \sum_{\tilde{k}\in I_-} V_\nu^{(D)}\left(\mathbf{r},\tilde{k}\right) b_{\tilde{k}} e^{-iE_{\tilde{k}} t}, \tag{4.14}$$

whereby (mind the summation convention)

$$\mathsf{H}_{\nu,\nu'}^{(D)} U_{\nu'}^{(D)}(\mathbf{r},k) = E_k U_\nu^{(D)}(\mathbf{r},k) \qquad \mathsf{H}_{\nu,\nu'}^{(D)} V_{\nu'}^{(D)}\left(\mathbf{r},\tilde{k}\right) = -\left|E_{\tilde{k}}\right| V_\nu^{(D)}\left(\mathbf{r},\tilde{k}\right). \tag{4.15}$$

In (4.14), the expansion coefficients, referred to as c_k and $b_{\tilde{k}}$, are c-numbers, obtainable from a prescribed amplitude $\Psi_\nu^{(D)}(\mathbf{r})$ at an initial time $t = 0$ making use of the orthonormality of those eigenfunctions. Mind the labels $k \in I_+$ and $\tilde{k} \in I_-$, counting the positive-energy eigenmodes $U_\nu^{(D)}(\mathbf{r},k)$ or rather the negative-energy eigenmodes $V_\nu^{(D)}\left(\mathbf{r},\tilde{k}\right)$ of the Dirac Hamiltonian, have in the presence of a (weak) electrostatic field $\mathscr{E}(\mathbf{r}) = -\nabla\Phi(\mathbf{r})$ possibly different codomains (for instance, if there exist bound states). Because a complete set of orthonormal eigenfunctions of the Dirac Hamiltonian comprises naturally the positive- *and* the negative-energy eigenstates *jointly*, the position operator in the Dirac theory, if defined by the operation of multiplication

$$\mathsf{x}_a\Psi_\nu^{(D)}(\mathbf{r},t) = r_a\Psi_\nu^{(D)}(\mathbf{r},t), \tag{4.16}$$

is *not* an operator defined over well-defined states of a *particle*, since in the evaluation of the expectation value $\left\langle \Psi^{(D)} \right| \mathsf{x}_a \left| \Psi^{(D)} \right\rangle$ of the position operator in a given Dirac state $\left| \Psi^{(D)} \right\rangle$ the positive-energy solutions *interfere* with the (seemingly unphysical) negative-energy solutions. The origin of this difficulty is the assumed consent of the Dirac amplitude $\Psi_\nu^{(D)}(\mathbf{r}, t)$ being a probability amplitude for *particles* just like in nonrelativistic Schrödinger quantum mechanics, which is plainly wrong, as has been first revealed by the analysis of the meaning of locality in quantum mechanics by Newton and Wigner [1]. For a brilliantly witty discussion of this point, already elucidated in pioneering work by Foldy and Wouthuysen [2], we refer to Costella and McKellar [3]. Indeed interpreting $\Psi_\nu^{(D)}(\mathbf{r}, t)$ as a probability amplitude gives cause to several well-known absurdities, for instance, the components of the "velocity" operator in the Heisenberg picture do not commute and have eigenvalues equal to $\pm c$, see [14, 15, 18]. Related to this is the concession that the phenomenon of the highly oscillatory "Zitterbewegung," sometimes discussed as an inevitable property of the relativistic electron, is actually not a physical property of a moving *particle* with attributes mass, charge, and spin [3].

4.3.1 The Newton–Wigner Amplitude and the Notion of Energy Separation

A physically correct probability amplitude, that in the style of the discussion given by Costella and McKellar [3] we refer to in what follows as the *Newton–Wigner* amplitude $\Psi_\nu^{(NW)}(\mathbf{r}, t)$, can be constructed from the four-component Dirac amplitude $\Psi_\nu^{(D)}(\mathbf{r}, t)$ by a suitable *unitary* transformation T, such that

$$\Psi_\nu^{(NW)}(\mathbf{r}, t) = \mathsf{T}_{\nu,\nu'} \Psi_{\nu'}^{(D)}(\mathbf{r}, t) \tag{4.17}$$

permits a meaningful expectation value, in particular, for the operators of position, velocity, spin, and also orbital angular momentum [2, 3].

To this end one looks first for a unitary transformation T requiring the transformed Dirac Hamiltonian

$$\mathsf{H}^{(NW)} \equiv \mathsf{T}\mathsf{H}^{(D)}\mathsf{T}^\dagger \tag{4.18}$$

be an *even* operator, so that

$$\tilde{\mathsf{H}}^{(NW)}\beta = \beta\tilde{\mathsf{H}}^{(NW)}. \tag{4.19}$$

Taken by itself the criterion (4.19) only ensures, minding our choice $\beta = diag\{1, 1, -1, -1\}$ in Dirac–Pauli representation that $\mathsf{H}^{(NW)}$ assumes a block-diagonal guise, but that doesn't warrant $\mathsf{H}^{(NW)}$ being *energy-separating*, the latter notion first introduced by Eriksen and Kolsrud [13]. Possibly the criterion is better understood introducing abstract bra- and ket-notation, so that $U_\mu^{(D)}(\mathbf{r}, k) = \left\langle \mathbf{r}, \mu | U_k^{(D)} \right\rangle$ and $V_\mu^{(D)}\left(\mathbf{r}, \tilde{k}\right) = \left\langle \mathbf{r}, \mu | V_{\tilde{k}}^{(D)} \right\rangle$. Introducing the spectral representations of the Dirac Hamiltonian and it's associated energy-sign operator $\Lambda^{(D)}$ in the basis of Dirac eigenstates,

$$\mathsf{H}^{(D)} = \sum_{k\in I_+} E_k \left|U_k^{(D)}\right\rangle\left\langle U_k^{(D)}\right| + \sum_{\tilde{k}\in I_-} (-\left|E_{\tilde{k}}\right|) \left|V_{\tilde{k}}^{(D)}\right\rangle\left\langle V_{\tilde{k}}^{(D)}\right| \tag{4.20}$$

$$\Lambda^{(D)} = \sum_{k} \left|U_k^{(D)}\right\rangle\left\langle U_k^{(D)}\right| - \sum_{\tilde{k}} \left|V_{\tilde{k}}^{(D)}\right\rangle\left\langle V_{\tilde{k}}^{(D)}\right| = \frac{\mathsf{H}^{(D)}}{\sqrt{\mathsf{H}^{(D)}\mathsf{H}^{(D)}}}, \tag{4.21}$$

it follows at once in terms of the unitary transformed Dirac eigenstates

$$\begin{aligned} \left|U_k^{(NW)}\right\rangle &= \mathsf{T}\left|U_k^{(D)}\right\rangle \\ \left|V_{\tilde{k}}^{(NW)}\right\rangle &= \mathsf{T}\left|V_{\tilde{k}}^{(D)}\right\rangle \end{aligned} \tag{4.22}$$

for the Dirac β-operator the representation

$$\beta = \mathsf{T}\Lambda^{(D)}\mathsf{T}^\dagger = \frac{\mathsf{H}^{(NW)}}{\sqrt{\mathsf{H}^{(NW)}\mathsf{H}^{(NW)}}}. \tag{4.23}$$

In reverse order, the *Newton–Wigner* Hamiltonian $\mathsf{H}^{(NW)}$ is defined as being one of a kind among all unitary transformed Hamiltonians with block-diagonal guise, satisfying additionally the identity [13]:

$$\mathsf{H}^{(NW)} = \beta\sqrt{\mathsf{H}^{(NW)}\mathsf{H}^{(NW)}}. \tag{4.24}$$

As a direct consequence of (4.24) being true then

$$\begin{aligned} (0_{4\times1})_\nu &= \left(\mathsf{H}^{(NW)} - \beta\sqrt{\mathsf{H}^{(NW)}\mathsf{H}^{(NW)}}\right)_{\nu,\nu'} U_{\nu'}^{(NW)}(\mathbf{r},k) \\ &= E_k\left(\hat{1}_{4\times4} - \beta\right)_{\nu,\nu'} U_{\nu'}^{(NW)}(\mathbf{r},k) \end{aligned} \tag{4.25}$$

$$\begin{aligned} (0_{4\times1})_\nu &= \left(\mathsf{H}^{(NW)} - \beta\sqrt{\mathsf{H}^{(NW)}\mathsf{H}^{(NW)}}\right)_{\nu,\nu'} V_{\nu'}^{(NW)}\left(\mathbf{r},\tilde{k}\right) \\ &= (-\left|E_{\tilde{k}}\right|)\left(\hat{1}_{4\times4} + \beta\right)_{\nu,\nu'} V_{\nu'}^{(NW)}\left(\mathbf{r},\tilde{k}\right). \end{aligned} \tag{4.26}$$

Assuming nonvanishing eigenvalues, $E_k \neq 0$ and $E_{\tilde{k}} \neq 0$, this entails at once, minding β being diagonal in Dirac–Pauli representation that $U_\nu^{(NW)}(\mathbf{r},k) \equiv 0$ regarding the lower components $\nu = 3, 4$ and further $V_\nu^{(NW)}\left(\mathbf{r},\tilde{k}\right) \equiv 0$ regarding the upper components $\nu = 1, 2$. This distinguishing feature indeed is the essence of the said *energy-separating* property of the Newton–Wigner representation, it being the representation in which the operator's position, velocity, orbital angular momentum, and spin of the free theory are agreeable to physical intuition just like in classical physics [2, 3, 19].

Perhaps, the physical meaning of the concept of energy separation becomes more comprehensible if one applies the unitary transformation T to the Dirac amplitude (4.14) and rewrites the Newton–Wigner amplitude (4.17) now with the transformed Dirac eigenstates (4.22) as

$$\Psi_\nu^{(NW)}(\mathbf{r}, t) = \sum_{k \in I_+} U_\nu^{(NW)}(\mathbf{r}, k)\, c_k e^{iE_k t} + \sum_{\tilde{k} \in I_-} V_\nu^{(NW)}\left(\mathbf{r}, \tilde{k}\right) b_{\tilde{k}} e^{-iE_{\tilde{k}} t}. \tag{4.27}$$

Expressly stated, $\Psi_\nu^{(NW)}(\mathbf{r}, t)$ assumes in consequence of $U_\nu^{(NW)}(\mathbf{r}, k) \equiv 0$ for $\nu = 3, 4$ and $V_\nu^{(NW)}(\mathbf{r}, k) \equiv 0$ for $\nu = 1, 2$ the guise

$$\Psi_\nu^{(NW)}(\mathbf{r}, t) = \begin{pmatrix} \psi_+(\mathbf{r}, t) \\ \psi_-(\mathbf{r}, t) \\ 0 \\ 0 \end{pmatrix}_\nu + \begin{pmatrix} 0 \\ 0 \\ \chi_+(\mathbf{r}, t) \\ \chi_-(\mathbf{r}, t) \end{pmatrix}_\nu . \tag{4.28}$$

With $\mathsf{H}^{(NW)}$ commuting with β in Pauli–Dirac representation, this entails an explicit 2×2 block matrix representation

$$\mathsf{H}^{(NW)} \equiv \begin{pmatrix} \mathsf{H}_{2\times 2}^{(e)}, & 0_{2\times 2} \\ 0_{2\times 2}, & \mathsf{H}_{2\times 2}^{(p)} \end{pmatrix}. \tag{4.29}$$

Then one readily confirms the resulting equations of motion governing the time evolution of the *two*-component amplitudes $\psi_\sigma(\mathbf{r}, t)$ and $\chi_\sigma(\mathbf{r}, t)$ in (4.28) are by construction forever propagating *without* any interference of positive- and negative-energy states, i.e.:

$$\begin{aligned} i\hbar\partial_t \psi_\sigma(\mathbf{r}, t) &= \left(\mathsf{H}_{2\times 2}^{(e)}\right)_{\sigma,\sigma'} \psi_{\sigma'}(\mathbf{r}, t) \\ i\hbar\partial_t \chi_\sigma(\mathbf{r}, t) &= \left(\mathsf{H}_{2\times 2}^{(p)}\right)_{\sigma,\sigma'} \chi_{\sigma'}(\mathbf{r}, t) \\ \sigma, \sigma' &\in \{+, -\}. \end{aligned} \tag{4.30}$$

These are Schrödinger–Pauli-type equations, whereas the amplitude $\psi_\sigma(\mathbf{r}, t)$ describes the electron as a particle (wave packet) with attributes mass m, spin $\pm\frac{\hbar}{2}$, and charge $-|e|$ being composed *exclusively* of positive-energy eigenfunctions $U_\mu^{(NW)}(\mathbf{r}, k)$. Correlating with this, as the amplitude $\chi_\sigma(\mathbf{r}, t)$ is composed solely of negative-energy eigenfunctions $V_\mu^{(NW)}\left(\mathbf{r}, \tilde{k}\right)$, it has been given a physical interpretation already by Dirac himself via charge conjugation, so that in effect the conjugated wavefunction $\chi^{\star}_{-\sigma}(\mathbf{r}, t)$ is describing the positron as an "anti-particle" (wave packet) with attributes mass m, spin $\mp\frac{\hbar}{2}$, and opposite charge $+|e|$. That way Dirac's hole picture implicitly already involved a quantum field theory context. The specific unitary transformation T, that leads from the Dirac Hamiltonian $\mathsf{H}^{(D)}$ to the energy-separating Hamiltonian $\mathsf{H}^{(NW)}$, indeed enables to bare the roots of Dirac's discovery of antimatter. A discussion of the meaning of the negative-energy states $\chi^{\star}_{-\sigma}(\mathbf{r}, t)$ can be found in Milonni's book [20], or else, for instance, in the appendix of our article [21, Appendix].

Last not least, while the Dirac Hamiltonian $\mathsf{H}^{(D)}$ is unique due to its linearity and minimal coupling to external potentials representing the electromagnetic fields, the Newton–Wigner Hamiltonian $\mathsf{H}^{(NW)}$ is the only one enabling a meaningful nonrelativistic limit in terms of particles and anti-particles moving as completely separated entities along their individual trajectories, as any bubble chamber track reveals. Unfortunately, a disadvantage of this very useful property of $\mathsf{H}^{(NW)}$ is that locality of the Dirac Hamiltonian $\mathsf{H}^{(D)}$, as it comprises only first-order differential operators, has been traded off for a *nonlocal* operator, with $\sqrt{\mathsf{H}^{(NW)}\mathsf{H}^{(NW)}}$ essentially being defined in terms of an integral kernel. But in view of obtaining a suitable starting point for approximations endeavoring to the quantum mechanics of nonrelativistic particles that flaw has little concernment.

4.4 The β-Flow

If a charge-carrying Dirac fermion moves in the presence of magnetostatic and electrostatic fields superposed, it is generally accepted to be difficult [5, 13, 22–24] obtaining the unitary transformation T, and in this way the Newton–Wigner Hamiltonian $\mathsf{H}^{(NW)}$ from a Dirac Hamiltonian $\mathsf{H}^{(D)}$ as stated in (4.12). Let us agree on terming operators $\mathcal{O}$ as being *odd* and operators $\mathcal{E}$ as being *even*, iff

$$\begin{aligned} \mathcal{O}\beta &= -\beta\mathcal{O} \\ \mathcal{E}\beta &= \beta\mathcal{E}. \end{aligned} \tag{4.31}$$

We aim in this section at constructing a flow striving from a general Hamiltonian

$$\mathsf{H}^{(D)} = \beta mc^2 + \mathcal{E} + \mathcal{O} \tag{4.32}$$

toward the corresponding Newton–Wigner Hamiltonian $\mathsf{H}^{(NW)}$. For example, a well-known extension of the Dirac Hamiltonian in external fields, as stated afore in (4.12), takes into account, besides mass m and charge q_e, further phenomenological attributes for the spin-$\frac{1}{2}$ "particles" like an intrinsic *anomalous* magnetic moment μ_M or even an intrinsic electric dipole moment d_E, see, for instance, [25]. In this case

$$\mathcal{O} = c\alpha_b \Pi_b + i\beta\alpha_b \left(\frac{\mu_M}{c} E_b^{(ext)} - c d_E B_b^{(ext)}\right) \tag{4.33}$$

$$\mathcal{E} = e\Phi(\mathsf{x}) - \frac{\Pi_b}{mc}\left(\mu_M B_b^{(ext)} + d_E E_b^{(ext)}\right). \tag{4.34}$$

Of course, with $d_E \neq 0$ then (spatial) parity is *not* conserved [26]. Whereas in standard QED for sure $d_E \equiv 0$ [26], instead in electroweak theory $d_E \neq 0$ appears quite reasonable. For a thorough discussion, see [19, 27].

In order to determine the exact solution to the nonlinear flow equation (4.4) with generator $\eta(s) = [\beta, \mathsf{H}(s)]$, the idea is to look for an operator $\mathsf{Z}(s)$ representing a continuous unitary transformation of the Dirac matrix β by solving the flow equation:

$$\begin{aligned}\frac{d}{ds}\mathsf{Z}(s) &= [\omega(s), \mathsf{Z}(s)] \\ \mathsf{Z}(0) &= \beta,\end{aligned} \tag{4.35}$$

and, in particular, to choose the generator $\omega(s)$ of that "beta-flow" in such a way that the limiting value $\mathsf{Z}(\infty)$ commutes with the original Dirac Hamiltonian (4.12), or else with (4.32)

$$\left[\tilde{\mathsf{H}}^{(D)}, \mathsf{Z}(\infty)\right] = 0_{4\times4}. \tag{4.36}$$

Along the line of reasoning presented afore in (4.4) a suitable anti-symmetric generator of such a beta-flow emerges as

$$\omega(s) = \left[\tilde{\mathsf{H}}^{(D)}, \mathsf{Z}(s)\right] \tag{4.37}$$

whereby for convenience

$$\tilde{\mathsf{H}}^{(D)} = \frac{1}{mc^2}\mathsf{H}^{(D)} = \beta + \tilde{\mathcal{E}} + \tilde{\mathcal{O}} \tag{4.38}$$

denotes a scaled Dirac Hamiltonian. If (4.35) could be solved for $\mathsf{Z}(s)$, then the generator $\omega(s)$ was known explicitly and the unitary transformation of β could be represented as

$$\mathsf{Z}(s) = \mathsf{V}(s)\,\beta\,\mathsf{V}^{\dagger}(s), \tag{4.39}$$

whereby the unitary operator $\mathsf{V}(s)$ solves

$$\begin{aligned}\frac{d}{ds}\mathsf{V}(s) &= \omega(s)\,\mathsf{V}(s) \\ \mathsf{V}(0) &= 1_{4\times4}.\end{aligned} \tag{4.40}$$

And because the transformation $\mathsf{V}(s)$ is unitary, of course, there holds

$$\beta\beta = \mathsf{Z}(s)\,\mathsf{Z}(s) = \mathsf{Z}(\infty)\,\mathsf{Z}(\infty) = 1_{4\times4}. \tag{4.41}$$

Consideration should be given to an ambiguity regarding the representation of $\mathsf{Z}(s)$ in (4.39) with such a unitary transformation $\mathsf{V}(s)$. Indeed, with $\mathsf{N}(s)$ another unitary operator with attributes

$$\mathsf{N}(s)\,\beta = \beta\,\mathsf{N}(s)\,, \tag{4.42}$$

then instead of (4.39) one finds for $\mathsf{Z}(s)$ as well the entirely equivalent representation:

$$\begin{aligned} \mathsf{Z}(s) &= \mathsf{U}(s)\,\beta\,\mathsf{U}^{\dagger}(s) \\ \mathsf{V}(s) &= \mathsf{U}(s)\,\mathsf{N}(s) \\ \mathsf{U}(0)\,\mathsf{N}(0) &= \mathsf{1}_{4\times 4}. \end{aligned} \tag{4.43}$$

So, even if $\mathsf{Z}(s)$ was known exactly, compliant with the representation (4.39) said unitary transformation $\mathsf{V}(s)$ in (4.43) cannot be determined any better up to an undetermined block-diagonal factor $\mathsf{N}(s)$.

4.4.1 *Exact Solution of Beta-Flow Equation*

With $\omega(s)$ as specified in (4.37) the flow equation (4.35) determining $\mathsf{Z}(s)$ reads

$$\frac{d}{ds}\mathsf{Z}(s) = \left[\left[\tilde{\mathsf{H}}^{(D)}, \mathsf{Z}(s)\right], \mathsf{Z}(s)\right]. \tag{4.44}$$

Evaluation of the double commutator gives

$$\left[\left[\tilde{\mathsf{H}}^{(D)}, \mathsf{Z}(s)\right], \mathsf{Z}(s)\right] = \tilde{\mathsf{H}}^{(D)}\mathsf{Z}(s)\,\mathsf{Z}(s) - 2\mathsf{Z}(s)\,\tilde{\mathsf{H}}^{(D)}\mathsf{Z}(s) + \mathsf{Z}(s)\,\mathsf{Z}(s)\,\tilde{\mathsf{H}}^{(D)}. \tag{4.45}$$

Because of (4.41) then (4.45) simplifies and the differential equation (4.44) determining $\mathsf{Z}(s)$ is

$$\begin{aligned} \frac{1}{2}\frac{d}{ds}\mathsf{Z}(s) &= \tilde{\mathsf{H}}^{(D)} - \mathsf{Z}(s)\,\tilde{\mathsf{H}}^{(D)}\mathsf{Z}(s) \\ \mathsf{Z}(0) &= \beta. \end{aligned} \tag{4.46}$$

Serendipitously this being a matrix Riccati equation [28], we can find an *exact* solution to this initial value problem in the guise

$$\mathsf{Z}(s) = \mathsf{W}(s)\,\beta\,\mathsf{W}^{-1}(s)\,, \tag{4.47}$$

with

$$\mathsf{W}(s) = \mathsf{C}(s)\,\beta + \mathsf{S}(s) \tag{4.48}$$

$$\mathsf{C}(s) = \cosh\left(2s\,\tilde{\mathsf{H}}^{(D)}\right)$$
$$\mathsf{S}(s) = \sinh\left(2s\,\tilde{\mathsf{H}}^{(D)}\right).$$

Despite $\beta\tilde{\mathsf{H}}^{(D)} \neq \tilde{\mathsf{H}}^{(D)}\beta$, there holds as an identity

$$\beta\mathsf{S}(s)\,\mathsf{S}(s) + \mathsf{C}(s)\,\mathsf{C}(s)\,\beta = \mathsf{S}(s)\,\mathsf{S}(s)\,\beta + \beta\mathsf{C}(s)\,\mathsf{C}(s), \tag{4.49}$$

and therefore

$$\mathsf{W}^{\dagger}(s)\,\mathsf{W}(s)\,\beta = \beta\,\mathsf{W}^{\dagger}(s)\,\mathsf{W}(s). \tag{4.50}$$

With that said

$$\mathsf{Z}(s) = \mathsf{Z}^{\dagger}(s), \tag{4.51}$$

and in this way

$$\mathsf{Z}(s)\,\mathsf{Z}^{\dagger}(s) = \mathsf{Z}^{\dagger}(s)\,\mathsf{Z}(s) = \mathsf{Z}(s)\,\mathsf{Z}(s) = 1_{4\times 4}, \tag{4.52}$$

thus validating in accord with the representation (4.39) the operator $\mathsf{Z}(s)$ being unitary (and involutive as well). With $\Lambda^{(D)}$ the energy-sign operator (4.21) in the basis of Dirac eigenstates, there holds

$$\lim_{s\to\infty}\mathsf{C}^{-1}(s)\,\mathsf{S}(s) = \lim_{s\to\infty}\frac{\tilde{\mathsf{H}}^{(D)}}{\sqrt{\tilde{\mathsf{H}}^{(D)}\tilde{\mathsf{H}}^{(D)}}}\tanh\left(2s\sqrt{\tilde{\mathsf{H}}^{(D)}\tilde{\mathsf{H}}^{(D)}}\right) = \Lambda^{(D)}, \tag{4.53}$$

and thus for $s \to \infty$ a meaningful limiting value of $\mathsf{Z}(s)$ identical to the energy-sign operator $\Lambda^{(D)}$ exists:

$$\begin{aligned}
\mathsf{Z}(\infty) &= \lim_{s\to\infty}\mathsf{Z}(s)\\
&= \lim_{s\to\infty}\left(\mathsf{C}(s)\left(\beta + \mathsf{C}^{-1}(s)\,\mathsf{S}(s)\right)\beta\left(\beta + \mathsf{C}^{-1}(s)\,\mathsf{S}(s)\right)^{-1}\mathsf{C}^{-1}(s)\right)\\
&= \lim_{s\to\infty}\left(\mathsf{C}(s)\left(\left(\beta + \Lambda^{(D)}\right)\beta\right)\left(\beta + \Lambda^{(D)}\right)^{-1}\mathsf{C}^{-1}(s)\right)\\
&= \lim_{s\to\infty}\left(\mathsf{C}(s)\left(\Lambda^{(D)}\left(\beta + \Lambda^{(D)}\right)\right)\left(\beta + \Lambda^{(D)}\right)^{-1}\mathsf{C}^{-1}(s)\right)\\
&= \Lambda^{(D)}.
\end{aligned} \tag{4.54}$$

4.4.2 Construction of Unitary Transformation $\mathsf{V}(s)$

Even though the operator $\mathsf{Z}(s)$ in (4.47) is the exact solution of the beta-flow (4.44), this doesn't concurrently determine the unitary transformation $\mathsf{V}(s)$ in the representation (4.39). Of course, $\mathsf{W}^{\dagger}(s)\,\mathsf{W}(s)$ being positive definite then a unitary operator

$\mathsf{U}^{(P)}(s)$ related to the polar (P) decomposition of the operator $\mathsf{W}(s)$ can be identified in the guise [29]

$$\mathsf{U}^{(P)}(s) = \mathsf{W}(s)\left(\mathsf{W}^{\dagger}(s)\,\mathsf{W}(s)\right)^{-\frac{1}{2}}. \tag{4.55}$$

And with help of (4.50) it follows then

$$\mathsf{Z}(s) = \mathsf{U}^{(P)}(s)\,\beta\left(\mathsf{U}^{(P)}(s)\right)^{\dagger}, \tag{4.56}$$

with $\mathsf{U}^{(P)}(s)$ (after re-arrangement of the square root term) given by

$$\mathsf{U}^{(P)}(s) = (\mathsf{C}(s)\,\beta + \mathsf{S}(s))\left((\beta\,\mathsf{C}(s) + \mathsf{S}(s))\,(\mathsf{C}(s)\,\beta + \mathsf{S}(s))\right)^{-\frac{1}{2}}. \tag{4.57}$$

But now a complicacy arises, as the identification of (4.57) with the searched for unitary transformation $\mathsf{V}(s)$, setting

$$\mathsf{V}(s) \equiv \mathsf{U}^{(P)}(s)\,\beta \tag{4.58}$$

in view of the posed initial value (4.40), actually applies only iff

$$\left[\beta, \sqrt{\tilde{\mathsf{H}}^{(D)}\tilde{\mathsf{H}}^{(D)}}\right] = 0_{4\times 4}. \tag{4.59}$$

This special constraint immediately implying $\mathsf{C}(s)\,\beta = \beta\,\mathsf{C}(s)$, only then a meaningful limit of $\mathsf{U}^{(P)}(s)$ for $s \to \infty$ is readily obtained

$$\mathsf{U}^{(P)}(\infty) = \frac{\beta + \Lambda^{(D)}}{\sqrt{\left(\beta + \Lambda^{(D)}\right)^{2}}}. \tag{4.60}$$

Unfortunately, the restriction (4.59) is not fitting in at all, say, with the presence of an external electrostatic potential $\Phi(\mathbf{r}) \neq 0$ in the Dirac Hamiltonian, as the square $\tilde{\mathsf{H}}^{(D)}\tilde{\mathsf{H}}^{(D)}$ then also comprises odd terms (anti-commuting with β). Thus, except in (special) cases when (4.59) applies, as concerns, for instance, the Dirac Hamiltonian $\mathsf{H}_0^{(D)}$ in a static external magnetic field *without* coupling to an electrostatic potential, for a general Dirac Hamiltonian $\tilde{\mathsf{H}}^{(D)}$ the limiting value of $\mathsf{U}^{(P)}(s)$ indeed remains indeterminable due to our ignorance how to find the limiting value of terms like $\mathsf{C}^{-1}(s)\,\beta\,\mathsf{S}(s)$ for $s \to \infty$. This symptom reveals the polar decomposition (4.55) being pointless if (4.59) ceases to be valid. Briefly speaking, that way we cannot find the searched for unitary transformation for a general Dirac Hamiltonian, even though the limiting value $\mathsf{Z}(s)$ for $s \to \infty$ has according to (4.54) already an assigned value.

Progress comes, accepting (for the moment being) a simplified way of writing $\mathsf{Z} \equiv \mathsf{Z}(s)$, from the observation

$$(\beta + \mathsf{Z})\,\beta = \mathsf{Z}\,(\beta + \mathsf{Z}) \tag{4.61}$$

$$\left[\beta, (\beta \mathsf{Z} + \mathsf{Z}\beta)\right] = 0_{4\times4} = \left[\mathsf{Z}, (\beta \mathsf{Z} + \mathsf{Z}\beta)\right],$$

so that

$$\begin{aligned} \mathsf{Z}(s) &= \mathsf{Z}(\beta + \mathsf{Z})(\beta + \mathsf{Z})^{-1} \\ &= (\beta + \mathsf{Z})\,\beta\,(\beta + \mathsf{Z})^{-1} \\ &= (\beta + \mathsf{Z})\,\beta\,(\beta + \mathsf{Z})^{-2}(\beta + \mathsf{Z}). \end{aligned} \tag{4.62}$$

The operator $\beta + \mathsf{Z}$ being Hermitian (and excluding *zero* as an eigenvalue of that operator), then for sure $(\beta + \mathsf{Z})^2$ is positive definite, so that

$$(\beta + \mathsf{Z})^{-2} = \left(\sqrt{(\beta + \mathsf{Z})^2}\sqrt{(\beta + \mathsf{Z})^2}\right)^{-1}. \tag{4.63}$$

On these grounds there follows now the representation

$$\mathsf{Z}(s) = \mathsf{U}(s)\,\beta\,\mathsf{U}^{\dagger}(s), \tag{4.64}$$

with $\mathsf{U}(s)$ being unitary and built-up in terms of the exact solution $\mathsf{Z}(s)$ as given in (4.47):

$$\mathsf{U}(s) = \frac{\beta + \mathsf{Z}(s)}{\sqrt{(\beta + \mathsf{Z}(s))^2}} \tag{4.65}$$

$$\mathsf{U}(0) = \beta. \tag{4.66}$$

With the known limiting value (4.54) then

$$\mathsf{U}(\infty) = \frac{\beta + \Lambda^{(D)}}{\sqrt{(\beta + \Lambda^{(D)})^2}}. \tag{4.67}$$

Remarkably enough, even though the operator $\mathsf{U}^{(\mathsf{P})}(s)$ introduced in (4.55) has only under the special premise (4.59) for $s \to \infty$ a definite limit (4.60), nonetheless that limiting value coincides with the generally valid result (4.67). Yet, in view of the posed initial value (4.40), the searched for unitary transformation is *not* $\mathsf{U}(s)$, but

$$\mathsf{Y}(s) \equiv \mathsf{U}(s)\,\beta = \frac{\beta + \mathsf{Z}(s)}{\sqrt{(\beta + \mathsf{Z}(s))^2}}\beta. \tag{4.68}$$

So in point of fact with $\mathsf{Y}(s)$ given in (4.68), we have

$$\mathsf{Z}(s) = \mathsf{Y}(s)\,\beta\,\mathsf{Y}^{\dagger}(s) \tag{4.69}$$

$$\mathsf{Y}(0) = 1_{4\times4}$$

$$\mathsf{Y}(s)\,\mathsf{Y}^{\dagger}(s) = \mathsf{1}_{4\times4} = \mathsf{Y}^{\dagger}(s)\,\mathsf{Y}(s)\,.$$

Notably, based on the identities

$$\begin{aligned} \mathsf{U}(s)\,\mathsf{U}(s) &= \mathsf{1}_{4\times4} \\ \beta\mathsf{U}(s) &= \mathsf{U}(s)\,\mathsf{Z}(s)\,, \end{aligned} \tag{4.70}$$

there holds for this particular operator (4.68) as well

$$\begin{aligned} \mathsf{Y}(s)\,\mathsf{Y}(s) &= \mathsf{U}(s)\,\beta\mathsf{U}(s)\,\beta \\ &= \mathsf{U}(s)\,\mathsf{U}(s)\,\mathsf{Z}(s)\,\beta \\ &= \mathsf{Z}(s)\,\beta. \end{aligned} \tag{4.71}$$

4.4.3 Retrieval of the Newton–Wigner Hamiltonian

The basic cause of the unitary transformation $\mathsf{Y}(s)$ stated in (4.68) being for $s \to \infty$ in fact energy-separating is the identity

$$\mathsf{Y}(\infty)\,\beta\mathsf{Y}^{\dagger}(\infty) = \mathsf{Z}(\infty) = \Lambda^{(D)}. \tag{4.72}$$

The seeked for unitary transformation T that maps the Dirac Hamiltonian $\mathsf{H}^{(D)}$ to the Newton–Wigner Hamiltonian $\mathsf{H}^{(NW)}$ we thus identify directly from (4.68) as

$$\mathsf{T} \equiv \mathsf{Y}^{\dagger}(\infty) = \beta\frac{\beta + \Lambda^{(D)}}{\sqrt{\left(\beta + \Lambda^{(D)}\right)^2}}. \tag{4.73}$$

Obviously there holds now

$$\beta\mathsf{T} = \mathsf{T}\Lambda^{(D)}. \tag{4.74}$$

Applying T to positive-energy eigenstates $\left|U_k^{(D)}\right\rangle$ of $\tilde{\mathsf{H}}^{(D)}$, respectively, applying T to negative-energy eigenstates $\left|V_{\tilde{k}}^{(D)}\right\rangle$ of $\tilde{\mathsf{H}}^{(D)}$, it is manifested that the unitary transformation T is as well energy-separating:

$$\begin{aligned} \beta\left(\mathsf{T}\left|U_k^{(D)}\right\rangle\right) &= \mathsf{T}\Lambda^{(D)}\left|U_k^{(D)}\right\rangle = +\mathsf{T}\left|U_k^{(D)}\right\rangle \\ \beta\left(\mathsf{T}\left|V_{\tilde{k}}^{(D)}\right\rangle\right) &= \mathsf{T}\Lambda^{(D)}\left|V_{\tilde{k}}^{(D)}\right\rangle = -\mathsf{T}\left|V_{\tilde{k}}^{(D)}\right\rangle. \end{aligned} \tag{4.75}$$

Indeed, due to (4.54), we have

$$\sqrt{\mathsf{H}^{(D)}\mathsf{H}^{(D)}} = \Lambda^{(D)}\mathsf{H}^{(D)} \tag{4.76}$$

$$
\begin{aligned}
&= \mathsf{Z}(\infty)\,\mathsf{H}^{(D)} \\
&= \left(\mathsf{Y}(\infty)\,\beta \mathsf{Y}^{\dagger}(\infty)\right)\mathsf{H}^{(D)} \\
&= \left(\mathsf{T}^{\dagger}\beta\,\mathsf{T}\right)\mathsf{H}^{(D)},
\end{aligned}
$$

and so it follows

$$
\begin{aligned}
\sqrt{\left(\mathsf{T}\mathsf{H}^{(D)}\mathsf{T}^{\dagger}\right)\left(\mathsf{T}\mathsf{H}^{(D)}\mathsf{T}^{\dagger}\right)} &= \mathsf{T}\left(\sqrt{\mathsf{H}^{(D)}\mathsf{H}^{(D)}}\right)\mathsf{T}^{\dagger} \\
&= \mathsf{T}\left(\mathsf{T}^{\dagger}\beta\mathsf{T}\right)\mathsf{H}^{(D)}\mathsf{T}^{\dagger} \\
&= \beta\left(\mathsf{T}\mathsf{H}^{(D)}\mathsf{T}^{\dagger}\right).
\end{aligned}
\tag{4.77}
$$

Writing $\Lambda^{(D)}\mathsf{H}^{(D)} = \sqrt{\mathsf{H}^{(D)}\mathsf{H}^{(D)}} = \mathsf{H}^{(D)}\Lambda^{(D)}$, then directly from (4.76) along the lines indicated in (4.77)

$$
\beta\left(\mathsf{T}\mathsf{H}^{(D)}\mathsf{T}^{\dagger}\right) = \left(\mathsf{T}\mathsf{H}^{(D)}\mathsf{T}^{\dagger}\right)\beta. \tag{4.78}
$$

Comparing (4.24) with (4.77) we thus identify, among all unitary transformations producing merely a 2×2—block-diagonalization of the Dirac–Pauli Hamiltonian, the even *and* energy-separating Newton–Wigner Hamiltonian being

$$
\mathsf{H}^{(NW)} = \mathsf{T}\mathsf{H}^{(D)}\mathsf{T}^{\dagger}, \tag{4.79}
$$

whereas T is the specific unitary transformation stated in (4.73).

4.4.4 Eriksen Transformation

In pioneering work Eriksen [5] fixed the unitary transformation U_E bearing his name, by *postulating*

$$
\mathsf{U}_E^{\dagger}\beta = \beta\mathsf{U}_E. \tag{4.80}
$$

From this he concluded

$$
\Lambda^{(D)} = \mathsf{U}_E^{\dagger}\beta\mathsf{U}_E = \beta\mathsf{U}_E^2, \tag{4.81}
$$

and (excluding -1 as an eigenvalue of $\beta\,\Lambda^{(D)}$) obtained [5][30]

$$
\mathsf{U}_E = \sqrt{\beta\,\Lambda^{(D)}} \equiv \frac{\frac{1}{2}\left(1_{4\times 4} + \beta\,\Lambda^{(D)}\right)}{\sqrt{1_{4\times 4} + \frac{1}{4}\left(\beta\,\Lambda^{(D)} + \Lambda^{(D)}\beta\, - 2 \times 1_{4\times 4}\right)}}. \tag{4.82}
$$

Comparing U_E with our result (4.73) we readily confirm the limiting value $\mathsf{T} = \lim_{s\to\infty} \mathsf{Y}^{\dagger}(s)$ being identical to Eriksen's transformation:

$$\mathsf{T} = \beta \frac{\beta + \Lambda^{(D)}}{\sqrt{\left(\beta + \Lambda^{(D)}\right)^2}} = \mathsf{U}_E. \tag{4.83}$$

4.4.5 A Link Between the Hamiltonian Flow and the Beta-Flow

As a general proposition, the solution $\mathsf{H}(s)$ to the (nonlinear) Hamiltonian flow equation (4.4) with generator $\eta(s) = [\beta, \mathsf{H}(s)]$ enables a representation in the guise

$$\mathsf{H}(s) = \mathsf{V}^{\dagger}(s)\, \tilde{\mathsf{H}}^{(D)} \mathsf{V}(s), \tag{4.84}$$

whereby $\mathsf{V}(s)$ is the afore-introduced unitary operator solving the initial value problem (4.40) with the generator of the beta-flow (4.35)

$$\omega(s) = \left[\tilde{\mathsf{H}}^{(D)}, \mathsf{Z}(s)\right]. \tag{4.85}$$

The validation of (4.84) follows readily calculating the derivative and minding subsequently (4.40), (4.85), and (4.39). Then

$$\begin{aligned}
\frac{d}{ds}\mathsf{H}(s) &= \mathsf{V}^{\dagger}(s)\left[\tilde{\mathsf{H}}^{(D)}, \omega(s)\right]\mathsf{V}(s) \\
&= \mathsf{V}^{\dagger}(s)\left(\left[\tilde{\mathsf{H}}^{(D)}, \left[\tilde{\mathsf{H}}^{(D)}, \mathsf{Z}(s)\right]\right]\right)\mathsf{V}(s) \\
&= \mathsf{V}^{\dagger}(s)\begin{pmatrix} \tilde{\mathsf{H}}^{(D)}\tilde{\mathsf{H}}^{(D)}\left(\mathsf{V}(s)\beta\mathsf{V}^{\dagger}(s)\right) \\ -2\tilde{\mathsf{H}}^{(D)}\left(\mathsf{V}(s)\beta\mathsf{V}^{\dagger}(s)\right)\tilde{\mathsf{H}}^{(D)} \\ +\left(\mathsf{V}(s)\beta\mathsf{V}^{\dagger}(s)\right)\tilde{\mathsf{H}}^{(D)}\tilde{\mathsf{H}}^{(D)} \end{pmatrix}\mathsf{V}(s) \\
&= \mathsf{H}(s)\mathsf{H}(s)\beta - 2\mathsf{H}(s)\beta\mathsf{H}(s) + \beta\mathsf{H}(s)\mathsf{H}(s) \\
&= \left[\left[\beta, \mathsf{H}(s)\right], \mathsf{H}(s)\right],
\end{aligned} \tag{4.86}$$

or else

$$\begin{aligned}
\frac{d}{ds}\mathsf{H}(s) &= \left[\eta(s), \mathsf{H}(s)\right] \\
\eta(s) &= \left[\beta, \mathsf{H}(s)\right] \\
\mathsf{H}(0) &= \tilde{\mathsf{H}}^{(D)}.
\end{aligned} \tag{4.87}$$

The ODE obtained for $\mathsf{H}(s)$ this way, with (scaled) initial data $\mathsf{H}(0) = \tilde{\mathsf{H}}^{(D)}$, coincides with the Hamiltonian flow (4.4), thus substantiating the assertion (4.84). Not unsuspected then, said generators, $\eta(s)$ for the Hamiltonian flow (4.87) and $\omega(s)$ for the associated beta-flow (4.35), are mutually connected by the same unitary transformation $\mathsf{V}(s)$, i.e., once $\omega(s)$ is known, then $\eta(s)$ is known and vice versa

$$\eta(s) = -\mathsf{V}^{\dagger}(s)\,\omega(s)\,\mathsf{V}(s)\,. \tag{4.88}$$

Indeed

$$\begin{aligned}\omega(s) &= \left[\mathsf{H}^{(D)}, \mathsf{Z}(s)\right]\\ &= \left[\mathsf{H}^{(D)}, \mathsf{V}(s)\,\beta\mathsf{V}^{\dagger}(s)\right]\\ &= -\mathsf{V}(s)\,\eta(s)\,\mathsf{V}^{\dagger}(s)\,.\end{aligned}$$

Notably, upon insertion of the afore-obtained *exact* solution $\mathsf{Z}(s)$ stated in (4.47) then the generator $\omega(s)$ being a *known* function of the flow parameter s, this fact enables to write an explicit (formal though) solution of the linear ODE (4.40) determining the unitary transformation $\mathsf{V}(s)$ as an s-ordered exponential

$$\mathsf{V}(s) = T_s \exp\left[\int_0^s ds'\omega\left(s'\right)\right] \tag{4.89}$$

or else known as the Dyson series [17]. Conversely, using unitarity (4.40) and re-expressing the generator of the beta-flow $\omega(s)$ in terms of $\eta(s)$ using the identity (4.88), it follows at once

$$\begin{aligned}0 &= \mathsf{V}^{\dagger}(s)\frac{d}{ds}\Bigg(\underbrace{\mathsf{V}(s)\,\mathsf{V}^{\dagger}(s)}_{=1}\Bigg)\\ &= \mathsf{V}^{\dagger}(s)\left(\frac{d}{ds}\mathsf{V}(s)\right)\mathsf{V}^{\dagger}(s) + \frac{d}{ds}\mathsf{V}^{\dagger}(s)\\ &= \left(\mathsf{V}^{\dagger}(s)\,\omega(s)\,\mathsf{V}(s)\right)\mathsf{V}^{\dagger}(s) + \frac{d}{ds}\mathsf{V}^{\dagger}(s)\\ &= -\eta(s)\,\mathsf{V}^{\dagger}(s) + \frac{d}{ds}\mathsf{V}^{\dagger}(s)\,,\end{aligned}$$

i.e., $\mathsf{V}^{\dagger}(s)$ solves the ODE

$$\begin{aligned}\frac{d}{ds}\mathsf{V}^{\dagger}(s) &= \eta(s)\,\mathsf{V}^{\dagger}(s)\\ \mathsf{V}^{\dagger}(0) &= 1_{4\times 4}.\end{aligned} \tag{4.90}$$

This fact implicates an equivalent representation for the adjoint (or else inverse) operator $\mathsf{V}^{\dagger}(s)$ in the guise of the Dyson series constructed with the generator $\eta(s)$ of the Hamiltonian flow (4.87):

$$\mathsf{V}^{\dagger}(s) = T_s \exp\left[\int_0^s ds' \eta(s')\right]. \tag{4.91}$$

Alas, because $\left[\beta, \sqrt{\tilde{\mathsf{H}}^{(D)}\tilde{\mathsf{H}}^{(D)}}\right] \neq 0$ for a general Dirac Hamiltonian $\tilde{\mathsf{H}}^{(D)}$ then as well $[\omega(s_1), \omega(s_2)] \neq 0$, alternatively $[\eta(s_1), \eta(s_2)] \neq 0$, which feature as a rule prevents an elementary calculation of $\mathsf{V}(s)$ in closed form.

Be that as it may, with a generator $\omega(s)$ of known identity (4.85) the explicit solution $\mathsf{V}(s)$ of the *linear* ODE (4.40) in the guise of the Dyson series (4.89) or else the Dyson series (4.91) build with $\eta(s)$ indeed provides an *exact* solution to the *nonlinear* Hamiltonian flow equation (4.87). Mind however, that due to the afore-mentioned ambiguity (4.43), even though $\mathsf{H}(s)$ as a solution to a double-bracket flow (4.87) by construction strives for $s \to \infty$ to a limiting value $\mathsf{H}(\infty)$ commuting with β, see (4.7), this feature taken by itself does *not* warrant as well $\mathsf{H}(\infty)$ being energy-separating.

4.4.6 The Newton–Wigner Hamiltonian for a Special Class of Dirac Hamiltonians

The afore-derived exact result (4.79) for the Newton–Wigner Hamiltonian is amenable to a substantial simplification for the particular class of Dirac Hamiltonians

$$\tilde{\mathsf{H}}_0^{(D)} = \beta + \tilde{\mathcal{E}}_0 + \tilde{\mathcal{O}}_0 \tag{4.92}$$

exhibiting the exceptional feature of a vanishing anti-commutator [19]

$$\left\{\tilde{\mathcal{O}}_0, \tilde{\mathcal{E}}_0\right\} \equiv 0. \tag{4.93}$$

Replacing now everywhere in the exact expression (4.47) the operator $\tilde{\mathsf{H}}^{(D)}$ by $\tilde{\mathsf{H}}_0^{(D)}$, one readily obtains proceeding directly from (4.47) the result

$$\begin{aligned}
\mathsf{Z}_0(s) &= \mathsf{t}_0(s)\,\Lambda_0^{(D)} \\
\mathsf{t}_0(s) &= \tanh\left(2s\sqrt{\tilde{\mathsf{H}}_0^{(D)}\tilde{\mathsf{H}}_0^{(D)}} + \operatorname{artanh}\left(\beta\Lambda_0^{(D)}\right)\right) \\
\Lambda_0 &= \frac{\tilde{\mathsf{H}}_0^{(D)}}{\sqrt{\tilde{\mathsf{H}}_0^{(D)}\tilde{\mathsf{H}}_0^{(D)}}} \\
\sqrt{\tilde{\mathsf{H}}_0^{(D)}\tilde{\mathsf{H}}_0^{(D)}} &\equiv \sqrt{\left(\beta + \tilde{\mathcal{E}}_0\right)^2 + \tilde{\mathcal{O}}_0^2}.
\end{aligned} \tag{4.94}$$

The associated generator $\omega_0(s)$ of the beta-flow assumes then the guise

$$\begin{aligned}\omega_0(s) &= \left[\tilde{\mathsf{H}}_0^{(D)}, \mathsf{Z}_0(s)\right] \\ &= \sqrt{\tilde{\mathsf{H}}_0^{(D)}\tilde{\mathsf{H}}_0^{(D)}}\left(\mathsf{t}_0^{\dagger}(s) - \mathsf{t}_0(s)\right).\end{aligned} \tag{4.95}$$

Besides being manifestly *odd*,

$$\beta\omega_0(s)\beta = -\omega_0(s) = \omega_0^{\dagger}(s), \tag{4.96}$$

that generator $\omega_0(s)$ has a *vanishing* commutator at different flow parameter values s_1, s_2 as well

$$[\omega_0(s_1), \omega_0(s_2)] = \mathsf{0}_{4\times 4}. \tag{4.97}$$

Because with (4.94) the functional dependence of $\omega_0(s)$ on the flow parameter s is known, now upon insertion of (4.94) into (4.95) an *exact* analytical expression for the unitary transformation $\mathsf{V}_0(s)$ can be given, so that

$$\mathsf{H}_0(s) = \mathsf{V}_0^{\dagger}(s)\tilde{\mathsf{H}}_0^{(D)}\mathsf{V}_0(s). \tag{4.98}$$

This is because the ODE defining $\mathsf{V}_0(s)$,

$$\begin{aligned}\frac{d}{ds}\mathsf{V}_0(s) &= \omega_0(s)\mathsf{V}_0(s) \\ \mathsf{V}_0(0) &= \mathsf{1}_{4\times 4}\end{aligned} \tag{4.99}$$

with $\omega_0(s)$ as stated in (4.95), can be solved due to (4.97) *exactly*

$$\mathsf{V}_0(s) = T_s \exp\left[\int_0^s ds'\omega_0(s')\right] \equiv \exp\left[\int_0^s ds'\omega_0(s')\right]. \tag{4.100}$$

The evaluation of the integral being elementary there follows without difficulty

$$\mathsf{V}_0(s) = \sqrt{\mathsf{Z}_0(s)\beta} = \frac{\beta + \mathsf{Z}_0(s)}{\sqrt{(\beta + \mathsf{Z}_0(s))^2}}\beta. \tag{4.101}$$

This outcome is in full accordance with the afore-derived general result (4.83), by replacing in (4.83) the operator $\tilde{\mathsf{H}}^{(D)}$ by $\tilde{\mathsf{H}}_0^{(D)}$. The limiting value of $\mathsf{Z}_0(s)$ for $s \to \infty$ as determined from (4.94) being now

$$\mathsf{Z}_0(\infty) = \Lambda_0^{(D)} = \frac{\tilde{\mathsf{H}}_0^{(D)}}{\sqrt{\tilde{\mathsf{H}}_0^{(D)}\tilde{\mathsf{H}}_0^{(D)}}}. \tag{4.102}$$

Identifying

$$\mathsf{T}_0 \equiv \mathsf{V}_0^\dagger(\infty) = \sqrt{\beta \Lambda_0^{(D)}} = \beta \frac{\beta + \Lambda_0^{(D)}}{\sqrt{\left(\beta + \Lambda_0^{(D)}\right)^2}}, \tag{4.103}$$

one readily sees using

$$\begin{aligned} \beta \mathsf{T}_0 &= \mathsf{T}_0 \Lambda_0^{(D)} \\ \Lambda_0^{(D)} \tilde{\mathsf{H}}_0^{(D)} &= \sqrt{\tilde{\mathsf{H}}_0^{(D)} \tilde{\mathsf{H}}_0^{(D)}} \end{aligned} \tag{4.104}$$

that (4.93) implies as well

$$\left[\mathsf{T}_0, \sqrt{\tilde{\mathsf{H}}_0^{(D)} \tilde{\mathsf{H}}_0^{(D)}}\right] = 0_{4\times 4}. \tag{4.105}$$

It is then straightforward to show with (4.104) the property

$$\begin{aligned} \tilde{\mathsf{H}}_0^{(NW)} &= \mathsf{T}_0 \tilde{\mathsf{H}}_0^{(D)} \mathsf{T}_0^\dagger \\ &= \beta \left(\beta \mathsf{T}_0\right) \tilde{\mathsf{H}}_0^{(D)} \mathsf{T}_0^\dagger \\ &= \beta \left(\mathsf{T}_0 \Lambda_0^{(D)}\right) \tilde{\mathsf{H}}_0^{(D)} \mathsf{T}_0^\dagger \\ &= \beta \mathsf{T}_0 \left(\sqrt{\tilde{\mathsf{H}}_0^{(D)} \tilde{\mathsf{H}}_0^{(D)}}\right) \mathsf{T}_0^\dagger. \end{aligned}$$

That way, with (4.105) and $\mathsf{T}_0 \mathsf{T}_0^\dagger = 1_{4\times 4}$ the Newton–Wigner Hamiltonian $\tilde{\mathsf{H}}_0^{(NW)}$ associated with a Dirac Hamiltonian $\tilde{\mathsf{H}}_0^{(D)}$ obeying to (4.93) assumes the special guise

$$\tilde{\mathsf{H}}_0^{(NW)} = \beta \sqrt{\tilde{\mathsf{H}}_0^{(D)} \tilde{\mathsf{H}}_0^{(D)}} = \beta \sqrt{\left(\beta + \tilde{\mathcal{E}}_0\right)^2 + \tilde{\mathcal{O}}_0^2}. \tag{4.106}$$

Without coupling to external fields, i.e., $\tilde{\mathcal{E}}_0 = 0$ and $\tilde{\mathcal{O}}_0 = \alpha_a \tilde{\mathsf{p}}_a$, this result coincides with what has been obtained by Foldy and Wouthuysen [2], and independently afore by Pryce [31] and also by Tani [32]. Choosing $\tilde{\mathcal{E}}_0 = 0$ and $\tilde{\mathcal{O}}_0 = \alpha_a \tilde{\Pi}_a$ then (4.106) describes as well the relativistic electron in the presence of a magnetic induction field, a result first obtained by Case [4], and using different methods by Eriksen [5]. Other examples for Hamiltonians $\tilde{\mathsf{H}}_0^{(D)}$ with the feature (4.93), so that (4.106) is valid, are considered in [19, 27, 33–35].

4.5 Manifestly Even Representation of the Newton–Wigner Hamiltonian as a Series of Iterated Commutators

The ensuing considerations apply to a general Dirac Hamiltonian (4.32) incorporating static external potentials and/or fields, as specified, for instance, in (4.33), (4.34). According to what has been said afore, the beta-flow equation (4.44) transforms the operator β into

$$\mathsf{Z}(s) = \mathsf{V}(s)\,\beta\,\mathsf{V}^{\dagger}(s)\,, \tag{4.107}$$

whereas due to (4.84) that same unitary operator $\mathsf{V}(s)$ concurrently serves as well to represent the solution to the Hamiltonian flow equation (4.87):

$$\mathsf{H}(s) = \mathsf{V}^{\dagger}(s)\,\tilde{\mathsf{H}}^{(D)}\mathsf{V}(s)\,. \tag{4.108}$$

Writing with anti-symmetric operators $\Omega_u(s)$ and $\Omega_g(s)$ now

$$\mathsf{V}(s) = e^{\Omega_u(s)}e^{\Omega_g(s)}, \tag{4.109}$$

whereby $\Omega_u(s)$ is odd and $\Omega_g(s)$ is even, we have

$$\begin{aligned} \Omega_u^{\dagger}(s) &= -\Omega_u(s) \\ \beta\Omega_u(s)\,\beta &= -\Omega_u(s) \end{aligned} \tag{4.110}$$

$$\begin{aligned} \Omega_g^{\dagger}(s) &= -\Omega_g(s) \\ \beta\Omega_g(s)\,\beta &= +\Omega_g(s)\,. \end{aligned} \tag{4.111}$$

Then it follows rewriting (4.108)

$$\mathsf{H}(s) = e^{-\Omega_g(s)}e^{-\Omega_u(s)}\tilde{\mathsf{H}}^{(D)}e^{\Omega_u(s)}e^{\Omega_g(s)}, \tag{4.112}$$

whereas from (4.107) and (4.109) we obtain with help of (4.111), (4.110) now

$$\mathsf{Z}(s) = e^{\Omega_u(s)}e^{\Omega_g(s)}\beta\,e^{-\Omega_g(s)}e^{-\Omega_u(s)} = e^{2\Omega_u(s)}\beta. \tag{4.113}$$

This representation for $\mathsf{Z}(s)$ being inherently consistent with (4.71) leads at once to the identifications

$$\begin{aligned} \mathsf{Y}(s) &= e^{\Omega_u(s)} \\ \mathsf{T} &= e^{-\Omega_u(\infty)}. \end{aligned} \tag{4.114}$$

That a factorization like (4.109) exists follows from a general result in Lie group theory on generalized polar decompositions, understanding the role of the operator β in relations like (4.111) and (4.110) essentially being equivalent to the action of an "involutive automorphism" [36],[37].

Because the limiting value $\mathsf{H}(\infty)$ of the Hamiltonian flow (4.4) obeys by construction to $\left[\beta, \mathsf{H}(\infty)\right] = 0$, now the searched for Newton–Wigner (NW) Hamiltonian arises with help of (4.112) in the guise

$$\begin{aligned} \tilde{\mathsf{H}}^{(NW)} &\equiv e^{+\Omega_g(\infty)} \mathsf{H}(\infty)\, e^{-\Omega_g(\infty)} && (4.115) \\ &= e^{-\Omega_u(\infty)} \tilde{\mathsf{H}}^{(D)} e^{\Omega_u(\infty)} && (4.116) \\ &= \mathsf{T}\tilde{\mathsf{H}}^{(D)}\mathsf{T}^{\dagger}. \end{aligned}$$

If $\mathsf{H}(\infty)$ was obtained, for instance, solving the Hamiltonian flow (4.87) perturbatively along the lines indicated in [11], then for sure $\mathsf{H}(\infty)$ is an even operator, but it is *not* guaranteed $\mathsf{H}(\infty)$ being energy-separating as well. Seen from another perspective, the unitary transformation performed with the even operator $e^{+\Omega_g(\infty)}$ in (4.115) can be regarded, once $\Omega_g(\infty)$ was known, as a "correction scheme" that converts the merely block-diagonal (even) limiting value $\mathsf{H}(\infty)$ of the Hamiltonian flow (4.87) into the *unique* energy-separating Newton–Wigner Hamiltonian.

Alternatively, $\tilde{\mathsf{H}}^{(NW)}$ may be obtained directly from the unitary transformation (4.116) performed solely with the odd operator $\Omega_u(\infty)$. Once $\Omega_u(\infty)$ is known, then the common BCH-expansion [38] leads to a representation as a series of commutators:

$$\tilde{\mathsf{H}}^{(NW)} = \tilde{\mathsf{H}}^{(D)} - \left[\Omega_u(\infty), \tilde{\mathsf{H}}^{(D)}\right] + \frac{1}{2!}\left[\Omega_u(\infty), \left[\Omega_u(\infty), \tilde{\mathsf{H}}^{(D)}\right]\right] + \ldots. \quad (4.117)$$

Here it is important to realize that a straightforward evaluation of this series of iterated commutators is needlessly complicated, because due to (4.116) the operator $\tilde{\mathsf{H}}^{(NW)}$ is unconditionally guaranteed to be *even*. A considerable simplification thus results rewriting (4.116) in the equivalent guise

$$\begin{aligned} \tilde{\mathsf{H}}^{(NW)} &= \frac{1}{2}\left(e^{-\Omega_u(\infty)}\tilde{\mathsf{H}}^{(D)}e^{\Omega_u(\infty)} + \beta e^{-\Omega_u(\infty)}\tilde{\mathsf{H}}^{(D)}e^{\Omega_u(\infty)}\beta\right) && (4.118) \\ &= \frac{1}{2}e^{-\Omega_u(\infty)}\tilde{\mathsf{H}}^{(D)}e^{\Omega_u(\infty)} + \frac{1}{2}e^{+\Omega_u(\infty)}\left(\beta\tilde{\mathsf{H}}^{(D)}\beta\right)e^{-\Omega_u(\infty)} \\ &= \frac{1}{2}e^{-\Omega_u(\infty)}\left(\beta + \tilde{\mathcal{O}} + \mathcal{E}\right)e^{\Omega_u(\infty)} + \frac{1}{2}e^{+\Omega_u(\infty)}\left(\beta - \tilde{\mathcal{O}} + \mathcal{E}\right)e^{-\Omega_u(\infty)} \\ &= \frac{1}{2}\left(\exp\left(-\mathrm{ad}_{\Omega_u(\infty)}\right) + \exp\left(+\mathrm{ad}_{\Omega_u(\infty)}\right)\right)\circ\left(\beta + \tilde{\mathcal{E}}\right) \\ &\quad -\frac{1}{2}\left(\exp\left(\mathrm{ad}_{\Omega_u(\infty)}\right) - \exp\left(-\mathrm{ad}_{\Omega_u(\infty)}\right)\right)\circ\tilde{\mathcal{O}}, \end{aligned}$$

whereby the symbol ad_X denotes here a most useful notation to describe iterated commutators as powers, see, for instance, [39]. With given operators X and $\mathcal{F}$ then

$$\begin{aligned} \mathrm{ad}_X \circ \mathcal{F} &= \left[X, \mathcal{F}\right] \\ (\mathrm{ad}_X)^2 \circ \mathcal{F} &= \mathrm{ad}_X \circ \left[X, \mathcal{F}\right] = \left[X, \left[X, \mathcal{F}\right]\right] \\ &\dots \\ (\mathrm{ad}_X)^n \circ \mathcal{F} &= \underbrace{\left[X, \dots \left[X, \left[X, \mathcal{F}\right]\right] \dots\right]}_{\text{n-fold}}. \end{aligned} \tag{4.119}$$

It follows introducing formal power series

$$\begin{aligned} \exp\left(\mathrm{ad}_X\right) \circ \mathcal{F} &= \sum_{j=0}^{\infty} \frac{1}{j!} \left(\mathrm{ad}_X\right)^j \circ \mathcal{F} \equiv e^X \mathcal{F} e^{-X} \\ \cosh\left(\mathrm{ad}_X\right) \circ \mathcal{F} &= \sum_{j=0}^{\infty} \frac{1}{(2j)!} \left(\mathrm{ad}_X\right)^{2j} \circ \mathcal{F} \\ \sinh\left(\mathrm{ad}_X\right) \circ \mathcal{F} &= \sum_{j=0}^{\infty} \frac{1}{(2j+1)!} \left(\mathrm{ad}_X\right)^{2j+1} \circ \mathcal{F}. \end{aligned} \tag{4.120}$$

That way an *explicit* representation for the (scaled) energy-separating Newton–Wigner Hamiltonian is obtained directly from (4.118) in the guise of a series of iterated commutators composed solely with the odd operator $\Omega_u(\infty)$, each term in this series expansion being *manifestly* even

$$\tilde{\mathsf{H}}^{(NW)} = \cosh\left(\mathrm{ad}_{\Omega_u(\infty)}\right) \circ \left(\beta + \tilde{\mathcal{E}}\right) - \sinh\left(\mathrm{ad}_{\Omega_u(\infty)}\right) \circ \tilde{\mathcal{O}}. \tag{4.121}$$

Provided a perturbative expansion for the operator $\Omega_u(\infty)$ could be found, then the associated perturbative expansion for the Newton–Wigner–Hamiltonian as represented by (4.121) is reduced to evaluating just a few commutators.

4.5.1 A Perturbative Expansion in the Style of Magnus for the Operators $\Omega_u(s)$ and $\Omega_g(s)$ Defining $\mathsf{V}(s) = e^{\Omega_u(s)} e^{\Omega_g(s)}$

A restriction of the Dirac Hamiltonian $\tilde{\mathsf{H}}^{(D)}$ in (4.32) to the low-energy sector of its spectrum suggests a weighting of the respective contributions of the gradient terms as set, for instance, by the small ratio of the Compton wave length $\lambda_C \equiv \frac{\hbar}{mc}$ to a typical atomic length of order of the Bohr radius $a_B \gg \lambda_C$. For reasons of bookkeeping we therefore write in terms of a dimensionless parameter κ (scaled units)

$$\tilde{\mathsf{H}}^{(D)} = \beta + \kappa\,\tilde{\mathcal{O}} + \kappa^2\,\tilde{\mathcal{E}} \equiv \tilde{\mathsf{H}}_0 + \kappa\,\tilde{\mathsf{H}}_1 + \kappa^2\,\tilde{\mathsf{H}}_2, \tag{4.122}$$

and agree to equate $\kappa = 1$ at the end of the calculations. Notably, here we consider the term $\tilde{\mathcal{E}}$ comprising the electrostatic Coulomb potential being small of second order, as its nonrelativistic expectation value is comparable to the expectation value of the kinetic energy term $\tilde{\mathcal{O}}\tilde{\mathcal{O}}$, this property being in agreement with the virial theorem, for instance [40].

In the ensuing our objective is to find a perturbation series representation of the Newton–Wigner Hamiltonian (4.121) in powers of κ. To this end we first aim at obtaining a perturbation expansion of the anti-Hermitian operators $\Omega_u\,(s)$ and $\Omega_g\,(s)$, these operators in fact being closely connected to the generator $\omega\,(s)$ of the beta-flow (4.35) or else to the generator $\eta\,(s)$ of the Hamiltonian flow (4.87).

For that purpose let us rewrite the identity (4.88) in the guise

$$e^{-\Omega_u(s)}\omega\,(s)\,e^{\Omega_u(s)} = -e^{\Omega_g(s)}\eta\,(s)\,e^{-\Omega_g(s)}. \tag{4.123}$$

As the generator $\eta\,(s)$ of the Hamiltonian flow is by construction an odd operator,

$$\begin{aligned} \eta\,(s) &= \left[\beta, \mathsf{H}\,(s)\right] \\ \beta\eta\,(s) + \eta\,(s)\,\beta &= 0_{4\times 4} \end{aligned} \tag{4.124}$$

and the commutator of an even operator with an odd operator is always odd as well, then the even part of the left-hand side in (4.123) should vanish identically,

$$e^{-\Omega_u(s)}\omega\,(s)\,e^{\Omega_u(s)} + \beta e^{-\Omega_u(s)}\omega\,(s)\,e^{\Omega_u(s)}\beta = -e^{\Omega_g(s)}\left(\eta\,(s) + \beta\eta\,(s)\,\beta\right)e^{-\Omega_g(s)} \equiv 0. \tag{4.125}$$

That way a (hidden) correlation between the even and odd parts of the generator $\omega\,(s)$ of the beta-flow is revealed

$$\omega\,(s) = -e^{2\Omega_u(s)}\left(\beta\,\omega\,(s)\,\beta\right)e^{-2\Omega_u(s)}. \tag{4.126}$$

In consequence of $\omega\,(s)$ being (in general) composed of even and odd terms,

$$\omega\,(s) = \omega_g\,(s) + \omega_u\,(s)\,, \tag{4.127}$$

then

$$\begin{aligned} \omega_g\,(s) + \omega_u\,(s) &= e^{2\Omega_u(s)}\left(-\omega_g\,(s) + \omega_u\,(s)\right)e^{-2\Omega_u(s)} \\ &= \exp\left(2\mathsf{ad}_{\Omega_u(s)}\right)\circ\left(-\omega_g\,(s) + \omega_u\,(s)\right), \end{aligned} \tag{4.128}$$

equivalently

$$\omega_g\,(s) = \tanh\left(\mathsf{ad}_{\Omega_u(s)}\right)\circ\omega_u\,(s)\,. \tag{4.129}$$

From this insight a useful relation connecting the generator $\eta(s)$ solely with the odd part $\omega_u(s)$ of the generator $\omega(s)$ emerges in the guise

$$\begin{aligned}
-e^{\Omega_g(s)}\eta(s)\,e^{-\Omega_g(s)} &= e^{-\Omega_u(s)}\omega(s)\,e^{\Omega_u(s)} \\
&= \exp\left(-\mathsf{ad}_{\Omega_u(s)}\right)\circ\left(\omega_g(s)+\omega_u(s)\right) \\
&= \exp\left(-\mathsf{ad}_{\Omega_u(s)}\right)\left(\tanh\left(\mathsf{ad}_{\Omega_u(s)}\right)+1\right)\circ\omega_u(s) \\
&= \frac{1}{\cosh\left(\mathsf{ad}_{\Omega_u(s)}\right)}\circ\omega_u(s)\,.
\end{aligned} \tag{4.130}$$

Consequently the ODE (4.40) together with (4.123) leads to

$$\begin{aligned}
\frac{d}{ds}\mathsf{V}(s) &= \frac{d}{ds}\left(e^{\Omega_u(s)}e^{\Omega_g(s)}\right) \\
&= \left(\frac{d}{ds}e^{\Omega_u(s)}\right)e^{\Omega_g(s)}+e^{\Omega_u(s)}\left(\frac{d}{ds}e^{\Omega_g(s)}\right) \\
&\overset{!}{=} \omega(s)\left(e^{\Omega_u(s)}e^{\Omega_g(s)}\right) \\
&= -e^{\Omega_u(s)}e^{\Omega_g(s)}\eta(s)\,.
\end{aligned} \tag{4.131}$$

Employing the well-known formula for the derivative of an exponential $e^{\Omega(s)}$ in case $[\Omega(s_1),\Omega(s_2)]\neq 0$, for instance [38, 39], now

$$\begin{aligned}
\frac{d}{ds}e^{\Omega(s)} &= \int_0^1 d\tau\, e^{\tau\,\Omega(s)}\frac{d}{ds}\Omega(s)\,e^{-\tau\,\Omega(s)}e^{\Omega(s)} \\
&= \left(\int_0^1 d\tau\, e^{\tau\,\mathsf{ad}_{\Omega(s)}}\circ\frac{d}{ds}\Omega(s)\right)e^{\Omega(s)} \\
&= \left(\frac{\exp\left(\mathsf{ad}_{\Omega(s)}\right)-1}{\mathsf{ad}_{\Omega(s)}}\circ\frac{d}{ds}\Omega(s)\right)e^{\Omega(s)}.
\end{aligned} \tag{4.132}$$

Up next (4.131) along with (4.130) is readily shown to be equivalent to

$$\begin{aligned}
-e^{\Omega_g(s)}\eta(s)\,e^{-\Omega_g(s)} &= \frac{1}{\cosh\left(\mathsf{ad}_{\Omega_u(s)}\right)}\circ\omega_u(s) \\
&= \frac{1-\exp\left(-\mathsf{ad}_{\Omega_u(s)}\right)}{\mathsf{ad}_{\Omega_u(s)}}\circ\frac{d}{ds}\Omega_u(s)+\frac{\exp\left(\mathsf{ad}_{\Omega_g(s)}\right)-1}{\mathsf{ad}_{\Omega_g(s)}}\circ\frac{d}{ds}\Omega_g(s)\,.
\end{aligned} \tag{4.133}$$

The left-hand side in (4.133) being manifestly odd, the *first* term on the right-hand side in (4.133) decomposes into even and odd parts

$$\frac{1-\exp\left(-\mathsf{ad}_{\Omega_u(s)}\right)}{\mathsf{ad}_{\Omega_u(s)}}\circ\frac{d}{ds}\Omega_u(s) \tag{4.134}$$

$$= \left(\frac{1 - \cosh\left(\mathrm{ad}_{\Omega_u(s)}\right)}{\mathrm{ad}_{\Omega_u(s)}} + \frac{\sinh\left(\mathrm{ad}_{\Omega_u(s)}\right)}{\mathrm{ad}_{\Omega_u(s)}} \right) \circ \frac{d}{ds}\Omega_u(s)$$

$$= \underbrace{\frac{1 - \cosh\left(\mathrm{ad}_{\Omega_u(s)}\right)}{\mathrm{ad}_{\Omega_u(s)}} \circ \frac{d}{ds}\Omega_u(s)}_{\text{even}} + \underbrace{\frac{\sinh\left(\mathrm{ad}_{\Omega_u(s)}\right)}{\mathrm{ad}_{\Omega_u(s)}} \circ \frac{d}{ds}\Omega_u(s)}_{\text{odd}},$$

whereas the *second* term on the right-hand side in (4.133) is manifestly even

$$\underbrace{\frac{\exp\left(\mathrm{ad}_{\Omega_g(s)}\right) - 1}{\mathrm{ad}_{\Omega_g(s)}} \circ \frac{d}{ds}\Omega_g(s)}_{\text{even}} = \sum_{n=0}^{\infty} \frac{1}{(n+1)!}\mathrm{ad}^n_{\Omega_g(s)} \circ \frac{d}{ds}\Omega_g(s). \tag{4.135}$$

A decomposition like (4.134) applies, because the commutator of an even operator with an even operator or else the commutator of an odd operator with an odd operator always results in an operator being *even*, whereas the commutator of an even operator with an odd operator and vice versa always results in an operator being *odd*.

Equating the even parts and the odd parts on either side of (4.133) regarded separately one obtains instead of (4.133) now two equations for $\frac{d}{ds}\Omega_u(s)$ and $\frac{d}{ds}\Omega_g(s)$:

$$\frac{\sinh\left(\mathrm{ad}_{\Omega_u(s)}\right)}{\mathrm{ad}_{\Omega_u(s)}} \circ \frac{d}{ds}\Omega_u(s) = \frac{1}{\cosh\left(\mathrm{ad}_{\Omega_u(s)}\right)} \circ \omega_u(s) \tag{4.136}$$

$$\frac{1 - \cosh\left(\mathrm{ad}_{\Omega_u(s)}\right)}{\mathrm{ad}_{\Omega_u(s)}} \circ \frac{d}{ds}\Omega_u(s) + \frac{\exp\left(\mathrm{ad}_{\Omega_g(s)}\right) - 1}{\mathrm{ad}_{\Omega_g(s)}} \circ \frac{d}{ds}\Omega_g(s) = 0. \tag{4.137}$$

Obviously, the first equation (4.136) directly determines $\frac{d}{ds}\Omega_u(s)$, as it is *decoupled* from the equation for $\frac{d}{ds}\Omega_g(s)$, whereas the determination of $\frac{d}{ds}\Omega_g(s)$ with the second equation (4.137) requires prior knowledge of $\frac{d}{ds}\Omega_u(s)$. Fortunately though, the determination of $\Omega_g(s)$ is for the purpose of determining the Newton–Wigner Hamiltonian $\mathrm{H}^{(NW)}$ redundant, as the representation (4.121) reveals at one glance.

Solving finally (4.136) for $\frac{d}{ds}\Omega_u(s)$ an equivalent ODE determining $\Omega_u(s)$ is obtained

$$\frac{d}{ds}\Omega_u(s) = \frac{2\mathrm{ad}_{\Omega_u(s)}}{\sinh\left(2\mathrm{ad}_{\Omega_u(s)}\right)} \circ \omega_u(s) \tag{4.138}$$

$$\Omega_u(0) = 0.$$

This being at first sight a scary nonlinear problem, in what follows it proves otherwise, noting that integration of (4.138) with respect to s leads to the integral equation:

$$\Omega_u(s) = \int_0^s ds' \frac{2\mathrm{ad}_{\Omega_u(s')}}{\sinh\left(2\mathrm{ad}_{\Omega_u(s')}\right)} \circ \omega_u\left(s'\right). \tag{4.139}$$

Provided the odd part $\omega_u(s)$ of the generator $\omega(s)$ of the beta-flow can be considered as being small, then that integral equation is amenable to a perturbative solution for the odd operator $\Omega_u(s)$ by the method of Picard iteration. Clearly, the described perturbative solution method is a close relative to the well-known Magnus series expansion, for a comprehensive review see [39].

4.5.2 Perturbation Series for Generator $\omega(s)$ of Beta-Flow

We are looking in the ensuing for a representation of the generator of the beta-flow (4.85) in the guise of a perturbation series:

$$\omega(s) = \left[\tilde{\mathsf{H}}^{(D)}, \mathsf{Z}(s)\right] = \sum_{n=1}^{\infty} \kappa^n \omega^{(n)}(s). \tag{4.140}$$

To this end first a corresponding perturbation series expansion progressing in powers of κ of the solution $\mathsf{Z}(s)$ to the initial value problem (4.46) is searched for. Insertion of

$$\mathsf{Z}(s) = \mathsf{Z}_0(s) + \kappa\,\mathsf{Z}_1(s) + \kappa^2\,\mathsf{Z}_2(s) + \kappa^3\,\mathsf{Z}_3(s) + \ldots \tag{4.141}$$

and identifying $\tilde{\mathsf{H}}_0 = \beta$, $\tilde{\mathsf{H}}_1 = \tilde{\mathcal{O}}$, $\tilde{\mathsf{H}}_2 = \tilde{\mathcal{E}}$ and $\tilde{\mathsf{H}}_n = 0_{4\times4}$ for $n \geq 3$ it then follows from (4.46)

$$\begin{aligned} \frac{1}{2}\frac{d}{ds}\mathsf{Z}_0(s) &= \tilde{\mathsf{H}}_0 - \mathsf{Z}_0(s)\,\tilde{\mathsf{H}}_0\mathsf{Z}_0(s) \\ \mathsf{Z}_0(0) &= \beta, \end{aligned} \tag{4.142}$$

and thus for $n = 1, 2, 3, \ldots$

$$\begin{aligned} \frac{1}{2}\frac{d}{ds}\mathsf{Z}_n(s) &= \tilde{\mathsf{H}}_n - \sum_{j=0}^{n} \mathsf{Z}_j(s)\,\tilde{\mathsf{H}}_0\mathsf{Z}_{n-j}(s) - \sum_{j=0}^{n-1} \mathsf{Z}_j(s)\,\tilde{\mathsf{H}}_1\mathsf{Z}_{n-1-j}(s) - \sum_{j=0}^{n-2} \mathsf{Z}_j(s)\,\tilde{\mathsf{H}}_2\mathsf{Z}_{n-2-j}(s) \\ \mathsf{Z}_n(s=0) &= 0_{4\times4}. \end{aligned} \tag{4.143}$$

With $\mathsf{Z}_0(s) = \beta$ being the unique solution to (4.142) it follows from (4.143)

$$\mathsf{Z}_1(s) = \left(1 - e^{-4s}\right)\tilde{\mathcal{O}} \tag{4.144}$$

$$\mathsf{Z}_2(s) = -\frac{1}{2}\left(1 - e^{-4s}\right)^2 \tilde{\mathcal{O}}\tilde{\mathcal{O}}\beta. \tag{4.145}$$

This brings up the conjecture of $\mathsf{Z}_{2n}(s)$ being even and $\mathsf{Z}_{2n+1}(s)$ being odd, i.e.:

$$Z_j(s)\,\beta = (-1)^j\,\beta Z_j(s)\,. \tag{4.146}$$

Proof by induction:

- In view of (4.144), (4.145) the statement is true for $j = 1, 2$.
- Induction hypothesis: (4.146) applies for $j = n$.
- Induction step: It is shown that (4.146) is correct for $j = n + 1$ solving a differential equation for the difference $Z_{n+1}(s)\,\beta - (-1)^{n+1}\,\beta Z_{n+1}(s)$. Insertion of (4.143) leads to

$$\frac{1}{2}\frac{d}{ds}\left(Z_{n+1}(s)\,\beta - (-1)^{n+1}\,\beta Z_{n+1}(s)\right)$$
$$= \begin{cases} -\left(\sum\limits_{j=0}^{n+1} Z_j(s)\,\tilde{H}_0 Z_{n+1-j}(s) + \sum\limits_{j=0}^{n} Z_j(s)\,\tilde{H}_1 Z_{n-j}(s) + \sum\limits_{j=0}^{n-1} Z_j(s)\,\tilde{H}_2 Z_{n-1-j}(s)\right)\beta \\ +(-1)^{n+1}\,\beta\left(\sum\limits_{j=0}^{n+1} Z_j(s)\,\tilde{H}_0 Z_{n+1-j}(s) + \sum\limits_{j=0}^{n} Z_j(s)\,\tilde{H}_1 Z_{n-j}(s) + \sum\limits_{j=0}^{n-1} Z_j(s)\,\tilde{H}_2 Z_{n-1-j}(s)\right). \end{cases}$$

Making use of (4.146) for $j = 1, 2, ..., n$ then with $\tilde{H}_0$ and $\tilde{H}_2$ both being even and $\tilde{H}_1$ being odd, all terms $Z_j(s)$ with $0 \le j \le n$ eventually clear away, in total leaving over

$$\frac{1}{2}\frac{d}{ds}\left(Z_{n+1}(s)\,\beta - (-1)^{n+1}\,\beta Z_{n+1}(s)\right) = -2\left(Z_{n+1}(s)\,\beta - (-1)^{n+1}\,\beta Z_{n+1}(s)\right).$$

In conformity with the posed initial values $Z_{n+1}(s) = 0_{4\times4}$ the solution to this homogenous linear differential equation is

$$Z_{n+1}(s)\,\beta - (-1)^{n+1}\,\beta Z_{n+1}(s) = e^{-4s}\left(Z_{n+1}(0)\,\beta - (-1)^{n+1}\,\beta Z_{n+1}(0)\right) \equiv 0,$$

thus proving (4.146) being true for $j = n + 1$ as well.

So then (4.146) being true implies that

$$\beta Z_{2n+1}(s) + Z_{2n+1}(s)\,\beta = 0_{4\times4} \tag{4.147}$$
$$\beta Z_{2n}(s) + Z_{2n}(s)\,\beta = 2\,Z_{2n}(s)\,\beta. \tag{4.148}$$

In view of the property $Z(s)\,Z(s) = 1_{4\times4}$ of the exact solution (4.47) the terms $Z_j(s)$ of the perturbation series expansion (4.141) should obey to the following constraints:

$$\begin{aligned} Z_0(s)\,Z_0(s) &= 1_{4\times4} \\ Z_0(s)\,Z_1(s) + Z_1(s)\,Z_0(s) &= 0_{4\times4} \\ &\dots\;\dots\;\dots \\ &n \ge 2 \\ Z_0(s)\,Z_n(s) + Z_n(s)\,Z_0(s) + \sum_{j=1}^{n-1} Z_j(s)\,Z_{n-j}(s) &= 0_{4\times4}. \end{aligned} \tag{4.149}$$

Notably, on the basis of (4.148) the even terms $\mathsf{Z}_{2n}(s)$ are obtainable directly from (4.149)

$$\mathsf{Z}_{2n}(s) = -\frac{1}{2}\sum_{j=1}^{2n-1} \mathsf{Z}_j(s)\,\mathsf{Z}_{2n-j}(s)\,\beta. \tag{4.150}$$

The labor of solving the differential equations (4.143) thus concerns the odd terms $\mathsf{Z}_{2n+1}(s)$ only. Setting $n = 3$ in (4.143) then, after insertion of $\mathsf{Z}_0(s) = \beta$ together with $\mathsf{Z}_1(s)$ and $\mathsf{Z}_2(s)$ as stated in (4.144) and (4.145), respectively:

$$\begin{aligned}\frac{1}{2}\frac{d}{ds}\mathsf{Z}_3(s) &= -\sum_{j=0}^{3}\mathsf{Z}_j(s)\,\tilde{\mathsf{H}}_0\mathsf{Z}_{3-j}(s) - \sum_{j=0}^{2}\mathsf{Z}_j(s)\,\tilde{\mathsf{H}}_1\mathsf{Z}_{2-j}(s) - \sum_{j=0}^{1}\mathsf{Z}_j(s)\,\tilde{\mathsf{H}}_2\mathsf{Z}_{1-j}(s)\\ &= -2\mathsf{Z}_3(s) - \left(1-e^{-4s}\right)\left(1-e^{-8s}\right)\tilde{\mathcal{O}}^3 - \left(1-e^{-4s}\right)\left[\tilde{\mathcal{O}},\tilde{\mathcal{E}}\right]\beta.\end{aligned} \tag{4.151}$$

Conforming with the posed initial value $\mathsf{Z}_3(s=0) = 0_{4\times 4}$ the unique solution to this inhomogeneous ordinary linear differential equation is

$$\begin{aligned}\mathsf{Z}_3(s) &= 2\int_0^s ds' e^{-4(s-s')}\left(-\left(1-e^{-4s'}\right)\left(1-e^{-8s'}\right)\tilde{\mathcal{O}}^3 - \left(1-e^{-4s'}\right)\left[\tilde{\mathcal{O}},\tilde{\mathcal{E}}\right]\beta\right) \\ &= \left(-\frac{1}{2} + \left(\frac{3}{4}+2s\right)e^{-4s} - \frac{1}{2}e^{-8s} + \frac{1}{4}e^{-12s}\right)\tilde{\mathcal{O}}^3 + \left(-\frac{1}{2} + \left(\frac{1}{2}+2s\right)e^{-4s}\right)\left[\tilde{\mathcal{O}},\tilde{\mathcal{E}}\right]\beta.\end{aligned} \tag{4.152}$$

Next from (4.150)

$$\begin{aligned}\mathsf{Z}_4(s) &= -\frac{1}{2}\Big(\mathsf{Z}_1(s)\,\mathsf{Z}_3(s) + \mathsf{Z}_3(s)\,\mathsf{Z}_1(s) + \mathsf{Z}_2(s)\,\mathsf{Z}_2(s)\Big)\beta \\ &= \begin{cases}\left(\frac{3}{8} - \left(\frac{3}{4}+2s\right)e^{-4s} + \left(\frac{1}{2}+2s\right)e^{-8s} - \frac{1}{4}e^{-12s} + \frac{1}{8}e^{-16s}\right)\tilde{\mathcal{O}}^4\beta + \\ + \left(\frac{1}{4} - \left(\frac{1}{2}+s\right)e^{-4s} + \left(\frac{1}{4}+s\right)e^{-8s}\right)\left[\tilde{\mathcal{O}},\left[\tilde{\mathcal{O}},\tilde{\mathcal{E}}\right]\right].\end{cases}\end{aligned} \tag{4.153}$$

The determination of $\mathsf{Z}_5(s)$ follows along similar lines of the determination of $\mathsf{Z}_3(s)$, but is more elaborate. Inserting the previous results $\mathsf{Z}_j(s)$ for $j \leq 4$ and using the identities

$$\begin{aligned}\tilde{\mathcal{O}}^2\left[\tilde{\mathcal{O}},\tilde{\mathcal{E}}\right] + \left[\tilde{\mathcal{O}},\tilde{\mathcal{E}}\right]\tilde{\mathcal{O}}^2 &= \frac{2}{3}\left[\tilde{\mathcal{O}}^3,\tilde{\mathcal{E}}\right] + \frac{1}{3}\left[\tilde{\mathcal{O}},\left[\tilde{\mathcal{O}},\left[\tilde{\mathcal{O}},\tilde{\mathcal{E}}\right]\right]\right]\\ \tilde{\mathcal{O}}\left[\tilde{\mathcal{O}},\tilde{\mathcal{E}}\right]\tilde{\mathcal{O}} &= \frac{1}{3}\left[\tilde{\mathcal{O}}^3,\tilde{\mathcal{E}}\right] - \frac{1}{3}\left[\tilde{\mathcal{O}},\left[\tilde{\mathcal{O}},\left[\tilde{\mathcal{O}},\tilde{\mathcal{E}}\right]\right]\right]\end{aligned} \tag{4.154}$$

straightforward calculations give

$$\frac{1}{2}\frac{d}{ds}\mathsf{Z}_5(s) = -\sum_{j=0}^{5}\mathsf{Z}_j(s)\,\tilde{\mathsf{H}}_0\mathsf{Z}_{5-j}(s) - \sum_{j=0}^{4}\mathsf{Z}_j(s)\,\tilde{\mathsf{H}}_1\mathsf{Z}_{4-j}(s) - \sum_{j=0}^{3}\mathsf{Z}_j(s)\,\tilde{\mathsf{H}}_2\mathsf{Z}_{3-j}(s)$$

$$= -2\mathsf{Z}_5(s) + \begin{pmatrix} \left(\frac{3}{4} - (1+2s)e^{-4s} - 4se^{-8s} + \left(\frac{1}{2}+6s\right)e^{-12s} - \frac{3}{4}e^{-16s} + \frac{1}{2}e^{-20s}\right)\tilde{\mathcal{O}}^5 \\ +\left(\frac{1}{4} - \left(\frac{1}{3}+\frac{1}{3}s\right)e^{-4s} - \left(\frac{1}{12}+s\right)e^{-8s} + \left(\frac{1}{6}+\frac{4}{3}s\right)e^{-12s}\right)\left[\tilde{\mathcal{O}},\left[\tilde{\mathcal{O}},\left[\tilde{\mathcal{O}},\tilde{\mathcal{E}}\right]\right]\right]\beta \\ +\left(\frac{1}{2} - \left(\frac{5}{12}+\frac{8}{3}s\right)e^{-4s} - \frac{1}{6}e^{-8s} + \left(\frac{1}{12}+\frac{2}{3}s\right)e^{-12s}\right)\left[\tilde{\mathcal{O}}^3,\tilde{\mathcal{E}}\right]\beta \\ +\left(\frac{1}{2} - \left(\frac{1}{2}+2s\right)e^{-4s}\right)\left[\left[\tilde{\mathcal{O}},\tilde{\mathcal{E}}\right],\tilde{\mathcal{E}}\right] \end{pmatrix}.$$

Conforming with the posed initial value $\mathsf{Z}_5(s=0) = 0_{4\times 4}$ the unique solution to this *inhomogeneous* ordinary linear differential equation is

$$\mathsf{Z}_5(s) = 2\int_0^s ds' e^{-4(s-s')} \times$$
$$\times \begin{pmatrix} \left(\frac{3}{4} - (1+2s')e^{-4s'} - 4s'e^{-8s'} + \left(\frac{1}{2}+6s'\right)e^{-12s'} - \frac{3}{4}e^{-16s'} + \frac{1}{2}e^{-20s'}\right)\tilde{\mathcal{O}}^5 \\ +\left(\frac{1}{4} - \left(\frac{1}{3}+\frac{1}{3}s'\right)e^{-4s'} - \left(\frac{1}{12}+s'\right)e^{-8s'} + \left(\frac{1}{6}+\frac{4}{3}s'\right)e^{-12s'}\right)\left[\tilde{\mathcal{O}},\left[\tilde{\mathcal{O}},\left[\tilde{\mathcal{O}},\tilde{\mathcal{E}}\right]\right]\right]\beta \\ +\left(\frac{1}{2} - \left(\frac{5}{12}+\frac{8}{3}s'\right)e^{-4s'} - \frac{1}{6}e^{-8s'} + \left(\frac{1}{12}+\frac{2}{3}s'\right)e^{-12s'}\right)\left[\tilde{\mathcal{O}}^3,\tilde{\mathcal{E}}\right]\beta \\ +\left(\frac{1}{2} - \left(\frac{1}{2}+2s'\right)e^{-4s'}\right)\left[\left[\tilde{\mathcal{O}},\tilde{\mathcal{E}}\right],\tilde{\mathcal{E}}\right] \end{pmatrix}$$
$$= \begin{pmatrix} \left(\frac{3}{8} - \left(\frac{5}{8}+2s+2s^2\right)e^{-4s} + \left(\frac{1}{2}+2s\right)e^{-8s} - \left(\frac{5}{16}+\frac{3}{2}s\right)e^{-12s} + \frac{1}{8}e^{-16s} - \frac{1}{16}e^{-20s}\right)\tilde{\mathcal{O}}^5 \\ +\left(\frac{1}{8} - \left(\frac{5}{24}+\frac{2}{3}s+\frac{1}{3}s^2\right)e^{-4s} + \left(\frac{1}{6}+\frac{1}{2}s\right)e^{-8s} - \left(\frac{1}{12}+\frac{1}{3}s\right)e^{-12s}\right)\left[\tilde{\mathcal{O}},\left[\tilde{\mathcal{O}},\left[\tilde{\mathcal{O}},\tilde{\mathcal{E}}\right]\right]\right]\beta \\ +\left(\frac{1}{4} - \left(\frac{1}{6}+\frac{4}{3}s\right)e^{-4s} - \frac{1}{24}e^{-8s} - \left(\frac{1}{24}+\frac{1}{6}s\right)e^{-12s}\right)\left[\tilde{\mathcal{O}}^3,\tilde{\mathcal{E}}\right]\beta \\ +\left(\frac{1}{4} - \left(\frac{1}{4}+s+2s^2\right)e^{-4s}\right)\left[\left[\tilde{\mathcal{O}},\tilde{\mathcal{E}}\right],\tilde{\mathcal{E}}\right] \end{pmatrix}. \tag{4.155}$$

The respective terms $\omega^{(n)}(s)$ of the searched perturbation series (4.140) are now readily identified

$$\begin{aligned} \omega^{(0)}(s) &= \left[\tilde{\mathsf{H}}_0, \mathsf{Z}_0(s)\right] \\ \omega^{(1)}(s) &= \left[\tilde{\mathsf{H}}_0, \mathsf{Z}_1(s)\right] + \left[\tilde{\mathsf{H}}_1, \mathsf{Z}_0(s)\right] \\ n &\geq 2 \\ \omega^{(n)}(s) &= \left[\tilde{\mathsf{H}}_0, \mathsf{Z}_n(s)\right] + \left[\tilde{\mathsf{H}}_1, \mathsf{Z}_{n-1}(s)\right] + \left[\tilde{\mathsf{H}}_2, \mathsf{Z}_{n-2}(s)\right]. \end{aligned} \tag{4.156}$$

After insertion of the afore-obtained results for $\mathsf{Z}_n(s)$, straightforward algebra leads eventually to the results

$$\begin{aligned} \omega^{(0)}(s) &= 0_{4\times 4} \\ \omega^{(1)}(s) &= 2e^{-4s}\tilde{\mathcal{O}}\beta \\ \omega^{(2)}(s) &= 0_{4\times 4} \\ \omega^{(3)}(s) &= \left(\left(\frac{1}{2} - 4s\right)e^{-4s} - \frac{1}{2}e^{-12s}\right)\tilde{\mathcal{O}}^3\beta - 4s\cdot e^{-4s}\left[\tilde{\mathcal{O}},\tilde{\mathcal{E}}\right] \\ \omega^{(4)}(s) &= \left(\left(-\frac{1}{2} + 2s\right)e^{-4s} + \frac{1}{2}e^{-8s}\right)\left[\tilde{\mathcal{O}}^2,\tilde{\mathcal{E}}\right]\beta \end{aligned}$$

$$\omega^{(5)}(s) = \begin{pmatrix} \left(\left(-\frac{1}{4}+4s^2\right)e^{-4s}+\left(\frac{1}{8}+3s\right)e^{-12s}+\frac{1}{8}e^{-20s}\right)\tilde{O}^5\beta \\ +\left(\left(-\frac{1}{12}+\frac{1}{3}s+\frac{2}{3}s^2\right)e^{-4s}-\frac{1}{12}e^{-8s}+\left(\frac{1}{6}+\frac{2}{3}s\right)e^{-12s}\right)\left[\tilde{O},\left[\tilde{O},\left[\tilde{O},\tilde{\mathcal{E}}\right]\right]\right] \\ +\left(\left(-\frac{5}{12}+\frac{2}{3}s\right)e^{-4s}+\frac{7}{12}e^{-8s}+\left(-\frac{1}{6}+\frac{1}{3}s\right)e^{-12s}\right)\left[\tilde{O}^3,\tilde{\mathcal{E}}\right] \\ +4s^2e^{-4s}\left[\left[\tilde{O},\tilde{\mathcal{E}}\right],\tilde{\mathcal{E}}\right]\beta \end{pmatrix}. \tag{4.157}$$

As inferred from (4.156) together with (4.146), apparently the even-numbered terms $\omega^{(2n)}(s)$ are *even* whereas the odd-numbered terms $\omega^{(2n+1)}(s)$ are *odd*, i.e.:

$$\beta\omega^{(2n)}(s) = \omega^{(2n)}(s)\,\beta \tag{4.158}$$

$$\beta\omega^{(2n+1)}(s) = -\omega^{(2n+1)}(s)\,\beta. \tag{4.159}$$

Accordingly the series expansion (4.140) representing the generator $\omega(s)$ decomposes into

$$\begin{aligned} \omega(s) &= \omega_g(s) + \omega_u(s) \\ \omega_g(s) &= \sum_{n=2}^{\infty} \kappa^{2n}\omega^{(2n)}(s) \\ \omega_u(s) &= \sum_{n=0}^{\infty} \kappa^{2n+1}\omega^{(2n+1)}(s)\,. \end{aligned} \tag{4.160}$$

Note that because $\omega^{(0)}(s) = \omega^{(2)}(s) \equiv 0$ the series determining the even part $\omega_g(s)$ starts at the first with the fourth-order term $\omega^{(4)}(s)$ only.

4.5.3 *Expansion of Generator $\Omega_u(s)$ of Eriksen Transformation*

In view of the perturbation series representation of

$$\omega_u(s) = \sum_{n=0}^{\infty} \kappa^{2n+1}\omega^{(2n+1)}(s) \tag{4.161}$$

it is natural adopting a corresponding perturbation series expansion for the generator $\Omega_u(s)$ of the Eriksen transformation:

$$\Omega_u(s) = \sum_{n=0}^{\infty} \kappa^{2n+1}\Omega^{(2n+1)}(s)\,. \tag{4.162}$$

With the expansion

$$\frac{2z}{\sinh(2z)} = 1 - \frac{2}{3}z^2 + \frac{14}{45}z^4 - \frac{124}{945}z^6 + O\left(z^8\right) \tag{4.163}$$

there follows in place of (4.139)

$$\Omega_u(s) = \int_0^s ds' \left(1 - \frac{2}{3}\left(\mathsf{ad}_{\Omega_u(s')}\right)^2 + \frac{14}{45}\left(\mathsf{ad}_{\Omega_u(s')}\right)^4 - \frac{124}{945}\left(\mathsf{ad}_{\Omega_u(s')}\right)^6 + O\left(\kappa^8\right)\right) \circ \omega_u\left(s'\right). \tag{4.164}$$

Inserting the series expansions (4.161) and (4.162) into (4.164) one finds in a straightforward manner comparing coefficients of κ^{2n+1} on either side of (4.164) a recursion relation determining the searched for terms $\Omega^{(2n+1)}(s)$. Obviously the first-order term is

$$\Omega^{(1)}(s) = \int_0^s ds'\omega^{(1)}\left(s'\right) = -\frac{1-e^{-4s}}{2}\beta\tilde{O}, \tag{4.165}$$

which immediately implies the vanishing of the commutator

$$\left[\Omega^{(1)}\left(s'\right), \omega^{(1)}\left(s'\right)\right] = 0. \tag{4.166}$$

Consequently all the commutator terms $\left(\mathsf{ad}_{\Omega_u(s')}\right)^{2n} \circ \omega_u\left(s'\right)$ in (4.164) are at least small of order $O\left(\kappa^{2n+3}\right)$. With that said the third-order term is

$$\begin{aligned}\Omega^{(3)}(s) &= \int_0^s ds'\omega^{(3)}\left(s'\right) \\ &= \left(-\frac{1}{4} + e^{-4s}\left(\frac{1}{4}+s\right)\right)\left[\tilde{O}, \tilde{\mathcal{E}}\right] + \left(\frac{1}{6} - e^{-4s}\left(\frac{1}{8}+s\right) - \frac{e^{-12s}}{24}\right)\beta\tilde{O}^3,\end{aligned} \tag{4.167}$$

whereas the (more arduous) evaluation of the fifth-order term gives

$$\begin{aligned}\Omega^{(5)}(s) &= \begin{cases}\int_0^s ds'\omega^{(5)}\left(s'\right) \\ -\frac{2}{3}\int_0^s ds'\left(\left[\Omega^{(1)}\left(s'\right), \left[\Omega^{(3)}\left(s'\right), \omega^{(1)}\left(s'\right)\right]\right] + \left[\Omega^{(1)}\left(s'\right), \left[\Omega^{(1)}\left(s'\right), \omega^{(3)}\left(s'\right)\right]\right]\right)\end{cases} \\ &= \begin{pmatrix} \left(\frac{1}{10} - \left(\frac{1}{16} + \frac{1}{2}s + s^2\right)e^{-4s} - \left(\frac{1}{32} + \frac{1}{4}s\right)e^{-12s} - \frac{1}{160}e^{-20s}\right)\tilde{O}^5\beta \\ +\left(\left(\frac{5}{144} - \left(\frac{1}{48} + \frac{2}{9}s + \frac{1}{6}s^2\right)e^{-4s} + \frac{1}{36}se^{-8s} - \left(\frac{1}{72} + \frac{1}{18}s\right)e^{-12s}\right)\right)\left[\tilde{O}, \left[\tilde{O}, \left[\tilde{O}, \tilde{\mathcal{E}}\right]\right]\right] \\ +\left(-\frac{7}{288} + \left(\frac{1}{16} + \frac{1}{18}s\right)e^{-4s} - \left(\frac{1}{32} + \frac{1}{9}s\right)e^{-8s} - \left(\frac{1}{144} + \frac{1}{36}s\right)e^{-12s}\right)\left[\tilde{O}^3, \tilde{\mathcal{E}}\right] \\ +\left(-\frac{1}{8} + \left(\frac{1}{8} + \frac{1}{2}s + s^2\right)e^{-4s}\right)\left[\tilde{\mathcal{E}}, \left[\tilde{O}, \tilde{\mathcal{E}}\right]\right]\beta \end{pmatrix}.\end{aligned} \tag{4.168}$$

Notably, the term $\Omega^{(5)}(s)$ will appear in the calculations for the first time if the expansion of $\tilde{\mathsf{H}}^{(NW)}$ according to the lines indicated in (4.121) aims at an accuracy *better* than κ^6.

We did not address the issue of the *convergence* of the Magnus-type expansion (4.162). Guided by physical intuition, with operators $\tilde{\mathcal{E}}$ and $\tilde{O}$ assigned to a Dirac Hamiltonian in external potentials like in (4.12), the described expansion method is conjectured being convergent for all electric field strengths $\mathscr{E} = \left\| \frac{i}{\hbar} \left[\tilde{O}, \tilde{\mathcal{E}} \right] \right\|$ vastly below the Schwinger critical field $\mathscr{E}_S = \frac{m}{|e|} \frac{c^2}{\lambda_C} \simeq 1.3 \times 10^{18} \left[\frac{V}{m} \right]$. Perhaps one could find along the lines discussed in [39] a sharp estimate for the radius of convergence of the expansion (4.162), establishing that way, for instance, a stability criterion for the existence of the Newton–Wigner Hamiltonian of the relativistic electron in the presence of external electrostatic fields.

4.6 The Relativistic Corrections to the Hamiltonian $\mathsf{H}^{(SP)}$ of Schrödinger–Pauli Quantum Mechanics

The task is now to evaluate the afore-given *explicit* representation (4.121) for the Newton–Wigner Hamiltonian as a series of iterated commutators by inserting the obtained series expansion for $\Omega_u(s)$ in (4.162) and to find that way the searched for perturbative terms $\Omega^{(2n+1)}(s)$. Let us agree in the ensuing on abbreviating the limiting value of the operator $\Omega_u(s)$ for $s \to \infty$ as

$$\Omega_u(\infty) \equiv \Omega_u = \kappa \Omega^{(1)} + \kappa^3 \Omega^{(3)} + \kappa^5 \Omega^{(5)} + \ldots, \tag{4.169}$$

whereby according to (4.165), (4.167), (4.168) the terms $\Omega^{(2n+1)} \equiv \Omega^{(2n+1)}(\infty)$ are given by

$$\begin{aligned}
\Omega^{(1)} &= \frac{1}{2}\tilde{O}\beta \\
\Omega^{(3)} &= -\frac{1}{6}\tilde{O}^3\beta - \frac{1}{4}\left[\tilde{O}, \tilde{\mathcal{E}}\right] \\
\Omega^{(5)} &= \frac{1}{10}\tilde{O}^5\beta + \frac{5}{144}\left[\tilde{O}, \left[\tilde{O}, \left[\tilde{O}, \tilde{\mathcal{E}}\right]\right]\right] - \frac{7}{288}\left[\tilde{O}^3, \tilde{\mathcal{E}}\right] - \frac{1}{8}\left[\tilde{\mathcal{E}}, \left[\tilde{O}, \tilde{\mathcal{E}}\right]\right]\beta,
\end{aligned} \tag{4.170}$$

and so on. Notably, these results apply for general *odd* operators $\tilde{O}$ and for general *even* operators $\tilde{\mathcal{E}}$ as defined afore in (4.33) and (4.34), respectively.

Restricting the expansion of $\tilde{\mathsf{H}}^{(NW)}$ to order $O\left(\kappa^8\right)$ it follows directly from (4.121)

$$\tilde{\mathsf{H}}^{(NW)} = \begin{cases} \left(\beta + \kappa^2\tilde{\mathcal{E}}\right) \\ +\frac{1}{2}\left(\mathrm{ad}_{\Omega_u}\right)^2 \circ \left(\beta + \kappa^2\tilde{\mathcal{E}}\right) + \frac{1}{24}\left(\mathrm{ad}_{\Omega_u}\right)^4 \circ \left(\beta + \kappa^2\tilde{\mathcal{E}}\right) + \frac{1}{720}\left(\mathrm{ad}_{\Omega_u}\right)^6 \circ \left(\beta + \kappa^2\tilde{\mathcal{E}}\right) \\ -\left(\mathrm{ad}_{\Omega_u}\right) \circ \kappa\tilde{O} - \frac{1}{6}\left(\mathrm{ad}_{\Omega_u}\right)^3 \circ \kappa\tilde{O} - \frac{1}{120}\left(\mathrm{ad}_{\Omega_u}\right)^5 \circ \kappa\tilde{O} \\ +O\left(\kappa^8\right) \end{cases} \tag{4.171}$$

At first sight the indicated accuracy $O\left(\kappa^8\right)$ holds true with the operator Ω_u being expanded up to *and* including the fifth-order term $\Omega^{(5)}$, because with the expansion

$$\mathrm{ad}_{\Omega_u} \circ \mathcal{F} = \sum_{j=0}^{\infty} \kappa^{2j+1} \mathrm{ad}_{\Omega^{(2j+1)}} \circ \mathcal{F} \tag{4.172}$$

then

$$\begin{aligned}
\left(\mathrm{ad}_{\Omega_u}\right)^6 \circ \beta &= \kappa^6 \left(\mathrm{ad}_{\Omega^{(1)}}\right)^6 \circ \beta + O\left(\kappa^8\right) \\
\left(\mathrm{ad}_{\Omega_u(\infty)}\right)^5 \circ \left(\kappa \tilde{O}\right) &= \kappa^6 \left(\mathrm{ad}_{\Omega^{(1)}}\right)^5 \circ \tilde{O} + O\left(\kappa^8\right) \\
\left(\mathrm{ad}_{\Omega_u}\right)^4 \circ \left(\kappa^2 \tilde{\mathcal{E}}\right) &= \kappa^6 \left(\mathrm{ad}_{\Omega^{(1)}}\right)^4 \circ \tilde{\mathcal{E}} + O\left(\kappa^8\right).
\end{aligned} \tag{4.173}$$

Yet the commutator terms at order κ^{2n} in the expansion (4.171) involving the operators $\Omega^{(2n-1)}$ *cancel* for $n \geq 2$, because of the identity

$$\frac{1}{2} \mathrm{ad}_{\Omega^{(1)}} \circ \mathrm{ad}_{\Omega^{(2n-1)}} \circ \beta + \frac{1}{2} \mathrm{ad}_{\Omega^{(2n-1)}} \circ \mathrm{ad}_{\Omega^{(1)}} \circ \beta = \mathrm{ad}_{\Omega^{(2n-1)}} \circ \tilde{O}. \tag{4.174}$$

This is fortunate, as it implies, that in order to achieve accuracy $O\left(\kappa^{2n}\right)$ for $n \geq 2$ only the operators $\Omega^{(1)}, \Omega^{(3)}, ..., \Omega^{(2n-3)}$ are required! So we obtain from (4.171) now the following expansion for the (scaled) Newton–Wigner Hamiltonian in the guise (here we set $\kappa = 1$, as no book keeping is required anymore):

$$\tilde{\mathsf{H}}^{(NW)} = \beta + \tilde{\mathsf{h}}^{(2)} + \tilde{\mathsf{h}}^{(4)} + \tilde{\mathsf{h}}^{(6)} + ... \tag{4.175}$$

where

$$\tilde{\mathsf{h}}^{(2)} = \tilde{\mathcal{E}} + \frac{1}{2}\left[\Omega^{(1)}, \left[\Omega^{(1)}, \beta\right]\right] - \left[\Omega^{(1)}, \tilde{O}\right] \tag{4.176}$$

$$\tilde{\mathsf{h}}^{(4)} = \begin{pmatrix} +\frac{1}{24}\left[\Omega^{(1)}, \left[\Omega^{(1)}, \left[\Omega^{(1)}, \left[\Omega^{(1)}, \beta\right]\right]\right]\right] \\ +\frac{1}{2}\left[\Omega^{(1)}, \left[\Omega^{(1)}, \tilde{\mathcal{E}}\right]\right] \\ -\frac{1}{6}\left[\Omega^{(1)}, \left[\Omega^{(1)}, \left[\Omega^{(1)}, \tilde{O}\right]\right]\right] \end{pmatrix} \tag{4.177}$$

$$\tilde{\mathsf{h}}^{(6)} = \begin{cases} \frac{1}{2}\left[\Omega^{(3)}, \left[\Omega^{(3)}, \beta\right]\right] \\ +\frac{1}{2}\left[\Omega^{(1)}, \left[\Omega^{(3)}, \tilde{\mathcal{E}}\right]\right] + \frac{1}{2}\left[\Omega^{(3)}, \left[\Omega^{(1)}, \tilde{\mathcal{E}}\right]\right] \\ \\ +\frac{1}{24}\left[\Omega^{(1)}, \left[\Omega^{(1)}, \left[\Omega^{(1)}, \left[\Omega^{(3)}, \beta\right]\right]\right]\right] + \frac{1}{24}\left[\Omega^{(1)}, \left[\Omega^{(1)}, \left[\Omega^{(3)}, \left[\Omega^{(1)}, \beta\right]\right]\right]\right] \\ +\frac{1}{24}\left[\Omega^{(1)}, \left[\Omega^{(3)}, \left[\Omega^{(1)}, \left[\Omega^{(1)}, \beta\right]\right]\right]\right] + \frac{1}{24}\left[\Omega^{(3)}, \left[\Omega^{(1)}, \left[\Omega^{(1)}, \left[\Omega^{(1)}, \beta\right]\right]\right]\right] \\ +\frac{1}{24}\left[\Omega^{(1)}, \left[\Omega^{(1)}, \left[\Omega^{(1)}, \left[\Omega^{(1)}, \tilde{\mathcal{E}}\right]\right]\right]\right] \\ +\frac{1}{720}\left[\Omega^{(1)}, \left[\Omega^{(1)}, \left[\Omega^{(1)}, \left[\Omega^{(1)}, \left[\Omega^{(1)}, \left[\Omega^{(1)}, \beta\right]\right]\right]\right]\right]\right] \\ \\ -\frac{1}{6}\left[\Omega^{(1)}, \left[\Omega^{(1)}, \left[\Omega^{(3)}, \tilde{O}\right]\right]\right] - \frac{1}{6}\left[\Omega^{(1)}, \left[\Omega^{(3)}, \left[\Omega^{(1)}, \tilde{O}\right]\right]\right] - \frac{1}{6}\left[\Omega^{(3)}, \left[\Omega^{(1)}, \left[\Omega^{(1)}, \tilde{O}\right]\right]\right] \\ -\frac{1}{120}\left[\Omega^{(1)}, \left[\Omega^{(1)}, \left[\Omega^{(1)}, \left[\Omega^{(1)}, \left[\Omega^{(1)}, \tilde{O}\right]\right]\right]\right]\right]. \end{cases} \tag{4.178}$$

and so on. Simplifying the commutators in (4.176), (4.177), (4.178) gives eventually

$$\tilde{\mathsf{h}}^{(2)} = \tilde{\mathcal{E}} + \frac{1}{2}\beta\tilde{\mathcal{O}}^2 \tag{4.179}$$

$$\tilde{\mathsf{h}}^{(4)} = -\frac{1}{8}\beta\tilde{\mathcal{O}}^4 - \frac{1}{8}\left[\tilde{\mathcal{O}},\left[\tilde{\mathcal{O}},\tilde{\mathcal{E}}\right]\right] \tag{4.180}$$

$$\tilde{\mathsf{h}}^{(6)} = \begin{cases} \frac{1}{16}\beta\tilde{\mathcal{O}}^6 \\ \frac{1}{32}\left[\tilde{\mathcal{O}}^3,\left[\tilde{\mathcal{O}},\tilde{\mathcal{E}}\right]\right] + \frac{1}{64}\left[\tilde{\mathcal{O}},\left(\mathcal{O}^2\left[\tilde{\mathcal{O}},\tilde{\mathcal{E}}\right] + \left[\tilde{\mathcal{O}},\tilde{\mathcal{E}}\right]\mathcal{O}^2\right)\right] \\ +\frac{1}{128}\left[\tilde{\mathcal{O}},\left[\tilde{\mathcal{O}},\left[\tilde{\mathcal{O}},\left[\tilde{\mathcal{O}},\tilde{\mathcal{E}}\right]\right]\right]\right] \\ +\frac{1}{16}\beta\left(\tilde{\mathcal{O}}\left[\left[\tilde{\mathcal{O}},\tilde{\mathcal{E}}\right],\tilde{\mathcal{E}}\right] + \left[\left[\tilde{\mathcal{O}},\tilde{\mathcal{E}}\right],\tilde{\mathcal{E}}\right]\tilde{\mathcal{O}}\right) \end{cases} . \tag{4.181}$$

The term $\tilde{\mathsf{h}}^{(2)}$ in (4.179) comprises the Hamiltonian of Schrödinger–Pauli quantum mechanics, well known to generate the motion of the *non*-relativistic electron in the presence of an electromagnetic field (scaled units). The term $\tilde{\mathsf{h}}^{(4)}$ in (4.180) coincides with the standard outcome for the leading order relativistic corrections obtained with the (laborious) step-by-step FW-transformation method [2]. It combines the kinetic energy correction, the Darwin term, and the spin–orbit interaction with the (static) external electric field.

Our findings pertaining to the higher order terms $\tilde{\mathsf{h}}^{(6)}$ enlisted in (4.181) are in accord with results obtained by Silenko's correction scheme [35] applied to the original step-by-step FW-transformation, and they also coincide with the table provided by deVries and Jonker [30], who obtained (though in a less practicable guise) their expansion in powers of $\kappa = \frac{v}{c}$ with the Pauli–Achieser–Berestezki elimination method (using computer algebra) and provided a proof of equivalence of their approach with the unitary transformation method of Eriksen.

The standard guise of the Hamiltonian of Schrödinger–Pauli (SP) quantum mechanics for electrons is readily obtained from the upper diagonal 2×2-block in (4.179) after bringing the units in. Identifying the even operator as $\mathcal{E} = e\Phi(\mathsf{x})$ and the odd operator as $\mathcal{O} = c\alpha_a\Pi_a$, then with $\mathsf{B}_a = (\mathrm{rot}\,\mathsf{A})_a$

$$\mathsf{H}^{(SP)} \equiv mc^2\,\tilde{\mathsf{h}}^{(2)}_{2\times 2} = mc^2 + e\Phi(\mathsf{x}) + \frac{1}{2m}\Pi_a\Pi_a - \frac{e\hbar}{2mc}\sigma_a^{(P)}\,\mathsf{B}_a. \tag{4.182}$$

Thereafter, the leading order relativistic corrections to the Hamiltonian $\mathsf{H}^{(SP)}$ for electrons moving in the presence of an electrostatic field $\mathsf{E}_a = -\nabla_a\Phi$ are obtained correspondingly from the upper diagonal 2×2-block in (4.180) in the guise

$$mc^2\,\tilde{\mathsf{h}}^{(4)}_{2\times 2} = \begin{cases} -\frac{1}{2mc^2}\left(\frac{1}{2m}\Pi_a\Pi_a - \frac{e\hbar}{2mc}\sigma_a^{(P)}\mathsf{B}_a\right)^2 \\ +\frac{1}{8}\left(\frac{\hbar}{mc}\right)^2\nabla^2 e\Phi - \frac{1}{8}\left(\frac{\hbar}{mc}e\mathsf{E}_a\frac{1}{mc}\Pi_{a'} + \frac{1}{mc}\Pi_{a'}\frac{\hbar}{mc}e\mathsf{E}_a\right)\epsilon_{aa'a''}\sigma_{a''}^{(P)}. \end{cases} \tag{4.183}$$

Because $\left[\Pi_{a'}, \mathsf{E}_a\right]\epsilon_{aa'a''} = i\hbar(\mathrm{rot}\,\mathsf{E})_{a''} = 0$, then $(\mathsf{E}_a\Pi_{a'} + \Pi_{a'}\mathsf{E}_a)\epsilon_{aa'a''} = 2\,\mathsf{E}_a\Pi_{a'}\epsilon_{aa'a''}$, thus implying the coefficients of the Darwin term and the spin–orbit interaction being weighted in a ratio of 1 : 2.

It is worth mentioning here that within the context of quantum electrodynamics (QED) there emerges a specific (small though) correction adding to the Darwin *and* spin–orbit terms in (4.183) that changes the ratio 1 : 2 of the respective weights slightly, so that agreement with the Lamb shift experiment is obtained. A worth reading presentation of the history and why indeed the fundamental physics of the Lamb shift is of great significance to the genesis of quantum field theory has been given by Milonni [20].

Inspired by the critics issued by P.A.M. Dirac in his lectures on quantum field theory [41, Sect. 7] on the use of the Schrödinger picture, that he exemplified with a simple model Hamiltonian *not* conserving particle number, we suggested recently an alternative derivation of the key effects of QED, which prevents use of the Schrödinger picture and consequently avoids S-matrix perturbation theory and/or Feynman diagrams altogether. Instead our approach is based on the perturbative solution of a Wegner flow equation with a suitably chosen generator that ensures the limiting value of that flow strives toward a particle number conserving Hamiltonian, in point of fact this Hamiltonian being unitary equivalent to the original non-particle number conserving Hamiltonian of QED, see [21, 42, 43].

4.6.1 Remark on Correction Schemes

Beyond order κ^4 results obtained by the step-by-step FW-transformation method are not energy-separating. This has been realized already early on by Eriksen and Kolsrud [13]. An additional unitary transformation is required, so that the results obtained were energy-separating, i.e., the results would agree with results obtained by the Eriksen method [5]. A general correction scheme for the original step-by-step FW-transformation [2] expanded to order κ^{2n} for $n \geq 3$ has been developed by Silenko [35].

Likewise, our result (4.115) has a natural interpretation of a *correction* scheme for the Hamiltonian flow equation method employed by Bylev and Pirner (BP) in [11] that brings the limiting value of that flow:

$$\tilde{\mathsf{H}}(\infty) = \beta + \tilde{\mathsf{h}}^{(2,BP)} + \tilde{\mathsf{h}}^{(4,BP)} + \tilde{\mathsf{h}}^{(6,BP)} + ... \tag{4.184}$$

now to an energy-separating guise. Notably, in view of (4.115) and (4.137), the expansion for $\Omega_g(s)$ commences with a *fourth*-order term only and progresses in powers of κ^{2n} for $n = 2, 3, ...$ So, setting thereafter $\kappa = 1$ then

$$\Omega_g(\infty) = \Omega^{(4)} + \Omega^{(6)} + ...,$$

whereby $\Omega^{(2n)}(s)$ is obtainable from (4.137) upon insertion of the odd terms $\Omega^{(2n+1)}(s)$, these odd terms enlisted for $n = 1, 2, 3$ in (4.165), (4.167), and (4.168), respectively.

Consequently, the earliest correction to $\tilde{\mathsf{H}}(\infty)$ so that there applies in accord with (4.115)

$$\tilde{\mathsf{H}}^{(NW)} = e^{-\Omega_g}\tilde{\mathsf{H}}(\infty)e^{\Omega_g} = \beta + \tilde{\mathsf{h}}^{(2)} + \tilde{\mathsf{h}}^{(4)} + \tilde{\mathsf{h}}^{(6)} + \tilde{\mathsf{h}}^{(8)} + ... \tag{4.185}$$

regards the sixth-order term. Because $\tilde{\mathsf{h}}^{(2n,BP)} \equiv \tilde{\mathsf{h}}^{(2n)}$ for $n = 1, 2$ it follows

$$\begin{aligned} \tilde{\mathsf{h}}^{(6)} &= \tilde{\mathsf{h}}^{(6,BP)} - \left[\Omega^{(4)}, \tilde{\mathsf{h}}^{(2)}\right] \\ \tilde{\mathsf{h}}^{(8)} &= \tilde{\mathsf{h}}^{(8,BP)} - \left[\Omega^{(6)}, \tilde{\mathsf{h}}^{(2)}\right] - \left[\Omega^{(4)}, \tilde{\mathsf{h}}^{(4)}\right], \end{aligned} \tag{4.186}$$

and so on.

On a final note, the above-suggested method to obtain $\tilde{\mathsf{H}}^{(NW)}$ directly from (4.121) together with the associated generators $\Omega^{(2n+1)}$ from (4.164) is obviously preferable to the (far more) cumbersome procedure where first a limiting value $\tilde{\mathsf{H}}(\infty)$ is obtained employing a step-by-step unitary transformation of the Dirac Hamiltonian not being energy-conserving and thereafter applying a correction scheme to achieve eventually energy separation.

4.6.2 Conclusion

The problem of deducing a perturbation series representation pertaining to the unitary transformation of a general Dirac Hamiltonian $\tilde{\mathsf{H}}^{(D)} = \beta + \tilde{\mathcal{O}} + \tilde{\mathcal{E}}$ to an even *and* manifestly energy-separating form $\tilde{\mathsf{H}}^{(NW)} = e^{-\Omega_u}\tilde{\mathsf{H}}^{(D)}e^{+\Omega_u}$ has been resolved with a new Magnus-type series expansion for the generator Ω_u of this transformation. Notably the series representing Ω_u progresses in *odd* powers of the Sommerfeld parameter $\kappa = \frac{v}{c} = \frac{\lambda_C}{a_B} = \frac{e^2}{4\pi\varepsilon_0\hbar c} \approx \frac{1}{137}$. The expansion applies to the nonrelativistic sector spanned by the low-energy eigenstates of $\tilde{\mathsf{H}}^{(D)}$, thus assuming the action of the operator $\tilde{\mathcal{O}}$ on such states produces a small result of first order in κ only, and in addition assuming too, the action of the operator $\tilde{\mathcal{E}}$ on such states produces an even smaller result of second-order κ^2. These assumptions certainly apply to the case of atomic physics, where $\tilde{\mathcal{O}} = \frac{\alpha_b \Pi_b}{mc}$ and $\tilde{\mathcal{E}} = \frac{e\Phi(\mathsf{x})}{mc^2}$, because in view of the virial theorem [40] the expectation value of the Coulomb potential $e\Phi$, say in a stationary state of the hydrogen atom and using a gauge where $\Phi(\mathsf{r}) \to 0$ for $r \to \infty$, compares in order of magnitude to the associated expectation value of the nonrelativistic kinetic energy $\frac{1}{2}\beta\tilde{\mathcal{O}}^2$.

The introduced expansion method, see (4.121) together with (4.139), constitutes the central result of this chapter. It fully implements a convenient energy-separating scheme to construct the expansion in powers of $\kappa = \frac{v}{c}$ for the Newton–Wigner Hamiltonian of the relativistic electron moving in static external electromagnetic fields.

In view of the step-by-step FW-transformation method [2] not leading to an energy-separating result for the unitary transformed Dirac Hamiltonian, it would be interesting to apply the introduced flow equation approach as well to bosons carrying mass and charge with spin $S = 0$ (Klein–Gordon) or spin $S = 1$ (Proca), in this way checking the results obtained early on by Case [4].

References

1. Newton TD, Wigner EP. Localized states for elementary particles. Rev Mod Phys. 1949;21:400–6.
2. Foldy LL, Wouthuysen SA. On the Dirac theory of spin 1/2 particles and its nonrelativistic limit. Phys Rev. 1950;78:29–36.
3. Costella JP, McKellar BHJ. The Foldy-Wouthuysen transformation. Am J Phys. 1995;63(12).
4. Case KM. Some generalizations of the Foldy-Wouthuysen transformation. Phys Rev. 1954;95:1323–8.
5. Eriksen E. Foldy-Wouthuysen transformation. Exact solution with generalization to the two-particle problem. Phys Rev. 1958;111:1011–6.
6. Wegner F. Flow equations for Hamiltonians. Ann Phys. 1994;506(2):77–91.
7. Wegner FJ. Flow equations for Hamiltonians. Phys Rep. 2001;348:77–89.
8. Kehrein S. The flow equation approach to many-particle systems. Springer Tracts Modern Phys. 2006;217.
9. Brockett RW. Dynamical systems that sort lists, diagonalize matrices and solve linear programming problems. Linear Algebra Appl. 1991;146:79–91.
10. Driessel KR, Hentzel IR, Wasin S. On preserving structured matrices using double bracket operators: tridiagonal and Toeplitz matrices. JP J Algebra, Number Theory Appl. 2001;1(2):87–114.
11. Bylev A, Pirner H. Hamiltonian flow equations for a Dirac particle in an external potential. Phys Lett B. 1998.
12. Pursey DL. On generalized Foldy-Wouthuysen transformations. Nucl Phys. 1958;8:595–601.
13. Eriksen E, Kolsrud M. Canonical transformations of Dirac's equation to even forms. Expansion in terms of the external fields. Nuovo Cimento Suppl. 1960;18:1–39.
14. Baym G. Lectures on Quantum Mechanics. Taylor and Francis Group; 1969.
15. Messiah A. Quantum Mechanics II. North-Holland; 1970.
16. Bjorken JD, Drell SD. Relativistic Quantum Mechanics. McGraw-Hill Book Company; 1964. ISBN 07-005493-2.
17. Itzykson C, Zuber J-B. Quantum Field Theory. McGraw-Hill; 1980. ISBN 0-07-032071-3.
18. Grainer W. Relativistic quantum mechanics. Springer; 1990. 978-3-540-99535-7.
19. Zhang P, Zou L, Silenko AJ. Position and spin in relativistic quantum mechanics. Phys Rev A. 2020;101(3).
20. Milonni PW. The Quantum Vacuum. An Introduction to Quantum Electrodynamics. Academic Press Inc; 1994.
21. Schopohl N, Cetin NS. Advances in Quantum Science I - Theoretical and Methodological Approaches, Chapter1: From the Relativistic Field Theory of Quantum Electrodynamics to a Hamiltonian for Electrons and Positrons as Point Particles with Attributes Mass, Charge and Spin. Springer; 2026.

22. Morpurgo G. A generalization of the Foldy-Wouthuysen transformation. Il Nuovo Cimento. 1960;XV:624–640.
23. Osche GR. Dirac and Dirac-Pauli equations in the Foldy-Wouthuysen representation. Phys Rev D. 1977;15:2181–5.
24. de Vries E. Foldy- Wouthuysen transformations and related problems. Fortschr Phys. 1970;18:149–82.
25. Barut AO. Electrodynamics and classical theory of fields and particles. New York: Dover Publications, Inc.; 1980.
26. Pauli W. Relativistic field theories of elementary particles. Rev Mod Phys. 1941;13:203–32.
27. Silenko AJ. Foldy-Wouthuysen transformation for relativistic particles in external fields. J Math Phys. 2003;44(0).
28. Levin JJ. On the matrix Riccati equation. Trans Am Math Soc. 1959;10:519–24.
29. Furuta T. On the polar decomposition of an operator. Acta Sci Math. 1983;46:261–8.
30. de Vries E, Jonker JE. Non-relativistic approximations of the Dirac Hamiltonian. Nucl Phys B. 1968;6:213–25.
31. Pryce MHL. The mass-centre in the restricted theory of relativity and its connexion with the quantum theory of elementary particles. Proc Roy Soc. 1948;A195:62.
32. Tani S. Connection between particle models and field theories. Prog Theor Phys Kyoto. 1951;VI:267.
33. Silenko AJ. Validation of the Eriksen method for the exact Foldy-Wouthuysen representation. Phys Part Nucl Lett. 2013;10(3):321–5.
34. Silenko AJ. General properties of the relativistic Foldy-Wouthuysen transformation and proof of validity of the Foldy-Wouthuysen Hamiltonian. Phys Rev A. 2016;91: 022103.
35. Silenko AJ. General properties of the Foldy-Wouthuysen transformation and applicability of the corrected original Foldy-Wouthuysen method. Phys Rev A. 2016;93: 022108.
36. Zanna A. Recurrence relations and convergence of the generalized polar decomposition on Lie groups. Math Comput. 2004;73:761–76.
37. Iserles A, Norsett SP. On the solution of linear differential equations in Lie groups. Phil Trans R Soc Lond. 2002;A357(1754):983–1019.
38. Wilcox RM. Exponential operators and parameter differentiation in quantum physics. J Math Phys. 1967;8:962–82.
39. Iserles A. Expansions that grow on trees. Not AMS. 2002;49:430–40.
40. Schwinger J. Quantum Mechanics. 2001.
41. Dirac PAM. Lectures on Quantum Field Theory. Academic Press London; 1966.
42. Schopohl N, Suzan Cetin N. Advances in Quantum Science I - Theoretical and Methodological Approaches, Chapter2: Obtaining Schwinger's Magnetic Moment Anomaly and the Lamb Shift Terms from a Particle Number Conserving Hamiltonian being Unitary Equivalent to the Relativistic Hamiltonian of Quantum Electrodynamics. Springer; 2026.
43. Schopohl N, Suzan Cetin N. Advances in Quantum Science I - Theoretical and Methodological Approaches, Chapter3: The Nonrelativistic Hamiltonian of Electrons and Positrons: a Derivation from Quantum Electrodynamics. Springer; 2026.

Chapter 5
Density Matrix of a Quantum System Taking into Account an Environment

Elena V. Orlenko and Fedor E. Orlenko

Abstract This chapter is devoted to a detailed exposition of the mathematical apparatus of the density matrix from first principles, applied to the description of quantum systems taking into account the stochastic influence of a thermalized environment. Applications of this apparatus relevant to modern nanotechnology physics are considered. Kubo's formula is derived for the conductivity and, in general, for the generalized susceptibility of a quantum system in response to an external electromagnetic field. The fluctuation–dissipation theorem is proved in detail. The semiclassical limit of the density matrix in the Wigner representation to the classical distribution function in phase space is considered. The Liouville-von Neumann equation for the density matrix transforms into the classical Liouville equation.

Keywords Stochastic influence of thermalization · Density matrix · Thermalized environment · Paramagnetism · Magnetic moment · Quantum fluctuation · Autocorrelation function · Generalized susceptibility · Fluctuation–dissipation theorem · Wigner representation

5.1 Introduction

The mathematical apparatus of the Density Matrix plays a fundamental role in the quantum theory of multi-particles systems. This method is applicable in the study of stationary states of polyatomic, multiparticle systems in a magnetic field, taking into account both spin polarization and magnetic polarization phenomena in systems of particles (atoms, molecules), characterized by a total mechanical moment and a corresponding gyromagnetic factor. This method is also applicable to dynamic

E. V. Orlenko (✉)
Higher School of Fundamental Physical Research, Theoretical Physic Department, Peter the Great Saint-Petersburg Polytechnic University, Saint Petersburg, Russia
e-mail: eorlenko@spbstu.ru

F. E. Orlenko
Higher School of Technology and Energetics, Saint Petersburg State University of Industrial Technologies and Design, Saint Petersburg, Russia

T. Onishi (ed.), *Advances in Quantum Science I*,
https://doi.org/10.1007/978-981-95-9717-8_5

problems in describing multiparticle quantum systems, during their thermalization. These are dynamic effects associated with the precession of the magnetic moment of particles with an arbitrary spin in an alternating magnetic field, these are effects associated with the conductivity of quantum materials in alternating electric fields, as well as the effects of electric polarization in an arbitrary electric field.

The founder of the density matrix theory is the American scientist of Hungarian origin John von Neumann (Johann Janos von Neumann), whose name is associated with the application of operator theory to quantum mechanics. He was one of the creators of the rigorous mathematical apparatus of quantum mechanics. He outlined his approach to the axiomatization of quantum mechanics in his work Matematische Gründlagen Quantenmechanik (Mathematical Grounds of Quantum Mechanics).

Despite the fact that statistical physics is one of the very well developed areas of knowledge in both theoretical and experimental aspects, most of the methodological elements related to the use of the theory in theoretical study are placed in separate articles. It makes difficult to understand these methods, especially for those who do not use them systematically or are acquainted with them. By now, quite a few good books have been published that describe the apparatus of the density matrix, for example [1–5]. However, many formulas and relationships that are important for understanding the apparatus and its practical application are omitted from these books. They are scattered across various reference publications or included in appendices to articles discussing individual issues, and are not always accessible to readers. Particularly large difficulties arise when different authors use different notations, definitions, and symbols for the same quantities, which can lead to difficulties in understanding and errors in calculations. Therefore, in this chapter, without pretending to be comprehensive, we will demonstrate how the methods under consideration allow one to obtain solutions to some selected simple problems, focusing our attention on the details of calculating.

Since our aim is to illustrate how the methods of the quantum theory of statistical phenomena are used in relatively simple problems, we will mainly discuss the magnetic properties of spin systems, the questions related with conductivity and electrical polarization of multilevel quantum systems.

5.2 Theoretical Foundations of the Density Matrix

5.2.1 Density Matrix Definition

Consider a quantum system surrounded by a macroscopic medium (a quantum system inside a macroscopic device), for example, an atomic system that makes up the so-called gas cell. In the future, terminologically, we will designate such a system as a quantum system + environment. We will denote the entire set of variables describing the particles that make up the quantum system, x, while the entire set of variables describing the particles of the medium as q. Since the particles of the medium are the same quantum objects as the particles themselves that make up the quantum system, the wave function describing the quantum state of the entire system as a whole (quantum system + environment) will be a function of the entire set of variables x (for particles of a quantum subsystem) and q (for particles belonged to an environment): $\Psi(q, x, t)$. Let us assume that an isolated quantum system in the stationary case. Without taking into account any interactions within it, this system has a known energy spectrum and a corresponding set of eigenvectors of states:

$$\hat{H}^0 \psi_n(x, t) = E_n \psi_n(xt),$$
$$\psi_n(x, t) = \psi_n(x) e^{-\frac{i}{\hbar} E_n t}. \tag{5.1a}$$

Here, the notation of quantum states is used both in the form of a wave function $\psi_n^*(x)$ and $\psi_n(x)$ in the coordinate representation (for example, for the stationary state n, corresponding to the eigenenergy E_n and in the wave vectors form of a bra-vector $\langle \psi_m|$ and a ket-vector $|\psi_n\rangle$:

$$\hat{H}^0 |\psi_n(t)\rangle = E_n |\psi_n(t)\rangle,$$
$$|\psi_n(t)\rangle = e^{-i\frac{E_n t}{\hbar}} |\psi_n\rangle,$$
$$\langle \psi_n(t)| = \langle \psi_n| e^{i\frac{E_n t}{\hbar}}. \tag{5.1b}$$

According to the Sturm–Liouville problem, this set of eigenstates $\{|\psi_n\rangle\}$ has the properties of orthonality, normality and completeness:

$$\langle \psi_m | \psi_n \rangle \equiv \langle m | n \rangle = \delta_{mn},$$
$$\sum_n |\psi_n\rangle \langle \psi_n| = \sum_n |n\rangle \langle n| = \hat{1},$$

(in the vector form) and in coordinate representation

$$\int \psi_m^*(x) \psi_n(x) dx = \delta_{mn},$$
$$\sum_n \psi_n(x) e^{-\frac{i}{\hbar} E_n t} e^{+\frac{i}{\hbar} E_n t} \psi_n^*(x') = \delta(x - x').$$

Then the specified set (5.1a, 5.1b) can be used to expand an arbitrary wave vector in this basis. Let us expand the complete wave function (wave vector) that describes the state of the entire system in terms of the specified set of states of the quantum subsystem:

$$\Psi(q,x,t) = \int dx'\delta(x-x')\Psi(q,x',t) =$$
$$= \int dx' \sum_n \psi_n(x)\psi_n^*(x')\Psi(q,x',t) =$$
$$= \sum_n \psi_n(x) \int \psi_n^*(x')\Psi(q,x',t)dx' = \sum_n \psi_n(x)C_n(q,t),$$
$$|\Psi(q,t)\rangle = \sum_n |\psi_n\rangle\langle\psi_n||\Psi(q,t)\rangle = \sum_n |\psi_n\rangle C_n(q,t), \tag{5.2}$$

where the expansion coefficients in quantum states are denoted as $C_n(q,t) = \langle\psi_n|\Psi(q,t)\rangle = \int \psi_n^*(x')\Psi(q,x',t)dx'$. During the writing the equations in the vector form, like as a scalar product of *bra*- and *ket*-vectors the variable x, isw not written, it is used only for the coordinate representation.

Then the normalization condition for the wave function (wave vector) of the entire system (quantum system + environment) will have the form:

In vector form (as an scalar product)

$$\int dq\langle\Psi(q,t|\Psi(q,t)\rangle = 1 =$$
$$= \int dq\left\{\left(\sum_m \langle\Psi(q,t)|\psi_m\rangle\langle\psi_m|\right)\left(\sum_n |\psi_n\rangle\langle\psi_n||\Psi(q,t)\rangle\right)\right\} =$$
$$= \sum_{n,m} \langle\psi_m|\psi_n\rangle \int dq\langle\psi_n|\Psi(q)\rangle\langle\Psi(q)|\psi_m\rangle =$$
$$= \sum_n \langle\psi_n\overline{|\Psi(q,t)\rangle\langle\Psi(q,t)}|\psi_n\rangle = \sum_n \overline{C_n(q,t)C_n^*(q,t)}. \tag{5.3}$$

Here, integration over environmental variables q is denoted by a line on top of the integrands:

$$\overline{C_n(q,t)C_n^*(q,t)} = \int dq\langle\psi_n|\Psi(q,t)\rangle\langle\Psi(q,t)|\psi_m\rangle =$$
$$= \langle\psi_n\overline{|\Psi(q,t)\rangle\langle\Psi(q,t)}|\psi_m\rangle. \tag{5.4}$$

This notation implies, in the future, instead of sequential integration over environmental variables q, averaging over its states. Due to the fact that the environment is infinitely large compared to the described quantum system, and the number of particles that make up this environment is also infinitely large, in principle we cannot give a detailed (regular) quantum–mechanical description for such a system (environment) and therefore have to resort to averaging over its states.

In the form of wave functions

$$
\begin{aligned}
&\iint dqdx\big(\Psi^*(q,x,t)\Psi(q,x,t)\big) = 1 = \\
&= \int dq \int dx \left\{ \int dx' \sum_n \Psi^*(q,x',t)\psi_n(x')\psi_n^*(x) \int dx'' \sum_m \psi_m(x)\psi_m^*(x'')\Psi(q,x'',t) \right\} = \\
&= \sum_{n,m} \int dx\big(\psi_n^*(x)\psi_m(x)\big) \int dq \Big[\int dx''\big(\psi_m^*(x'')\Psi(q,x'',t)\big) \int dx'\big(\Psi^*(q,x',t)\psi_n(x')\big) \Big] = \\
&= \sum_{n,m} \delta_{nm} \left[\overline{\int dx''\big(\psi_m^*(x'')\Psi(q,x'',t)\big) \int dx'\big(\psi_n(x')\Psi^*(q,x',t)\big)} \right] = \\
&= \sum_n \overline{C_n(q,t)C_n^*(q,t)}\Big|_q = \sum_n w_n(t), \\
&C_n(q,t)C_n^*(q,t) = P_n(q,t), \\
&w_n(t) = \overline{P_n(q,t)}\Big|_q .
\end{aligned}
\tag{5.5}
$$

The expansion coefficients introduced in (5.2) have a probabilistic meaning according to the principle of superposition in quantum mechanics. The product of the coefficients $C_n(q,t)C_n^*(q,t) = P_n(q,t)$ directly has the meaning of the probability of detecting a quantum subsystem in the state n, provided that the particles of the "environment" have a set of coordinates q. When integration ("reduction") is carried out over this product of coefficients $w_n = \overline{C_n(q,t)C_n^*(q,t)} = \overline{P_n(q,t)}\big|_q$ over the specified set of environmental variables q, it acquires the meaning of the probability of finding a quantum subsystem in a quantum state n.

Thus $C_n(q)C_n^*(q) = P_n(q)$, it makes sense the probability of realization of the n-th quantum state, subject to a certain configuration q (state) of the environment. When integrating over all possible configurations q, and in fact, averaging over the states of the environment, this value $w_n = \overline{P_n(q)}$ becomes the probability of realizing the quantum state n of the quantum subsystem, taking into account the stochastic influence of the environment on it. Let us immediately note that the spread of the states of the environment, which is assumed to be macroscopic, consisting of a large number of particles ($N \sim N_A = 6\cdot 10^{23} >> 1$), is small, since the relative fluctuation of the parameters characterizing such an environment is of the order of $\frac{\overline{\Delta A}}{\overline{A}} \sim \frac{1}{\sqrt{N}}$, that is, it is characterized by stable parameters.

Let us now assume that the quantum system is characterized by a physical quantity L whose operator $\hat{L}$ acts only on the variables of the quantum system. Then the average value of this quantity, taking into account the stochastic impact on the quantum system of the environment, will by definition be equal to

$$
\begin{aligned}
\tilde{L}(t) &= \int dq \langle \Psi(q,t)|\hat{L}|\Psi(q,t)\rangle \\
&= \sum_{n,m} \int dq \langle \Psi(q,t)|\,\psi_m\rangle\langle\psi_m|\hat{L}|\psi_n\rangle\langle\psi_n\,|\Psi(q,t)\rangle =
\end{aligned}
$$

$$
\begin{aligned}
&= \sum_{n,m} \langle \psi_n \overline{|\Psi(q,t)\rangle\langle\Psi(q,t)|} \psi_m \rangle \langle \psi_m | \hat{L} | \psi_n \rangle \\
&= \sum_{n,m} \overline{C_n(q,t) C_m^*(q,t)} \langle m | \hat{L} | n \rangle = \\
&= \sum_{n,m} \langle n | \hat{\rho}(t) | m \rangle \langle m | \hat{L} | n \rangle = \sum_n \langle n | \hat{\rho}(t) \hat{L} | n \rangle = Sp\big(\hat{\rho}(t)\hat{L}\big), \\
&\overline{C_n(q,t) C_m^*(q,t)} = \langle n | \hat{\rho}(t) | m \rangle, \\
&\langle n | \hat{\rho}(t) | n \rangle = \overline{C_n(q,t) C_n^*(q,t)} = w_n(t) \Rightarrow \\
&\sum_n \langle n | \hat{\rho}(t) | n \rangle = \sum_n w_n(t) = 1, \Rightarrow Sp\hat{\rho}(t) = 1.
\end{aligned} \tag{5.6}
$$

Here the expression:

$$
\sum_{n,m} \langle \psi_n \overline{|\Psi(q,t)\rangle\langle\Psi(q,t)|} \psi_m \rangle \langle \psi_m | \hat{L} | \psi_n \rangle = \sum_{n,m} \overline{C_n(q,t) C_m^*(q,t)} \langle m | \hat{L} | n \rangle
$$

has an algebraic form, which formally corresponds to a product of two matrix, a matrix $\left[\overline{C_n(q,t)C_m^*(q,t)}\right] = [\langle n|\hat{\rho}(t)|m\rangle]$, corresponding to the any statistical operator, while it contains the statistical average on the environmental variable q and a matrix of the physical quantity L, $\left[\langle \psi_m|\hat{L}|\psi_n\rangle\right] \equiv \left[\langle m|\hat{L}|n\rangle\right]$, moreover, this expression is a sum of a diagonal elements of the resulted matrix:

$$
\sum_{n,m} \langle n | \hat{\rho}(t) | m \rangle \langle m | \hat{L} | n \rangle = \sum_n \langle n | \hat{\rho}(t) \hat{L} | n \rangle = Sp\big(\hat{\rho}(t)\hat{L}\big),
$$

that means Trace (Tr), ore, in German abbreviator, Spur (Sp). Thus, we have introduced a statistical operator $\hat{\rho}(t)$ whose matrix element in a given n-representation has the form $\overline{C_n(q,t)C_m^*(q,t)}\big|_q = \langle n|\hat{\rho}(t)|m\rangle$, where the diagonal elements have the meaning of the probability of realizing the n-th quantum state of a quantum subsystem under the influence of the stochastic influence of the surrounding macroscopic medium.

5.2.2 Coordinate Representation of a Statistical Operator

As is known from quantum mechanics, in order for a given operator of a physical quantity $\hat{L}$, represented as a matrix $\left(\langle n|\hat{L}|m\rangle\right) = (L_{nm})$ in some n-representation of the eigenvectors of the state of some operator $\hat{H}^0$, see Eq. (5.1), to translate into a coordinate representation, it is necessary to carry out the following unitary transformation:

$$\sum_n \sum_m |n\rangle\langle n|\hat{L}|m\rangle\langle m| \equiv \sum_n \sum_m \psi_n(x) L_{nm} \psi_m^*(x') =$$
$$= \sum_n \sum_m \psi_n(x) \left(\int \psi_n^*(y) \hat{L}(y) \psi_m(y) dy \right) \psi_m^*(x') =$$
$$= \int dy \left\{ \left(\sum_n \psi_n(x) \psi_n^*(y) \right) \hat{L}(y) \left(\sum_m \psi_m(y) \psi_m^*(x') \right) \right\} =$$
$$= \int dy \{ \delta(x-y) \hat{L}(y) \delta(y-x') \} = \hat{L}(x) \delta(x-x'). \tag{5.7}$$

We remember that where the specified set of state vectors has the properties of orthogonallity and normality and completeness, and once again made sure that any operator of a physical quantity in the coordinate representation is always diagonal. We note here that from the matrix element L_{nm} we have obtained the matrix element $L(xx')$ of the same operator, but in the coordinate representation.

Let us now carry out the same unitary transformation of the statistical operator whose matrix elements in the n-representation are known to us:

$$\sum_n \sum_m |n\rangle\langle n|\hat{\rho}(t)|m\rangle\langle m| \equiv \sum_n \sum_m \psi_n(x) \rho_{nm}(t) \psi_m^*(x') =$$
$$= \sum_n \sum_m \psi_n(x) \overline{C_n(q,t) C_m^*(q,t)} \psi_m^*(x') =$$
$$= \sum_n \sum_m \psi_n(x) \overline{\left[\int dy (\psi_n^*(y) \Psi(q,y,t)) \int dy' (\psi_m(y') \Psi^*(q,y',t)) \right]} \psi_m^*(x') =$$
$$= \int dy \int dy' \left\{ \left(\sum_n \psi_n(x) \psi_n^*(y) \right) \overline{\Psi(q,y,t) \Psi^*(q,y',t)} \left(\sum_m \psi_m(y') \psi_m^*(x') \right) \right\} =$$
$$= \int dy \int dy' \left\{ \delta(x-y) \overline{\Psi(q,y,t) \Psi^*(q,y',t)} \delta(y'-x') \right\} =$$
$$= \overline{\Psi(q,x,t) \Psi^*(q,x',t)} = \rho_{xx'}(t) \equiv \rho(xx',t), \tag{5.8}$$

where do we get the matrix $(\rho_{xx'}(t))$ in the coordinate representation. Here, in addition to the diagonal matrix elements there are still non-zero off-diagonal elements $\rho_{xx'}(t)$, which have the meaning of quantum correlations of particle density between points x and x'. We remind, that the diagonal matrix elements $\rho_{xx}(t) = \overline{|\Psi(qx,t)|^2}$ mean the probability density, to find a quantum system with a set of coordinates x, or, if we are talking about one particle, at the point x. If this diagonal matrix element is multiplied by the total number of particles N in a quantum system, then this will give the density of particles at point x.

5.2.3 The Equation for Density Matrix

The environment, like the quantum subsystem, consists of atoms, molecules, which are quantum objects. Suppose that we have the opportunity to take into account all the interactions of the particles that make up the medium and the quantum system, while taking into account the possible interactions of particles with external non-stationary fields, then the Hamiltonian $\hat{H}(q, x, t)$ of the complete system can be arranged as follows:

$$\hat{H}(q, x, t) = \hat{H}(x, t) + \hat{Q}(q, x, t), \tag{5.9}$$

where $\hat{H}(x, t)$ describes the quantum system itself, taking into account the interaction of particles that make up only the quantum system, between themselves and with an external time-dependent field. The operator $\hat{Q}(q, x, t)$ describes the particles of the environment, taking into account their interactions not only with each other, but also with the particles of the quantum system and the external field. Then the Schrödinger equation for the total wave function "quantum system + medium" in the non-relativistic limit will have the form:

$$i\hbar\frac{\partial\Psi(x, q, t)}{\partial t} = \left(\hat{H}(x, t) + \hat{Q}(q, x, t)\right)\Psi(x, q, t). \tag{5.10}$$

We use the expansion (5.2) of the wave function in terms of the complete set of stationary states (5.1), and then we have

$$\begin{aligned}
&\Psi(x, q, t) = \sum_n \psi_n(x)C_n(q, t),\\
&i\hbar\sum_n \psi_n(x)\dot{C}_n(q, t),\\
&= \left(\hat{H}(x, t) + \hat{Q}(q, x, t)\right)\sum_n \psi_n(x)C_n(q, t),\\
&\dot{C}_n(q, t) \equiv \frac{\partial}{\partial t}C_n(q, t).
\end{aligned} \tag{5.11}$$

Next, we multiply Eq. (5.11) on the left by the wave function $\psi_m^*(x)$ and integrate over x, then we get the same Schrödinger equation in the n-representation:

$$\begin{aligned}
&i\hbar\dot{C}_m(q, t) = \sum_n \left(\hat{H}(t) + \hat{Q}(q, t)\right)_{mn}C_n(q, t),\\
&i\hbar\dot{C}_m(q, t) = \sum_n (H_{mn}(t) + Q_{mn}(q, t))C_n(q, t),
\end{aligned} \tag{5.12}$$

and for the complex conjugate equation:

$$-i\hbar\dot{C}_m^*(q,t)=\sum_n C_n^*(q,t)(H_{nm}(t)+Q_{nm}(q,t)), \tag{5.13}$$

where the Hermitian property of the energy operators is used:

$$H_{mn}^*=H_{nm},\ Q_{mn}^*=Q_{nm}.$$

Let us now take the partial time derivative of the matrix element of the statistical operator $\rho_{nm}(t)=\overline{C_n(q,t)C_m^*(q,t)}$ and use the Schrödinger equations obtained in Eqs. (5.12), (5.13) written in the proper energy n-representation:

$$\begin{aligned}
&\frac{\partial}{\partial t}\rho_{nm}(t)=\frac{\partial}{\partial t}\overline{C_n(q,t)C_m^*(q,t)}\\
&=\overline{\left(\frac{\partial}{\partial t}C_n(q,t)\right)C_m^*(q,t)}+\overline{C_n(q,t)\left(\frac{\partial}{\partial t}C_m^*(q,t)\right)}=\\
&=\sum_k\left(\frac{-i}{\hbar}\right)\overline{(H_{nk}(t)+Q_{nk}(q,t))C_kq,t)C_m^*(q,t)}+\\
&+\sum_k\left(\frac{i}{\hbar}\right)\overline{C_n(q,t)C_k^*(q,t)(H_{km}(t)+Q_{km}(q,t))}=\\
&=\frac{-i}{\hbar}\sum_k\left(H_{nk}(t)\overline{C_k(q,t)C_m^*(q,t)}-\overline{C_n(q,t)C_k^*(q,t)}H_{km}(t)\right)-\\
&-\frac{i}{\hbar}\sum_k\overline{\left(Q_{nk}(q,t)C_k(q,t)C_m^*(q,t)-C_n(q,t)C_k^*(q,t)Q_{km}(q,t)\right)}\\
&=\frac{-i}{\hbar}\sum_k(H_{nk}(t)\rho_{km}(t)-\rho_{nk}(t)H_{km}(t))-\\
&-\frac{i}{\hbar}\sum_k\overline{\left(Q_{nk}(q,t)C_k(q,t)C_m^*(q,t)-C_n(q,t)C_k^*(q,t)Q_{km}(q,t)\right)}.
\end{aligned}\tag{5.14}$$

Let us study the resulting second term (5.14):

$$\begin{aligned}
&\frac{i}{\hbar}\sum_k\overline{\left(Q_{nk}(q,t)C_k(q,t)C_m^*(q,t)-C_n(q,t)C_k^*(q,t)Q_{km}(q,t)\right)}=\\
&=\frac{i}{\hbar}\sum_k\overline{\langle\psi_n|\hat{Q}(q,t)|\psi_k\rangle\langle\psi_k|\,\Psi(q,t)\rangle\langle\Psi(q,t)|\psi_m\rangle}-\\
&-\sum_k\langle\psi_n|\,\Psi(q,t)\rangle\langle\Psi(q,t)|\psi_k\rangle\,\langle\psi_k|\hat{Q}(q,t)|\psi_m\rangle=\\
&=\frac{i}{\hbar}\left(\begin{array}{l}\overline{\langle\psi_n|\hat{Q}(q,t)|\Psi(q,t)\rangle\langle\Psi(q,t)|\psi_m\rangle}-\\ -\overline{\langle\psi_n|\,\Psi(q,t)\rangle\langle\Psi(q,t)|\hat{Q}(q,t)|\psi_m\rangle}\end{array}\right)=
\end{aligned}$$

$$= -\frac{i}{\hbar}\langle\psi_n|\overline{\left(|\Psi(q,t)\rangle\langle\Psi(q,t)|\hat{Q}(q,t) - \hat{Q}(q,t)|\Psi(q,t)\rangle\langle\Psi(q,t)|\right)}|\psi_m\rangle \to 0. \tag{5.15}$$

Both matrix elements included in expression (5.15) express the average value of the energy of the environment, taking into account the degree of influence of the quantum subsystem on it by taking into account the transitions of the quantum system between states n and m. Since the environment is considered to be infinitely large compared to the quantum subsystem, the degree of influence of the quantum system on the environment is negligible. Consequently, the matrix elements taken from the energy operator of the environment, taking into account the influence of the quantum subsystem on it, can be considered almost identical, and since their contributions enter the expression with different signs, they cancel each other out.

Thus, we have the equation for the density matrix obtained from (5.14) in its own energy representation:

$$\frac{\partial}{\partial t}\rho_{nm}(t) = \frac{-i}{\hbar}\sum_k (H_{nk}(t)\rho_{km}(t) - \rho_{nk}(t)H_{km}(t)),$$

$$\langle\psi_n|\frac{\partial\hat{\rho}(t)}{\partial t}|\psi_m\rangle = \frac{-i}{\hbar}\langle\psi_n|\hat{H}\hat{\rho} - \hat{\rho}\hat{H}|\psi_m\rangle,$$

$$\langle\psi_n|\frac{\partial\hat{\rho}(t)}{\partial t} + \frac{i}{\hbar}[\hat{H}\hat{\rho}]|\psi_m\rangle = 0 \tag{5.16}$$

or in general form:

$$\frac{\partial\hat{\rho}(t)}{\partial t} + \frac{i}{\hbar}[\hat{H}\hat{\rho}] = 0. \tag{5.17}$$

This equation is named after Liouville-von Neumann, or Liouville's quantum equation.

5.2.4 *The Equation for the Density Matrix in Coordinate Representation*

We write Eq. (5.16) in the coordinate representation as

$$i\hbar\frac{\partial\rho_{xx'}(t)}{\partial t} = \int dx''(H_{xx''}\rho_{x''x'} - \rho_{x'x''}H_{x''x'}), \tag{5.18}$$

where according to Eq. (5.7) the matrix element $H_{xx'}$ has the form:

$$H_{xx'} = \hat{H}(x)\delta(x - x') = \left(-\frac{\hbar^2}{2m}\frac{\partial^2}{\partial x^2} + U(x)\right)\delta(x - x').$$

According to the rule of matrix multiplication, the right side of the Eq. (5.18) is rewritten as an integral over the "running" index x''. Consider it more detailed

$$\int dx''(H_{xx''}\rho_{x''x'} - \rho_{xx''}H_{x''x'}) =$$

$$= \int dx''\left(-\frac{\hbar^2}{2m}\right)\left(\frac{\partial^2}{\partial x^2}\delta(x-x'')\rho_{x''x'} - \rho_{xx''}\frac{\partial^2}{\partial x''^2}\delta(x''-x')\right) +$$

$$+\int dx''(U(x)\delta(x-x'')\rho_{x''x'} - \rho_{xx''}U(x'')\delta(x''-x')) =$$

$$= \left(-\frac{\hbar^2}{2m}\right)\frac{\partial^2}{\partial x^2}\int dx''(\delta(x-x'')\rho_{x''x'}) +$$

$$+\frac{\hbar^2}{2m}\int \rho_{xx''}\frac{\partial^2}{\partial x''^2}\delta(x''-x')dx'' + \left(U(x)\rho_{xx'} - \rho_{xx'}U(x')\right) =$$

$$= \left\{-\frac{\hbar^2}{2m}\frac{\partial^2}{\partial x^2}\rho_{xx'} + \left(U(x)-U(x')\right)\rho_{xx'}\right\} +$$

$$+\frac{\hbar^2}{2m}\left(\cancel{\rho_{xx''}\frac{\partial}{\partial x''}\delta(x''-x')}\Big|_{-\infty}^{+\infty} - \int\frac{\partial}{\partial x''}\rho_{xx''}\frac{\partial}{\partial x''}\delta(x''-x')dx''\right) =$$

$$= \left\{-\frac{\hbar^2}{2m}\frac{\partial^2}{\partial x^2}\rho_{xx'} + \left(U(x)-U(x')\right)\rho_{xx'}\right\} +$$

$$+\frac{\hbar^2}{2m}\left(\cancel{-\frac{\partial^2}{\partial x''^2}\rho_{x'x''}\delta(x''-x)}\Big|_{-\infty}^{+\infty} + \int\frac{\partial^2}{\partial x''^2}\rho_{xx''}\delta(x''-x')dx''\right) =$$

$$= \left\{-\frac{\hbar^2}{2m}\frac{\partial^2}{\partial x^2}\rho_{xx'} + \left(U(x)-U(x')\right)\rho_{xx'}\right\} + \frac{\hbar^2}{2m}\frac{\partial^2}{\partial x'^2}\rho_{xx'} =$$

$$= \left\{-\frac{\hbar^2}{2m}\left(\frac{\partial^2}{\partial x^2} - \frac{\partial^2}{\partial x'^2}\right)\rho_{xx'} + \left(U(x)-U(x')\right)\rho_{xx'}\right\}, \Rightarrow$$

Hence, instead of Eq. (5.18) we have the following

$$i\hbar\frac{\partial\rho_{xx'}(t)}{\partial t} = -\frac{\hbar^2}{2m}\left(\frac{\partial^2}{\partial x^2} - \frac{\partial^2}{\partial x'^2}\right)\rho_{xx'} + \left(U(x) - U(x')\right)\rho_{xx'}. \tag{5.19}$$

5.2.5 A Stationary Case. Equilibrium System

If there is no explicit dependence on time in the interaction operator describing the quantum system, then the energy of the quantum system is an integral of motion, therefore, the explicit dependence on time also disappears for the density matrix operator, that is $\frac{\partial \hat{\rho}}{\partial t} = 0$, and, therefore, the following commutation relation holds for the density operator: $\left[\hat{H}\hat{\rho}\right] = 0$. The latter formally means the existence of a common basis of state eigenvectors that diagonalize both operators, Hamiltonian and a Density Matrix. It is easy to show

$$\begin{aligned}
&\exists\{|n\rangle\}:\hat{H}|n\rangle = E_n|n\rangle\\
&\langle m|n\rangle = \delta_{mn},\\
&\sum_k |k\rangle\langle k| = \hat{\mathbf{1}}.\\
&[\hat{H}\hat{\rho}] = 0, \Rightarrow\\
&\langle m|[\hat{H}\hat{\rho}]|n\rangle = 0,\\
&\langle m|\hat{H}\hat{\mathbf{1}}\hat{\rho} - \hat{\rho}\hat{\mathbf{1}}\hat{H}|n\rangle =\\
&= \sum_k \left(\langle m|\hat{H}|k\rangle\langle k|\hat{\rho}|n\rangle - \langle m|\hat{\rho}|k\rangle\langle k|\hat{H}|n\rangle\right) =\\
&= \sum_k H_{mk}\rho_{k_n} - \sum_k \rho_{mk}H_{kn} =\\
&= \sum_k (E_k\langle m||k\rangle\langle k|\hat{\rho}|n\rangle - \langle m|\hat{\rho}|k\rangle\langle k||n\rangle E_n) =\\
&= (E_m - E_n)\langle m|\hat{\rho}|n\rangle = 0,\\
&E_m \neq E_n, \Rightarrow \langle m|\hat{\rho}|n\rangle = 0,\\
&\langle n|\hat{\rho}|n\rangle = \rho_n \neq 0, \Rightarrow \langle m|\hat{\rho}|n\rangle = \rho_n\langle m|n\rangle = 0.
\end{aligned} \tag{5.20}$$

That means a functional dependence of the Density Matrix operator on the Hamiltonian $\hat{\rho} = \wp(\hat{H})$.

Let's start with a "pure" ensemble. Suppose there is a complex quantum system. This system consists of many quasi-independent quantum subsystems. For an example, the electrons belonged to the different none-interactive atoms, or molecules. Then the total Hamiltonian $\hat{H}$ of whole quantum system have a form of sum of the Hamiltonians $\hat{H}_\nu$ describing each none-interactive subsystems, that is $\hat{H} = \sum_\nu \hat{H}_\nu$. In other words, this operator is additive. Then the wave function of the whole system is factorized as: $\Psi(x_1, x_2, ...x_N) = \prod_{\nu=1}^{N} \psi_\nu(x_\nu)$, where $\psi_\nu(x_\nu)$ is the function of subsystem ν. We mean under the (x_ν) the set of variables describing the ν-th subsystem, then we have

$$\begin{aligned}
\hat{H}\Psi(x_1, x_2, ...x_N) &= \left(\sum_\nu \hat{H}_\nu\right)\prod_{\nu=1}^{N}\psi_\nu(x_\nu)\\
&= \left(\sum_\nu E_\nu\right)\Psi(x_1, x_2, ...x_N).
\end{aligned} \tag{5.21}$$

Now consider the "mixed" ensemble of the above complex quantum system. Taking into account an environment and performing the environmental average over the wave function product (see Eq. (5.8))

$$\overline{\Psi(x_1, x_2, ...x_N, q)\Psi^*(x'_1, x'_2, ...x'_N, q)}_q = \rho_{(x_1, x_2, ...x_N)(x'_1, x'_2, ...x'_N)}, \tag{5.22}$$

then we have for the coordinate representation following expression:

$$\begin{aligned}\rho_{x_1 x_2 ...x_N x'_1 x'_2 ...x'_N} &= \overline{\Psi(x_1, x_2, ...x_N q)\Psi^*(x'_1, x'_2, ...x'_N q)} = \\ &= \overline{\prod_{\nu=1}^{N} \psi_\nu(x_\nu q) \prod_{\eta=1}^{N} \psi^*_\eta(x'_\eta q)} = \prod_{\nu=1}^{N} \overline{\psi_\nu(x_\nu q)\psi^*_\nu(x'_\nu q)} \\ &= \prod_{\nu=1}^{N} \rho_{x_\nu x'_\nu}.\end{aligned} \tag{5.23}$$

So then, for the general operator form we write

$$\begin{aligned}\hat{\rho} &= \prod_{\nu=1}^{N} \hat{\rho}_\nu, \Rightarrow \\ \ln \hat{\rho} &= \sum_\nu \ln \hat{\rho}_\nu.\end{aligned} \tag{5.24}$$

We will assume that in the case of an equilibrium system, the density matrix describing both the system as a whole and individual quasi-independent subsystems should have the same functional dependence on the energy operator (Hamiltonian) describing the entire system or subsystem, respectively: $\hat{\rho} = \wp(\hat{H})$, $\hat{\rho}_\nu = \wp(\hat{H}_\nu)$. Keeping in mind the additivity property of the Hamiltonian, (or the additivity of the Hamiltonian) as well as the logarithm of the density matrix, we will look for this connection between the logarithm of the density matrix and the Hamiltonian in linear form:

$$\begin{aligned}\ln \hat{\rho} &= A + B\hat{H}, \\ \ln \hat{\rho}_\nu &= A_\nu + B_\nu \hat{H}_\nu,\end{aligned} \tag{5.25}$$

where

$$\begin{aligned}\hat{H} &= \sum_\nu \hat{H}_\nu, \\ \ln \hat{\rho} &= \sum_\nu \ln \hat{\rho}_\nu, \\ \Rightarrow A + B\hat{H} &= \sum_\nu \left(A_\nu + B_\nu \hat{H}_\nu\right).\end{aligned} \tag{5.26}$$

It is obvious that the dependence of the logarithm of the density matrix on the Hamiltonian can only be linear, while the quadratic$\hat{H}^2$, cubic $\hat{H}^3$ and higher order Hamiltonians do not satisfy the additivity principle.

The equation cannot be satisfied for any possible coefficients, since (see the following equalities) $\ln \hat{\rho} = \sum_\nu \ln \hat{\rho}_\nu$ if

$$\ln \hat{\rho}_\nu = A_\nu + B_\nu \hat{H}_\nu + C_\nu \hat{H}_\nu^2 + D_\nu \hat{H}_\nu^3 + \dots :$$

$$\begin{aligned}
\ln \hat{\rho} &= A + B\hat{H} + C\hat{H}^2 + D\hat{H}^3 + \dots \\
&= \sum_\nu \left(A_\nu + B_\nu \hat{H}_\nu + C_\nu \hat{H}_\nu^2 + D_\nu \hat{H}_\nu^3 + \dots\right) \\
C\hat{H}^2 &= \sum_\nu C_\nu \hat{H}_\nu^2, \\
D\hat{H}^3 &= \sum_\nu D_\nu \hat{H}_\nu^3, \dots
\end{aligned}$$

The equality Eq. (5.25) is satisfied only for the linear case, if the sought coefficients A and A_ν, as well as B and B_ν, correspond to the following relations:

$$\begin{cases} A = \sum_\nu A_\nu, \\ B = B_\nu. \end{cases} \tag{5.27}$$

Then the density matrix of the equilibrium system will have the form: $\hat{\rho} = \exp\left(A + B\hat{H}\right)$. Taking into account the diagonality of the matrix in its own energy representation and the normalization condition, we see that

$$\begin{aligned}
1 &= Sp\hat{\rho} = \sum_n^\infty \rho_{nn} = Sp\left(\exp\left(A + B\hat{H}\right)\right) = \\
&= e^A \sum_n^\infty \langle n|\left(\exp\left(B\hat{H}\right)\right)|n\rangle = e^A \sum_n^\infty (\exp(BE_n)).
\end{aligned} \tag{5.28}$$

The normalization condition can be fulfilled only if the parameter B is negative, we denote it as B = (-1/T), where T is considered to be the temperature on the energy scale (the exponent must be dimensionless). Assuming that the semi-classical limit to the distribution function in the canonical Gibbs form during the transition of a quantum system must be fulfilled, while the environment over which the averaging is performed is macroscopic and thermalized, playing the role of a "bath," the parameter upon the onset of equilibrium between the subsystems and appearing in the equilibrium distribution function of the thermalized system is temperature. Then in the final form we have the expression for the density matrix of the equilibrium system:

$$\hat{\rho} = e^{A}\exp\left(-\frac{\hat{H}}{T}\right),$$

$$Sp\hat{\rho} = e^{A}Sp\left(\exp\left(-\frac{\hat{H}}{T}\right)\right) = 1,$$

$$\Rightarrow e^{A} = \frac{1}{Sp\left(\exp\left(-\frac{\hat{H}}{T}\right)\right)}.$$

$$\hat{\rho} = \frac{\exp\left(-\frac{\hat{H}}{T}\right)}{Sp\left(\exp\left(-\frac{\hat{H}}{T}\right)\right)}. \tag{5.29}$$

Let us denote $Sp\left(\exp\left(-\frac{\hat{H}}{T}\right)\right) = z$, where z is a small statistical sum, which determines a free energy F, $F = -T\ln z$, then density matrix of the equilibrium system will have the form:

$$\hat{\rho} = \frac{\exp\left(-\frac{\hat{H}}{T}\right)}{z} = \exp\left(\frac{F-\hat{H}}{T}\right). \tag{5.30}$$

Note: Keeping in mind the additivity property of the Hamiltonian, the conservation law of the number of particles N $\left[\hat{H}\hat{\rho}\right] = 0$, $\left[\hat{H}\hat{N}\right] = 0$ (where N is an additive quantity $\hat{N} = \sum_{\nu}\hat{N}_{\nu}$ and an integral of motion), and the logarithm of the density matrix, we will look for this connection between the logarithm of the density matrix and the Hamiltonian in linear form:

$$\ln\hat{\rho} = A + B\hat{H} + C\hat{N},$$

$$\ln\hat{\rho}_{\nu} = A_{\nu} + B_{\nu}\hat{H}_{\nu} + C_{\nu}\hat{N}_{\nu},$$

$$\hat{H} = \sum_{\nu}\hat{H}_{\nu}, \hat{N} = \sum_{\nu}\hat{N}_{\nu}$$

$$\ln\hat{\rho} = \sum_{\nu}\ln\hat{\rho}_{\nu}, \Rightarrow A + B\hat{H} + C\hat{N}$$

$$= \sum_{\nu}\left(A_{\nu} + B_{\nu}\hat{H}_{\nu} + C_{\nu}\hat{N}_{\nu}\right). \tag{5.31}$$

The last equality is satisfied if the sought coefficients A and A_{ν}, as well as B and B_{ν}, C and C_{ν}, correspond to the following relations:

$$A = \sum_{\nu}(A_{\nu}),$$

$$B = B_{\nu}, C = C_{\nu}. \tag{5.32}$$

Then the density matrix of the equilibrium system, taking into account the conservation of the number of particles, will have the form: $\hat{\rho} = \exp(A + B\hat{H} + C\hat{N})$, taking into account the diagonality of the matrix in its own energy representation and the normalization condition, we see that

$$\begin{aligned} Sp\hat{\rho} &= \sum_{r}^{\infty} \rho_{rr} = Sp(\exp(A + B\hat{H} + C\hat{N},)) = \\ &= e^{A} \sum_{r}^{\infty} \langle r|(\exp(B\hat{H}))|r\rangle \\ &= e^{A} \sum_{r}^{\infty} (\exp(BE_r + CN_r)) = 1, \\ B &= -\frac{1}{T}, C = \frac{\mu}{T}. \end{aligned} \tag{5.33}$$

The normalization condition can be fulfilled only if the parameter B is negative, we will denote it as B = (-1/T), where T will be considered the temperature on the energy scale (the exponent must be dimensionless), and the parameter C is nothing more than the chemical potential divided by the temperature C = μ/T. Then in the final form we have an expression for the density matrix of the equilibrium system:

$$\begin{aligned} &\hat{\rho} = e^{A} \exp\left(-\frac{\hat{H} - \mu\hat{N}}{T}\right), \\ &Sp\hat{\rho} = e^{A} Sp\left(\exp\left(\frac{\mu\hat{N} - \hat{H}}{T}\right)\right) = 1, \\ &\Rightarrow e^{A} = \frac{1}{Sp\left(\exp\left(\frac{\mu\hat{N} - \hat{H}}{T}\right)\right)}, \\ &Sp\left(\exp\left(\frac{\mu\hat{N} - \hat{H}}{T}\right)\right) = Z \\ &\hat{\rho} = \frac{\exp\left(\frac{\mu\hat{N} - \hat{H}}{T}\right)}{Sp\left(\exp\left(\frac{\mu\hat{N} - \hat{H}}{T}\right)\right)} = \frac{\exp\left(\frac{\mu\hat{N} - \hat{H}}{T}\right)}{Z} \\ &= \exp\left(\frac{\Omega + \mu\hat{N} - \hat{H}}{T}\right). \\ &\Omega = -T \ln Z. \end{aligned} \tag{5.34}$$

Here, Z is grand statistical sum and Ω is a grand (high) thermodynamic potential.

5.2.6 Interaction Representation for a Density Matrix

Quantum mechanics uses three time representations that are related by unitary transformations: the Schrödinger representation, the Heisenberg representation, and the interaction representation. The original representation is the Schrödinger representation, which contains the equation for the wave function (for non-relativistic systems, this is the Schrödinger equation):

$$i\hbar\frac{\partial}{\partial t}|\Psi(t)\rangle = \hat{H}|\Psi(t)\rangle. \tag{5.35}$$

Here the time dependence remains with the wave function, while the operators of physical quantities do not contain an explicit time dependence. Let there be a set of eigenvectors of the state of some stationary Schrödinger problem

$$\begin{aligned}
&\hat{H}^0|\psi_n(t)\rangle = E_n|\psi_n(t)\rangle,\\
&|\psi_n(t)\rangle = |\psi_n\rangle e^{-\frac{i}{\hbar}E_n t}\\
&\langle\psi_m|\,\psi_n\rangle = \delta_{mn},\\
&\sum_n |\psi_n\rangle\langle\psi_n| = \hat{1}.
\end{aligned} \tag{5.36}$$

Then the expansion of the vector in basis (5.35) will have the form:

$$|\Psi(t)\rangle = \sum_n |\psi_n(t)\rangle\langle\psi_n(t)|\,\Psi(t)\rangle = \sum_n |\psi_n\rangle\langle\psi_n|\,\Psi(t)\rangle. \tag{5.37}$$

The quantum–mechanical average value of a physical quantity L described by the operator $\hat{L}$ for system (5.35) will have the form:

$$\overline{L}(t) = \langle\Psi(t)|\hat{L}|\Psi(t)\rangle = \sum_{m,n} \langle\Psi(t)\,|\psi_m\rangle\langle\psi_m|\hat{L}|\psi_n\rangle\langle\psi_n|\,\Psi(t)\rangle. \tag{5.38}$$

If the Hamiltonian in (5.35) is explicitly independent of time, then formally, the operator of the time evolution of the wave function will be the operator $\hat{U} = e^{\frac{-i}{\hbar}\hat{H}t}$:

$$\begin{aligned}
&|\Psi(t)\rangle = e^{\frac{-i}{\hbar}\hat{H}t}|\Psi(0)\rangle,\\
&i\hbar\frac{\partial}{\partial t}|\Psi(t)\rangle = \hat{H}|\Psi(t)\rangle.
\end{aligned} \tag{5.39}$$

This operator is unitary because its Hermitian adjoint operator is its inverse:

$$\begin{aligned}
&\hat{U} = e^{\frac{-i}{\hbar}\hat{H}t},\ \hat{U}^\dagger = e^{\frac{i}{\hbar}\hat{H}^\dagger t},\\
&\hat{H}^\dagger = \hat{H}, \Rightarrow
\end{aligned}$$

$$\hat{U}^{\dagger} = e^{\frac{i}{\hbar}\hat{H}^{\dagger}t} = e^{\frac{i}{\hbar}\hat{H}t} = \hat{U}^{-1},$$
$$\hat{U}\hat{U}^{\dagger} = \hat{1}. \tag{5.40}$$

Let some wave vector of the initial state $|\Psi(0)\rangle$ (explicitly independent of time), which we denote as $|\Psi(0)\rangle \equiv |\Phi\rangle$, can be obtained from $|\Psi(t)\rangle = \hat{U}|\Phi\rangle$ (39) by a unitary transformation

$$|\Phi\rangle = \hat{U}^{-1}|\Psi(t)\rangle. \tag{5.41}$$

The average value of a physical quantity should not depend on the representation of the state vectors, therefore, for the operator of a physical quantity there should exist a following transformation:

$$\overline{L} = \langle\Psi(t)|\hat{L}_{\Psi}|\Psi(t)\rangle = \langle\Phi|\hat{U}^{-1}\hat{L}_{\Psi}\hat{U}|\Phi\rangle = \langle\Phi|\hat{\Lambda}_{\Phi}|\Phi\rangle,$$
$$\hat{\Lambda}_{\Phi} = \hat{U}^{-1}\hat{L}_{\Psi}\hat{U} = e^{\frac{i}{\hbar}\hat{H}t}\hat{L}_{\Psi}e^{-\frac{i}{\hbar}\hat{H}t}, \tag{5.42}$$

that is, the operator $\hat{\Lambda}_{\Phi}$ is the operator of the same physical quantity L, but in a new, Φ- representation. In this case, it is Heisenberg's representation, in which the wave vector is independent of time, while the operator of the physical quantity contains explicit time dependence.

If now the Hamiltonian that describes some system can be broken down into two terms of the form:

$$\hat{H}(t) = \hat{H}^0 + \hat{V}(t). \tag{5.43}$$

In this case, another time-unitary transformation of the operators of physical quantities and wave vectors can be introduced, translating them into the so-called interaction representation:

$$|\Psi_{Sch}(t)\rangle = e^{\frac{-i}{\hbar}\hat{H}^0 t}|\Phi_{\text{int}}(t)\rangle$$
$$|\Phi_{\text{int}}(t)\rangle = e^{\frac{i}{\hbar}\hat{H}^0 t}|\Psi_{Sch}(t)\rangle$$
$$\overline{L}(t) = \langle\Psi(t)|\hat{L}|\Psi(t)\rangle = \langle\Phi_{\text{int}}(t)|e^{\frac{i}{\hbar}\hat{H}^0 t}\hat{L}e^{\frac{-i}{\hbar}\hat{H}^0 t}|\Phi_{\text{int}}(t)\rangle,$$
$$\hat{\Lambda}_{\text{int}}(t) = e^{\frac{i}{\hbar}\hat{H}^0 t}\hat{L}e^{\frac{-i}{\hbar}\hat{H}^0 t}. \tag{5.44}$$

Let us assume that we need to solve the equation for the density matrix for the case when the Hamiltonian can be represented as the sum of two contributions (5.43), and the density matrix of the unperturbed system is known to us and corresponds to a certain equilibrium state of the system (5.30)

$$\hat{\rho}^0 = \frac{\exp\left(-\frac{\hat{H}^0}{T}\right)}{Sp\left(\exp\left(-\frac{\hat{H}^0}{T}\right)\right)} = \exp\left(\frac{F - \hat{H}^0}{T}\right).$$

Then, if the interaction operator contains a small parameter and this operator can be considered a perturbation, then we could solve the Liouville-von Neumann equation using the method of successive approximations:

$$\frac{\partial \hat{\rho}(t)}{\partial t} + \frac{i}{\hbar}\left[\left(\hat{H}^0 + \hat{V}(t)\right)\hat{\rho}(t)\right] = 0,$$
$$\hat{\rho}(t) = \hat{\rho}^0 + \hat{\rho}^{(1)} + \hat{\rho}^{(2)} + \ldots, \tag{5.45}$$

Then, as a first approximation, we leave in the equation terms that are not higher than the first order of smallness:

$$\frac{\partial (\hat{\rho}^0 + \hat{\rho}^{(1)})}{\partial t} + \frac{i}{\hbar}\left[\left(\hat{H}^0 + \hat{V}(t)\right)\left(\hat{\rho}^0 + \hat{\rho}^{(1)}\right)\right] = 0,$$
$$\frac{\partial (\hat{\rho}^{(1)}(t)}{\partial t} = -\frac{i}{\hbar}\left[\hat{H}^0\hat{\rho}^0\right] - \frac{i}{\hbar}\left[\hat{H}^0\hat{\rho}^{(1)}(t)\right] - \frac{i}{\hbar}\left[\hat{V}(t)\hat{\rho}^0\right],$$
$$\left[\hat{H}^0\hat{\rho}^0\right] = 0, \Rightarrow$$
$$\hat{\rho}^{(1)}(t) = -\frac{i}{\hbar}\int_{-\infty}^{t}\left[\hat{H}^0\hat{\rho}^{(1)}(t')\right]dt' - \frac{i}{\hbar}\int_{-\infty}^{t}\left[\hat{V}(t')\hat{\rho}^0\right]dt'. \tag{5.46}$$

As a result, already at the first iteration, we arrive at the integral equation for the first correction of the density matrix. It is known that the procedure for solving an integral equation is much more complicated than a differential equation, therefore let's try to implement the same algorithm, but in the interaction representation. To do this, we must obtain the equation for the density matrix in this representation.

$$\frac{\partial \hat{\rho}(t)}{\partial t} + \frac{i}{\hbar}\left[\left(\hat{H}^0 + \hat{V}(t)\right)\hat{\rho}(t)\right] = 0,$$
$$\hat{\rho}_{\text{int}}(t) = e^{\frac{i}{\hbar}\hat{H}^0 t}\hat{\rho}(t)e^{-\frac{i}{\hbar}\hat{H}^0 t},$$
$$\frac{\partial \hat{\rho}_{\text{int}}(t)}{\partial t} = \frac{i}{\hbar}\hat{H}^0 e^{\frac{i}{\hbar}\hat{H}^0 t}\hat{\rho}(t)e^{-\frac{i}{\hbar}\hat{H}^0 t}$$
$$+ e^{\frac{i}{\hbar}\hat{H}^0 t}\frac{\partial \hat{\rho}(t)}{\partial t}e^{-\frac{i}{\hbar}\hat{H}^0 t} - e^{\frac{i}{\hbar}\hat{H}^0 t}\hat{\rho}(t)e^{-\frac{i}{\hbar}\hat{H}^0 t}\frac{i}{\hbar}\hat{H}^0 =$$
$$= \frac{i}{\hbar}\left[\hat{H}^0\hat{\rho}_{\text{int}}(t)\right] - e^{\frac{i}{\hbar}\hat{H}^0 t}\frac{i}{\hbar}\left[\left(\hat{H}^0 + \hat{V}(t)\right)\hat{\rho}(t)\right]e^{-\frac{i}{\hbar}\hat{H}^0 t} =$$
$$= \frac{i}{\hbar}\left[\hat{H}^0\hat{\rho}_{\text{int}}(t)\right] - \frac{i}{\hbar}e^{\frac{i}{\hbar}\hat{H}^0 t}\left[\hat{H}^0\hat{\rho}(t)\right]e^{-\frac{i}{\hbar}\hat{H}^0 t} - \frac{i}{\hbar}e^{\frac{i}{\hbar}\hat{H}^0 t}\left[\hat{V}(t)\hat{\rho}(t)\right]e^{-\frac{i}{\hbar}\hat{H}^0 t},$$
$$\frac{i}{\hbar}e^{\frac{i}{\hbar}\hat{H}^0 t}\left[\hat{H}^0\hat{\rho}(t)\right]e^{-\frac{i}{\hbar}\hat{H}^0 t} = \frac{i}{\hbar}e^{\frac{i}{\hbar}\hat{H}^0 t}\left(\hat{H}^0\hat{\rho}(t) - \hat{\rho}(t)\hat{H}^0\right)e^{-\frac{i}{\hbar}\hat{H}^0 t} =$$
$$= \frac{i}{\hbar}\left(\hat{H}^0 e^{\frac{i}{\hbar}\hat{H}^0 t}\hat{\rho}(t)e^{-\frac{i}{\hbar}\hat{H}^0 t} - e^{\frac{i}{\hbar}\hat{H}^0 t}\hat{\rho}(t)e^{-\frac{i}{\hbar}\hat{H}^0 t}\hat{H}^0\right)$$
$$= \frac{i}{\hbar}\left[\hat{H}^0\hat{\rho}_{\text{int}}(t)\right], \Rightarrow$$

$$\frac{\partial \hat{\rho}_{\text{int}}(t)}{\partial t} = \frac{i}{\hbar}\big[\hat{H}^0 \hat{\rho}_{\text{int}}(t)\big] - \frac{i}{\hbar}\big[\hat{H}^0 \hat{\rho}_{\text{int}}(t)\big]$$
$$- \frac{i}{\hbar} e^{\frac{i}{\hbar}\hat{H}^0 t}\big[\hat{V}(t)\hat{\rho}(t)\big]e^{-\frac{i}{\hbar}\hat{H}^0 t},$$
$$\frac{\partial \hat{\rho}_{\text{int}}(t)}{\partial t} = -\frac{i}{\hbar} e^{\frac{i}{\hbar}\hat{H}^0 t}\big(\hat{V}(t)\hat{1}\hat{\rho}(t) - \hat{\rho}(t)\hat{1}\hat{V}(t)\big)e^{-\frac{i}{\hbar}\hat{H}^0 t} =$$
$$= -\frac{i}{\hbar} e^{\frac{i}{\hbar}\hat{H}^0 t}\Big(\hat{V}(t)e^{-\frac{i}{\hbar}\hat{H}^0 t}e^{\frac{i}{\hbar}\hat{H}^0 t}\hat{\rho}(t) - \hat{\rho}(t)e^{-\frac{i}{\hbar}\hat{H}^0 t}e^{\frac{i}{\hbar}\hat{H}^0 t}\hat{V}(t)\Big)e^{-\frac{i}{\hbar}\hat{H}^0 t} =$$
$$= -\frac{i}{\hbar}\big[\hat{V}_{\text{int}}(t)\hat{\rho}_{\text{int}}(t)\big]. \tag{5.47}$$

Thus, the equation for the density matrix in the interaction representation is as follows:

$$\frac{\partial \hat{\rho}_{\text{int}}(t)}{\partial t} - \frac{i}{\hbar}\big[\hat{V}_{\text{int}}(t)\hat{\rho}_{\text{int}}(t)\big] = 0,$$
$$\hat{V}_{\text{int}}(t) = e^{\frac{i}{\hbar}\hat{H}^0 t}\hat{V}(t)e^{-\frac{i}{\hbar}\hat{H}^0 t}. \tag{5.48}$$

In this representation, the implementation of the method of successive approximations (within the framework of perturbation theory) looks more convenient. Let non-stationary fields with small amplitudes of intensities be imposed on a system in an equilibrium state, i.e., the perturbation operator $\hat{V}(t)$ will contain a small parameter, then we will look for the density matrix in the form of an expansion in terms of the small parameter: $\hat{\rho}(t) = \hat{\rho}^0 + \hat{\rho}^{(1)} + \hat{\rho}^{(2)} + ..$, while writing the equation in the interaction representation (5.48):

$$\frac{\partial \hat{\rho}_{\text{int}}(t)}{\partial t} = -\frac{i}{\hbar}\big[\hat{V}_{\text{int}}(t)\hat{\rho}_{\text{int}}(t)\big]$$
$$\hat{\rho}_{\text{int}}(t) = \hat{\rho}^0_{\text{int}} + \hat{\rho}^{(1)}_{\text{int}} + \hat{\rho}^{(2)}_{\text{int}} + \dots$$
$$\hat{\rho}^0_{\text{int}} = e^{\frac{i}{\hbar}\hat{H}^0 t}\hat{\rho}^0 e^{-\frac{i}{\hbar}\hat{H}^0 t} = e^{\frac{i}{\hbar}\hat{H}^0 t}\frac{e^{-\frac{\hat{H}^0}{T}}}{Sp\Big(e^{-\frac{\hat{H}^0}{T}}\Big)}$$
$$e^{-\frac{i}{\hbar}\hat{H}^0 t} = \frac{e^{-\frac{\hat{H}^0}{T}}}{Sp\Big(e^{-\frac{\hat{H}^0}{T}}\Big)} \equiv \hat{\rho}^0. \tag{5.49}$$

In the first approximation we have

$$\frac{\partial \hat{\rho}^{(1)}_{\text{int}}(t)}{\partial t} = -\frac{i}{\hbar}\big[\hat{V}_{\text{int}}(t)\hat{\rho}^0_{\text{int}}\big], \_ \Rightarrow$$
$$\hat{\rho}^{(1)}_{\text{int}}(t) = -\frac{i}{\hbar}\int_{-\infty}^{t}\big[\hat{V}_{\text{int}}(t')\hat{\rho}^0_{\text{int}}\big]dt'. \tag{5.50}$$

In the second approximation:

$$\frac{\partial\hat{\rho}_{\text{int}}^{(1)}(t)}{\partial t}+\frac{\partial\hat{\rho}_{\text{int}}^{(2)}}{\partial t}=-\frac{i}{\hbar}\left[\hat{V}_{\text{int}}(t)\left(\hat{\rho}_{\text{int}}^{0}+\hat{\rho}_{\text{int}}^{(1)}\right)\right]=$$
$$=-\frac{i}{\hbar}\left[\hat{V}_{\text{int}}(t)\left(\hat{\rho}_{\text{int}}^{0}\right)\right]-\frac{i}{\hbar}\left[\hat{V}_{\text{int}}(t)\left(\hat{\rho}_{\text{int}}^{(1)}\right)\right],$$
$$\frac{\partial\hat{\rho}_{\text{int}}^{(1)}(t)}{\partial t}=-\frac{i}{\hbar}\left[\hat{V}_{\text{int}}(t)\hat{\rho}_{\text{int}}^{0}\right],\Rightarrow$$
$$\frac{\partial\hat{\rho}_{\text{int}}^{(2)}(t)}{\partial t}=-\frac{i}{\hbar}\left[\hat{V}_{\text{int}}(t)\hat{\rho}_{\text{int}}^{(1)}(t)\right],\Rightarrow$$
$$\hat{\rho}_{\text{int}}^{(2)}(t)=-\frac{i}{\hbar}\int\limits_{-\infty}^{t}\left[\hat{V}_{\text{int}}(t')\hat{\rho}_{\text{int}}^{(1)}(t')\right]dt'=$$
$$=\left(-\frac{i}{\hbar}\right)^{2}\int\limits_{-\infty}^{t}\int\limits_{-\infty}^{t_2}\left[\hat{V}_{\text{int}}(t_2)\left[\hat{V}_{\text{int}}(t_1)\hat{\rho}_{\text{int}}^{0}\right]\right]dt_1dt_2,\ \ldots \tag{5.51}$$

Then, we have a solution for the n-th approximation in the form:

$$\hat{\rho}_{\text{int}}^{(n)}(t)=\left(-\frac{i}{\hbar}\right)^{n}\int\limits_{-\infty}^{t}\int\limits_{-\infty}^{t_{n-1}}\ldots\int\limits_{-\infty}^{t_2}\left[\hat{V}_{\text{int}}(t_{n-1})\left[\hat{V}_{\text{int}}(t_{n-2})\ldots\left[\hat{V}_{\text{int}}(t_1)\hat{\rho}_{\text{int}}^{0}\right]\ldots\right]\right]dt_1dt_2\ldots dt_{n-2}dt_{n-1}.$$

Here the moments of time are: $t_1\leq t_2\leq\ldots\leq t_{n-2}\leq t_{n-1}\leq t$. Introducing the chronological ordering operator, we have

$$\hat{\rho}_{\text{int}}^{(1)}(t)=-\frac{i}{\hbar}\int\limits_{-\infty}^{t}\left[\hat{V}_{\text{int}}(t')\hat{\rho}_{\text{int}}^{0}\right]dt',$$

$$\hat{\rho}_{\text{int}}^{(2)}(t)=-\frac{i}{\hbar}\int\limits_{-\infty}^{t}\left[\hat{V}_{\text{int}}(t')\hat{\rho}_{\text{int}}^{(1)}(t')\right]dt'$$
$$=\left(-\frac{i}{\hbar}\right)^{2}\int\limits_{-\infty}^{t}\int\limits_{-\infty}^{t'}\left[\hat{V}_{\text{int}}(t')\left[\hat{V}_{\text{int}}(t_1)\hat{\rho}_{\text{int}}^{0}\right]\right]dt_1dt'=$$
$$=\frac{1}{2}\left(-\frac{i}{\hbar}\right)^{2}\hat{T}\int\limits_{-\infty}^{t}\int\limits_{-\infty}^{t}\left[\hat{V}_{\text{int}}(t')\left[\hat{V}_{\text{int}}(t_1)\hat{\rho}_{\text{int}}^{0}\right]\right]dt_1dt_2,$$

$$\hat{\rho}_{\text{int}}^{(n)}(t)=\left(-\frac{i}{\hbar}\right)^{n}\int\limits_{-\infty}^{t}\int\limits_{-\infty}^{t_{n-1}}\ldots\int\limits_{-\infty}^{t_2}\left[\hat{V}_{\text{int}}(t_{n-1})\left[\hat{V}_{\text{int}}(t_{n-2})\ldots\left[\hat{V}_{\text{int}}(t_1)\hat{\rho}_{\text{int}}^{0}\right]\ldots\right]\right]$$

$$dt_1 dt_2 ... dt_{n-2} dt_{n-1} =$$

$$= \frac{1}{n!}\left(-\frac{i}{\hbar}\right)^n \hat{T} \int_{-\infty}^{t} \int_{-\infty}^{t} ... \int_{-\infty}^{t} \left[\hat{V}_{\text{int}}(t_{n-1})\left[\hat{V}_{\text{int}}(t_{n-2})...\left[\hat{V}_{\text{int}}(t_1)\hat{\rho}^0_{\text{int}}\right]...\right]\right]$$

$$dt_1 dt_2 ... dt_{n-2} dt_{n-1}. \tag{5.52}$$

Then we have a general solution in the form:

$$\hat{\rho}_{\text{int}}(t) = \sum_n \frac{1}{n!}\left(-\frac{i}{\hbar}\right)^n \hat{T} \int_{-\infty}^{t} \int_{-\infty}^{t} ... \int_{-\infty}^{t} \left[\hat{V}_{\text{int}}(t_{n-1})\left[\hat{V}_{\text{int}}(t_{n-2})...\left[\hat{V}_{\text{int}}(t_1)\hat{\rho}^0_{\text{int}}\right]...\right]\right]$$

$$dt_1 dt_2 ... dt_{n-2} dt_{n-1} =$$

$$= \hat{T} e^{-\frac{i}{\hbar}\int_{-\infty}^{t} \hat{V}_{\text{int}}(t')dt'} \hat{\rho}^0_{\text{int}} e^{\frac{i}{\hbar}\int_{-\infty}^{t} \hat{V}_{\text{int}}(t'')dt''}. \tag{5.53}$$

5.3 Physical Applications of the Density Matrix Apparatus

5.3.1 Paramagnetism of a System of Particles with an Arbitrary Proper Mechanical Moment J Interaction

A gas of particles with their own mechanical moment of magnitude j and corresponding to it eigen magnetic momentum $m = gj = 2\mu_B j$ is placed in a constant uniform magnetic field of intensity $\boldsymbol{H}$. We find the magnetization of a unit volume of this gas in thermodynamic equilibrium at temperature T, consider the concentration to be a constant n, consider the gyromagnetic factor of the particle to be equal to g (for electrons, protons, neutrons, this factor is equal to $g = 2\mu_B$, where the Bohr magneton is $\mu_B = \frac{|e|\hbar}{2mc}$, where the denominator is the mass of an electron, proton, or neutron, respectively). For atoms, the gyromagnetic factor is the Lande factor, which is calculated. Let $\mathbf{H}||\mathbf{z}$, then a magnetization of the volume unit with the particles concentration n will be $M_z = n\overline{m}_z$. Our task is to calculate the average value of the projection of the magnetic moment per particle for the case of finite temperatures, assuming the system to be thermalized. We write the operator of projection of the magnetic moment onto the axis as $\hat{m}_z = g\hat{j}_z$. According to expression (5.6) of Chap. 1, the average value of a physical quantity, taking into account the thermalization of the system, can be calculated using the density matrix as $\overline{m}_z = Sp(\hat{m}_z\hat{\rho})$. The density matrix describing a quantum system in thermodynamic equilibrium with the environment has the form (5.29), (5.30) of Chap. 1 is.

$$\hat{\rho} = \frac{e^{-\frac{\hat{H}}{T}}}{Sp\left(e^{-\frac{\hat{H}}{T}}\right)}.$$

The Hamiltonian describing the interaction of the magnetic moment of a particle with a magnetic field has the form $\hat{H} = -\hat{m}_z H_z = -g\hat{j}_z H_z$, where eigen mechanical moment and Hamiltonian in eigent representation have the set of values:

$$\begin{aligned} &\hat{j}_z|m\rangle = m|m\rangle \\ &m : -j, -j+1 \ldots j-1, j \\ &\hat{H}|m\rangle = -gmH_z|m\rangle. \end{aligned} \tag{5.54}$$

The calculation of the $\overline{m}_z$ is following:

$$\begin{aligned} &\overline{m}_z = Sp(\hat{m}_z\hat{\rho}) = \frac{Sp\left(\hat{m}_z e^{-\frac{\hat{H}}{T}}\right)}{Sp\left(e^{-\frac{\hat{H}}{T}}\right)}, \\ &\hat{m}_z|m\rangle = g\hat{j}_z|m\rangle = gm|m\rangle, \\ &\hat{H} = -\hat{m}_z H_z = -g\hat{j}_z H_z, \\ &\hat{H}|m\rangle = -g\hat{j}_z H_z|m\rangle = -gmH_z|m\rangle. \end{aligned} \tag{5.55}$$

Then we have

$$\begin{aligned} &\overline{m}_z = Sp(\hat{m}_z\hat{\rho}) = \frac{\sum\limits_{m=-j}^{m=j} mg e^{\frac{gH_z}{T}m}}{\sum\limits_{m=-j}^{m=j} e^{\frac{gH_z}{T}m}} = \\ &= g\frac{\frac{\partial}{\partial\beta}\sum\limits_{m=-j}^{m=j} e^{\beta m}}{\sum\limits_{m=-j}^{m=j} e^{\beta m}} = g\frac{\partial}{\partial\beta}\ln\left(\sum_{m=-j}^{m=j} e^{\beta m}\right), \\ &\left(\frac{gH_z}{T}\right) = \beta. \end{aligned} \tag{5.56}$$

The sum calculation is performed by following way:

$$\sum_{m=-j}^{m=j} e^{\beta m} = e^{-\beta j}\sum_{m=-j}^{m=j} e^{\beta m} e^{\beta j} = e^{-\beta j}\sum_{m=-j}^{m=j} e^{\beta(m+j)}$$

$$
\begin{aligned}
&= e^{-\beta j} \sum_{n=(m+j)=0}^{n=2j} e^{\beta n} = \\
&= e^{-\beta j} \frac{e^{(2j+1)\beta} - 1}{e^{\beta} - 1} = \frac{e^{(j+1)\beta} - e^{-\beta j}}{e^{\beta} - 1} = \\
&= \frac{e^{(j+\frac{1}{2})\beta} - e^{-\beta(j+\frac{1}{2})}}{e^{\beta/2} - e^{-\beta/2}} = \frac{2sh(j+\frac{1}{2})\beta}{2sh(\beta/2)}.
\end{aligned}
\tag{5.57}
$$

Then, we have for the magnetization

$$
\begin{aligned}
\tilde{m}_z &= g\frac{\partial}{\partial\beta}\ln\left(\sum_{m=-j}^{m=j} e^{\beta m}\right) = g\frac{\partial}{\partial\beta}\left(\ln\left[sh\left(j+\frac{1}{2}\right)\beta\right] - \ln[sh(\beta/2)]\right) = \\
&= g\left(\left(j+\frac{1}{2}\right)cth\left(j+\frac{1}{2}\right)\beta - \frac{1}{2}cth(\beta/2)\right), \\
M_z &= ng\left\{\left(j+\frac{1}{2}\right)cth\left(\left(j+\frac{1}{2}\right)\frac{gH_z}{T}\right) - \frac{1}{2}cth\left(\frac{gH_z}{2T}\right)\right\}.
\end{aligned}
\tag{5.58}
$$

For the case of "mesa" particles, the mechanical moment of which is much greater than unity $j >> 1$, and the mesoscopic value of the magnetic moment $jg = M >> \mu_B$, for example, when a ferromagnet shaving is placed in a dielectric, the so-called lyotropic materials, we have

$$
\begin{aligned}
M_z &= n\left\{(jg)cth\left((jg)\frac{H_z}{T}\right) - \frac{g}{2}cth\left(\frac{gjH_z}{2jT}\right)\right\} = \\
&= n\left\{Mcth\left(M\frac{H_z}{T}\right) - \frac{g}{2}\frac{1}{th\left(\frac{MH_z}{2Tj}\right)}\right\} \approx \\
&= n\left\{Mcth\left(M\frac{H_z}{T}\right) - \frac{g}{2}\frac{1}{\left(\frac{MH_z}{2Tj}\right)}\right\} \\
&= n\left\{Mcth\left(M\frac{H_z}{T}\right) - \frac{gj}{2}\frac{1}{\left(\frac{MH_z}{2T}\right)}\right\} = \\
&= Mn\left\{cth\left(\frac{MH_z}{T}\right) - \frac{1}{\left(\frac{MH_z}{T}\right)}\right\} = MnL\left(\frac{MH_z}{T}\right), \\
&]j \sim 1000, \_\_M \sim 10^{-17}\_\_H_z \sim 10^6\Gamma c\_\_MH_z \sim 10^{-11}
\end{aligned}
\tag{5.59}
$$

where $L(x) = cthx - \frac{1}{x}$ is Langevin function.

5.3.2 Equation of Precession of the Proper Magnetic Moment for an Arbitrary Spin of Particles in a Magnetic Field

A gas of particles with their own mechanical moment of magnitude j is placed in a time-dependent uniform magnetic field of intensity $H(t)$. Our task is to find an equation of motion for the average value $\frac{\partial}{\partial t}\overline{\mathbf{m}}(t)$. Here $\overline{\mathbf{m}}(t)$ is the average magnetic moment of one particle. As above, the operator of the magnetic moment of particle is $\hat{\mathbf{m}} = g\hat{\mathbf{j}}$, where $\hat{\mathbf{j}} = \sum_{\alpha=x,y,z} \hat{j}_\alpha \mathbf{e}_\alpha$ is an operator of mechanical moment. Then, we can write for operators of magnetic moments projections as $\begin{array}{c}\hat{m}_\alpha = g\hat{j}_\alpha, \\ (\alpha : x, y, z).\end{array}$.

The statistically averaged physical value of magnetic moment projection would be calculated by using density matrix (see Eq. (5.6) Chapt.1) as $\overline{m}_\alpha = Sp(\hat{m}_\alpha \hat{\rho}(t))$. In Schrödinger representation, operators of physical values are not time dependent, whole dependence of time is in the density matrix operator, where the equation (see Chapt1, Eq. (5.17)) is

$$\frac{\partial}{\partial t}\hat{\rho} = -\frac{i}{\hbar}[\hat{H}(t)\hat{\rho}],$$
$$\hat{H}(t) = -\hat{\mathbf{m}} \cdot H(t).$$

The calculation of time dependence of the each projection of the magnetic moment will be done by the following way:

$$\begin{aligned}
\frac{\partial}{\partial t}\overline{m}_\alpha(t) &= Sp\left(\hat{m}_\alpha \frac{\partial}{\partial t}\hat{\rho}\right) = -\frac{i}{\hbar}Sp\big(\hat{m}_\alpha[\hat{H}(t)\hat{\rho}]\big) = \\
&= \frac{i}{\hbar}Sp(\hat{m}_\alpha[\hat{\mathbf{m}} \cdot H(t)\hat{\rho}]) = \frac{i}{\hbar}Sp \sum_{\beta=x,y,z} \big(\hat{m}_\alpha H_\beta(t)[\hat{m}_\beta \hat{\rho}]\big) = \\
&= \frac{i}{\hbar}\sum_\beta H_\beta(t) Sp\big(\hat{m}_\alpha[\hat{m}_\beta\hat{\rho}]\big) = \frac{i}{\hbar}\sum_{\beta=x,y,z} H_\beta(t) Sp\big([\hat{m}_\alpha\hat{m}_\beta]\hat{\rho}\big) = \\
&= \frac{i}{\hbar}\sum_\beta H_\beta(t) Sp\big([g\hat{j}_\alpha g\hat{j}_\beta]\hat{\rho}\big) = \frac{i}{\hbar}g^2\sum_\beta H_\beta(t) Sp\big([\hat{j}_\alpha\hat{j}_\beta]\hat{\rho}\big) = \\
&= \frac{i}{\hbar}g^2\sum_\beta H_\beta(t) Sp\big(i\varepsilon_{\alpha\beta\gamma}\hat{j}_\gamma\hat{\rho}\big).
\end{aligned} \tag{5.60}$$

Here we used the commutation rules for operators of mechanical moments projections $[\hat{j}_\alpha\hat{j}_\beta] = i\varepsilon_{\alpha\beta\gamma}\hat{j}_\gamma$, where $\varepsilon_{\alpha\beta\gamma}$-Levi–Civita symbols. Then, after taking trace (Sp), we have

$$\frac{\partial}{\partial t}\overline{m}_\alpha(t) = \frac{-g^2}{\hbar}\sum_\beta H_\beta(t)\varepsilon_{\alpha\beta\gamma}\overline{j}_\gamma =$$

$$= \frac{-g}{\hbar}\sum_{\beta} H_{\beta}(t)\varepsilon_{\alpha\beta\gamma}\overline{m}_{\gamma} = \frac{-g}{\hbar}(H(t)\times\overline{\mathbf{m}})_{\alpha}$$

$$\equiv \frac{g}{\hbar}(\overline{\mathbf{m}}\times H(t))_{\alpha} = (\overline{\mathbf{m}}\times\boldsymbol{\Omega})_{\alpha},$$

$$\boldsymbol{\Omega}(t) = \frac{g}{\hbar}H(t), \Rightarrow \frac{\partial}{\partial t}\overline{\mathbf{m}} = (\overline{\mathbf{m}}\times\boldsymbol{\Omega}(t)).$$

There is a Bloch Equation describing a precession of the average value of the eigent magnetic moment of quantum particle in the arbitrary magnetic field: $\frac{\partial}{\partial t}\overline{\mathbf{m}} = (\overline{\mathbf{m}}\times\boldsymbol{\Omega}(t))$. $\boldsymbol{\Omega}(t)$-precession frequency of the magnetic moment vector, the frequency vector is co-directed with the magnetic field vector.

5.3.3 *Kubo's Formula for Conductivity*

Let us consider a certain conducting sample, the current density operator in which we denote by $\hat{\mathbf{j}} = q\hat{\mathbf{v}}$. The sample is not a superconductor, so in the absence of an external field, in the equilibrium state of the system, the current density j = 0. Let this conducting sample be placed in an electric field of intensity $\mathbf{E}(t) = \mathbf{e}_x E_0 \cos\omega t$, then denoting the vector potential as $\mathbf{A}(t)$, the operator of interaction of current carriers in the sample with the field will have the form:

$$\begin{aligned}
\hat{V}(t) &= -\frac{1}{c}\mathbf{A}(t)\cdot\hat{\mathbf{j}},\\
rot\mathbf{A}(t) &= H(t)\\
-\frac{1}{c}\frac{\partial\mathbf{A}(t)}{\partial t} &= E(t) = \mathbf{e}_x E_0\cos\omega t,\\
\mathbf{A}(t) &= -\mathbf{e}_x\frac{c}{\omega}E_0\sin\omega t,\\
\hat{V}(t) &= \hat{\mathbf{j}}\cdot\mathbf{e}_x\frac{E_0}{\omega}\sin\omega t = j_x\frac{E_0}{\omega}\sin\omega t.
\end{aligned} \tag{5.61}$$

Our task is to obtain in analytical form in general the response of a conducting sample to an electric field, that is, to determine the conductivity of the system at a given potential, taking into account all possible quantum transitions in the system at a finite temperature. It is required to determine the conductivity of a given isotropic sample at a given temperature T. To begin with, we define the general structure of the matrix of the current density operator in its own energy representation, where

$$\hat{H}^0|n\rangle = E_n|n\rangle,$$

the set of states $\{|n\rangle\}$ has the properties of orthogonality $\langle n|m\rangle = \delta_{nm}$ and completeness $\sum_n |n\rangle\langle n| = \hat{1}$. In the Schrödinger representation, the current density operator does not contain an explicit time dependence and is defined as the velocity operator multiplied by the charge. In turn, the velocity operator is the operator of the time derivative of the radius vector (or its projection onto a given axis)

$$\hat{j}_x = e\hat{v}_x = e\frac{i}{\hbar}\left[\hat{H}^0\hat{x}\right].$$

Then an arbitrary matrix element of the current density operator is defined as

$$\begin{aligned}
(j_x)_{nm} &= e\frac{i}{\hbar}\left[\hat{H}^0\hat{x}\right]_{nm} \\
&= e\frac{i}{\hbar}\{E_n x_{nm} - x_{nm}E_m\} = ei\omega_{nm}x_{nm}, \\
\omega_{nm} &= \frac{E_n - E_m}{\hbar}. \Rightarrow \\
j_{nn} &= 0 \\
j_{nm} &\neq 0.
\end{aligned} \tag{5.62}$$

We consider a conducting sample in which, in the absence of an external electric field, the current density should be zero, that is,

$$\begin{aligned}
\tilde{j}_x &= Sp\left(\hat{j}_x\hat{\rho}^0\right) = 0, \\
\hat{\rho}^0 &= \frac{e^{-\frac{\hat{H}^0}{T}}}{Sp\left(e^{-\frac{\hat{H}^0}{T}}\right)}, \\
\left(\hat{\rho}^0\right)_{mn} &= \frac{e^{-\frac{E_n}{T}}}{Sp\left(e^{-\frac{\hat{H}^0}{T}}\right)}\delta_{mn} = \frac{e^{-\frac{E_n}{T}}}{\sum_k e^{-\frac{E_k}{T}}}\delta_{mn}, \\
Sp\left(\hat{j}_x\hat{\rho}^0\right) &= \sum_{n,m} j_{nm}\rho^0_{mn} = \frac{\sum_n j_{nn}e^{-\frac{E_n}{T}}}{\sum_k e^{-\frac{E_k}{T}}} = 0, \Rightarrow \\
j_{nn} &= 0 \\
j_{nm} &\neq 0.
\end{aligned} \tag{5.63}$$

Thus, in the general case, the structure of the current density matrix is determined in which all diagonal elements will be equal to zero, and the off-diagonal ones will be different from zero. Let us now find the average value of the current density taking into account the interaction with the external electric field (5.61). To do this, we solve the equation for the density matrix in the first approximation (see (5.50))

$$\frac{\partial\hat{\rho}^{(1)}_{\text{int}}(t)}{\partial t} = -\frac{i}{\hbar}\left[\hat{V}_{\text{int}}(t)\hat{\rho}^0_{\text{int}}\right],$$

$$\hat{\rho}_{\text{int}}^{(1)}(t) = -\frac{i}{\hbar}\int_{-\infty}^{t}\left[\hat{V}_{\text{int}}(t')\hat{\rho}_{\text{int}}^{0}\right]dt'.$$

To obtain a solution, we need to translate the perturbation operator describing the interaction of the current density with the field into an interaction representation:

$$\begin{aligned}
&\hat{V}_{\text{int}}(t) = e^{\frac{i}{\hbar}\hat{H}^0 t}\hat{V}(t)e^{-\frac{i}{\hbar}\hat{H}^0 t} = \hat{j}_{x\text{int}}\frac{E_0}{\omega}\sin\omega t\\
&j_{\text{int_}nm} = \langle n|e^{\frac{i}{\hbar}\hat{H}^0 t}\hat{j}_{x\text{int}}e^{-\frac{i}{\hbar}\hat{H}^0 t}|m\rangle =\\
&= \sum_{k,k'}\langle n|e^{\frac{i}{\hbar}\hat{H}^0 t}|k\rangle\langle k|\hat{j}_{x\text{int}}|k'\rangle\langle k'|e^{-\frac{i}{\hbar}\hat{H}^0 t}|m\rangle =\\
&= \sum_{k,k'}e^{\frac{i}{\hbar}E_k t}\langle n||k\rangle\langle k|\hat{j}_{x\text{int}}|k'\rangle e^{-\frac{i}{\hbar}E_{k'}t}\langle k'||m\rangle =\\
&= j_{nm}e^{\frac{i}{\hbar}E_n t}e^{-\frac{i}{\hbar}E_m t} = j_{nm}e^{i\omega_{nm}t},\\
&\langle n||k\rangle = \delta_{nk},\ \langle k'||m\rangle = \delta_{k'm},\\
&\omega_{nm} = \frac{i}{\hbar}(E_n - E_m).
\end{aligned} \tag{5.64}$$

$$\begin{aligned}
&\rho_{\text{int_}nm}^{(1)}(t) = -\frac{i}{\hbar}\int_{-\infty}^{t}\left[\hat{V}_{\text{int}}(t')\hat{\rho}_{\text{int}}^{0}\right]_{nm}dt',\\
&\left[\hat{V}_{\text{int}}(t')\hat{\rho}_{\text{int}}^{0}\right]_{nm} = \langle n|\left[\hat{V}_{\text{int}}(t')\hat{\rho}_{\text{int}}^{0}\right]|m\rangle\\
&= \langle n|\hat{V}_{\text{int}}(t')\hat{\rho}_{\text{int}}^{0}|m\rangle - \langle n|\hat{\rho}_{\text{int}}^{0}\hat{V}_{\text{int}}(t')|m\rangle =\\
&= \sum_{k}\langle n|\hat{V}_{\text{int}}(t')|k\rangle\langle k|\hat{\rho}_{\text{int}}^{0}|m\rangle\\
&- \sum_{k}\langle n|\hat{\rho}_{\text{int}}^{0}|k\rangle\langle k|\hat{V}_{\text{int}}(t')|m\rangle =\\
&= \langle n|\hat{V}_{\text{int}}(t')|m\rangle\rho_{mm}^{0} - \rho_{nn}^{0}\langle n|\hat{V}_{\text{int}}(t')|m\rangle =\\
&= \langle n|\hat{V}_{\text{int}}(t')|m\rangle(\rho_{mm}^{0} - \rho_{nn}^{0}) = (\rho_{mm}^{0} - \rho_{nn}^{0})j_{nm}e^{i\omega_{nm}t'}\frac{E_0}{\omega}\sin\omega t',\\
&\rho_{\text{int_}nm}^{(1)}(t) = -\frac{i}{\hbar}(\rho_{mm}^{0} - \rho_{nn}^{0})j_{nm}\frac{E_0}{\omega}\int_{-\infty}^{t}e^{i\omega_{nm}t'}\sin\omega t'dt' =\\
&= -\frac{i}{\hbar}\left(e^{\frac{F-E_m}{T}} - e^{\frac{F-E_n}{T}}\right)j_{nm}\frac{E_0}{\omega}\int_{-\infty}^{t}e^{i\omega_{nm}t'}\sin\omega t'dt'.
\end{aligned} \tag{5.65}$$

Then, the average value of the current density, in the first approximation, will have the form:

$$\begin{aligned}\tilde{j}_x(t) &= Sp\big(\hat{j}_x(\hat{\rho}^0 + \hat{\rho}^{(1)}(t)\big)\\ &= Sp\Big(\hat{j}_{x\_\mathrm{int}}\hat{\rho}^{(1)}_{\mathrm{int}}(t)\Big) = \sum_{m,n} j_{\mathrm{int}\_mn}(t)\rho^{(1)}_{\mathrm{int}\_nm}(t) =\\ &= \sum_{m,n} j_{mn}e^{i\omega_{mn}t}\left(-\frac{i}{\hbar}\right)(\rho^0_{mm} - \rho^0_{nn})j_{nm}\frac{E_0}{\omega}\int\limits_{-\infty}^{t} e^{i\omega_{nm}t'}\sin\omega t' dt' =\\ &= \left(-\frac{i}{\hbar}\right)\frac{E_0}{\omega}\sum_{m,n}|j_{mn}|^2(\rho^0_{mm} - \rho^0_{nn})e^{i\omega_{mn}t}\int\limits_{-\infty}^{t} e^{i\omega_{nm}t'}\sin\omega t' dt'.\end{aligned} \tag{5.66}$$

Let us consider the integral in detail with respect to the fact, that there are a lot of transitions between states in the conductive sample, that are in resonance with the external field, so $\omega_{nm} \approx \omega$, then

$$\begin{aligned}&e^{i\omega_{mn}t}\int\limits_{-\infty}^{t} e^{i\omega_{nm}t'}\sin\omega t' dt'\\ &= \frac{1}{2i}\left(e^{i\omega_{mn}t}\int\limits_{-\infty}^{t} e^{i\omega_{nm}t'}e^{i\omega t'}dt' - e^{i\omega_{mn}t}\int\limits_{-\infty}^{t} e^{i\omega_{nm}t'}e^{-i\omega t'}dt'\right) =\\ &= \frac{1}{2i}\left(e^{i\omega t}\underline{e^{-i\omega t}e^{i\omega_{mn}t}}\int\limits_{-\infty}^{t} e^{i(\omega+\omega_{nm})t'}dt' - \underline{\underline{e^{i\omega_{mn}t}e^{i\omega t}}}e^{-i\omega t}\int\limits_{-\infty}^{t} e^{i(\omega_{nm}-\omega)t'}dt'\right) =\\ &= \frac{1}{2i}\left(e^{i\omega t}\int\limits_{-\infty}^{t} e^{i(\omega+\omega_{nm})(t'-t)}dt' - e^{-i\omega t}\int\limits_{-\infty}^{t} e^{i(\omega_{nm}-\omega)(t'-t)}dt'\right) = \big(t' - t = \tau\big)\\ &= \frac{1}{2i}\left(e^{i\omega t}\int\limits_{-\infty}^{0} e^{i(\omega+\omega_{nm})\tau}d\tau - e^{-i\omega t}\int\limits_{-\infty}^{0} e^{i(\omega_{nm}-\omega)\tau}d\tau\right) =\\ &= \frac{1}{4i}\left(e^{i\omega t}\mathrm{Re}\int\limits_{-\infty}^{\infty} e^{i(\omega+\omega_{nm})\tau}d\tau - e^{-i\omega t}\mathrm{Re}\int\limits_{-\infty}^{\infty} e^{i(\omega_{nm}-\omega)\tau}d\tau\right) =\\ &= \frac{1}{2i}\big(e^{i\omega t}\pi\delta(\omega+\omega_{nm}) - e^{-i\omega t}\pi\delta(\omega-\omega_{nm})\big).\end{aligned} \tag{5.67}$$

Then, we have for the average value of the current density in the sample

$$\tilde{j}_x(t) = \left(-\frac{\pi}{2\hbar}\right)\frac{E_0}{\omega}\sum_{m,n}|j_{mn}|^2(\rho^0_{mm} - \rho^0_{nn})\big(e^{i\omega t}\delta(\omega+\omega_{nm}) - e^{-i\omega t}\delta(\omega-\omega_{nm})\big) \equiv$$

$$\equiv \left(-\frac{\pi}{2\hbar}\right)\frac{E_0}{\omega}\big(\sum_{m,n} |j_{mn}|^2(\rho^0_{mm}-\rho^0_{nn})e^{i\omega t}\delta(\omega+\omega_{nm})$$

$$-\sum_{m,n} |j_{mn}|^2(\rho^0_{nn}-\rho^0_{mm})\big(e^{-i\omega t}\delta(\omega-\omega_{mn})\big)\big) \equiv$$

$$\equiv \left(-\frac{\pi}{2\hbar}\right)\frac{E_0}{\omega}\sum_{m,n} |j_{mn}|^2(\rho^0_{mm}-\rho^0_{nn})\delta(\omega+\omega_{nm})$$

$$+e^{-i\omega t}\sum_{m,n} |j_{mn}|^2(\rho^0_{mm}-\rho^0_{nn})(\delta(\omega+\omega_{nm}))) =$$

$$= \left(-\frac{\pi}{\hbar\omega}\right)\frac{E_0}{2}\left(e^{i\omega t}+e^{-i\omega t}\right)\sum_{m,n} |j_{mn}|^2(\rho^0_{mm}-\rho^0_{nn})(\delta(\omega+\omega_{nm})) =$$

$$= \left(-\frac{\pi}{\hbar\omega}\right)E_0\cos(\omega t)\sum_{m,n} |j_{mn}|^2(\rho^0_{mm}-\rho^0_{nn})(\delta(\omega+\omega_{nm})). \tag{5.68}$$

And for a conductivity

$$\begin{aligned}
\mathrm{Re}\sigma &= \left(\frac{-\pi}{\hbar\omega}\right)\sum_{m,n} |j_{mn}|^2(\rho^0_{mm}-\rho^0_{nn})\delta(\omega+\omega_{nm}) \\
&= \left(-\frac{\pi}{\hbar\omega}\right)\sum_{m,n} |j_{mn}|^2\rho^0_{nn}\left(e^{\frac{E_n-E_m}{T}}-1\right)\delta(\omega+\omega_{nm}) = \\
&= \left(-\frac{\pi}{\hbar\omega}\right)\left(e^{\frac{-\hbar\omega}{T}}-1\right)\sum_{m,n} |j_{mn}|^2\rho^0_{nn}\delta(\omega+\omega_{nm}) \\
&= \left(\frac{\pi}{\hbar\omega}\right)\left(1-e^{\frac{-\hbar\omega}{T}}\right)\sum_{m,n} |j_{mn}|^2\rho^0_{nn}\delta(\omega+\omega_{nm}).
\end{aligned} \tag{5.69}$$

The resulting relationship is the Kubo formula for conductivity:

$$\begin{aligned}
\mathrm{Re}\sigma &= \left(\frac{\pi}{\hbar\omega}\right)\left(1-e^{\frac{-\hbar\omega}{T}}\right)\sum_{m,n} |j_{mn}|^2\rho^0_{nn}\delta(\omega+\omega_{nm}), \\
\rho^0_{nn} &= \frac{e^{-\frac{E_n}{T}}}{\sum_k e^{-\frac{E_k}{T}}} = \frac{e^{-\frac{E_n}{T}}}{z} = e^{\frac{F-E_n}{T}}, \\
F &= -T\ln z.
\end{aligned} \tag{5.70}$$

It is clear that the main reason for the occurrence of current in a conducting system is the finite probability of interlevel transitions in a quantum system, caused by the interaction of the dipole moments of transitions with an external field; it is this interaction that causes quantum transitions in the system, which leads to the appearance of a “response” of the system to the influence of an external field. The internal characteristic of a system that demonstrates the system's ability to respond to external influences, its susceptibility, in this case is conductivity. For the case of high temperatures ($\frac{\hbar\omega}{T} << 1$) the formula (5.70) is transformed to the form:

$$\begin{aligned} \mathrm{Re}\sigma &= \left(\frac{\pi}{\hbar\omega}\right)\left(\frac{\hbar\omega}{T}\right)\sum_{m,n}|j_{mn}|^2\rho^0_{nn}\delta(\omega+\omega_{nm}) \\ &= \frac{\pi}{T}\sum_{m,n}|j_{mn}|^2\rho^0_{nn}\delta(\omega+\omega_{nm}). \end{aligned} \tag{5.71}$$

5.4 Quantum Fluctuations of Physical Quantities

We have shown above that the ability of a conducting system to respond to an external field is determined by the presence of an internal characteristic in the system—a physical quantity (in our case, it was the current density) that determines the interaction of the system with this field (with the vector potential, Eq. (5.61)). This interaction excites interlevel quantum transitions in the system, which lead to the appearance of an uncompensated average value of the current density that responds to the external field, whereas the average value of this quantity in the absence of an external field is zero. Interlevel transitions excited by an external field are the driving mechanism for inducing a response in the system that is proportional to the field value with a proportionality coefficient that is the conductivity. In the absence of an external field, the system is "noisy" at a finite temperature; it has a finite value of current density fluctuations, while the average value of the current density itself in the absence of a field is zero. If we take into account that thermal fluctuations (noise) of this physical quantity are caused by the same interlevel transitions as in the case of a field, with the same probability of these transitions at the same values of the matrix elements of this quantity, then it is not surprising that the spectral density of noise can be related to the conductivity of the system under the action of an external harmonic field. This is the position we are going to prove in this section.

5.4.1 Autocorrelation Function of a Physical Quantity in a Quantum System

Consider a quantum system in equilibrium with a thermalized environment in the absence of external field effects. The physical quantity (e.g., current density, dipole moment, or magnetic moment of an atom) that characterizes this system and is described by the operator has a mean value equal to zero in the absence of an external field. In the Heisenberg representation, the operator of this physical quantity will contain an explicit time dependence: $\hat{\alpha}(t) = e^{\frac{i}{\hbar}\hat{H}^0 t}\hat{\alpha}e^{-\frac{i}{\hbar}\hat{H}^0 t}$, where $\hat{H}^0$ is Hamiltonian describing a quantum system in the absence of external fields. Then for the specified physical quantity, we can introduce an autocorrelation function, by analogy with classical systems:

$$K(\tau) = \frac{1}{2}\left(\overline{\hat{\alpha}(t)\hat{\alpha}(t+\tau) + \hat{\alpha}(t+\tau)\hat{\alpha}(t)}\right) = \\ = Sp\left(\hat{\rho}^0 \frac{1}{2}(\hat{\alpha}(t)\hat{\alpha}(t+\tau) + \hat{\alpha}(t+\tau)\hat{\alpha}(t))\right). \tag{5.72}$$

Here the overline means averaging over the equilibrium ensemble. For the operator $\hat{\alpha}(t)$ we define the Fourier transformation:

$$\hat{\alpha}(t) = \frac{1}{\sqrt{2\pi}} \int_{-\infty}^{\infty} \hat{\alpha}(\omega) e^{-i\omega t} d\omega, \\ \hat{\alpha}(\omega) = \frac{1}{\sqrt{2\pi}} \int_{-\infty}^{\infty} \hat{\alpha}(t) e^{i\omega t} dt. \tag{5.73}$$

Taking these transformations into account, expression (5.72) can be rewritten as

$$K(\tau) = Sp\left(\hat{\rho}^0 \frac{1}{2}(\hat{\alpha}(t)\hat{\alpha}(t+\tau) + \hat{\alpha}(t+\tau)\hat{\alpha}(t))\right) = \\ = Sp\left\{\hat{\rho}^0 \frac{1}{2}\left(\frac{1}{\sqrt{2\pi}} \int_{-\infty}^{\infty} \hat{\alpha}(\omega) e^{-i\omega t} d\omega \frac{1}{\sqrt{2\pi}} \int_{-\infty}^{\infty} \hat{\alpha}(\omega') e^{-i\omega'(t+\tau)} d\omega' + \right.\right. \\ \left.\left. + \frac{1}{\sqrt{2\pi}} \int_{-\infty}^{\infty} \hat{\alpha}(\omega') e^{-i\omega'(t+\tau)} d\omega' \frac{1}{\sqrt{2\pi}} \int_{-\infty}^{\infty} \hat{\alpha}(\omega) e^{-i\omega t} d\omega\right)\right\} = \\ = \frac{1}{2}\frac{1}{2\pi} Sp\left\{\hat{\rho}^0\left(\left(\int_{-\infty}^{\infty} d\omega \int_{-\infty}^{\infty} \hat{\alpha}(\omega)\hat{\alpha}(\omega') e^{-i(\omega+\omega')t} e^{-i\omega'\tau} d\omega' + \right.\right.\right. \\ \left.\left.\left. + \int_{-\infty}^{\infty} d\omega' \int_{-\infty}^{\infty} \hat{\alpha}(\omega')\hat{\alpha}(\omega) e^{-i\omega'\tau} e^{-i(\omega+\omega')t} d\omega\right)\right)\right\} = \\ = \frac{1}{2\pi}\frac{1}{2} \int_{-\infty}^{\infty} d\omega \int_{-\infty}^{\infty} d\omega' \left\{Sp\left(\hat{\rho}^0\left(\hat{\alpha}(\omega)\hat{\alpha}(\omega') + \hat{\alpha}(\omega')\hat{\alpha}(\omega)\right)\right) e^{-i(\omega+\omega')t} e^{-i\omega'\tau}\right\}. \tag{5.74}$$

Let us calculate in detail the trace of the product of operators:

$$Sp\big(\hat{\rho}^0\big(\hat{\alpha}(\omega)\hat{\alpha}(\omega') + \hat{\alpha}(\omega')\hat{\alpha}(\omega)\big)\big) =$$
$$= \sum_{n,k} \rho^0_{nk} \sum_m \big(\alpha_{km}(\omega)\alpha_{mn}(\omega') + \alpha_{km}(\omega')\alpha_{mn}(\omega)\big).$$
$$\rho^0_{nk} = e^{\frac{F-E_k}{T}}\delta_{nk}, \Rightarrow$$
$$\sum_{n,k} \rho^0_{nk} \sum_m \big(\alpha_{km}(\omega)\alpha_{mn}(\omega') + \alpha_{km}(\omega')\alpha_{mn}(\omega)\big) =$$
$$= \sum_n \rho^0_{nn} \sum_m \big(\alpha_{nm}(\omega)\alpha_{mn}(\omega') + \alpha_{km}(\omega')\alpha_{mn}(\omega)\big). \tag{5.75}$$

Here the set of states $\{|n\rangle\}$ that are eigenstates of the Hamiltonian describing a quantum system in the absence of external fields $\hat{H}^0|n\rangle = E_n|n\rangle$. They have the properties of orthogonality $\langle n|\, m\rangle = \delta_{nm}$ and completeness $\sum_n |n\rangle\langle n| = \hat{1}$. For the matrix element of the Fourier transform of the operator $\hat{\alpha}(\omega)$ we have

$$\alpha_{nm}(\omega) = \frac{1}{\sqrt{2\pi}} \int_{-\infty}^{\infty} \alpha_{nm}(t)e^{i\omega t}dt = \frac{1}{\sqrt{2\pi}} \int_{-\infty}^{\infty} \alpha_{nm}e^{i\omega_{nm}t}e^{i\omega t}dt =$$
$$= \frac{\alpha_{nm}}{\sqrt{2\pi}} \int_{-\infty}^{\infty} e^{i(\omega+\omega_{nm})t}dt = \frac{\alpha_{nm}}{\sqrt{2\pi}} 2\pi\,\delta(\omega + \omega_{nm}), \Rightarrow$$
$$\alpha_{nm}(\omega) = \alpha_{nm}\sqrt{2\pi}\,\delta(\omega + \omega_{nm}). \tag{5.76}$$

Hence,

$$\sum_n \rho^0_{nn} \sum_m \big(\alpha_{nm}(\omega)\alpha_{mn}(\omega') + \alpha_{nm}(\omega')\alpha_{mn}(\omega)\big) =$$
$$= 2\pi \sum_n \rho^0_{nn} \sum_m \big(\alpha_{nm}\delta(\omega + \omega_{nm})\alpha_{mn}\delta\big(\omega' + \omega_{mn}\big) + \alpha_{nm}\delta\big(\omega' + \omega_{nm}\big)\alpha_{mn}\delta(\omega + \omega_{mn})\big) =$$
$$= 2\pi \sum_{n,m} \rho^0_{nn}|\alpha_{nm}|^2 \big(\delta(\omega + \omega_{nm})\delta\big(\omega' + \omega_{mn}\big) + \delta\big(\omega' + \omega_{nm}\big)\delta(\omega + \omega_{mn})\big) =$$
$$= 2\pi \sum_{n,m} \rho^0_{nn}|\alpha_{nm}|^2 \big(\delta(\omega + \omega_{nm})\delta\big(\omega' + \omega\big) + \delta\big(\omega' + \omega\big)\delta(\omega + \omega_{mn})\big) =$$
$$= 2\pi\,\delta\big(\omega' + \omega\big) \sum_{n,m} \rho^0_{nn}|\alpha_{nm}|^2 (\delta(\omega + \omega_{nm}) + \delta(\omega + \omega_{mn})) =$$
$$= 2\pi\,\delta\big(\omega' + \omega\big) \sum_{n,m} \rho^0_{nn}|\alpha_{nm}|^2 \delta(\omega + \omega_{nm})$$
$$+ 2\pi\,\delta\big(\omega' + \omega\big) \sum_{n\rightleftarrows m} \rho^0_{mm}|\alpha_{nm}|^2 \delta(\omega + \omega_{nm}) =$$
$$= 2\pi\,\delta\big(\omega' + \omega\big) \sum_{n,m} \big(\rho^0_{nn} + \rho^0_{mm}\big)|\alpha_{nm}|^2 \delta(\omega + \omega_{nm}). \tag{5.77}$$

Substituting the obtained expression (5.77) into the integral (5.74), we have the following expression for the autocorrelation function

$$K(\tau)=\frac{1}{2\pi}\frac{1}{2}\int\limits_{-\infty}^{\infty}d\omega\int\limits_{-\infty}^{\infty}Sp\big(\hat{\rho}^0\big(\hat{\alpha}(\omega)\hat{\alpha}(\omega')+\hat{\alpha}(\omega')\hat{\alpha}(\omega)\big)\big)e^{-i(\omega+\omega')t}e^{-i\omega'\tau}d\omega'=$$

$$=\frac{1}{2\pi}\frac{1}{2}\int\limits_{-\infty}^{\infty}d\omega\int\limits_{-\infty}^{\infty}2\pi\delta\big(\omega'+\omega\big)\sum_{n,m}\big(\rho_{nn}^0+\rho_{mm}^0\big)|\alpha_{nm}|^2\delta(\omega+\omega_{nm})e^{-i(\omega+\omega')t}e^{-i\omega'\tau}d\omega'=$$

$$=\frac{1}{2}\int\limits_{-\infty}^{\infty}d\omega\sum_{n,m}\big(\rho_{nn}^0+\rho_{mm}^0\big)|\alpha_{nm}|^2\delta(\omega+\omega_{nm})\int\limits_{-\infty}^{\infty}\delta(\omega'+\omega)e^{-i(\omega+\omega')t}e^{-i\omega'\tau}d\omega'=$$

$$=\frac{1}{2}\int\limits_{-\infty}^{\infty}d\omega e^{i\omega\tau}\sum_{n,m}\big(\rho_{nn}^0+\rho_{mm}^0\big)|\alpha_{nm}|^2\delta(\omega+\omega_{nm}). \tag{5.78}$$

If we set the time interval $\tau\rightarrow 0$ in the autocorrelation function in expressions (5.72), (5.74) and (5.78) to zero, we obtain an expression for the mean value of the square of the physical quantity α:

$$K(\tau)|_{\tau\rightarrow 0}=\overline{\alpha^2}=D=\frac{1}{2}\int\limits_{-\infty}^{\infty}d\omega\left\{\sum_{n,m}\big(\rho_{nn}^0+\rho_{mm}^0\big)|\alpha_{nm}|^2\delta(\omega+\omega_{nm})\right\}. \tag{5.79}$$

Taking into account that in the absence of external fields in a system in thermodynamic equilibrium, the average value of the physical quantity $\overline{\alpha}=Sp(\hat{\rho}^0\hat{\alpha})=0$ is equal to zero, then the dispersion of this quantity will be $D=\overline{\alpha^2}-\overline{\alpha}^2\equiv\overline{\alpha^2}$, then spectral density of the dispersion will be

$$D_\omega=\overline{\alpha_\omega^2}=\frac{1}{2}\sum_{n,m}\big(\rho_{nn}^0+\rho_{mm}^0\big)|\alpha_{nm}|^2\delta(\omega+\omega_{nm})=$$

$$=\frac{1}{2}\sum_{n,m}\left(e^{\frac{F-E_n}{T}}+e^{\frac{F-E_m}{T}}\right)|\alpha_{nm}|^2\delta(\omega+\omega_{nm})=$$

$$=\frac{1}{2}\sum_{n,m}e^{\frac{F-E_n}{T}}\left(1+e^{\frac{E_n-E_m}{T}}\right)|\alpha_{nm}|^2\delta(\omega+\omega_{nm})=$$

$$=\frac{1}{2}\left(1+e^{-\frac{\hbar\omega}{T}}\right)\sum_{n,m}e^{\frac{F-E_n}{T}}|\alpha_{nm}|^2\delta(\omega+\omega_{nm}). \tag{5.80}$$

If we consider the current density in a conducting sample as a physical quantity, then the spectral density of current noise is written as

$$\overline{j_\omega^2}=\frac{1}{2}\left(1+e^{-\frac{\hbar\omega}{T}}\right)\sum_{n,m}e^{\frac{F-E_n}{T}}|j_{nm}|^2\delta(\omega+\omega_{nm}). \tag{5.81}$$

Comparing which with expression (5.69) for conductivity, we notice

$$
\begin{aligned}
\overline{j_{\omega}^{2}} &= \frac{1}{2}\frac{\left(1+e^{-\frac{\hbar\omega}{T}}\right)}{\left(1-e^{\frac{-\hbar\omega}{T}}\right)}\left(\frac{\hbar\omega}{\pi}\right)\mathrm{Re}\sigma = \\
&= \left(\frac{\hbar\omega}{2\pi}\right)\mathrm{Re}\sigma\, cth\frac{\hbar\omega}{2T} \\
&= \left(\frac{\hbar\omega}{\pi}\right)\mathrm{Re}\sigma\left(\frac{1}{2}+\frac{1}{e^{\frac{\hbar\omega}{T}}-1}\right).
\end{aligned}
\tag{5.82}
$$

Since conductivity σ is a quantity reciprocal to impedance Z (total resistance), a complex quantity in which the real part R is the active resistance, and the imaginary part is the difference between the capacitive and inductive resistance, which we denote by X, then

$$
\mathrm{Re}\sigma = \mathrm{Re}\left(\frac{1}{R+iX}\right) = \frac{R}{|Z|^{2}}. \tag{5.83}
$$

Substituting this expression (5.83) into Eq. (5.82) and taking into account Ohm's law that the product of the current density and the absolute value of the impedance gives the potential, we have

$$
\overline{j_{\omega}^{2}}|Z|^{2} = \overline{\varphi_{\omega}^{2}} = \left(\frac{\hbar\omega}{2\pi}\right)\mathrm{R}cth\frac{\hbar\omega}{2T}. \tag{5.84}
$$

In case of high temperatures $\frac{\hbar\omega}{T} << 1$ expression (5.84) becomes

$$
\overline{\varphi_{\omega}^{2}} = \left(\frac{\hbar\omega}{2\pi}\right)\mathrm{R}\left(\frac{2T}{\hbar\omega}\right) = \frac{TR}{\pi} \tag{5.85}
$$

which is the so-called Nyquist formula for the spectral density of potential noise. At the given high temperatures, this is "white" noise, when the spectral density of the noise is the same for all frequencies and is proportional to the temperature. The classical limit corresponds to the Nyquist formula for thermal noise of the EMF on an active resistance in an open circuit (in the absence of currents). (Johnson–Nyquist noise.)

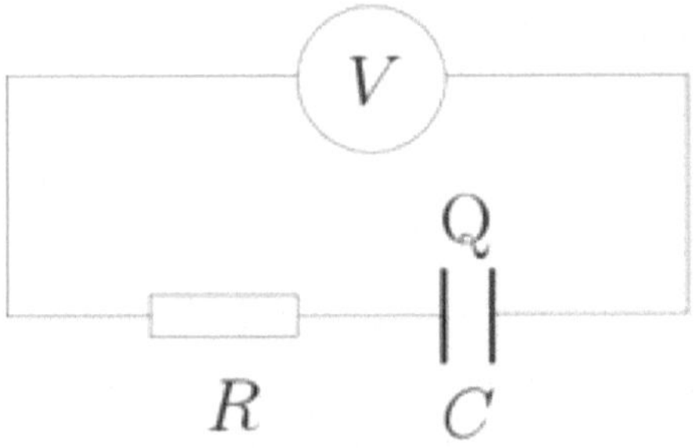

5.4.2 Fluctuation–Dissipation Theorem

Let us consider a certain quantum system characterized by a set of eigenstates of the energy operator $\hat{H}^0$ that correspond to the eigenvalues of the energy E_n and satisfy the conditions of orthonormality and completeness:

$$\hat{H}^0|n\rangle = E_n|n\rangle,$$
$$\langle m|n\rangle = \delta_{mn},$$
$$\sum_n |n\rangle\langle n| = \hat{1}. \tag{5.86}$$

The system is in equilibrium with the environment at a given temperature T. The system is affected by an external field, the force characteristic of which (generalized intensity) has the form:

$$\mathbf{C}(t) = \frac{1}{2}\left(\mathbf{C_0}(q)e^{-i\omega t} + \mathbf{C_0^*}(q)e^{i\omega t}\right). \tag{5.87}$$

This field causes a response from the system in the form of the appearance of the average value of a physical quantity $\boldsymbol{\alpha}$, which "reacts" to the appearance of the field, interacting with it:

$$\tilde{\alpha}(t) = \frac{1}{2}\left(\eta\mathbf{C_0}(q)e^{-i\omega t} + \eta * \mathbf{C_0^*}(q)e^{i\omega t}\right). \tag{5.88}$$

Here, the parameter η (admittance) is phenomenologically introduced, which is the generalized susceptibility of the system to the influence of an external field. The operator of this physical quantity in the Schrödinger representation is denoted by $\hat{\alpha}$, then the operator describing the interaction of the quantum system with the field will have the form:

$$\hat{V}(t) = -\hat{\alpha} \cdot \mathbf{C}(t). \tag{5.89}$$

We will consider the field amplitude to be sufficiently small (weak fields), so much so that the contribution from the indicated interaction will be much less than the characteristic distances between energy levels, that is, we will consider the interaction operator to be a perturbation. The interaction of a quantum system with a field causes quantum transitions, the probability of which per unit time (in the first approximation for the perturbation) has the form:

$$\begin{aligned} \hat{V}(t) &= -\hat{\alpha}C(t) \\ w_{nm} &= \frac{2\pi}{\hbar}|V_{nm}|^2[\delta(E_n - E_m + \hbar\omega) + \delta(E_m - E_n + \hbar\omega)] = \\ &= \frac{2\pi}{\hbar}|\alpha_{nm}|^2\frac{1}{2}|C_0(q)|^2[\delta(\hbar\omega_{nm} + \hbar\omega) + \delta(\hbar\omega_{mn} + \hbar\omega)] = \\ &= \frac{\pi}{\hbar^2}|\alpha_{nm}|^2|C_0(q)|^2[\delta(\omega_{nm} + \omega) + \delta(\omega_{mn} + \omega)]. \end{aligned} \tag{5.90}$$

Then the energy Q absorbed by the system per unit of time from the source, spent on the excitation of interlevel transitions, will have the form:

$$\begin{aligned} \tilde{Q} &= \sum_n \rho_{nn}^0 \sum_m w_{nm}\hbar\omega_{mn} = \\ &= \frac{\pi}{\hbar^2}|C_0(q)|^2 \sum_n \rho_{nn}^0 \sum_m |\alpha_{nm}|^2\hbar\omega_{mn} \\ &[\delta(\omega_{nm} + \omega) + \delta(\omega_{mn} + \omega)]. \end{aligned} \tag{5.91}$$

Taking into account the property of the delta function $\delta(\omega_{nm} + \omega) \neq 0$, if $-\omega_{nm} = \omega_{mn} = \omega$, then

$$\begin{aligned} &\frac{\pi}{\hbar^2}|C_0(q)|^2 \sum_n \rho_{nn}^0 \sum_m |\alpha_{nm}|^2\hbar\omega_{mn}[\delta(\omega_{nm} + \omega) + \delta(\omega_{mn} + \omega)] = \\ &= \frac{\pi}{\hbar^2}|C_0(q)|^2\hbar\omega \sum_n \rho_{nn}^0 \sum_m |\alpha_{nm}|^2[\delta(\omega_{nm} + \omega) - \delta(\omega_{mn} + \omega)]. \end{aligned} \tag{5.92}$$

Let us split expression (5.92) into two separate terms, and in the second of them we will replace the equivalent indices $n \leftrightarrow m$, then

$$\begin{aligned} &\frac{\pi}{\hbar^2}|C_0(q)|^2\hbar\omega \sum_n \rho_{nn}^0 \sum_m |\alpha_{nm}|^2[\delta(\omega_{nm} + \omega) - \delta(\omega_{mn} + \omega)] = \\ &= \frac{\pi}{\hbar}|C_0(q)|^2\omega\left[\sum_n \rho_{nn}^0 \sum_m |\alpha_{nm}|^2\delta(\omega_{nm} + \omega) - \sum_m \rho_{mm}^0 \sum_n |\alpha_{mn}|^2\delta(\omega_{nm} + \omega)\right] = \\ &= \frac{\pi}{\hbar}|C_0(q)|^2\omega \sum_{n,m}[\rho_{nn}^0 - \rho_{mm}^0]|\alpha_{nm}|^2\delta(\omega_{nm} + \omega). \end{aligned} \tag{5.93}$$

Taking into account expression (5.70) for the matrix element of the density matrix describing the equilibrium system, we finally have

$$\begin{aligned}\tilde{Q} &= \frac{\pi}{\hbar}|C_0(q)|^2\omega\sum_{n,m}\left[\rho^0_{nn}-\rho^0_{mm}\right]|\alpha_{nm}|^2\delta(\omega_{nm}+\omega) = \\ &= \frac{\pi}{\hbar}|C_0(q)|^2\omega\sum_{n,m}\left[e^{\frac{F-E_n}{T}}-e^{\frac{F-E_m}{T}}\right]|\alpha_{nm}|^2\delta(\omega_{nm}+\omega) = \\ &= \frac{\pi}{\hbar}|C_0(q)|^2\omega\sum_{n,m}e^{\frac{F-E_n}{T}}\left[1-e^{\frac{E_n-E_m}{T}}\right]|\alpha_{nm}|^2\delta(\omega_{nm}+\omega) = \\ &= \frac{\pi}{\hbar}|C_0(q)|^2\omega\left[1-e^{\frac{-\hbar\omega}{T}}\right]\sum_{n,m}e^{\frac{F-E_n}{T}}|\alpha_{nm}|^2\delta(\omega_{nm}+\omega). \end{aligned} \tag{5.94}$$

On the other hand, phenomenologically, the energy consumed by the system per unit of time goes to inducing the system's response in the form (5.88). Then the energy consumed by the system from the source per unit of time (equal to the power of the source)

$$\begin{aligned}\tilde{Q} &= \overline{-\dot{\mathbf{C}}(t)\cdot\tilde{\alpha}(t)} = \\ &= \overline{-\frac{1}{2}\left(-i\omega\mathbf{C}_0(q)e^{-i\omega t}+i\omega\mathbf{C}^*_0(q)e^{i\omega t}\right)\cdot\frac{1}{2}\left(\eta\mathbf{C}_0(q)e^{-i\omega t}+\eta*\mathbf{C}^*_0(q)e^{i\omega t}\right)} = \\ &= \frac{i\omega}{4}|C_0(q)|^2(\eta*-\eta) = \frac{i\omega}{4}|C_0(q)|^2(-2i)\mathrm{Im}\eta = \frac{\omega}{2}|C_0(q)|^2\mathrm{Im}\eta. \end{aligned} \tag{5.95}$$

Now, comparing expressions (5.95) and (5.94) for the consumed power, we can write for the imaginary part of the generalized susceptibility

$$\begin{aligned}&\frac{\omega}{2}|C_0(q)|^2\mathrm{Im}\eta = \frac{\pi}{\hbar}|C_0(q)|^2 \\ &\omega\left[1-e^{\frac{-\hbar\omega}{T}}\right]\sum_{n,m}e^{\frac{F-E_n}{T}}|\alpha_{nm}|^2\delta(\omega_{nm}+\omega), \\ &\mathrm{Im}\eta = \frac{2\pi}{\hbar}\left[1-e^{\frac{-\hbar\omega}{T}}\right]\sum_{n,m}e^{\frac{F-E_n}{T}}|\alpha_{nm}|^2\delta(\omega_{nm}+\omega). \end{aligned} \tag{5.96}$$

Comparing this expression (5.96) now with expression (5.80) for the spectral density of dispersion (square of fluctuation), we can come to the conclusion:

$$\begin{aligned}&\overline{\alpha^2_\omega} = \frac{\left(1+e^{-\frac{\hbar\omega}{T}}\right)}{2\left(1-e^{\frac{-\hbar\omega}{T}}\right)}\frac{\hbar}{2\pi}\mathrm{Im}\eta = \\ &= \frac{\hbar}{2\pi}\mathrm{Im}\eta\frac{1}{2}cth\frac{\hbar\omega}{2T} = \frac{\hbar}{2\pi}\mathrm{Im}\eta\left(\frac{1}{2}+\frac{1}{e^{\frac{\hbar\omega}{T}}-1}\right). \end{aligned} \tag{5.97}$$

The resulting relationship expresses the essence of the fluctuation–dissipation theorem [6], which is as follows.

The ability of a system to respond to an external field is determined by the presence in the system of an internal characteristic, a physical quantity that determines the interaction of the system with this field. The specified interaction excites interlevel quantum transitions in the system, which lead to the appearance of an uncompensated average value of this physical quantity, responding to an external field, whereas the average value of this quantity in the absence of an external field is zero. Interlevel transitions excited by an external field are the driving mechanism for inducing a response in the system proportional to the magnitude of the field with a proportionality coefficient representing the admittance. In the absence of an external field, a system at a finite temperature is "noisy"; it has a finite value of fluctuations of a physical quantity, the average value of which in the absence of a field is equal to zero. If we take into account that thermal fluctuations (noise) of this physical quantity are caused by the same interlevel transitions with the same probability of these transitions with the same values of matrix elements of this quantity, then it is not surprising that the spectral density of noise is associated with the admittance of the system under an external harmonic field. It is interesting to note the practical side of this theorem. The higher the value of the generalized susceptibility of the system to the external field, the more noise this system makes at a finite temperature. However, we are talking about the spectral density of noise, which is still proportional to the Planck distribution function, which is the number of photons in a given mode (with a given frequency) at a finite temperature. Where does the Planck distribution function for thermal photons come from when an electron system in thermal equilibrium is considered? In this case, the "electronic" system, which is in thermal equilibrium at a finite temperature, plays the role of the notorious abstract "black body," from the internal surfaces of which the number of emitted photons is equal to the number of simultaneously absorbed photons. The emission and absorption of photons by the electron system during single-photon transitions (in the linear approximation) determines those very transitions that lead to thermal noise of a physical quantity. In this case, the electron subsystem at a finite temperature is in equilibrium with the photon subsystem.

5.5 Wigner Representation

5.5.1 Unitary Transformation

The density matrix in coordinate representation has the form (in the one-dimensional case) (8)

$$\overline{\Psi(q,x)\Psi^*(q,x')} = \rho_{xx'} \equiv \rho(xx').$$

For the multidimensional case (22) (a multiparticle system located in real, three-dimensional space):

$$\overline{\Psi(q, x_{1\alpha}, ...x_{N\alpha})\Psi^*(q, x'_{1\alpha}...x'_{N\alpha})} = \rho(x_{1\alpha}...x_{N\alpha}, x'_{1\alpha}...x'_{N\alpha}),$$

We describe a quantum system (for example, an atomic gas, the atoms of which perform a thermal chaotic motion), while the "working" characteristics of the atoms remain quantum, such as atomic terms, between which quantum transitions can occur with a change in quantum numbers, spin states, etc. The movement of the atom itself is characterized by the de Broglie wavelength, which determines the length of quantum correlations in the movement of atoms. Let us designate

$$\begin{aligned} r &= x - x', \\ Q &= \frac{x + x'}{2}, \Rightarrow \\ x &= Q + \frac{r}{2}, x' = Q - \frac{r}{2}. \end{aligned} \tag{5.98}$$

Similarly, for a multiparticle (N-particle) system we have

$$\begin{aligned} r_{i\alpha} &= x_{i\alpha} - x'_{i\alpha}, \\ Q_{i\alpha} &= \frac{x_{i\alpha} + x'_{i\alpha}}{2}. \\ \alpha &\to (x, y, z). \end{aligned} \tag{5.99}$$

For the one-dimensional case, we introduce the following integral transformation of the density matrix:

$$\begin{aligned} f_W(PQ, t) &= \int_{-\infty}^{\infty} e^{-\frac{iPx}{\hbar}} \rho(xx't) e^{\frac{iPx'}{\hbar}} d(x - x') = \\ &= \int_{-\infty}^{\infty} e^{-\frac{iPr}{\hbar}} \rho\left(Q + \frac{r}{2}, Q - \frac{r}{2}\right) dr, \\ \rho(xx') &= \frac{1}{2\pi\hbar} \int_{-\infty}^{\infty} f_W(PQ) e^{\frac{iPr}{\hbar}} dP = \\ &= \frac{1}{2\pi\hbar} \int_{-\infty}^{\infty} e^{\frac{iPx}{\hbar}} f_W\left(P, \frac{x + x'}{2}\right) e^{-\frac{iPx'}{\hbar}} dP. \end{aligned} \tag{5.100}$$

Similarly, for a multiparticle (N-particle) system we have the transformation:

$$f_W(P_{1\alpha}...P_{N\alpha}Q_{1\alpha}...Q_{N\alpha}) = \int\limits_{-\infty}^{\infty} ... \int\limits_{-\infty}^{\infty} e^{-\frac{iP_{1\alpha}r_{1\alpha}}{\hbar}} ... e^{-\frac{iP_{N\alpha}r_{N\alpha}}{\hbar}} \rho(x_{1\alpha}...x_{N\alpha}x'_{1\alpha}...x'_{N\alpha}))dr_{1\alpha}...dr_{N\alpha},$$

$$\rho(x_{1\alpha}...x_{N\alpha}x'_{1\alpha}...x'_{N\alpha}) = \frac{1}{(2\pi\hbar)^{3N}} \int\limits_{-\infty}^{\infty} ... \int\limits_{-\infty}^{\infty} f_W(P_{1\alpha}...P_{N\alpha}, Q_{1\alpha}...Q_{N\alpha}) e^{\frac{iP_{1\alpha}r_{1\alpha}}{\hbar}} dP_{1\alpha}...dP_{N\alpha}. \tag{5.101}$$

Let us now consider some physical quantity A represented by a matrix in coordinate representation; for a matrix element in the one-dimensional case we have

$$A(xx') = \hat{A}(x)\delta(x - x'),$$

$$\hat{A}_W(P, Q) = \int\limits_{-\infty}^{\infty} e^{-\frac{iPx}{\hbar}} A(xx') e^{\frac{iPx'}{\hbar}} d(x - x')$$

$$= \int\limits_{-\infty}^{\infty} e^{-\frac{iPr}{\hbar}} \hat{A}\left(Q + \frac{r}{2}\right)\delta(r)dr = \hat{A}_W(Q). \tag{5.102}$$

Then the statistical average value of this physical quantity will have the form:

$$\overline{A} = Sp(\hat{\rho}\hat{A}) = \int\limits_{-\infty}^{\infty} dx \int\limits_{-\infty}^{\infty} dx' (\rho(xx')A(x'x)) =$$

$$= \int\limits_{-\infty}^{\infty} dx \int\limits_{-\infty}^{\infty} dx' \left(\left(\frac{1}{2\pi\hbar} \int\limits_{-\infty}^{\infty} f_W(PQ) e^{\frac{iPr}{\hbar}} dP\right) A(x'x)\right). \tag{5.103}$$

Taking into account (5.98), we have for the transformation to new variables, where the Jacobian of the transformations is defined as

$$dxdx' = |J|drdQ = \left|\frac{\partial(x, x')}{\partial(r, Q)}\right| drdQ =$$

$$= \begin{vmatrix} \frac{\partial x}{\partial r} & \frac{\partial x'}{\partial r} \\ \frac{\partial x}{\partial Q} & \frac{\partial x'}{\partial Q} \end{vmatrix} drdQ = \begin{vmatrix} \frac{1}{2} & -\frac{1}{2} \\ 1 & 1 \end{vmatrix} drdQ = drdQ. \tag{5.104}$$

We can replace the integrals over x and x' in (5.103) with integrals over the variables r and Q:

$$\tilde{A} = \int\limits_{-\infty}^{\infty} dx \int\limits_{-\infty}^{\infty} dx' (\rho(xx')A(x'x)) =$$

$$= \int_{-\infty}^{\infty} dx \int_{-\infty}^{\infty} dx' \left(\left(\frac{1}{2\pi\hbar} \int_{-\infty}^{\infty} f_W(PQ) e^{\frac{iPr}{\hbar}} dP \right) A(x'x) \right) =$$

$$= \int_{-\infty}^{\infty} dQ \int_{-\infty}^{\infty} dr \left(\left(\frac{1}{2\pi\hbar} \int_{-\infty}^{\infty} f_W(PQ) e^{\frac{iPr}{\hbar}} dP \right) A(x'x) \right) =$$

$$= \int_{-\infty}^{\infty} dQ \int_{-\infty}^{\infty} f_W(PQ) dP \frac{1}{2\pi\hbar} \int_{-\infty}^{\infty} dr \left(e^{\frac{iPr}{\hbar}} A(x'x) \right) =$$

$$= \frac{1}{2\pi\hbar} \int_{-\infty}^{\infty} \hat{A}_W(Q) dQ \int_{-\infty}^{\infty} f_W(PQ) dP =$$

$$= \frac{1}{2\pi\hbar} \int_{-\infty}^{\infty} dQ \int_{-\infty}^{\infty} dP \left(f_W(P, Q) \hat{A}_W(Q) \right). \tag{5.105}$$

The expression for the average statistical value of the quantity A of a quantum system formally coincides with the averaged quantity A using the distribution function of the classical ensemble defined in a two-dimensional phase space. In the case of a system of N particles, averaging similar to (5.105) will be reduced to calculating the corresponding integral over the volume of the 6N-dimensional phase space:

$$\bar{A} = Sp\hat{\rho}\hat{A} = \int_{-\infty}^{\infty} dx_{1\alpha} \ldots \int_{-\infty}^{\infty} dx_{N\alpha}$$

$$\int_{-\infty}^{\infty} dx'_{1\alpha} \ldots \int_{-\infty}^{\infty} dx'_{N\alpha} \left(\rho(x_{1\alpha} \ldots x_{N\alpha}, x'_{1\alpha} \ldots x'_{N\alpha}) A(x'_{1\alpha} \ldots x'_{N\alpha}, x_{1\alpha} \ldots x_{N\alpha},) \right),$$

$$\rho\left(x_{1\alpha} \ldots x_{N\alpha} x'_{1\alpha} \ldots x'_{N\alpha}\right) = \frac{1}{(2\pi\hbar)^{3N}}$$

$$\int_{-\infty}^{\infty} \ldots \int_{-\infty}^{\infty} f_W(P_{1\alpha} \ldots P_{N\alpha}, Q_{1\alpha} \ldots Q_{N\alpha}) e^{\frac{iP_{1\alpha} r_{1\alpha}}{\hbar}} \ldots e^{\frac{iP_{N\alpha} r_{N\alpha}}{\hbar}} dP_{1\alpha} \ldots dP_{N\alpha}.$$

$$\bar{A} = \int_{-\infty}^{\infty} dx_{1\alpha} \ldots \int_{-\infty}^{\infty} dx_{N\alpha}$$

$$\int_{-\infty}^{\infty} dx'_{1\alpha} \ldots \int_{-\infty}^{\infty} dx'_{N\alpha}$$

$$(\frac{1}{(2\pi\hbar)^{3N}}\int\limits_{-\infty}^{\infty}...\int\limits_{-\infty}^{\infty}f_W(P_{1\alpha}...P_{N\alpha},Q_{1\alpha}...Q_{N\alpha})e^{\frac{iP_{1\alpha}r_{1\alpha}}{\hbar}}...e^{\frac{iP_{N\alpha}r_{N\alpha}}{\hbar}}dP_{1\alpha}...dP_{N\alpha})$$

$$A(x'_{1\alpha}...x'_{N\alpha},x_{1\alpha}...x_{N\alpha},))=$$

$$=\int\limits_{-\infty}^{\infty}dQ_{1\alpha}...\int\limits_{-\infty}^{\infty}dQ_{N\alpha}$$

$$\int\limits_{-\infty}^{\infty}dr_{1\alpha}...\int\limits_{-\infty}^{\infty}dr_{N\alpha}$$

$$(\left(\frac{1}{(2\pi\hbar)^{3N}}\int\limits_{-\infty}^{\infty}...\int\limits_{-\infty}^{\infty}f_W(P_{1\alpha}...P_{N\alpha},Q_{1\alpha}...Q_{N\alpha})e^{\frac{iP_{1\alpha}r_{1\alpha}}{\hbar}}...e^{\frac{iP_{N\alpha}r_{N\alpha}}{\hbar}}dP_{1\alpha}...dP_{N\alpha}\right)$$

$$A\left(Q_{1\alpha}-\frac{r_{1\alpha}}{2}...Q_{N\alpha}-\frac{r_{1N}}{2}\right)\delta(r_{1\alpha}...r_{N\alpha}))=$$

$$=\int\limits_{-\infty}^{\infty}dQ_{1\alpha}...\int\limits_{-\infty}^{\infty}dQ_{N\alpha}\frac{1}{(2\pi\hbar)^{3N}}$$

$$\int\limits_{-\infty}^{\infty}...\int\limits_{-\infty}^{\infty}dP_{1\alpha}...dP_{N\alpha}f_W(P_{1\alpha}...P_{N\alpha},Q_{1\alpha}...Q_{N\alpha})$$

$$\int\limits_{-\infty}^{\infty}dr_{1\alpha}...\int\limits_{-\infty}^{\infty}dr_{N\alpha}A\left(x'_{1\alpha}...x'_{N\alpha},x_{1\alpha}...x_{N\alpha}\right)e^{\frac{iP_{1\alpha}r_{1\alpha}}{\hbar}}...e^{\frac{iP_{N\alpha}r_{N\alpha}}{\hbar}}=$$

$$=\frac{1}{(2\pi\hbar)^{3N}}\int\limits_{-\infty}^{\infty}dQ_{1\alpha}...\int\limits_{-\infty}^{\infty}dQ_{N\alpha}\int\limits_{-\infty}^{\infty}...$$

$$\int\limits_{-\infty}^{\infty}dP_{1\alpha}...dP_{N\alpha}f_W(P_{1\alpha}...P_{N\alpha},Q_{1\alpha}...Q_{N\alpha})\hat{A}_W(Q_{1\alpha}...Q_{N\alpha}). \tag{5.106}$$

5.5.2 Equation for the Density Matrix in Wigner Representation

In the coordinate representation (one-dimensional case), the equation for the matrix element has the form (5.19):

$$i\hbar\frac{\partial \rho_{xx'}(t)}{\partial t} = -\frac{\hbar^2}{2m}\left(\frac{\partial^2}{\partial x^2} - \frac{\partial^2}{\partial x'^2}\right)\rho_{xx'} + \left(U(x) - U(x')\right)\rho_{xx'}.$$

In order to perform an integral transformation into the Wigner representation, it is necessary to perform an integral transformation

$$\begin{aligned} i\hbar \int_{-\infty}^{\infty} e^{-\frac{i\mathrm{Pr}}{\hbar}} \frac{\partial \rho(xx't)}{\partial t} dr = \\ -\frac{\hbar^2}{2m}\int_{-\infty}^{\infty} e^{-\frac{i\mathrm{Pr}}{\hbar}}\left(\frac{\partial^2}{\partial x^2} - \frac{\partial^2}{\partial x'^2}\right)\rho(xx')dr \\ + \int_{-\infty}^{\infty} e^{-\frac{i\mathrm{Pr}}{\hbar}}\left(U(x) - U(x')\right)\rho(xx')dr. \end{aligned} \tag{5.107}$$

Let's consider each term separately. The first (left part):

$$(1) \int_{-\infty}^{\infty} e^{-\frac{i\mathrm{Pr}}{\hbar}} \frac{\partial \rho(xx', t)}{\partial t} dr = \frac{\partial}{\partial t}\int_{-\infty}^{\infty} e^{-\frac{i\mathrm{Pr}}{\hbar}} \rho(xx', t)dr = \frac{\partial}{\partial t} f_W(PQ, t). \tag{5.108}$$

The second is

$$\begin{aligned} &(2) \int_{-\infty}^{\infty} e^{-\frac{i\mathrm{Pr}}{\hbar}}\left(\frac{\partial^2}{\partial x^2} - \frac{\partial^2}{\partial x'^2}\right)\rho(xx')dr = \\ &= \int_{-\infty}^{\infty} e^{-\frac{i\mathrm{Pr}}{\hbar}}\left(\frac{1}{2}\frac{\partial}{\partial Q} + \frac{\partial}{\partial r} + \frac{1}{2}\frac{\partial}{\partial Q} - \frac{\partial}{\partial r}\right)\left(\frac{1}{2}\frac{\partial}{\partial Q} + \frac{\partial}{\partial r} - \frac{1}{2}\frac{\partial}{\partial Q} + \frac{\partial}{\partial r}\right)\rho(xx')dr = \\ &= 2\int_{-\infty}^{\infty} e^{-\frac{i\mathrm{Pr}}{\hbar}} \frac{\partial}{\partial Q}\frac{\partial}{\partial r}\rho(xx')dr = 2\frac{\partial}{\partial Q}\int_{-\infty}^{\infty} e^{-\frac{i\mathrm{Pr}}{\hbar}} \frac{\partial}{\partial r}\rho(xx')dr = \\ &= 2\frac{\partial}{\partial Q}\left\{\rho(Q + \frac{r}{2}, Q - \frac{r}{2}, t)e^{-\frac{i\mathrm{Pr}}{\hbar}}\Big|_{-\infty}^{\infty} - \left(-\frac{i}{\hbar}P\right)\int_{-\infty}^{\infty} e^{-\frac{i\mathrm{Pr}}{\hbar}}\rho(xx')dr\right\} = \\ &= 2\left(\frac{i}{\hbar}P\right)\frac{\partial f_W(PQ, t)}{\partial Q}. \end{aligned} \tag{5.109}$$

Here the relation (5.98) and the Jacobian of the transformations were taken into account

$$x = Q + \frac{r}{2}, x' = Q - \frac{r}{2},$$

$$
\begin{aligned}
&Q=\frac{x+x'}{2},\ r=x-x'\\
&\frac{\partial}{\partial x}=\frac{\partial}{\partial Q}\frac{\partial Q}{\partial x}+\frac{\partial}{\partial r}\frac{\partial r}{\partial x}=\frac{1}{2}\frac{\partial}{\partial Q}+\frac{\partial}{\partial r},\\
&\frac{\partial^2}{\partial x^2}=\left(\frac{1}{2}\frac{\partial}{\partial Q}+\frac{\partial}{\partial r}\right)^2\\
&\frac{\partial}{\partial x'}=\frac{\partial}{\partial Q}\frac{\partial Q}{\partial x'}+\frac{\partial}{\partial r}\frac{\partial r}{\partial x'}=\frac{1}{2}\frac{\partial}{\partial Q}-\frac{\partial}{\partial r},\\
&\frac{\partial^2}{\partial x'^2}=\left(\frac{1}{2}\frac{\partial}{\partial Q}-\frac{\partial}{\partial r}\right)^2
\end{aligned}
$$

$$
\frac{\partial^2}{\partial x^2}-\frac{\partial^2}{\partial x'^2}=\left(\frac{1}{2}\frac{\partial}{\partial Q}+\frac{\partial}{\partial r}\right)^2-\left(\frac{1}{2}\frac{\partial}{\partial Q}-\frac{\partial}{\partial r}\right)^2=2\frac{\partial^2}{\partial Q\partial r}. \tag{5.110}
$$

Thus, the kinetic energy operator acting on the density matrix in the Wigner representation has the form

$$
\boxed{\hat{T}_W f_W=-\frac{i\hbar}{m}P\frac{\partial f_W(PQ,t)}{\partial Q}}. \tag{5.111}
$$

The third, the contribution of the potential energy

$$
\begin{aligned}
&(3)\int_{-\infty}^{\infty}e^{-\frac{iPr}{\hbar}}\left(U(x)-U(x')\right)\rho(xx')dr=\\
&=\int_{-\infty}^{\infty}e^{-\frac{iPr}{\hbar}}\left(U\left(Q+\frac{r}{2}\right)\right)\rho(xx')dr\\
&-\int_{-\infty}^{\infty}e^{-\frac{iPr}{\hbar}}U\left(Q-\frac{r}{2}\right)\rho(xx')dr=\\
&=\left\{U_W\left(Q+\frac{1}{2}i\hbar\frac{\partial}{\partial P}\right)-U_W\left(Q-\frac{1}{2}i\hbar\frac{\partial}{\partial P}\right)\right\}f_W(P,Q,t).
\end{aligned} \tag{5.112}
$$

Here the first integral I considered in details as follows:

$$
\begin{aligned}
I&=\int_{-\infty}^{\infty}e^{-\frac{iPr}{\hbar}}\left(U\left(Q+\frac{r}{2}\right)\right)\rho(xx')dr\\
&=\int_{-\infty}^{\infty}e^{-\frac{iPr}{\hbar}}\left(\sum_{\nu}\frac{1}{\nu!}\left(\frac{r}{2}\right)^{\nu}\frac{\partial^{\nu}}{\partial Q^{\nu}}U(Q)\right)\rho(xx')dr=
\end{aligned}
$$

$$
\begin{aligned}
&= \int_{-\infty}^{\infty} e^{-\frac{iPr}{\hbar}} \left(U(Q) + \frac{r}{2}\frac{\partial U(y)}{\partial y}\bigg|_{y=Q} + \frac{1}{2}\left(\frac{r}{2}\right)^2 \frac{\partial^2}{\partial Q^2} U(Q)\right. \\
&\left. + \frac{1}{3!}\left(\frac{r}{2}\right)^3 \frac{\partial^3}{\partial Q^3} U(Q) + \ldots + \frac{1}{\nu!}\left(\frac{r}{2}\right)^\nu \frac{\partial^\nu}{\partial Q^\nu} U(Q) + \ldots\right)\rho(xx')dr = \\
&= \sum_\nu \frac{1}{\nu!}\frac{\partial^\nu U(Q)}{\partial Q^\nu} \int_{-\infty}^{\infty} e^{-\frac{iPr}{\hbar}} \left(\frac{r}{2}\right)^\nu \rho(xx')dr \\
&= \sum_\nu \frac{1}{\nu!}\frac{\partial^\nu U(Q)}{\partial Q^\nu} \left(-\frac{\hbar}{2i}\right)^\nu \frac{\partial^\nu}{\partial P^\nu} \int_{-\infty}^{\infty} e^{-\frac{iPr}{\hbar}} \rho(xx')dr = \\
&= \sum_\nu \frac{1}{\nu!}\frac{\partial^\nu U(Q)}{\partial Q^\nu} \left(\frac{i\hbar}{2}\right)^\nu \frac{\partial^\nu f_W(P,Q,t)}{\partial P^\nu} \\
&= U\left(Q + \frac{1}{2} i\hbar \frac{\partial}{\partial P}\right) f_W(P,Q,t).
\end{aligned} \tag{5.113}
$$

and the second integral II considered by the same way is

$$
\begin{aligned}
II &= \int_{-\infty}^{\infty} e^{-\frac{iPr}{\hbar}} U\left(Q - \frac{r}{2}\right)\rho(xx')dr \\
&= U\left(Q - \frac{1}{2} i\hbar \frac{\partial}{\partial P}\right) f_W(P,Q,t).
\end{aligned} \tag{5.114}
$$

By collecting expressions (1), (2), and (3) together, (5.108), (5.109), and (5.112), we obtain an equation for the density matrix in the Wigner representation, which will have the form:

$$
\begin{aligned}
& i\hbar \frac{\partial}{\partial t} f_W(PQ,t) = \\
& = -\frac{\hbar^2}{2m} 2\left(\frac{i}{\hbar}P\right) \frac{\partial f_W(PQ,t)}{\partial Q} \\
& + \left\{U\left(Q + \frac{1}{2} i\hbar \frac{\partial}{\partial P}\right) - U\left(Q - \frac{1}{2} i\hbar \frac{\partial}{\partial P}\right)\right\} f_W(P,Q,t), \\
& \frac{\partial}{\partial t} f_W(PQ,t) = -\frac{1}{m} P \frac{\partial f_W(PQ,t)}{\partial Q} \\
& + \frac{1}{i\hbar}\left\{U\left(Q + \frac{1}{2} i\hbar \frac{\partial}{\partial P}\right) - U\left(Q - \frac{1}{2} i\hbar \frac{\partial}{\partial P}\right)\right\} f_W(P,Q,t)
\end{aligned} \tag{5.115}
$$

or

$$\boxed{\frac{\partial}{\partial t}f_W(PQ,t)+\frac{P}{m}\frac{\partial f_W(PQ,t)}{\partial Q}+\frac{i}{\hbar}\left\{U\left(Q+\frac{1}{2}i\hbar\frac{\partial}{\partial P}\right)-U\left(Q-\frac{1}{2}i\hbar\frac{\partial}{\partial P}\right)\right\}f_W(P,Q,t)=0.} \tag{5.116}$$

In the semiclassical limit $\frac{i\hbar\partial}{\partial P}=\hat{x}\sim\lambda\!\!\!^{-}\ll\Delta Q\sim l_{fp}$, where l_{fp} is a mean free pass, then Eq. (5.115) is transformed by a following way:

$$\begin{aligned}
&\frac{\partial}{\partial t}f_W(PQ,t)+\frac{P}{m}\frac{\partial f_W(PQ,t)}{\partial Q}+\\
&+\frac{i}{\hbar}\left\{\left(U(Q)+\frac{1}{2}i\hbar\frac{\partial f_W(P,Q,t)}{\partial P}\frac{\partial U(Q)}{\partial Q}\right)-\left(U(Q)+\frac{1}{2}i\hbar\frac{\partial f_W(P,Q,t)}{\partial P}\frac{\partial U(Q)}{\partial Q}\right)\right\}\\
&f_W(P,Q,t)=0,\\
&\frac{\partial}{\partial t}f_W(PQ,t)+\frac{P}{m}\frac{\partial f_W(PQ,t)}{\partial Q}-\frac{\partial U(Q)}{\partial Q}\frac{\partial f_W(P,Q,t)}{\partial P}=0,\\
&H(P,Q)=\frac{P^2}{2m}+U(Q)\\
&\frac{\partial}{\partial t}f_W(PQ,t)+\frac{\partial H(P,Q)}{\partial P}\frac{\partial f_W(P,Q,t)}{\partial Q}-\frac{\partial H(P,Q)}{\partial Q}\frac{\partial f_W(P,Q,t)}{\partial P}\\
&=\frac{\partial}{\partial t}f_W(PQ,t)+\{H(P,Q)f_W(P,Q,t)\}=0.
\end{aligned} \tag{5.117}$$

$$\boxed{\frac{\partial}{\partial t}f_W(PQ,t)+\{H(P,Q)f_W(P,Q,t)\}=0.}$$

Then the Hamiltonian becomes a classical Hamiltonian function $H(P,Q)=\frac{P^2}{2m}+U(Q)$, depending on the coordinate Q of the point particle and its momentum P. Here we introduce the standard notation for Poisson brackets:

$$\{H(P,Q)f_W(P,Q,t)\}=\frac{\partial H(P,Q)}{\partial P}\frac{\partial f_W(P,Q,t)}{\partial Q}-\frac{\partial H(P,Q)}{\partial Q}\frac{\partial f_W(P,Q,t)}{\partial P}. \tag{5.118}$$

We obtain the classical Liouville equation for the one-dimensional case:

$$\frac{\partial}{\partial t}f_W(PQ,t)+\{H(P,Q)f_W(P,Q,t)\}=0. \tag{5.119}$$

5.6 Summary of the Density Matrix Theory

The density matrix theory is a powerful tool for describing a wide range of phenomena in quantum systems occurring in the presence of a thermalized environment with which the quantum system is in thermodynamic equilibrium. When describing quantum systems, one has to take into account the presence of a macroscopic device in which the quantum system under study is located. As a rule, it is difficult in an experiment, and in other cases simply impossible, to ensure the isolation of the quantum object of study, therefore the results of the measurements are certainly affected by stochastic factors on the part of the device, which in general must be taken into account in the theoretical description of the phenomenon or process under study. The processes and phenomena occurring in a quantum system themselves, in the presence of the influence of the "device," can differ radically from those processes occurring in an isolated quantum system. Therefore, a mathematical tool is required that allows describing such a quantum system in "difficult weather conditions." Such a mathematical apparatus, which allows us to correctly describe a quantum system, taking into account the stochastic influence of the environment, is the Density Matrix Theory, briefly outlined in this chapter.

Acknowledgements We would like to express our gratitude to my teacher Anna Maruchek, a brilliant pianist, for the inspiration and creative atmosphere she provided while writing this chapter.

References

1. von Neumann J., In: Wheeler NA editors. Mathematical foundations of quantum mechanics. New Edition. Translated by Robert T. Beyer. Princeton University Press; 2018. ISBN 9781400889921. Pages: 328
2. Kubo R, Toda M, Hashitsume N. "Statistical Physics II" Nonequilibrium Statistical Mechanics, Springer, Berlin, Heidelberg , New-York, Tokyo; 1985.
3. Kvasnikov IA. Thermodynamics and statistical physics. V.3: The theory of non-equilibrium systems; 1987. ISBN: 5354017025 ISBN-13(EAN): 9785354017027, London.
4. Akhiezer AI, Peletminskii SV. Methods of statistical physics, translated by M. Schukin, Bib ID: 1665456, Oxford ; New York : Pergamon Press, xv, 450 p.; 1981. ISBN: 0080250408
5. Landau LD, Lifschitz EM. Vol. 5—Statistical Physics part 1 Third Edition, Pergamon Press Oxford, New-York, Toronto, Sydney, Paris Frankfurt; 1980.
6. Callen HB, Welton TA. Fluctuation-dissipation theorem. Phys Rev. 1951;83:34–39.
7. Callen HB, Welton RF. Phys Rev. 1952;86:702.
8. Nyquist H. Phys Rev. 1928;32:110.
9. Kubo R. Rep Prog Phys 1966;29, Part I, 255.
10. Abragam A. The principal of nuclear magnetism. Oxford: Clarendon Press; 1962.

Chapter 6
Quantum Entanglement of Quantum Particles in High Energy Scattering—Formulation and Evaluation

Shigenori Seki

Abstract We consider entanglement between two outgoing particles of the final states in two-particle scatterings. An S-matrix theory is often useful in order for the analyses of phenomena in particle scatterings. There are an elastic channel and inelastic channels in the two-particle scatterings. We formulate the entanglement entropy of the two-particle final states in each elastic and inelastic channel in the context of the S-matrix theory. Then we concretely evaluate the entanglement entropy in terms of experimental data for two examples of hadron scattering, i.e., proton-proton scattering and proton-neutron scattering. In the proton-neutron scattering, the inelastic channel produces more entanglement than the elastic one.

Keywords Quantum entanglement · Particle scattering · Quantum field theory · S-matrix theory · Mandelstam variables · Partial wave expansion · Elastic scattering · Inelastic scattering · Proton-proton scattering · Proton-neutron scattering

6.1 Introduction

6.1.1 What Is Quantum Entanglement?

Entanglement is one of the most important phenomena in quantum theory. It shows a kind of correlation in a multi-body system. Let us consider spin-$\frac{1}{2}$ states in a two-body (A and B) system. The quantum state spaces of A and B are the Hilbert spaces, $\mathcal{H}_{\rm A}$ and $\mathcal{H}_{\rm B}$, respectively. A Hilbert space in quantum mechanics is a space of state vectors with an inner product equipped. Thus the two-body states are in the Hilbert space $\mathcal{H}_{\rm A} \otimes \mathcal{H}_{\rm B}$. $\otimes$ stands for the direct product of $\mathcal{H}_{\rm A}$ and $\mathcal{H}_{\rm B}$ in linear algebra. Each Hilbert space $\mathcal{H}_{\rm A,B}$ is spanned by two vectors, $\{|\uparrow\rangle_{\rm A,B}, |\downarrow\rangle_{\rm A,B}\}$, and the inner products are defined as

S. Seki (✉)
Institute for Fundamental Sciences, Setsunan University, Neyagawa, Osaka, Japan
e-mail: shigenori.seki@setsunan.ac.jp

T. Onishi (ed.), *Advances in Quantum Science I*,
https://doi.org/10.1007/978-981-95-9717-8_6

$$_A\langle\uparrow|\uparrow\rangle_A = {}_B\langle\downarrow|\downarrow\rangle_B = 1, \quad {}_A\langle\uparrow|\downarrow\rangle_A = {}_B\langle\uparrow|\downarrow\rangle_B = 0. \tag{6.1}$$

A state in this two-body system is described as

$$|\psi\rangle = \sum_{m,n=\uparrow,\downarrow} c_{mn}|m\rangle_A \otimes |n\rangle_B. \tag{6.2}$$

When c_{mn} is separable, i.e., $c_{mn} = c_m c_n$, $|\psi\rangle$ is an unentangled state. On the other hand, when c_{mn} is inseparable, $|\psi\rangle$ is an entangled state. For example,

$$|\psi_u\rangle = \frac{1}{\sqrt{2}}(|\uparrow\rangle_A \otimes |\uparrow\rangle_B + |\downarrow\rangle_A \otimes |\uparrow\rangle_B) \tag{6.3}$$

is unentangled, but

$$|\psi_e\rangle = \frac{1}{\sqrt{2}}(|\uparrow\rangle_A \otimes |\downarrow\rangle_B + |\downarrow\rangle_A \otimes |\uparrow\rangle_B) \tag{6.4}$$

is entangled.

We introduce a density matrix, $\rho = |\psi\rangle\langle\psi|$, which is on $\mathcal{H}_A \otimes \mathcal{H}_B$. Then the reduced density matrix ρ_A is given by $\rho_A = \mathrm{tr}_B\rho$. tr_B stands for taking the trace only over the Hilbert space $\mathcal{H}_B$ (not over $\mathcal{H}_A$), and similarly tr_A for taking the trace only over $\mathcal{H}_A$. Therefore ρ_A is on $\mathcal{H}_A$. In terms of the reduced density matrix, the entanglement entropy (EE), S_{EE}, is defined as

$$S_{EE} = -\mathrm{tr}_A\rho_A \ln \rho_A. \tag{6.5}$$

In the case of (6.4), its entanglement entropy is equal to $\ln 2$.[1] Note that the density matrix ρ is, in general, normalized so that $\mathrm{tr}_A\mathrm{tr}_B\rho = 1$.

[1] In the spin-$\frac{1}{2}$ system, $\mathrm{tr}_B\rho$ means

$$\mathrm{tr}_B\rho = {}_B\langle\uparrow|\rho|\uparrow\rangle_B + {}_B\langle\downarrow|\rho|\downarrow\rangle_B.$$

When $\rho = |\psi_e\rangle\langle\psi_e|$, the reduced density matrix becomes

$$\begin{aligned}
\rho_A = \mathrm{tr}_B\rho &= \frac{1}{2}\,{}_B\langle\uparrow|(|\uparrow\rangle_A \otimes |\downarrow\rangle_B + |\downarrow\rangle_A \otimes |\uparrow\rangle_B)({}_A\langle\uparrow| \otimes {}_B\langle\downarrow| + {}_A\langle\downarrow| \otimes {}_B\langle\uparrow|)|\uparrow\rangle_B \\
&+ \frac{1}{2}\,{}_B\langle\downarrow|(|\uparrow\rangle_A \otimes |\downarrow\rangle_B + |\downarrow\rangle_A \otimes |\uparrow\rangle_B)({}_A\langle\uparrow| \otimes {}_B\langle\downarrow| + {}_A\langle\downarrow| \otimes {}_B\langle\uparrow|)|\downarrow\rangle_B \\
&= \frac{1}{2}(|\uparrow\rangle_A\, {}_B\langle\uparrow|\downarrow\rangle_B + |\downarrow\rangle_A\, {}_B\langle\uparrow|\uparrow\rangle_B)({}_A\langle\uparrow|\,{}_B\langle\downarrow|\uparrow\rangle_B + {}_A\langle\downarrow|\,{}_B\langle\uparrow|\uparrow\rangle_B) \\
&+ \frac{1}{2}(|\uparrow\rangle_A\, {}_B\langle\downarrow|\downarrow\rangle_B + |\downarrow\rangle_A\, {}_B\langle\downarrow|\uparrow\rangle_B)({}_A\langle\uparrow| \otimes {}_B\langle\downarrow|\downarrow\rangle_B + {}_A\langle\downarrow| \otimes {}_B\langle\uparrow|\downarrow\rangle_B) \\
&= \frac{1}{2}|\downarrow\rangle_A\,{}_A\langle\downarrow| + \frac{1}{2}|\uparrow\rangle_A\,{}_A\langle\uparrow|.
\end{aligned}$$

Then the entanglement entropy becomes

$$-\mathrm{tr}_A\rho_A \ln \rho_A = -\frac{1}{2}\ln\frac{1}{2}\cdot \mathrm{tr}_A(|\downarrow\rangle_A\,{}_A\langle\downarrow| + |\uparrow\rangle_A\,{}_A\langle\uparrow|) = -\frac{1}{2}\ln\frac{1}{2}\cdot 2 = \ln 2.$$

In this article, we shall study the entanglement between two particles which is produced by a scattering process.

6.2 S-Matrix Theory for Particle Scattering

We use the S-matrix theory [1, 2] in order to formulate the entanglement entropy of two-particle final states in scatterings.

6.2.1 Momentum Hilbert Space and S-Matrix

Let us consider the collision of two incoming particles, A_0 and B_0. There are three kinds of channels depending on outgoing particles as follows:

$$\mathrm{A}_0\mathrm{B}_0 \to \begin{cases} \mathrm{A}_0\mathrm{B}_0 & \text{an elastic channel} \\ \mathrm{A}_i\mathrm{B}_i \quad (i = 1, 2, \ldots, n) & \text{two-particle inelastic channels} \\ X & \text{multi-particle inelastic channels,} \end{cases} \tag{6.6}$$

where $(\mathrm{A}_i, \mathrm{B}_i) \neq (\mathrm{A}_0, \mathrm{B}_0)$. Since the final states in the multi-particle inelastic channels consist of many combinations of out-going particles, we symbolically describe them as X. We focus on the elastic channel, $\mathrm{A}_0\mathrm{B}_0 \to \mathrm{A}_0\mathrm{B}_0$, and the two-particle inelastic channel, $\mathrm{A}_0\mathrm{B}_0 \to \mathrm{A}_1\mathrm{B}_1$. The state of A_i with momentum $\mathbf{p}$ and the one of B_i with momentum $\mathbf{q}$ are in the momentum Hilbert spaces, $\mathcal{H}_{\mathrm{A}_i}$ and $\mathcal{H}_{\mathrm{B}_i}$, respectively, i.e.,

$$|\mathbf{p}\rangle_{\mathrm{A}_i} \in \mathcal{H}_{\mathrm{A}_i}, \quad |\mathbf{q}\rangle_{\mathrm{B}_i} \in \mathcal{H}_{\mathrm{B}_i}. \tag{6.7}$$

These Hilbert spaces are given by Fock spaces. In a quantum field theory, the one-particle state of A_i with momentum $\mathbf{p}$ is given by a creation operator acting on a vacuum state. Similarly the one-particle state of B_i with momentum $\mathbf{q}$ is given by a creation operator for B_i. The Fock space consists of these states with infinite number of various momenta, $\mathbf{p}$ and $\mathbf{q}$. Therefore, we note that the volume of the momentum Hilbert spaces is infinite. The two-particle states of $\mathrm{A}_i\mathrm{B}_i$ $(i = 0, 1)$ is denoted by

$$|\mathbf{p}, \mathbf{q}\rangle_i \equiv |\mathbf{p}\rangle_{\mathrm{A}_i} \otimes |\mathbf{q}\rangle_{\mathrm{B}_i} \in \mathcal{H}_{\mathrm{A}_i} \otimes \mathcal{H}_{\mathrm{B}_i}. \tag{6.8}$$

The inner products of the states are defined as

$${}_I\langle\mathbf{p}|\mathbf{q}\rangle_J = \delta_{IJ}\sqrt{2E_{I\mathbf{p}}\,2E_{J\mathbf{q}}}\,\delta^{(3)}(\mathbf{p} - \mathbf{q}) \quad (I, J = \mathrm{A}_0, \mathrm{B}_0, \mathrm{A}_1, \mathrm{B}_1). \tag{6.9}$$

δ_{IJ} is a Kronecker delta,

$$\delta_{IJ} = \begin{cases} 1 & (I = J) \\ 0 & (I \neq J), \end{cases} \tag{6.10}$$

and $\delta^{(3)}(\mathbf{p} - \mathbf{q})$ is a three-dimensional Dirac delta function as

$$\delta^{(3)}(\mathbf{a}) = \delta(a_1)\delta(a_2)\delta(a_3) \quad (\mathbf{a} = (a_1, a_2, a_3)), \tag{6.11}$$

where $\delta(a_i)$ (i = 1, 2, 3) is a usual (one-dimensional) Dirac Delta function. $E_{I\mathbf{p}}$ is the energy of particle I with momentum $\mathbf{p}$ and mass m_I, that is, $E_{I\mathbf{p}} = \sqrt{m_I^2 + \mathbf{p}^2}$. Hence, the inner product of the two-particle states becomes

$$\begin{aligned} {}_i\langle \mathbf{p}, \mathbf{q} | \mathbf{p}', \mathbf{q}' \rangle_j &= {}_{\mathrm{A}_i}\langle \mathbf{p} | \mathbf{p}' \rangle_{\mathrm{A}_j} \, {}_{\mathrm{B}_i}\langle \mathbf{q} | \mathbf{q}' \rangle_{\mathrm{B}_j} \\ &= \delta_{ij} \sqrt{2E_{\mathrm{A}_i\mathbf{p}} 2E_{\mathrm{A}_j\mathbf{q}}} \sqrt{2E_{\mathrm{B}_i\mathbf{p}} 2E_{\mathrm{B}_j\mathbf{q}}} \delta^{(3)}(\mathbf{p} - \mathbf{p}') \delta^{(3)}(\mathbf{q} - \mathbf{q}') \end{aligned} \tag{6.12}$$

The complete set on orthogonal basis is

$$\mathbf{1} = \sum_{i=0}^{n} \int \frac{d^3\mathbf{p}\, d^3\mathbf{q}}{2E_{\mathrm{A}_i\mathbf{p}} 2E_{\mathrm{B}_i\mathbf{q}}} |\mathbf{p}, \mathbf{q}\rangle_i \, {}_i\langle \mathbf{p}, \mathbf{q}| + \int dX |X\rangle\langle X|, \tag{6.13}$$

where $\mathbf{1}$ is an identity matrix. $|X\rangle$ means all final states in the multi-particle inelastic channels of (6.6). $\int dX$ is the integration over all the momenta of particles appearing in those final states.

We set the momenta of incoming particles, A_0 and B_0, in (6.6) to $\mathbf{k}$ and $\mathbf{l}$, so that the initial state is described as

$$|\mathrm{ini}\rangle = |\mathbf{k}, \mathbf{l}\rangle_0. \tag{6.14}$$

Once the initial state is fixed, the final state is given by an S-matrix, S, as $S|\mathrm{ini}\rangle = S|\mathbf{k}, \mathbf{l}\rangle_0$. The S-matrix implies infinite time evolution. Therefore, the S-matrix satisfies the unitarity condition,

$$S^\dagger S = \mathbf{1}, \tag{6.15}$$

in order for preserving the completeness of the basis of states. We also use the transition T-matrix, T, defined by $S = \mathbf{1} + 2iT$. Then the unitarity condition (6.15) is rewritten as

$$T - T^\dagger = 2iT^\dagger T. \tag{6.16}$$

This condition can be rewritten in terms of T-matrix elements,

$$\begin{aligned} & {}_i\langle \mathbf{p}, \mathbf{q}|T|\mathbf{k}, \mathbf{l}\rangle_j - {}_i\langle \mathbf{p}, \mathbf{q}|T^\dagger|\mathbf{k}, \mathbf{l}\rangle_j \\ &= 2i\, {}_i\langle \mathbf{p}, \mathbf{q}|T^\dagger T|\mathbf{k}, \mathbf{l}\rangle_j \\ &= 2i \sum_{k=0}^{n} \int \frac{d^3\mathbf{p}'\, d^3\mathbf{q}'}{2E_{\mathrm{A}_k\mathbf{p}'}\, 2E_{\mathrm{B}_k\mathbf{q}'}}\, {}_i\langle \mathbf{p}, \mathbf{q}|T^\dagger|\mathbf{p}', \mathbf{q}'\rangle_k\, {}_k\langle \mathbf{p}', \mathbf{q}'|T|\mathbf{k}, \mathbf{l}\rangle_j \\ &\quad +2i \int dX\, {}_i\langle \mathbf{p}, \mathbf{q}|T^\dagger|X\rangle\langle X|T|\mathbf{k}, \mathbf{l}\rangle_j, \end{aligned} \tag{6.17}$$

in which we used (6.13) for the second equality.

In the S-matrix and T-matrix elements, the energy-momentum is conserved from the initial state $|\mathbf{k}, \mathbf{l}\rangle_j$ to the final state $|\mathbf{p}, \mathbf{q}\rangle_i$. Therefore, it is convenient to extract factors of energy-momentum conservation from the S-matrix and T-matrix elements [1],

$${}_i\langle \mathbf{p}, \mathbf{q}|T|\mathbf{k}, \mathbf{l}\rangle_j = \delta(E_i(\mathbf{p}, \mathbf{q}) - E_j(\mathbf{k}, \mathbf{l}))\delta^{(3)}(\mathbf{p}+\mathbf{q}-\mathbf{k}-\mathbf{l})\, {}_i\langle \mathbf{p}, \mathbf{q}|\hat{T}|\mathbf{k}, \mathbf{l}\rangle_j \tag{6.18}$$

$${}_i\langle \mathbf{p}, \mathbf{q}|S|\mathbf{k}, \mathbf{l}\rangle_j = \delta(E_i(\mathbf{p}, \mathbf{q}) - E_j(\mathbf{k}, \mathbf{l}))\delta^{(3)}(\mathbf{p}+\mathbf{q}-\mathbf{k}-\mathbf{l})\, {}_i\langle \mathbf{p}, \mathbf{q}|\hat{S}|\mathbf{k}, \mathbf{l}\rangle_j \tag{6.19}$$

where $E_i(\mathbf{p}, \mathbf{q}) = E_{\mathrm{A}_i\mathbf{p}} + E_{\mathrm{B}_i\mathbf{q}}$. $\hat{S}$ and $\hat{T}$ are the modified S and T matrices by extracting the energy-momentum conservation factors, and $\hat{S} = \mathbf{1} + 2i\hat{T}$.

6.2.2 Partial Wave Expansion

In the analyses of particle scatterings, partial wave expansion is often useful [1, 2]. In this subsection, we adopt a center-of-mass frame. If one particle has momentum $\mathbf{p}$, the other has $-\mathbf{p}$ in the two-particle initial and final states (see Fig. 6.1). Therefore, such states are denoted by

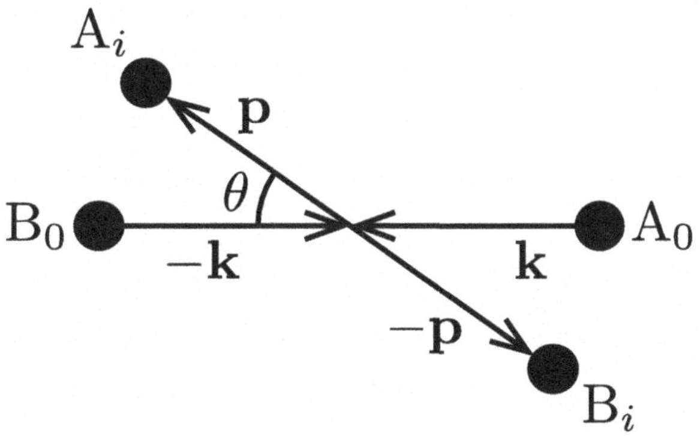

Fig. 6.1 The elastic and two-particle inelastic scattering, $\mathrm{A}_0\mathrm{B}_0 \to \mathrm{A}_i\mathrm{B}_i$ ($i = 0, 1, \ldots, n$). θ is a scattering angle

$$|\mathbf{p}\rangle\rangle_i \equiv |\mathbf{p}, -\mathbf{p}\rangle_i. \tag{6.20}$$

The initial state is $|\text{ini}\rangle = |\mathbf{k}\rangle\rangle_0$. For the elastic and two-particle inelastic channels, the scattering angle is denoted by θ. Here we introduce Mandelstam variables, s and t[2]:

$$s = (E_{\text{A}_0\mathbf{k}} + E_{\text{B}_0\mathbf{k}})^2 = (E_{\text{A}_i\mathbf{p}} + E_{\text{B}_i\mathbf{p}})^2, \tag{6.22}$$
$$t = 2k^2(\cos\theta - 1) \quad (k \equiv |\mathbf{k}|). \tag{6.23}$$

$\sqrt{s}$ is called the center-of-mass energy.

The inner product (6.12) and the complete set (6.13) become

$$_i\langle\langle\mathbf{p}|\mathbf{q}\rangle\rangle_j = \delta_{ij} 2E_{\text{A}_i\mathbf{p}} 2E_{\text{B}_i\mathbf{p}} \delta^{(3)}(\mathbf{p}-\mathbf{q})\delta^{(3)}(0), \tag{6.24}$$
$$\mathbf{1} = \sum_{i=0}^{n} \int \frac{d^3\mathbf{p}}{2E_{\text{A}_i\mathbf{p}} 2E_{\text{B}_i\mathbf{q}}\delta^{(3)}(0)} |\mathbf{p}\rangle\rangle_i\, {}_i\langle\langle\mathbf{p}| + \int dX |X\rangle\langle X|. \tag{6.25}$$

Although $\delta^{(3)}(0)$ diverges, we leave it at this stage. It will be regularized later (see the Sects. 6.3.2 and 6.3.3).

The unitarity condition (6.17) is rewritten in the center-of-mass frame,

$$\begin{aligned} & {}_i\langle\langle\mathbf{p}|\hat{T}|\mathbf{k}\rangle\rangle_j - {}_i\langle\langle\mathbf{p}|\hat{T}^\dagger|\mathbf{k}\rangle\rangle_j \\ &= 2i \sum_{h=0}^{n} \int \frac{d^3\mathbf{q}}{2E_{\text{A}_h\mathbf{q}} 2E_{\text{B}_h\mathbf{q}}} {}_i\langle\langle\mathbf{p}|\hat{T}^\dagger|\mathbf{q}\rangle\rangle_h\, \delta(E_h(\mathbf{p}, -\mathbf{p}) - E_j(\mathbf{k}, -\mathbf{k}))\, {}_h\langle\langle\mathbf{q}|\hat{T}|\mathbf{k}\rangle\rangle_j \\ &+ 2i F_{ij}(\mathbf{p}, \mathbf{k}), \end{aligned} \tag{6.26}$$

---

[2] The energy-momenta of the in-coming particles, A_0, B_0, and the out-going particles, A_i, B_i, are denoted by $p_{\text{A}_0} = (E_{\text{A}_0\mathbf{k}}, \mathbf{k})$, $p_{\text{B}_0} = (E_{\text{B}_0\mathbf{k}}, -\mathbf{k})$, $p_{\text{A}_i} = (E_{\text{A}_i\mathbf{p}}, \mathbf{p})$ and $p_{\text{B}_i} = (E_{\text{B}_i\mathbf{p}}, -\mathbf{p})$ respectively. The definition of the Mandelstam variables, s and t, are

$$s \equiv -(p_{\text{A}_0} + p_{\text{B}_0})^2, \quad t \equiv -(p_{\text{A}_0} - p_{\text{A}_i})^2. \tag{6.21}$$

Then the s is calculated as

$$s = -\{-(E_{\text{A}_0\mathbf{k}} + E_{\text{B}_0\mathbf{k}})^2 + (\mathbf{k} + (-\mathbf{k}))^2\} = (E_{\text{A}_0\mathbf{k}} + E_{\text{B}_0\mathbf{k}})^2 = (E_{\text{A}_i\mathbf{p}} + E_{\text{B}_i\mathbf{p}})^2.$$

The last equality is due to the energy conservation in the scattering process.

Since we are considering high energy scatterings, the absolute values of momenta are much larger than the masses. Therefore, we can neglect the contribution of masses, so that $p_{\text{A}_0} \approx (k, \mathbf{k})$, $p_{\text{B}_0} \approx (k, -\mathbf{k})$, $p_{\text{A}_i} \approx (p, \mathbf{p})$ and $p_{\text{B}_i} \approx (p, -\mathbf{p})$, where $k = |\mathbf{k}|$ and $p = |\mathbf{p}|$. Then $s \approx (k+k)^2 = 4k^2 \approx 4p^2$, while the Mandelstam variable t is given by

$$t \approx -(k-p)^2 + (\mathbf{k}-\mathbf{p})^2 = -2kp + 2\mathbf{k}\cdot\mathbf{p} = 2k^2(\cos\theta - 1),$$

where we used $\mathbf{k}\cdot\mathbf{p} = kp\cos\theta$ (see Fig. 6.1). At the same time.

where

$$F_{ij}(\mathbf{p}, \mathbf{k}) = \int dX\, {}_i\langle\!\langle \mathbf{p}|\hat{T}^\dagger|X\rangle \delta(E_X - E_j(\mathbf{k}, -\mathbf{k}))\delta^{(3)}(\mathbf{p}_X - \mathbf{k} - (-\mathbf{k}))\langle X|\hat{T}|\mathbf{k}\rangle\!\rangle_j. \tag{6.27}$$

E_X and $\mathbf{p}_X$ are the total energy and momentum of multi-particle state $|X\rangle$. $F_{ij}(\mathbf{p}, \mathbf{k})$ is known as an overlap matrix [2]. All the contributions of multi-particle inelastic channels X are collected into the overlap matrix.

We expand the T-matrix element and the overlap matrix by partial waves,

$$ {}_i\langle\!\langle \mathbf{p}|\hat{T}|\mathbf{k}\rangle\!\rangle_j = \frac{\sqrt{s}}{\pi\sqrt{pk}}\sum_{\ell=0}^{\infty}(2\ell+1)\tau_{ij}^{\ell}P_\ell(\cos\theta), \tag{6.28}$$

$$F_{ij}(\mathbf{p}, \mathbf{k}) = \frac{\sqrt{s}}{2\pi\sqrt{pk}}\sum_{\ell=0}^{\infty}(2\ell+1)\phi_{ij}^{\ell}P_\ell(\cos\theta). \tag{6.29}$$

P_ℓ is a Legendre polynomial (see Appendix). τ_{ij}^{ℓ} and ϕ_{ij}^{ℓ} are called partial wave modes which, in other words, are defined by Eqs. (6.28) and (6.29) due to the relation (6.119). Since the T-matrix elements (6.28) are related to the scattering amplitudes through

$$\begin{aligned} {}_i\langle\!\langle \mathbf{p}|\hat{S}|\mathbf{k}\rangle\!\rangle_0 &= \frac{\sqrt{s}}{\pi\sqrt{pk}}\sum_{\ell=0}^{\infty}(2\ell+1)(1+2i\tau_{i0}^{\ell})P_\ell(\cos\theta) \\ &= \frac{2\sqrt{s}}{\pi\sqrt{pk}}\left(\delta(1-\cos\theta) + \frac{i}{16\pi}\mathcal{A}_i\right). \end{aligned} \tag{6.30}$$

The scattering amplitudes are

$$\mathcal{A}_i = 16\pi\sum_{\ell=0}^{\infty}(2\ell+1)\tau_{i0}^{\ell}P_\ell(\cos\theta). \tag{6.31}$$

$\mathcal{A}_0$ is the elastic amplitude, while $\mathcal{A}_i$ ($i = 1, 2, \ldots, n$) are the two-particle inelastic amplitudes. $\delta(1-\cos\theta)$ in (6.30) comes from the infinite sum of Legendre polynomials:

$$2\delta(1-\cos\theta) = \sum_{\ell=0}^{\infty}(2\ell+1)P_\ell(\cos\theta). \tag{6.32}$$

By using the scattering amplitudes (6.31), we describe the elastic differential cross section for $i = 0$ and the two-particle inelastic ones for $i = 1, 2, \ldots, n$ as

$$\frac{d\sigma_i}{dt} = \frac{|\mathcal{A}_i|^2}{256\pi k^4} = \frac{\pi}{k^4}\left|\sum_{\ell=0}^{\infty}(2\ell+1)\tau_{i0}^{\ell}P_\ell(\cos\theta)\right|^2. \tag{6.33}$$

Note that $d\sigma_i/d\cos\theta = 2k^2 d\sigma_i/dt$ from (6.23). Then the cross sections are calculated from (6.33),

$$\sigma_i = \int_{-1}^{1} d\cos\theta \, \frac{d\sigma_i}{d\cos\theta} = \frac{4\pi}{k^2} \sum_{\ell=0}^{\infty} (2\ell+1) \left|\tau_{i0}^{\ell}\right|^2, \tag{6.34}$$

where we used the orthogonality of Legendre polynomials:

$$\int_{-1}^{1} d\cos\theta \; P_\ell(\cos\theta) P_{\ell'}(\cos\theta) = \frac{2}{2\ell+1}\delta_{\ell\ell'}. \tag{6.35}$$

The cross section for the multi-particle inelastic channels is derived from the overlap matrix,

$$\sigma_i^X = \frac{2\pi}{k^2} \sum_{\ell=0}^{\infty} (2\ell+1)\phi_{i0}^{\ell}. \tag{6.36}$$

Let us reconsider the unitarity condition (6.26) in terms of the partial wave modes. Substituting (6.28) and (6.29) into (6.26), we obtain

$$\mathrm{Im}\tau_{ij}^{\ell} = \sum_{h=0}^{n} \tau_{ih}^{\ell *} \tau_{hj}^{\ell} + \frac{1}{2}\phi_{ij}^{\ell}. \tag{6.37}$$

(See Appendix about the detailed derivation.) From this equation, we can describe the total cross section as

$$\sigma_{\text{tot}} = \frac{4\pi}{k^2} \sum_{\ell=0}^{\infty} (2\ell+1)\mathrm{Im}\tau_{00}^{\ell} = \sum_{i=0}^{n} \sigma_i + \sigma_0^X. \tag{6.38}$$

6.3 Formulation of Quantum Entanglement Entropy

We shall formulate the entanglement entropy of two-particle final states [3, 5].

6.3.1 *Final States and Density Matrices*

The final state $S|\text{ini}\rangle (= S|\mathbf{k}, \mathbf{l}\rangle_0)$ includes all the channels. Since we are interested in two-particle final states, we need to the projection of $S|\text{ini}\rangle$ onto the two-particle states. We especially concentrate on the elastic channel $A_0B_0 \to A_0B_0$ and the two-particle inelastic channel $A_0B_0 \to A_1B_1$. Referring the complete set (6.13), we introduce the projection operator onto $\mathcal{H}_{A_0} \otimes \mathcal{H}_{B_0}$ for the elastic channel:

$$\int \frac{d^3\mathbf{p}\, d^3\mathbf{q}}{2E_{\mathrm{A_0}\mathbf{p}}\, 2E_{\mathrm{B_0}\mathbf{q}}} |\mathbf{p}, \mathbf{q}\rangle_{0\,0}\langle \mathbf{p}, \mathbf{q}|, \tag{6.39}$$

and define the final state in the elastic channel as

$$\begin{aligned} |\mathrm{fin}\rangle_{\mathrm{el}} &= \left(\int \frac{d^3\mathbf{p}\, d^3\mathbf{q}}{2E_{\mathrm{A_0}\mathbf{p}}\, 2E_{\mathrm{B_0}\mathbf{q}}} |\mathbf{p}, \mathbf{q}\rangle_{0\,0}\langle \mathbf{p}, \mathbf{q}| \right) S|\mathbf{k}, \mathbf{l}\rangle_0 \\ &= \int \frac{d^3\mathbf{p}\, d^3\mathbf{q}}{2E_{\mathrm{A_0}\mathbf{p}}\, 2E_{\mathrm{B_0}\mathbf{q}}} |\mathbf{p}, \mathbf{q}\rangle_{0\,0}\langle \mathbf{p}, \mathbf{q}|(\mathbf{1} + 2iT)|\mathbf{k}, \mathbf{l}\rangle_0. \end{aligned} \tag{6.40}$$

In the same way, we introduce the projection operator onto $\mathcal{H}_{\mathrm{A_1}} \otimes \mathcal{H}_{\mathrm{B_1}}$ for the inelastic channel:

$$\int \frac{d^3\mathbf{p}\, d^3\mathbf{q}}{2E_{\mathrm{A_1}\mathbf{p}}\, 2E_{\mathrm{B_1}\mathbf{q}}} |\mathbf{p}, \mathbf{q}\rangle_{1\,1}\langle \mathbf{p}, \mathbf{q}|, \tag{6.41}$$

and the final state in the two-particle inelastic channel is defined as

$$\begin{aligned} |\mathrm{fin}\rangle_{\mathrm{inel}} &= \left(\int \frac{d^3\mathbf{p}\, d^3\mathbf{q}}{2E_{\mathrm{A_1}\mathbf{p}}\, 2E_{\mathrm{B_1}\mathbf{q}}} |\mathbf{p}, \mathbf{q}\rangle_{1\,1}\langle \mathbf{p}, \mathbf{q}| \right) S|\mathbf{k}, \mathbf{l}\rangle_0 \\ &= \int \frac{d^3\mathbf{p}\, d^3\mathbf{q}}{2E_{\mathrm{A_1}\mathbf{p}}\, 2E_{\mathrm{B_1}\mathbf{q}}} |\mathbf{p}, \mathbf{q}\rangle_{1\,1}\langle \mathbf{p}, \mathbf{q}|2iT|\mathbf{k}, \mathbf{l}\rangle_0. \end{aligned} \tag{6.42}$$

The identity matrix $\mathbf{1}$ in S contributes to $|\mathrm{fin}\rangle_{\mathrm{el}}$, but not to $|\mathrm{fin}\rangle_{\mathrm{inel}}$, because ${}_1\langle \mathbf{p}, \mathbf{q}|\mathbf{1}|\mathbf{k}, \mathbf{l}\rangle_0$ vanishes.

The density matrices for those two-particle final states are defined as

$$\rho_{\mathrm{el}} = \frac{1}{\mathcal{N}_{\mathrm{el}}} |\mathrm{fin}\rangle_{\mathrm{el}\;\mathrm{el}}\langle \mathrm{fin}|, \quad \rho_{\mathrm{inel}} = \frac{1}{\mathcal{N}_{\mathrm{inel}}} |\mathrm{fin}\rangle_{\mathrm{inel}\;\mathrm{inel}}\langle \mathrm{fin}|, \tag{6.43}$$

where $\mathcal{N}_{\mathrm{el}}$ and $\mathcal{N}_{\mathrm{inel}}$ are normalization factors, and will be determined later. From those density matrices, we obtain the reduced density matrices,

$$\begin{aligned} \rho_{\mathrm{A_0}} &\equiv \mathrm{tr}_{\mathrm{B_0}} \rho_{\mathrm{el}} = \int \frac{d^3\mathbf{r}}{2E_{\mathrm{B_0}\mathbf{r}}}\, {}_{\mathrm{B_0}}\langle \mathbf{r}|\rho_{\mathrm{el}}|\mathbf{r}\rangle_{\mathrm{B_0}} \\ &= \frac{1}{\mathcal{N}_{\mathrm{el}}} \int \frac{d^3\mathbf{r}}{2E_{\mathrm{B_0}\mathbf{r}}} \int \frac{d^3\mathbf{p}\, d^3\mathbf{q}}{2E_{\mathrm{A_0}\mathbf{p}}\, 2E_{\mathrm{B_0}\mathbf{q}}}\, |\mathbf{p}\rangle_{\mathrm{A_0}}\;{}_{\mathrm{B_0}}\langle \mathbf{r}|\mathbf{q}\rangle_{\mathrm{B_0}}\;{}_0\langle \mathbf{p}, \mathbf{q}|S|\mathbf{k}, \mathbf{l}\rangle_0 \\ &\quad \times \int \frac{d^3\mathbf{p}'\, d^3\mathbf{q}'}{2E_{\mathrm{A_0}\mathbf{p}'}\, 2E_{\mathrm{B_0}\mathbf{q}'}} \left({}_0\langle \mathbf{p}', \mathbf{q}'|S|\mathbf{k}, \mathbf{l}\rangle_0 \right)^*\; {}_{\mathrm{B_0}}\langle \mathbf{q}'|\mathbf{r}\rangle_{\mathrm{B_0}}\;{}_{\mathrm{A_0}}\langle \mathbf{p}'| \\ &= \frac{1}{\mathcal{N}_{\mathrm{el}}} \int \frac{d^3\mathbf{p}}{2E_{\mathrm{A_0}\mathbf{p}}} \frac{\delta(E_{\mathrm{A_0}\mathbf{p}} + E_{\mathrm{B_0}\mathbf{k}+\mathbf{l}-\mathbf{p}} - E_{\mathrm{A_0}\mathbf{k}} - E_{\mathrm{B_0}\mathbf{l}})\delta(0)}{2E_{\mathrm{A_0}\mathbf{p}}\, 2E_{\mathrm{B_0}\mathbf{k}+\mathbf{l}-\mathbf{p}}} \\ &\quad \times \left| {}_0\langle \mathbf{p}, \mathbf{k}+\mathbf{l}-\mathbf{p}|\hat{S}|\mathbf{k}, \mathbf{l}\rangle_0 \right|^2 |\mathbf{p}\rangle_{\mathrm{A_0}}\;{}_{\mathrm{A_0}}\langle \mathbf{p}|, \end{aligned} \tag{6.44}$$

and, in the same way,

$$\begin{aligned}\rho_{A_1} &\equiv \mathrm{tr}_{B_1}\rho_{\mathrm{inel}} = \int \frac{d^3\mathbf{r}}{2E_{B_1\mathbf{r}}}\,{}_{B_1}\langle\mathbf{r}|\rho_{\mathrm{inel}}|\mathbf{r}\rangle_{B_1}\\ &= \frac{1}{\mathcal{N}_{\mathrm{inel}}}\int \frac{d^3\mathbf{p}}{2E_{A_1\mathbf{p}}}\frac{\delta(E_{A_1\mathbf{p}} + E_{B_1\mathbf{k}+\mathbf{l}-\mathbf{p}} - E_{A_0\mathbf{k}} - E_{B_0\mathbf{l}})\delta(0)}{2E_{A_1\mathbf{p}}\,2E_{B_1\mathbf{k}+\mathbf{l}-\mathbf{p}}}\\ &\quad\times 4\left|{}_1\langle\mathbf{p},\mathbf{k}+\mathbf{l}-\mathbf{p}|\hat{T}|\mathbf{k},\mathbf{l}\rangle_0\right|^2|\mathbf{p}\rangle_{A_1}\,{}_{A_1}\langle\mathbf{p}|.\end{aligned} \tag{6.45}$$

Hereafter, we use the center-of-mass frame again. The reduced density matrix for the elastic channel (6.44) is rewritten in simpler expression as

$$\rho_{A_0} = \frac{1}{\mathcal{N}_{\mathrm{el}}}\int \frac{d^3\mathbf{p}}{2E_{A_0\mathbf{p}}}\frac{\delta(p-k)\delta(0)}{4p\sqrt{s}}\left|{}_0\langle\langle\mathbf{p}|\hat{S}|\mathbf{k}\rangle\rangle_0\right|^2|\mathbf{p}\rangle_{A_0}\,{}_{A_0}\langle\mathbf{p}|, \tag{6.46}$$

because the delta function for energy conservation is described as

$$\delta(E_{A_0\mathbf{p}} + E_{B_0\mathbf{p}} - \sqrt{s}) = \frac{2E_{A_0\mathbf{p}}\,2E_{B_0\mathbf{p}}}{4p\sqrt{s}}\delta(p-k), \tag{6.47}$$

where the center-of-mass energy $\sqrt{s} = E_{A_0\mathbf{k}} + E_{B_0\mathbf{k}}$. The reduced density matrix for the two-particle inelastic channel (6.45) is rewritten as

$$\rho_{A_1} = \frac{1}{\mathcal{N}_{\mathrm{inel}}}\int \frac{d^3\mathbf{p}}{2E_{A_1\mathbf{p}}}\frac{\delta(p-k')\delta(0)}{p\sqrt{s}}\left|{}_1\langle\langle\mathbf{p}|\hat{T}|\mathbf{k}\rangle\rangle_0\right|^2|\mathbf{p}\rangle_{A_1}\,{}_{A_1}\langle\mathbf{p}|, \tag{6.48}$$

where

$$k' = \sqrt{\frac{(m_{A_1}^2 + m_{B_1}^2 - s)^2 - 4m_{A_1}^2 m_{B_1}^2}{4s}} \tag{6.49}$$

(see also (6.130)). Note that $\sqrt{s}$ is also equal to $E_{A_1k'} + E_{B_1k'}$.

The normalization factors, $\mathcal{N}_{\mathrm{el}}$ and $\mathcal{N}_{\mathrm{inel}}$, are determined so that

$$\mathrm{tr}_{A_0}\mathrm{tr}_{B_0}\rho_{\mathrm{el}} = \mathrm{tr}_{A_0}\rho_{A_0} = 1, \quad \mathrm{tr}_{A_1}\mathrm{tr}_{B_1}\rho_{\mathrm{inel}} = \mathrm{tr}_{A_1}\rho_{A_1} = 1. \tag{6.50}$$

Substituting (6.46) and (6.48) into (6.50), we calculate the normalization factor for the elastic channel,

$$\mathcal{N}_{\mathrm{el}} = \delta(0)\delta^{(3)}(0)\int d^3\mathbf{p}\,\frac{\delta(p-k)}{4p\sqrt{s}}\left|{}_0\langle\langle\mathbf{p}|\hat{S}|\mathbf{k}\rangle\rangle_0\right|^2, \tag{6.51}$$

and the one for the two-particle inelastic channel,

$$\mathcal{N}_{\text{inel}} = \delta(0)\delta^{(3)}(0) \int d^3\mathbf{p}\, \frac{\delta(p-k')}{p\sqrt{s}} \left| {}_1\langle\!\langle \mathbf{p}|\hat{T}|\mathbf{k}\rangle\!\rangle_1 \right|^2. \tag{6.52}$$

6.3.2 Quantum Entanglement Entropy

It is often difficult to directly compute the trace of $\rho_A \ln \rho_A$ in (6.5). However it is easier to compute the trace of $(\rho_A)^n$. This quantity is associated with Rényi entropy,

$$S_{\text{RE}} = \frac{1}{1-n} \ln \text{tr}_A(\rho_A)^n. \tag{6.53}$$

We can compute the entanglement entropy S_{EE} through the Rényi entropy S_{RE}, i.e.,

$$S_{\text{EE}} = \lim_{n\to 1} S_{\text{RE}} = -\lim_{n\to 1} \frac{\partial}{\partial n} \text{tr}_A(\rho_A)^n, \tag{6.54}$$

because

$$\begin{aligned} \lim_{n\to 1} S_{\text{RE}} &= -\lim_{n\to 1} \frac{\ln \text{tr}_A(\rho_A)^n - \ln \text{tr}_A(\rho_A)^1}{n-1} \\ &= -\lim_{n\to 1} \frac{\partial}{\partial n} \ln \text{tr}_A(\rho_A)^n = -\lim_{n\to 1} \frac{\frac{\partial}{\partial n}\text{tr}_A(\rho_A)^n}{\text{tr}_A(\rho_A)^n} \\ &= -\lim_{n\to 1} \frac{\partial}{\partial n} \text{tr}_A(\rho_A)^n \end{aligned} \tag{6.55}$$

due to the normalization of the density matrix, i.e., $\text{tr}_A\rho_A = 1$, furthermore,

$$-\lim_{n\to 1} \frac{\partial}{\partial n} \text{tr}_A(\rho_A)^n = -\lim_{n\to 1} \text{tr}_A(\rho_A)^n \ln \rho_A = -\text{tr}_A \rho_A \ln \rho_A. \tag{6.56}$$

This is equal to the entanglement entropy S_{EE} by definition (see Eq. (6.5)).

Elastic Channel

Let us compute the entanglement entropy S_{el} of the final state in the elastic channel, $A_0B_0 \to A_0B_0$. From (6.46) and (6.51), we obtain

$$\text{tr}_{A_0}(\rho_{A_0})^n = \int d^3\mathbf{p}\, \delta^{(3)}(0) \left(\frac{\delta(p-k)\delta(0)}{\mathcal{N}_{\text{el}} \cdot 4p\sqrt{s}} \left| {}_0\langle\!\langle \mathbf{p}|\hat{S}|\mathbf{k}\rangle\!\rangle_0 \right|^2 \right)^n. \tag{6.57}$$

We denote the partial wave modes for $\hat{S}$ by $s^{\ell}_{ij} = \delta_{ij} + 2i\tau_{ij}$. Equation (6.57) is rewritten in terms of the partial wave modes as

$$\mathrm{tr}_{\mathrm{A}_0}(\rho_{\mathrm{A}_0})^n = \int_{-1}^{1} d\cos\theta \left(\frac{\delta(0)}{2\pi k^2 \delta^{(3)}(0)}\right)^{n-1} (\mathcal{P}_{\mathrm{el}}(\cos\theta))^n, \tag{6.58}$$

$$\mathcal{P}_{\mathrm{el}}(\cos\theta) \equiv \frac{\left|\sum_{\ell=0}^{\infty}(2\ell+1)s^{\ell}_{00}P_{\ell}(\cos\theta)\right|^2}{2\sum_{\ell=0}^{\infty}(2\ell+1)\left|s^{\ell}_{00}\right|^2}. \tag{6.59}$$

Since the scattering process has azimuthal symmetry with respect to the spherical coordinates of $\mathbf{p}$, the three-dimensional integration $\int d^3\mathbf{p}$ is reduced to $2\pi\int_{-1}^{1} d\cos\theta \int_0^{\infty} dp\, p^2$. When there is such azimuthal symmetry, the three-dimensional Dirac delta function is expanded in terms of Legendre polynomials as

$$\delta^{(3)}(\mathbf{p}-\mathbf{k}) = \frac{\delta(p-k)}{4\pi k^2}\sum_{\ell=0}^{\infty}(2\ell+1)P_{\ell}(\cos\theta), \quad \cos\theta = \frac{\mathbf{p}\cdot\mathbf{k}}{pk}. \tag{6.60}$$

By taking the limit, $\mathbf{p} \to \mathbf{k}$, we obtain

$$\frac{4\pi k^2 \delta^{(3)}(0)}{\delta(0)} = \sum_{\ell=0}^{\infty}(2\ell+1)P_{\ell}(1) = \sum_{\ell=0}^{\infty}(2\ell+1). \tag{6.61}$$

This corresponds to the infinite volume of momentum Hilbert space, V, i.e.,

$$V \equiv \sum_{\ell=0}^{\infty}(2\ell+1). \tag{6.62}$$

V will be regularized in Sect. 6.3.3. By the use of V, we describe (6.58) as

$$\mathrm{tr}_{\mathrm{A}_0}(\rho_{\mathrm{A}_0})^n = \left(\frac{V}{2}\right)^{1-n}\int_{-1}^{1} d\cos\theta\, (\mathcal{P}_{\mathrm{el}}(\cos\theta))^n. \tag{6.63}$$

Consequently we obtain the entanglement entropy for the elastic channel via (6.54),

$$S_{\mathrm{el}}\left(\equiv -\lim_{n\to 1}\frac{\partial}{\partial n}\mathrm{tr}_{\mathrm{A}_0}(\rho_{\mathrm{A}_0})^n\right) = \ln\frac{V}{2} - \int_{-1}^{1} d\cos\theta\, \mathcal{P}_{\mathrm{el}}(\cos\theta)\ln\mathcal{P}_{\mathrm{el}}(\cos\theta). \tag{6.64}$$

Since (6.59) is expressed by the partial wave modes, we can interpret them to the cross sections (6.33), (6.34) and (6.38), and the infinite volume (6.62), so that we obtain

$$\mathcal{P}_{\rm el}(\cos\theta) = \delta(1-\cos\theta)\left(1 - \frac{\sigma_{\rm el}}{\frac{\pi}{k^2}V - (\sigma_{\rm tot} - \sigma_{\rm el})}\right) + \frac{2k^2}{\frac{\pi}{k^2}V - (\sigma_{\rm tot} - \sigma_{\rm el})}\frac{d\sigma_{\rm el}}{dt}, \tag{6.65}$$

where $d\sigma_{\rm el}/dt$ and $\sigma_{\rm el}$ are respectively equivalent to $d\sigma_0/dt$ in (6.33) and σ_0 in (6.34). $\mathcal{P}_{\rm el}(\cos\theta)$ satisfies

$$\int_{-1}^{1} d\cos\theta\, \mathcal{P}_{\rm el}(\cos\theta) = 1. \tag{6.66}$$

(6.64) and (6.66) are reminiscent of Shannon entropy and probability density.

Two-Particle Inelastic Channel

In the previous subsection, we have obtained the entanglement entropy of the final state in the elastic channel. In the same way, we shall formulate the entanglement entropy of the final state in the two-particle inelastic channel, $A_0B_0 \to A_1B_1$. From (6.48) and (6.52) we calculate $\mathrm{tr}_{A_1}(\rho_{A_1})^n$ by the use of the partial wave modes,

$$\begin{aligned}\mathrm{tr}_{A_1}(\rho_{A_1})^n &= \int d^3\mathbf{p}\,\delta^{(3)}(0)\left(\frac{\delta(p-k')\delta(0)}{\mathcal{N}_{\rm inel}\cdot p\sqrt{s}}\left|{}_1\langle\!\langle \mathbf{p}|\hat{T}|\mathbf{k}\rangle\!\rangle_0\right|^2\right)^n \\ &= \left(\frac{V}{2}\right)^{1-n}\int_{-1}^{1} d\cos\theta\,(\mathcal{P}_{\rm inel}(\cos\theta))^n,\end{aligned} \tag{6.67}$$

with

$$\mathcal{P}_{\rm inel}(\cos\theta) \equiv \frac{\left|\sum_{\ell=0}^{\infty}(2\ell+1)\tau_{10}^{\ell}P_\ell(\cos\theta)\right|^2}{2\sum_{\ell=0}^{\infty}(2\ell+1)\left|\tau_{10}^{\ell}\right|^2}. \tag{6.68}$$

There is τ_{10}^{ℓ} in (6.68) instead of s_{00}^{ℓ} in (6.59), because the contribution of 1 in $1 + 2i\tau_{10}^{\ell}(= s_{10}^{\ell})$ does not exist in the inelastic channel. Then we obtain the entanglement entropy of the final state in the inelastic channel,

$$S_{\rm inel}\left(\equiv -\lim_{n\to 1}\frac{\partial}{\partial n}\mathrm{tr}_{A_1}(\rho_{A_1})^n\right) = \ln\frac{V}{2} - \int_{-1}^{1} d\cos\theta\,\mathcal{P}_{\rm inel}(\cos\theta)\ln\mathcal{P}_{\rm inel}(\cos\theta). \tag{6.69}$$

(6.68) is rewritten in terms of the cross sections as

$$\mathcal{P}_{\rm inel}(\cos\theta) = \frac{2k^2}{\sigma_{\rm inel}}\frac{d\sigma_{\rm inel}}{dt}, \tag{6.70}$$

where $d\sigma_{\rm inel}/dt$ and $\sigma_{\rm inel}$ are respectively equivalent to $d\sigma_1/dt$ in (6.33) and σ_1 in (6.34). $\mathcal{P}_{\rm inel}(\cos\theta)$ also satisfies

$$\int_{-1}^{1} d\cos\theta\,\mathcal{P}_{\rm inel}(\cos\theta) = 1. \tag{6.71}$$

6.3.3 Volume Regularization

Since there appears the infinite volume of momentum Hilbert space, V, in the elastic and inelastic entanglement entropy, (6.64) and (6.69), it is necessary to regularize it.

Let us see the elastic channel. There is the term proportional to $\delta(1-\cos\theta)$ in the S-matrix element (6.30). It implies the non-interaction term, which originates from $\mathbf{1}$ in $S=\mathbf{1}+2iT$. The first term in (6.65) is also proportional to $\delta(1-\cos\theta)$, that is to say, it is the contribution of non-interacting scattering. Indeed, $\delta(1-\cos\theta)$ has support only at $\cos\theta=1$, i.e., the scattering angle θ is equal to zero. Therefore, in order to remove the contribution of the non-interacting term to the entanglement entropy, we regularize V so that the first term in (6.65) vanishes, namely,

$$V=\frac{k^2}{\pi}\sigma_{\text{tot}}. \tag{6.72}$$

Currently we do not know the concrete way to realize such regularization. Hence we call it an "ideal" volume regularization. Substituting (6.72) into (6.64) and (6.65), we obtain the regularized entanglement entropy for the elastic channel,

$$\begin{aligned}\tilde{S}_{\text{el}} &= \ln\frac{k^2\sigma_{\text{tot}}}{2\pi}-\int_{-1}^{1}d\cos\theta\,\mathcal{P}_{\text{el}}(\cos\theta)\ln\mathcal{P}_{\text{el}}(\cos\theta),\\ \mathcal{P}_{\text{el}}(\cos\theta) &= \frac{1}{\sigma_{\text{el}}}\frac{d\sigma_{\text{el}}}{d\cos\theta}.\end{aligned} \tag{6.73}$$

The entanglement entropy for the elastic channel (6.64) and the one for the inelastic channel (6.69) have a common term, $\ln(V/2)$. Therefore we use the same volume regularization (6.72) for both of the elastic and inelastic entanglement entropy. Then the entanglement entropy for the inelastic channel becomes

$$\begin{aligned}\tilde{S}_{\text{inel}} &= \ln\frac{k^2\sigma_{\text{tot}}}{2\pi}-\int_{-1}^{1}d\cos\theta\,\mathcal{P}_{\text{inel}}(\cos\theta)\ln\mathcal{P}_{\text{inel}}(\cos\theta),\\ \mathcal{P}_{\text{inel}}(\cos\theta) &= \frac{1}{\sigma_{\text{inel}}}\frac{d\sigma_{\text{inel}}}{d\cos\theta}.\end{aligned} \tag{6.74}$$

We note that (6.70) does not depend on V from the beginning, and $\mathcal{P}_{\text{el}}(\cos\theta)$ after the regularization in (6.73) has the same form as $\mathcal{P}_{\text{inel}}(\cos\theta)$ in (6.74).

6.4 Evaluation of Quantum Entanglement Entropy

In this section, we concretely evaluate the regularized entanglement entropy for the hadron scatterings by the use of experimental data [4, 5]. In calculating the entanglement entropy, one of the Mandelstam variables, t, is more useful than $\cos\theta$.

Therefore, we rewrite (6.73) and (6.74) as

$$\tilde{S}_{\rm el} = -\int_{-4k^2}^{0} dt\, \mathcal{P}_{\rm el}(t) \ln\left(\frac{4\pi}{\sigma_{\rm tot}}\mathcal{P}_{\rm el}(t)\right), \quad \mathcal{P}_{\rm el}(t) = \frac{1}{\sigma_{\rm el}}\frac{d\sigma_{\rm el}}{dt}, \tag{6.75}$$

$$\tilde{S}_{\rm inel} = -\int_{-4k^2}^{0} dt\, \mathcal{P}_{\rm inel}(t) \ln\left(\frac{4\pi}{\sigma_{\rm tot}}\mathcal{P}_{\rm inel}(t)\right), \quad \mathcal{P}_{\rm inel}(t) = \frac{1}{\sigma_{\rm inel}}\frac{d\sigma_{\rm inel}}{dt}, \tag{6.76}$$

where

$$P_{\rm el}(\cos\theta) = 2k^2\mathcal{P}_{\rm el}(t), \quad P_{\rm inel}(\cos\theta) = 2k^2\mathcal{P}_{\rm inel}(t) \tag{6.77}$$

due to (6.23). Note that $\mathcal{P}_{\rm el}(t)$ and $\mathcal{P}_{\rm inel}(t)$ satisfy

$$\int_{-4k^2}^{0} dt\, \mathcal{P}_{\rm el}(t) = 1, \quad \int_{-4k^2}^{0} dt\, \mathcal{P}_{\rm inel}(t) = 1. \tag{6.78}$$

Hence we can regard $\mathcal{P}_{\rm el}(t)$ and $\mathcal{P}_{\rm inel}(t)$ as some kind of "probability density".

6.4.1 Proton-Proton Elastic Scattering

Let us evaluate the entanglement entropy for the final state of elastic proton-proton scattering. We can find the experimental data of cross sections given by Tevatron [6, 7] and CERN Large Hadron Collider (LHC) [8–11] (see Table 6.1).

In order to compute the entanglement entropy (6.73), we need the differential cross section as a function of t. Here we use the diffraction peak model, in which the scattering amplitude is described as

$$\mathcal{A}_0(s,t) = is\sigma_{\rm tot}e^{\frac{1}{2}Bt}. \tag{6.79}$$

B is a slope parameter. Since the energy level in Tevatron and LHC is sufficiently high, we can assume $s \approx 4k^2$ from (6.22) (see the footnote 2). Then the differential cross section becomes

$$\frac{d\sigma_{\rm el}}{dt} = \frac{\sigma_{\rm tot}^2}{16\pi}e^{Bt}. \tag{6.80}$$

Table 6.1 Experimental data of cross sections by Tevatron and LHC, and the elastic entanglement entropy for each center-of-mass energy

$\sqrt{s}$ [GeV]	$\sigma_{\rm tot}$ [mb]	$\sigma_{\rm el}$ [mb]	$\tilde{S}_{\rm el}$
1800	72.1	16.6	0.9176
7000	98.58	25.43	1.031
8000	101.7	27.1	1.063
13000	110.6	31.0	1.114

We calculate the elastic cross section,

$$\sigma_{el} = \int_{-\infty}^{0} dt \, \frac{d\sigma_{el}}{dt} = \frac{\sigma_{tot}^2}{16\pi B}, \tag{6.81}$$

because we can assume $-4k^2 \approx \infty$ at the high energy. From (6.81), the slope parameter is described in terms of the cross sections as

$$B = \frac{\sigma_{tot}^2}{16\pi\sigma_{el}}. \tag{6.82}$$

The differential cross section (6.80) with (6.82) thus becomes

$$\frac{d\sigma_{el}}{dt} = \frac{\sigma_{tot}^2}{16\pi} \exp\left(\frac{\sigma_{tot}^2}{16\pi\sigma_{el}} t\right). \tag{6.83}$$

By using (6.80) and the data of the cross sections, we evaluate the regularized entanglement entropy for the elastic channel (6.75). The results of $\tilde{S}_{el}$ are shown in Table 6.1. The elastic scattering process yields more entanglement at higher center-of-mass energy.

Cut-Off Regularization

For the proton-proton elastic scattering, let us try to consider another concrete (but not ideal) regularization scheme, that is, the cut-off regularization based on the impact-parameter representation (see Fig. 6.2) [4].

The scattering amplitude (6.31) is rewritten in the impact-parameter representation,

$$\mathcal{A}_0 = 32\pi k^2 \int_0^{\infty} b db \, \tau(b) J_0(b\sqrt{-t}), \tag{6.84}$$

where b is an impact parameter and J_0 is the Bessel function of the first kind (see Appendix). $\tau(b)$ is defined by this equation, namely,

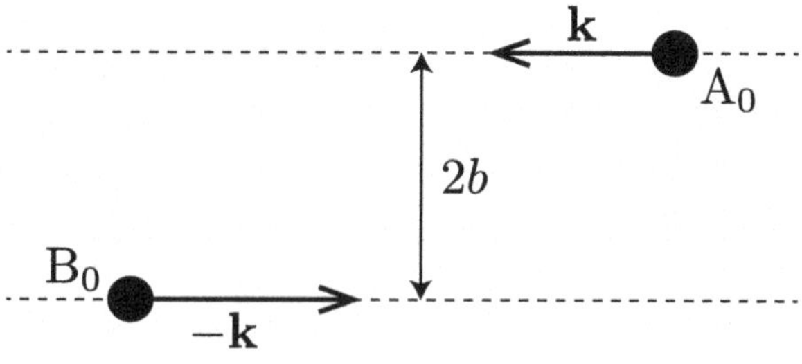

Fig. 6.2 The incoming particles A_0 and B_0 with the impact parameter b

$$\tau(b) = \frac{1}{32\pi k^2}\int_0^\infty \sqrt{-t}d\sqrt{-t}\,\mathcal{A}_0(s,t)J_0(b\sqrt{-t}) = i\frac{\sigma_{\text{tot}}}{8\pi B}e^{-\frac{b^2}{2B}}, \tag{6.85}$$

where we used (6.79). At very large impact parameter, the two particles are far from each other and almost non-interactive. Therefore we suppress the integration range of large b in (6.84).

By introducing a cut-off function $c(b)$, the scattering amplitude is modified as

$$\hat{\mathcal{A}} = 32\pi k^2 \int_0^\infty bdb\, c(b)\tau(b)J_0(b\sqrt{-t}). \tag{6.86}$$

This leads to the changes of the cross sections;

$$\frac{d\hat{\sigma}_{\text{el}}}{dt} = 4\pi\left|\int_0^\infty bdb\, c(b)\tau(b)J_0(b\sqrt{-t})\right|^2, \tag{6.87}$$

$$\hat{\sigma}_{\text{tot}} = 8\pi\int_0^\infty bdb\, c^2(b)\text{Im}\tau(b), \tag{6.88}$$

$$\hat{\sigma}_{\text{el}} = \int_{-\infty}^0 dt\,\frac{d\hat{\sigma}_{\text{el}}}{dt} = 8\pi\int_0^\infty bdb\, c^2(b)|\tau(b)|^2. \tag{6.89}$$

In the following, we shall study two kinds of concrete cut-off functions.

Step-function cut-off

First we consider the step-function cut-off given by

$$c(b) = \begin{cases} 1 & (b \le 2\Lambda) \\ 0 & (b > 2\Lambda), \end{cases} \tag{6.90}$$

where Λ is a cut-off parameter. By using (6.85), the cross sections become

$$\frac{d\hat{\sigma}_{\text{el}}}{dt} = 4\pi\left|\int_0^{2\Lambda} bdb\,\tau(b)J_0(b\sqrt{-t})\right|^2 = \frac{\sigma_{\text{tot}}^2}{16\pi B^2}\left(\int_0^{2\Lambda} bdb\, e^{-\frac{b^2}{2B}}J_0(b\sqrt{-t})\right)^2, \tag{6.91}$$

$$\hat{\sigma}_{\text{tot}} = 8\pi\int_0^{2\Lambda} bdb\,\text{Im}\tau(b) = \sigma_{\text{tot}}\left(1 - e^{-\frac{2}{B}\Lambda^2}\right), \tag{6.92}$$

$$\hat{\sigma}_{\text{el}} = 8\pi\int_0^{2\Lambda} bdb\,|\tau(b)|^2 = \frac{\sigma_{\text{tot}}^2}{16\pi B}\left(1 - e^{-\frac{4}{B}\Lambda^2}\right). \tag{6.93}$$

Since the impact parameter b and the angular momentum ℓ are related by $b = \ell/k$, the cut-off of ℓ is given by $\ell \le 2\Lambda k$. The infinite volume (6.62) is then regularized as

$$V = \sum_{\ell=0}^{2\Lambda k}(2\ell+1) = 2k^2 \int_0^{2\Lambda} bdb = 4k^2\Lambda^2. \tag{6.94}$$

We determine the cut-off parameter Λ by (6.72) with (6.92), i.e.,

$$\frac{4\pi\Lambda^2}{\sigma_{\text{tot}}} = 1 - e^{-\frac{2}{B}\Lambda^2}. \tag{6.95}$$

We then numerically calculate the regularized entanglement entropy (6.75) in terms of (6.92), (6.93) and Λ fixed by (6.95).

Gaussian cut-off

Next we consider the Gaussian cut-off function:

$$c(b) = \exp\left(-\frac{1}{2}\frac{b^2}{4\Lambda^2}\right), \tag{6.96}$$

which modifies the cross sections as

$$\frac{d\hat{\sigma}_{\text{el}}}{dt} = \frac{\sigma_{\text{tot}}^2}{16\pi}\left(1+\frac{B}{4\Lambda^2}\right)^{-2} \exp\left(\frac{B}{1+\frac{B}{4\Lambda^2}}t\right), \tag{6.97}$$

$$\hat{\sigma}_{\text{tot}} = \sigma_{\text{tot}}\left(1+\frac{B}{2\Lambda^2}\right)^{-1}, \tag{6.98}$$

$$\hat{\sigma}_{\text{el}} = \frac{\sigma_{\text{tot}}^2}{16\pi B}\left(1+\frac{B}{4\Lambda^2}\right)^{-1}. \tag{6.99}$$

The infinite volume (6.62) becomes

$$V = 2k^2\int_0^\infty bdb\, c^2(b) = 4k^2\Lambda^2. \tag{6.100}$$

Since this V has the same expression as (6.94), we numerically calculate Λ from (6.72), (6.98) and (6.100), i.e.,

$$\frac{4\pi\Lambda^2}{\sigma_{\text{tot}}} = \left(1+\frac{B}{2\Lambda^2}\right)^{-1}. \tag{6.101}$$

Thus the regularized entanglement entropy (6.75) is described as

$$\hat{S}_{\text{el}} = 1 - \ln\frac{4\pi B\left(1+\frac{B}{2\Lambda^2}\right)}{\sigma_{\text{tot}}\left(1+\frac{B}{4\Lambda^2}\right)}. \tag{6.102}$$

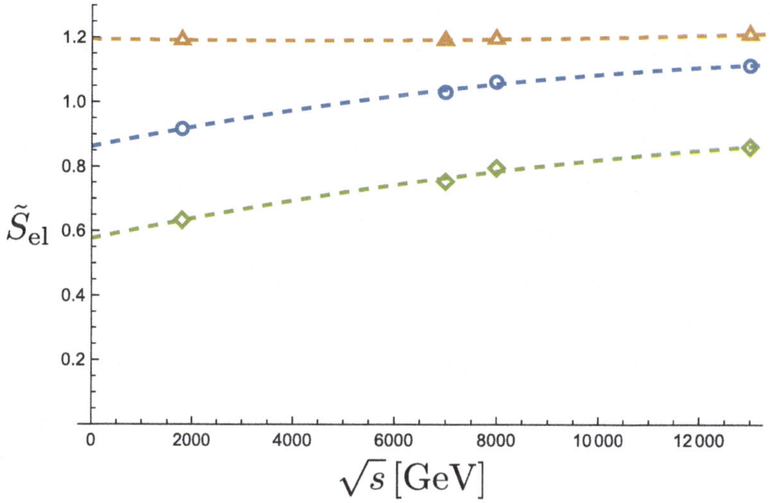

Fig. 6.3 The regularized entanglement entropy $\tilde{S}_{\text{el}}$ on the center-of-mass energy $\sqrt{s}$ [GeV] in the ideal volume regularization (circle), the step-function regularization (triangle) and the Gaussian regularization (square)

Finally let us compare the entanglement entropy numerically computed with the experimental data (Table 6.1) in the ideal volume regularization, the step-function cut-off regularization and the Gaussian cut-off regularization. The results are depicted in Fig. 6.3. Comparing with the ideal volume regularization, the cut-off regularization does not sufficiently avoid the contribution by the non-interacting part of scattering process. On the other hand, the Gaussian cut-off regularization over-erases such contribution. However, all the three regularizations show the common behavior that the entanglement entropy increases as the center-of-mass energy in the higher energy region.

6.4.2 *Proton-Neutron Scattering*

In this subsection, we focus on the following two channels:

$$pn \to \begin{cases} pn & \text{an elastic channel} \\ np & \text{an inelastic channel.} \end{cases} \tag{6.103}$$

in the proton-neutron scattering (Fig. 6.4). Of course there exist any other two-particle inelastic channels and multi-particle inelastic channels. Note that we distinguish pn

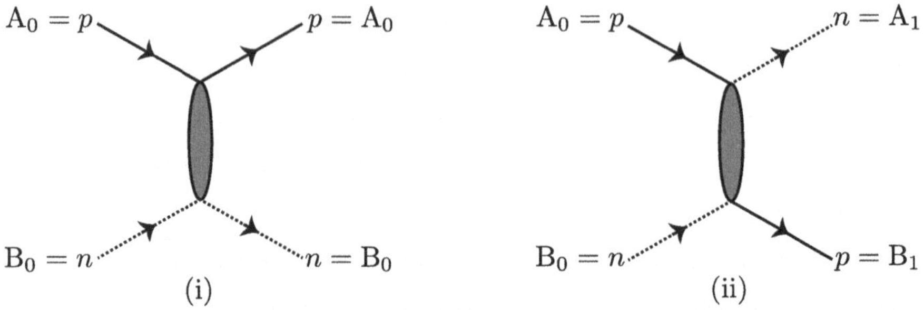

Fig. 6.4 (i) The elastic channel, $pn \to pn$. (ii) The inelastic channel, $pn \to np$

and np. Following our definition of the two-particle state (6.8), $|\mathbf{p}\rangle_{\text{proton}} \otimes |\mathbf{q}\rangle_{\text{netron}}$ is different from $|\mathbf{q}\rangle_{\text{netron}} \otimes |\mathbf{p}\rangle_{\text{proton}}$. Actually, the inner products (6.12) of these states are

$$\left({}_{\text{proton}}\langle \mathbf{p}| \otimes {}_{\text{neutron}}\langle \mathbf{q}|\right)\left(|\mathbf{q}'\rangle_{\text{neutron}} \otimes |\mathbf{p}'\rangle_{\text{proton}}\right) = {}_{\text{proton}}\langle \mathbf{p}|\mathbf{q}\rangle_{\text{neutron}}\, {}_{\text{neutron}}\langle \mathbf{q}|\mathbf{p}\rangle_{\text{proton}} = 0, \tag{6.104}$$

although

$$\left({}_{\text{proton}}\langle \mathbf{p}| \otimes {}_{\text{neutron}}\langle \mathbf{q}|\right)\left(|\mathbf{p}'\rangle_{\text{proton}} \otimes |\mathbf{q}'\rangle_{\text{neutron}}\right) = {}_{\text{proton}}\langle \mathbf{p}|\mathbf{p}'\rangle_{\text{proton}}\, {}_{\text{neutron}}\langle \mathbf{q}|\mathbf{q}'\rangle_{\text{neutron}}$$
$$= 2E_{\text{protonp}}\, 2E_{\text{neutronq}}\, \delta^{(3)}(\mathbf{p} - \mathbf{p}')\delta^{(3)}(\mathbf{q} - \mathbf{q}'). \tag{6.105}$$

In other words, for the incoming particles, $(A_0, B_0) = (p, n)$, the two pairs of the outgoing particles, $(A_0, B_0) = (p, n)$ and $(A_1, B_1) = (n, p)$, are different. The former is in the elastic channel and the latter is in the inelastic one.

In certain manner, ones say both of the two channels (6.103) are elastic phenomenologically. However, we here focus on the forward scattering with t-channel exchange (see also (6.75) and (6.76)), and we can also study those channels separately in experimental research. The elastic channel is neutral current exchange, while the inelastic channel is charged current exchange (Fig. 6.4).

In the following, we shall evaluate the entanglement entropy under the ideal volume regularization in terms of the experimental data by CERN Intersecting Storage Ring (ISR) [12–19] (see also Appendix).

Quantum Entanglement Entropy for the Elastic Channel: $pn \to pn$

Although the integration range of t in (6.75) and (6.76) is $-4k^2 \leq t \leq 0$, we assume $-4k^2 \to -\infty$ for convenience in computing. Indeed this assumption is good enough approximation, because we are here interested in only high energy scatterings.

Let us evaluate the entanglement entropy (6.75). We can find the data of total cross section σ_{tot} in [12, 13]. These references have analyzed the data for proton-proton

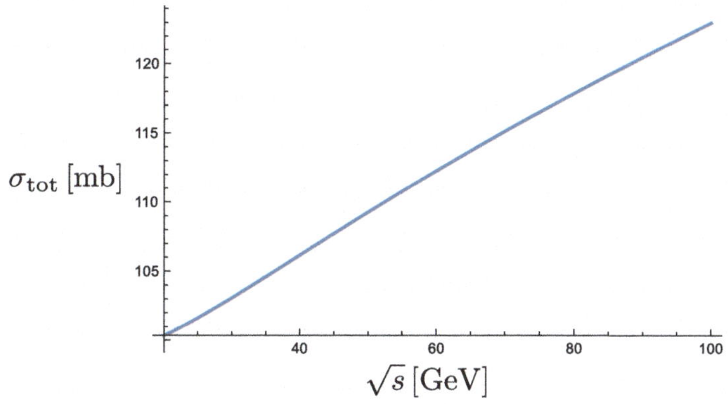

Fig. 6.5 The total cross section σ_{tot} depending on the center-of-mass energy $\sqrt{s}$

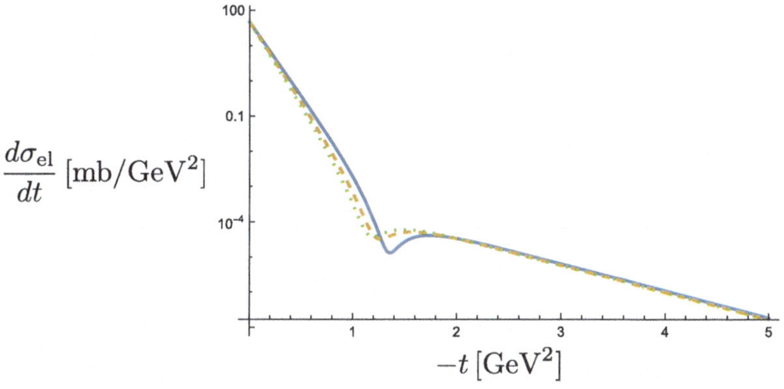

Fig. 6.6 The elastic differential cross section $\frac{d\sigma_{el}}{dt}$ [mb/GeV^2] depending on $-t$ [GeV^2] at $\sqrt{s}$ =30 (solid line), 60 (dashed line) and 90 (dotted line) [GeV]

scattering. However, since the mass difference between a proton and a neutron is negligible enough at high energy, we can borrow their data for the analysis of proton-neutron scattering. The plots of the total cross section is shown in Fig. 6.5. This data will be used also in the analysis of entanglement in the inelastic channel.

The elastic differential cross section $\frac{d\sigma_{el}}{dt}$ at, for example, $\sqrt{s} = 30, 60, 90$ GeV is plotted in Fig. 6.6. Note that the horizontal axis is $-t$, because one of the Mandelstam variables, t, is negative. The elastic differential cross section exponentially decreases as $-t$, and does not indicate a remarkable $\sqrt{s}$-dependence. We can see that the fitting function for the differential cross section given by [12, 13] has two slopes.

We numerically calculate the elastic cross section (6.34), i.e.,

$$\sigma_{el} = \int_{-\infty}^{0} \frac{d\sigma_{el}}{dt}, \tag{6.106}$$

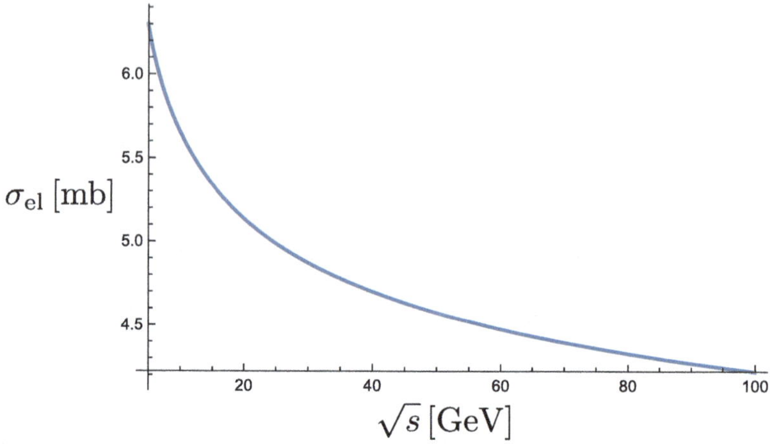

Fig. 6.7 The elastic cross section $\sigma_{\rm el}$ [mb] depending on $\sqrt{s}$ [GeV]

and plot it as the function of the center-of-mass energy in Fig. 6.7. The elastic cross section decreases as the center-of-mass energy $\sqrt{s}$.

Finally we shall numerically calculate the entanglement entropy (6.75). Before the calculation, we pay attention to the integration kernel in (6.75):

$$\mathcal{D}_{\rm el} \equiv -\mathcal{P}_{\rm el}(t) \ln \left(\frac{4\pi}{\sigma_{\rm tot}} \mathcal{P}_{\rm el}(t) \right). \tag{6.107}$$

One can recognize this with a kind of entanglement entropy distribution. $\mathcal{D}_{\rm el}$ at, for example, $\sqrt{s} = 30, 60, 90$ GeV are depicted in Fig. 6.8. All the three examples show

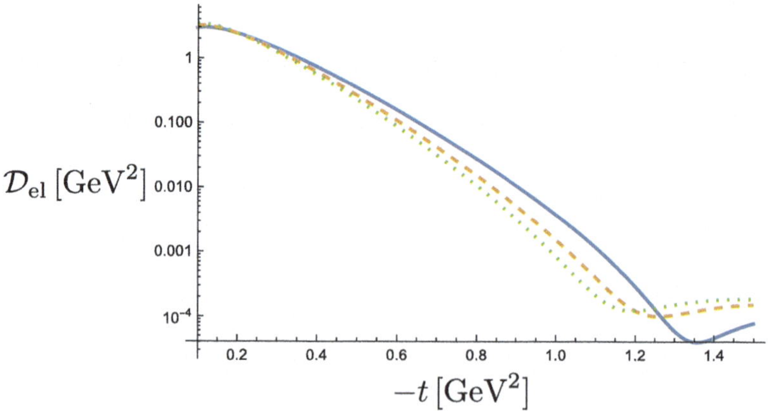

Fig. 6.8 The elastic entanglement entropy distribution $\mathcal{D}_{\rm el}$ [GeV^2] depending on $-t$ [GeV^2] at $\sqrt{s}$ =30 (solid line), 60 (dashed line) and 90 (dotted line) [GeV]

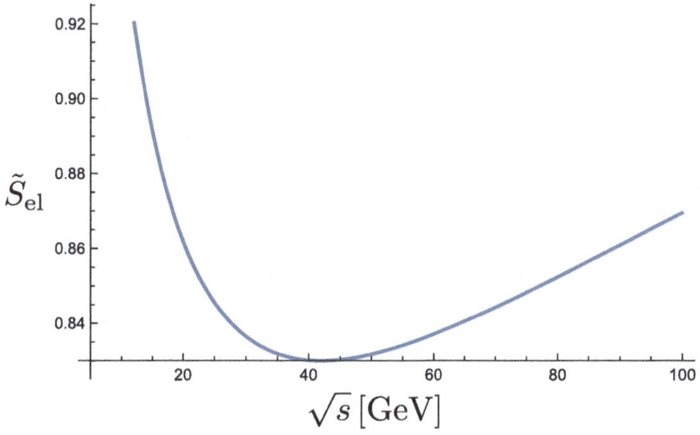

Fig. 6.9 The elastic entanglement entropy $\tilde{S}_{\rm el}$ depending on the center-of-mass energy $\sqrt{s}$ [GeV]

that the elastic entanglement entropy density $\mathcal{D}_{\rm el}$ decreases as $-t$ increases. We have magnified the plot of $\mathcal{D}_{\rm el}$ in the small $-t$ region, because, as we have seen, this region is exponentially dominant for $\tilde{S}_{\rm el}$. We compute the entanglement entropy,

$$\tilde{S}_{\rm el} = \int_{-\infty}^{0} dt\, \mathcal{D}_{\rm el}, \tag{6.108}$$

and it is depicted in Fig. 6.9. $\tilde{S}_{\rm el}$ has a minimum value, 0.830042, at $\sqrt{s} = 42.1609$ GeV. In the low energy region, $\sqrt{s} < 42.1609$ GeV, as $\sqrt{s}$ increases, the entanglement entropy decreases, in other words, the final state becomes less entangled as $\sqrt{s}$. On the other hand, in the high energy region, the entanglement entropy increases as $\sqrt{s}$, that is to say, the higher energy, more strongly the particles interact.

Quantum Entanglement Entropy for the Inelastic Channel: $pn \rightarrow np$

Since we can find the fitting function of the inelastic differential cross section in [18, 19], we plot it at, for example, $\sqrt{s} = 30, 60, 90$ GeV in Fig. 6.10. Since we can see in Fig. 6.10 that the inelastic differential cross section in the large $-t$ region is exponentially small as same as the elastic one is, we take a limit of the integration range, $-4k^2 \leq t \leq 0$, to $-\infty < t \leq 0$ for convenience. This is sufficiently good approximation. Then we numerically compute the inelastic cross section,

$$\sigma_{\rm inel} = \int_{-\infty}^{0} \frac{d\sigma_{\rm inel}}{dt}. \tag{6.109}$$

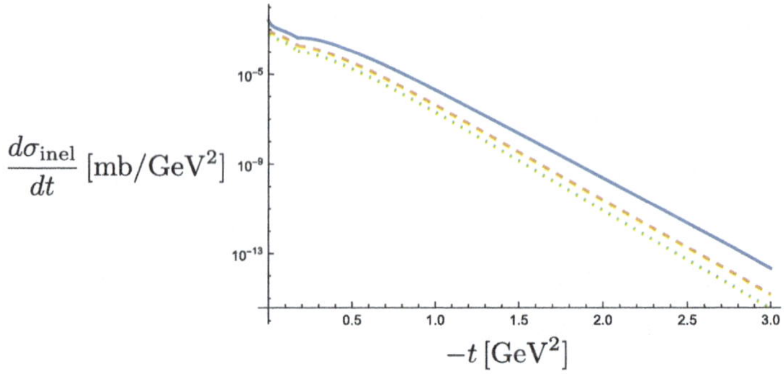

Fig. 6.10 The inelastic differential cross section $\frac{d\sigma_{\text{inel}}}{dt}$ [mb/GeV^2] depending on $-t$ [GeV^2] at $\sqrt{s}$ =30 (solid line), 60 (dashed line) and 90 (dotted line) [GeV]

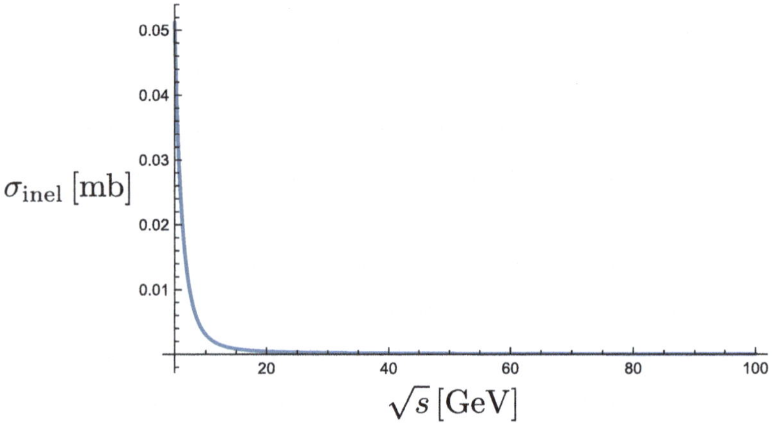

Fig. 6.11 The inelastic cross section σ_{inel} [mb] depending on the center-of-mass energy $\sqrt{s}$ [GeV]

This is plotted in Fig. 6.11. From (6.76), the entanglement entropy distribution in the inelastic channel, $\mathcal{D}_{\text{inel}}$, is defined as

$$\mathcal{D}_{\text{inel}} \equiv -\mathcal{P}_{\text{inel}}(t) \ln \left(\frac{4\pi}{\sigma_{\text{tot}}} \mathcal{P}_{\text{inel}}(t) \right). \tag{6.110}$$

We draw $\mathcal{D}_{\text{inel}}$ at, for example, $\sqrt{s} = 30$, 60, 90 GeV in Fig. 6.12. In all the cases of $\sqrt{s}$, the inelastic entanglement entropy density decreases as $-t$ in similar to the elastic entanglement entropy density. We then compute the inelastic entanglement entropy,

$$\tilde{S}_{\text{inel}} = \int_{-\infty}^{0} dt\, \mathcal{D}_{\text{inel}}. \tag{6.111}$$

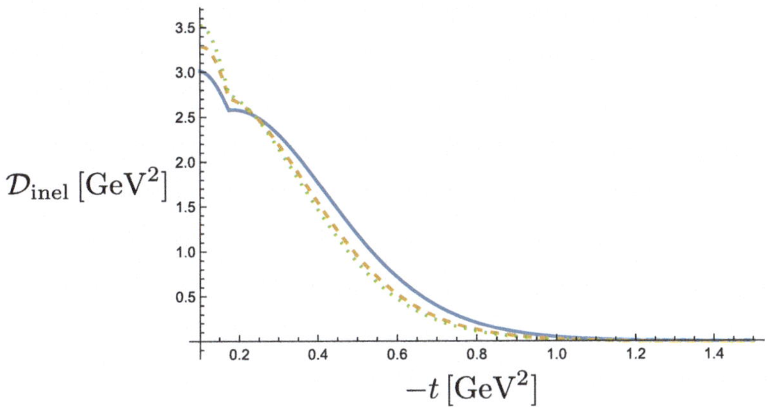

Fig. 6.12 The elastic entanglement entropy distribution $\mathcal{D}_{\rm inel}$ [GeV^2] depending on $-t$ [GeV^2] at $\sqrt{s}$ =30 (solid line), 60 (dashed line) and 90 (dotted line) [GeV]

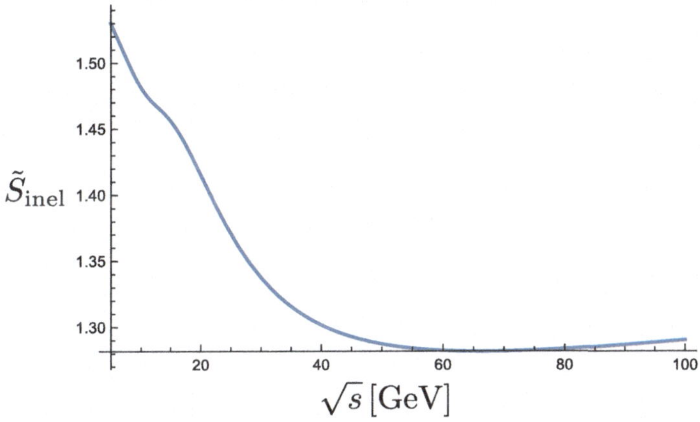

Fig. 6.13 The inelastic entanglement entropy $\tilde{S}_{\rm inel}$ depending on the center-of-mass energy $\sqrt{s}$ [GeV]

This is depicted in Fig. 6.13. $\tilde{S}_{\rm inel}$ has a minimum value, 1.28188, at $\sqrt{s} = 66.3979$ GeV. The inelastic entanglement entropy decreases as $\sqrt{s}$ in the region of $\sqrt{s} < 66.3979$ GeV, while it increases in the region of $\sqrt{s} > 66.3979$ GeV. Thus one can see the inelastic entanglement entropy has similar behavior to the elastic one.

Elastic Versus Inelastic

Let us compare the entanglement entropy in the elastic channel with the one in the inelastic channel. The two plots of entanglement entropy are drawn together in Fig. 6.14. The inelastic entanglement entropy is always larger than the elastic one.

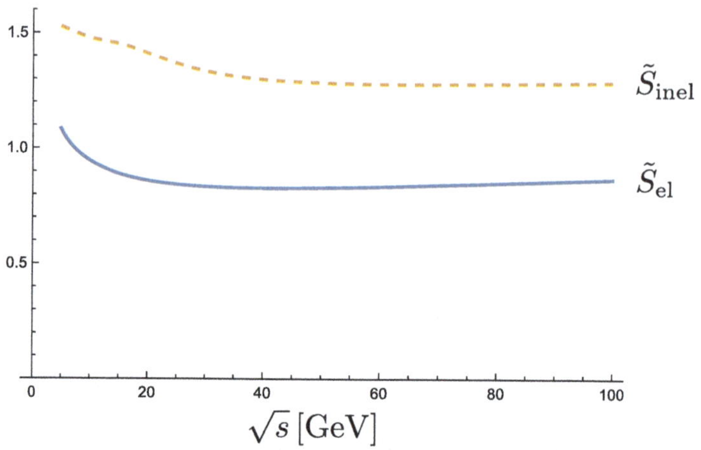

Fig. 6.14 The entanglement entropy in the elastic channel $\tilde{S}_{\rm el}$ (solid line) and the one in the inelastic channel $\tilde{S}_{\rm inel}$ (dashed line) depending on the center-of-mass energy $\sqrt{s}$ [GeV]

Let us recall the diffraction peak model, in which the t-dependence factor in the differential cross section (6.80) is proportional to e^{Bt}. Therefore we assume the elastic and inelastic differential cross sections as

$$\frac{d\sigma_{\rm el}}{dt} = C_{\rm el} e^{B_{\rm el} t}, \quad \frac{d\sigma_{\rm inel}}{dt} = C_{\rm inel} e^{B_{\rm inel} t}, \tag{6.112}$$

where $B_{\rm el}$ and $B_{\rm inel}$ are slopes. $C_{\rm el}$ and $C_{\rm inel}$ depend on s but not on t. Then the cross sections become

$$\sigma_{\rm el} = \int_{-\infty}^{0} dt \, \frac{d\sigma_{\rm el}}{dt} = \frac{C_{\rm el}}{B_{\rm el}}, \quad \sigma_{\rm inel} = \int_{-\infty}^{0} dt \, \frac{d\sigma_{\rm inel}}{dt} = \frac{C_{\rm inel}}{B_{\rm inel}}. \tag{6.113}$$

Therefore the "probability density" (6.75) and (6.76) become

$$\mathcal{P}_{\rm el}(t) = B_{\rm el} e^{B_{\rm el} t}, \quad \mathcal{P}_{\rm inel}(t) = B_{\rm inel} e^{B_{\rm inel} t}, \tag{6.114}$$

$C_{\rm el}$ and $C_{\rm inel}$ dependence disappear. The entanglement entropy distribution (6.107) and (6.110) become

$$\mathcal{D}_{\rm el} = -B_{\rm el} e^{B_{\rm el} t} \left(\ln \frac{4\pi B_{\rm el}}{\sigma_{\rm tot}} + B_{\rm el} t \right), \tag{6.115}$$

$$\mathcal{D}_{\rm inel} = -B_{\rm inel} e^{B_{\rm inel} t} \left(\ln \frac{4\pi B_{\rm inel}}{\sigma_{\rm tot}} + B_{\rm inel} t \right). \tag{6.116}$$

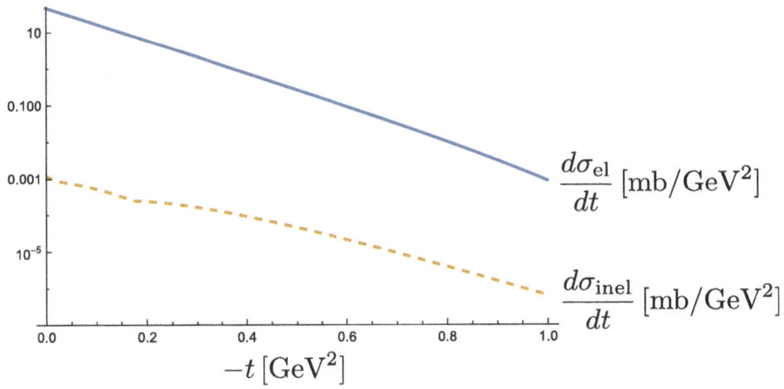

Fig. 6.15 The differential cross sections at small $-t$ in the elastic and inelastic channels

Consequently the entanglement entropy (6.108) and (6.111) are described as

$$\tilde{S}_{\mathrm{el}} = 1 - \ln \frac{4\pi}{\sigma_{\mathrm{tot}}} - \ln B_{\mathrm{el}}, \quad \tilde{S}_{\mathrm{inel}} = 1 - \ln \frac{4\pi}{\sigma_{\mathrm{tot}}} - \ln B_{\mathrm{inel}}. \tag{6.117}$$

Therefore the discrepancy between $\tilde{S}_{\mathrm{el}}$ and $\tilde{S}_{\mathrm{inel}}$ at a fixed center-of-mass energy originates from the slopes, B_{el} and B_{inel}. Let us compare the differential cross sections in the elastic and inelastic channels. In Fig. 6.15, they are depicted at $\sqrt{s} = 50\,\mathrm{GeV}$ for instance. We concentrate on the region of very small $-t$, because the differential cross sections are exponentially small at large $-t$. As we can see in Fig. 6.15, the slope B_{el} in the elastic channel is larger than the slope B_{inel} in the inelastic channel. This is the reason why the inelastic entanglement entropy $\tilde{S}_{\mathrm{inel}}$ is larger than the elastic one $\tilde{S}_{\mathrm{el}}$.

6.5 Summary

We have formulated the entanglement entropy of the final state in two-particle scatterings in the context of the S-matrix theory. Since in our formulation there appears a divergence due to the infinite Hilbert space volume, we have suggested to regularize it by the "ideal" volume regularization.

We have concretely evaluated the entanglement entropy in the proton-proton elastic scattering in terms of the experimental data. We have tried the cut-off regularization and the Gaussian regularization in addition to the volume regularization. The cut-off and Gaussian regularizations provide us with the regularized entanglement entropies that are different from the one by the volume regularization. However all the regularizations show the same behavior of the entanglement entropy, that is, the entanglement entropy decreases at low energy and increases at higher energy.

We have also evaluated the entanglement entropies in the proton-neutron elastic and inelastic scatterings in the scheme of volume regularization. Both the elastic and inelastic entanglement entropies have common property, that is, they decrease in low energy and increase in high energy as the center-of-mass energy. Furthermore, we have found that the inelastic entanglement entropy is always larger than the elastic one. Employing the diffraction peak model, we have shown that this relation is interpreted to the relation between the elastic and inelastic slopes.

Appendix

Legendre Polynomials

Legendre polynomials $P_n(x)$ are defined as the coefficients in the Taylor series as follows:

$$\frac{1}{\sqrt{1-2xt+t^2}} = \sum_{\ell=0}^{\infty} P_\ell(x)t^\ell. \tag{6.118}$$

The Legendre polynomials satisfy $P_\ell(1) = 1$, and the orthogonality;

$$\int_{-1}^{1} P_\ell(x)P_{\ell'}(x)\,dx = \frac{2}{2\ell+1}\delta_{\ell\ell'}. \tag{6.119}$$

Bessel Function of the First Kind

Bessel functions of the first kind $J_\alpha(x)$ is one of the solutions $y(x)$ of Bessel's differential equation:

$$x^2\frac{d^2y}{dx^2} + x\frac{dy}{dx} + (x^2-\alpha^2)y = 0. \tag{6.120}$$

$J_\alpha(x)$ is described in terms of an infinite sum as

$$J_\alpha(x) = \sum_{n=0}^{\infty} \frac{(-1)^n}{n!\,\Gamma(n+\alpha+1)}\left(\frac{x}{2}\right)^{2n+\alpha}, \tag{6.121}$$

where $\Gamma(n+\alpha+1)$ is the Gamma function, which is defined by

$$\Gamma(z)=\int_0^{\infty} t^{z-1}e^{-t}\,dt \quad (\mathrm{Re}(z)>0). \tag{6.122}$$

$\Gamma(n)=(n-1)!$ for an integer n.

$J_n(x)$ (n is an integer) can be also described in terms of an integral as

$$J_n(x)=\frac{1}{\pi}\int_0^{\pi}\cos(n\theta-x\sin\theta)d\theta. \tag{6.123}$$

The Derivation of Unitarity Condition in Partial Waves

The left hand side of (6.26) is equal to

$${}_i\langle\!\langle \mathbf{p}|\hat{T}|\mathbf{k}\rangle\!\rangle_j-({}_j\langle\!\langle \mathbf{k}|\hat{T}|\mathbf{p}\rangle\!\rangle_i)^*=\frac{\sqrt{s}}{\pi\sqrt{pk}}\sum_{\ell=0}^{\infty}(2\ell+1)(\mathrm{Im}\tau_{ij}^{\ell})P_\ell(\cos\theta)\,. \tag{6.124}$$

${}_i\langle\!\langle \mathbf{p}|\hat{T}|\mathbf{k}\rangle\!\rangle_j$ is equal to ${}_j\langle\!\langle \mathbf{k}|\hat{T}|\mathbf{p}\rangle\!\rangle_i$ due to the time reversal symmetry in scattering processes. In order to compute the first term in the right hand side of (6.26), we set the three-dimensional Cartesian coordinates so that $\mathbf{k}/k=(0,0,1)$, and $\mathbf{p}/p=(\sin\theta,0,\cos\theta)$. $\mathbf{q}$ is in general described as $\mathbf{q}/q=(\sin\theta'\cos\varphi',\sin\theta'\sin\varphi',\cos\theta')$. Then the angle between $\mathbf{k}$ and $\mathbf{q}$ and the one between $\mathbf{q}$ and $\mathbf{p}$ are respectively given by

$$\frac{\mathbf{k}\cdot\mathbf{q}}{kq}=\cos\theta',\quad \frac{\mathbf{q}\cdot\mathbf{p}}{qp}=\sin\theta\sin\theta'\cos\varphi'+\cos\theta\cos\theta'\equiv\cos\omega. \tag{6.125}$$

In order to calculate the first term in the right hand side of (6.26), we show a formula of Legendre polynomials. If $\cos\omega=\cos\theta\cos\theta'+\sin\theta\sin\theta'\cos(\varphi-\varphi')$, we can rewrite a Legendre polynomial $P_n(\cos\omega)$ as the sum of products of Legendre polynomials,

$$\begin{aligned}P_n(\cos\omega)=&\;P_n(\cos\theta)P_n(\cos\theta')\\&+2\sum_{m=1}^{n}\frac{(n-m)!}{(n+m)!}P_n^m(\cos\theta)P_n^m(\cos\theta')\cos(m(\varphi-\varphi')).\end{aligned} \tag{6.126}$$

Then the first term in the right hand side of (6.26) is equal to

$$
\begin{aligned}
&2i\sum_{h=0}^{n}\int\frac{d^3\mathbf{q}}{2E_{\mathrm{A}_h\mathbf{q}}2E_{\mathrm{B}_h\mathbf{q}}}{}_i\langle\!\langle\mathbf{p}|\hat{T}^\dagger|\mathbf{q}\rangle\!\rangle_h\frac{2E_{\mathrm{A}_h\mathbf{q}}2E_{\mathrm{B}_h\mathbf{q}}}{4q\sqrt{s}}\delta(q-q')\,{}_h\langle\!\langle\mathbf{q}|\hat{T}|\mathbf{k}\rangle\!\rangle_j \\
&=2i\sum_{j=0}^{n}\int_0^\infty dq\int_0^\pi d\theta'\int_0^{2\pi}d\varphi' q^2\sin\theta'\frac{\sqrt{s}}{\pi\sqrt{pq}}\sum_{\ell=0}^{\infty}(2\ell+1)\tau_{ij}^{\ell *}P_\ell(\cos\omega) \\
&\quad\times\frac{\delta(q-q')}{4q\sqrt{s}}\frac{\sqrt{s}}{\pi\sqrt{qk}}\sum_{\ell'=0}^{\infty}(2\ell'+1)\tau_{j0}^{\ell'}P_{\ell'}(\cos\theta') \\
&=2i\sum_{h=0}^{n}\int_0^\infty dq\int_{-1}^{1}d\cos\theta'\int_0^{2\pi}d\varphi'\frac{\sqrt{s}\delta(q-q')}{4\pi^2\sqrt{pk}} \\
&\quad\times\sum_{\ell,\ell'=0}^{\infty}(2\ell+1)(2\ell'+1)\tau_{ih}^{\ell *}\tau_{hj}^{\ell'}P_\ell(\cos\omega)P_{\ell'}(\cos\theta') \\
&=\frac{\sqrt{s}}{\pi\sqrt{pk}}\sum_{\ell=0}^{\infty}(2\ell+1)\left(\sum_{h=0}^{n}\tau_{ih}^{\ell *}\tau_{hj}^{\ell}\right)P_\ell(\cos\theta),
\end{aligned}
\tag{6.127}
$$

where we used (6.126).

q' is derived as follows. The center-of-mass energy of the state $|\mathbf{k}\rangle\!\rangle_j$ is

$$
\sqrt{s}=E_{\mathrm{A}_j\mathbf{k}}+E_{\mathrm{B}_j\mathbf{k}}=\sqrt{m_{\mathrm{A}_j}^2+k^2}+\sqrt{m_{\mathrm{B}_j}^2+k^2}
\tag{6.128}
$$

This energy is conserved through the scattering process. Therefore, the energy conservation provides with the delta function, $\delta((E_{\mathrm{A}_h\mathbf{q}}+E_{\mathrm{B}_h\mathbf{q}})-(E_{\mathrm{A}_j\mathbf{k}}+E_{\mathrm{B}_j\mathbf{k}}))=\delta(E_{\mathrm{A}_h\mathbf{q}}+E_{\mathrm{B}_h\mathbf{q}}-\sqrt{s})$. By solving $E_{\mathrm{A}_h\mathbf{q}}+E_{\mathrm{B}_h\mathbf{q}}-\sqrt{s}=0$ with respect to q, we obtain

$$
\begin{aligned}
&\delta((E_{\mathrm{A}_h\mathbf{q}}+E_{\mathrm{B}_h\mathbf{q}})-(E_{\mathrm{A}_j\mathbf{k}}+E_{\mathrm{B}_j\mathbf{k}})) \\
&=\delta\left(\frac{q\sqrt{s}}{E_{\mathrm{A}_h\mathbf{q}}E_{\mathrm{B}_h\mathbf{q}}}(q-q')+O((q-q')^2)\right)=\frac{2E_{\mathrm{A}_h\mathbf{q}}2E_{\mathrm{B}_h\mathbf{q}}}{4q\sqrt{s}}\delta(q-q')
\end{aligned}
\tag{6.129}
$$

where

$$
q'=\sqrt{\frac{\left(m_{\mathrm{A}_j}^2+m_{\mathrm{B}_j}^2-s\right)^2-4m_{\mathrm{A}_j}^2m_{\mathrm{B}_j}^2}{4s}}.
\tag{6.130}
$$

Fitting Functions for Proton-Neutron Scattering from Experimental Data

The function of s for total cross section in proton-proton scattering is given by [12, 13]. Since the mass difference between a proton and a neutron is small and the electromagnetic force is weak, that function is available also for proton-neutron scattering. [12, 13] give us

$$\sigma_{\text{tot}} = -\frac{1}{s}\text{Im}(M_+(s) - M_-(s)), \tag{6.131}$$

where

$$\text{Im}M_+(s) = -s\big(163 - 20.4\ln s + 1.75(\ln s)^2\big), \tag{6.132}$$
$$\text{Im}M_-(s) = -73.1\, s^{0.48}. \tag{6.133}$$

Since the elastic channel of proton-neutron scattering is neutral current exchange, the elastic differential cross section for the proton-proton scattering given by [17] is available for the elastic proton-neutron scattering, so that

$$\frac{d\sigma_{\text{el}}}{dt} = \left| i\, 6.88 e^{(2.69+0.306\ln s - i\,0.153\pi)t} + \left(\frac{10.3}{s} - i\,0.035\right) e^{0.89t}\right|^2. \tag{6.134}$$

[18, 19] give us the inelastic differential cross section of proton-neutron scattering:

$$\frac{d\sigma_{\text{inel}}}{dt} = \frac{2\pi}{s^2} 0.389\left(|\varphi_1|^2 + |\varphi_2|^2 + |\varphi_3|^2 + |\varphi_4|^2 + |\varphi_5|^2\right), \tag{6.135}$$

where

$$\varphi_i = \varphi_i^c + \varphi_i^v \quad (i = 1, 2, \ldots, 5). \tag{6.136}$$

φ_i^c are as follows:

$$\varphi_1^c = 15\frac{t}{s - 4m^2}\frac{m^2}{s}(-2.2)e^{5t}, \tag{6.137}$$

$$\varphi_2^c = 15\left\{\frac{t}{t - m_\pi^2} + \frac{4t}{m^2} - 0.9 - \frac{0.45t}{s - 4m^2} + \frac{0.65t}{s - 4m^2}\frac{m^2}{s} + \frac{0.65t}{s - 4m^2}\left(\frac{m^2}{s} + 0.5\right)\right\}e^{5t}, \tag{6.138}$$

$$\varphi_3^c = 0, \tag{6.139}$$

$$\varphi_4^c = 15\left\{\frac{t}{t - m_\pi^2} + \frac{4t}{m^2} + \frac{t}{s - 4m^2}\left(1.55\frac{m^2}{s} + 0.325\right)\right\} e^{5t}, \tag{6.140}$$

$$\varphi_5^c = 15\frac{m^2}{s - 4m^2}\sqrt{-\frac{t}{4m^2}\frac{s - 4m^2 + t}{s}}\, 2.2e^{5t}, \tag{6.141}$$

where proton (or neutron) mass $m = 0.939\,\text{GeV}$ and pion mass $m_\pi = 0.139\,\text{GeV}$. φ_i^v are

$$\varphi_1^v = \varphi_3^v = \beta_{\text{eff}}(t)s^{\alpha_{\text{eff}}(t)}, \tag{6.142}$$

$$\varphi_2^v = -\varphi_4^v = 17t\beta_{\text{eff}}(t)s^{\alpha_{\text{eff}}(t)}, \tag{6.143}$$

$$\varphi_5^v = -\sqrt{-17t}\,\beta_{\text{eff}}(t)s^{\alpha_{\text{eff}}(t)}, \tag{6.144}$$

with

$$\alpha_{\text{eff}}(t) = \begin{cases} 0.74 + 0.67t & (|t| < 0.175) \\ 0.63 + 0.04t & (|t| > 0.175), \end{cases} \tag{6.145}$$

$$\beta_{\text{eff}}(t) = -0.12e^{5.5t}. \tag{6.146}$$

References

1. Van Hove L. A phenomenological discussion of inelastic collisions at high energies. Nuovo Cim. 1963;28:2344.
2. Bialas A, Van Hove L. Two-particle and many-particle channels in high-energy collisions. Nuovo Cim. 1965;38:1385.
3. Peschanski R, Seki S. Entanglement entropy of scattering particles. Phys Lett B. 2016;758:89.
4. Peschanski R, Seki S. Evaluation of entanglement entropy in high energy elastic scattering. Phys Rev D. 2019;100(7):076012.
5. Peschanski R, Seki S. Entanglement in Elastic and Inelastic Two-particle Scatterings at High Energy. Phys Rev D. 2026;113(10):105023.
6. Amos NA et al. [E710 Collaboration], Measurement of the $\bar{p}p$ total cross-section at $\sqrt{s} = 1.8$ TeV. Phys Rev Lett. 1989;63:2784.
7. Amos NA et al. [E-710 Collaboration], A luminosity independent measurement of the $\bar{p}p$ total cross-section at $\sqrt{s} = 1.8$ tev. Phys Lett. B 1990;243:158.
8. Antchev G, et al. [TOTEM Collaboration], Measurement of proton-proton elastic scattering and total cross-section at $\sqrt{s} = 7$ TeV. Europhys Lett. 2013;101:21002.
9. Antchev G, et al. [TOTEM Collaboration], Luminosity-independent measurement of the proton-proton total cross section at $\sqrt{s} = 8$ TeV. Phys Rev Lett. 2013;111(1):012001.
10. Antchev G, et al. [TOTEM Collaboration], Evidence for non-exponential elastic proton-proton differential cross-section at low $|t|$ and $\sqrt{s} = 8$ TeV by TOTEM. Nucl Phys B. 2015;899:527.
11. G. Antchev et al. [TOTEM Collaboration], First measurement of elastic, inelastic and total cross-section at $\sqrt{s} = 13$ TeV by TOTEM and overview of cross-section data at LHC energies. Eur Phys J C. 2019;79(2):103.
12. Block MM, Cahn R. Forward hadronic pp and $\bar{p}p$ elastic scattering amplitudes: analysis of existing data and extrapolations to collider energies. Phys Lett B. 1983;120B:224.
13. Block MM, Cahn R. Near forward p-p and $\bar{p}$-p elastic scattering: slope analysis of existing data. Phys Lett B. 1983;120B:229.
14. Kreisler MN, Davis MB, Longo MJ, O'Brien DD. Neutron-proton charge exchange scattering from 8-GeV/c to 29-GeV/c. Nucl Phys B. 1975;84:3.
15. Kwak N, Lohrmann E, Nagy E, Regler M, Schmidt-Parzefall W, Schubert KR, Winter K, Brandt A, Dibon H, Flugge G, et al. Experimental results on large angle elastic pp scattering at the CERN ISR. Phys Lett B. 1975;58:233.

16. Dehaven CE, Ayre CA, Gustafson HR, Jones LW, Longo MJ, Ramana Murthy PV, Roberts TJ, Whalley MR. Neutron-proton differential cross sections in the range 70 to 400 GeV/c. Nucl Phys B. 1979;148:1–17.
17. Phillips RJN, Barger VD. Model independent analysis of the structure in p p scattering. Phys Lett B. 1973;46:412.
18. Bouquet A, Diu B. The energy dependence of neutron-proton charge exchange. Nuovo Cim A. 1978;43:53.
19. Bizard G, Diu B. Phenomenological FIT of the n p charge-exchange cross-section at intermediate and high-energy. Nuovo Cim A. 1975;25:467.

Part II
Methodological Development in Quantum Chemistry

Chapter 7
A Unified Semiclassical Framework for Ultrafast Competitive Electron Transfer in Multiredox Molecular Systems

Serguei V. Feskov and Anatoly I. Ivanov

Abstract Ultrafast multistage electron transfer (ET) in molecular systems with multiple redox centers is fundamental to photochemical energy conversion, including processes in natural photosynthesis, molecular optoelectronics, and organic photovoltaics. These systems often operate under nonequilibrium conditions, where solvent relaxation, intramolecular vibrations, and competing ET pathways jointly determine reaction kinetics and product yields. In this chapter, we present a unified semiclassical framework for modeling ultrafast competitive ET in multiredox compounds embedded in polar environments with complex relaxation dynamics. The approach constructs diabatic free energy surfaces (FESs) in a multidimensional coordinate space that integrates both polarization and relaxation components of the environment within a unified representation. Electron dynamics are described using a stochastic point-transition method that captures the coupling between nonadiabatic quantum transitions and classical nuclear motion. The formalism generalizes and unifies several established semiclassical models—including the Najbar-Tachiya, Zusman-Beratan, and Sumi-Marcus approaches—and supports efficient simulation of multistage ET cascades. As an application, we investigate ultrafast charge separation in donor–acceptor–acceptor (DA_1A_2) triads, showing how hot charge shift to a secondary acceptor can suppress nonequilibrium charge recombination. Numerical simulations reveal how reorganization energies, vibrational coupling, molecular geometry, and bending angle collectively influence ET efficiency. The proposed framework offers a general and scalable tool for the rational design of photofunctional molecular systems.

Keywords Electron transfer · Macromolecular compounds · Photochemistry · Ultrafast reactions · Nonequilibrium processes

---

S. V. Feskov (✉)
Volgograd State University, Universitetsky prosp., Volgograd, Russia
e-mail: serguei.feskov@volsu.ru

A. I. Ivanov
Volgograd State University, Universitetsky Prosp., Volgograd, Russia
e-mail: anatoly.ivanov@volsu.ru

T. Onishi (ed.), *Advances in Quantum Science I*,
https://doi.org/10.1007/978-981-95-9717-8_7

7.1 Introduction

Electron transfer (ET) is a fundamental process in chemical physics that underpins a wide range of technologies, including solar energy conversion, molecular electronics, and electrocatalysis [1–15]. In many applications, achieving fast and directional ET is important, as it competes with energy-dissipative or chemically unproductive pathways that limit overall efficiency and selectivity. This is particularly true for photoinduced ET, where charge-transfer reactions occur on femtosecond to picosecond timescales and are strongly influenced by solvent polarization and intramolecular vibrational relaxation.

7.1.1 Nonequilibrium Nuclear Configurations

In multistage ET processes, nonequilibrium nuclear configurations typically result from electronic transitions, such as optical excitation or preceding charge transfer steps, that shift the system away from thermal equilibrium. These nonequilibrium states can substantially influence the kinetics and pathways of subsequent ET events by altering the energy landscape and the coupling between electronic and nuclear degrees of freedom [16–23]. As a result, nonequilibrium ET often deviates significantly from classical thermally activated behavior, necessitating theoretical approaches that account for the dynamic evolution of the nuclear environment. These effects have been extensively investigated through ultrafast spectroscopic experiments and advanced theoretical modeling [24–30].

7.1.2 Multiredox Systems

The challenge of controlling photoinduced ET on ultrafast timescales is particularly important in organic photovoltaic and photocatalytic systems, where efficient charge separation (CS) must outcompete recombination pathways. In multiredox macromolecular systems, such as molecular triads and extended donor–acceptor cascades, multiple ET steps can occur either sequentially or in parallel [31, 32]. In such architectures, accurately predicting and optimizing ET efficiency requires models that incorporate both nuclear relaxation dynamics and electronic couplings across multiple intermediate states [26, 33].

7.1.3 Photosynthetic Reaction Centers

Natural photosynthetic reaction centers (RCs) exemplify the remarkable efficiency of multistage ET, where photoinduced charge separation proceeds through a series of ET events with near-unity quantum yields [34, 35]. These biological systems motivate the design of artificial compounds that replicate multistage directional ET. However, achieving this in synthetic systems requires theoretical models that go beyond single-step, equilibrium-based descriptions—models that can capture nonequilibrium effects, multistate competition, and environmental complexity.

7.1.4 Symmetry-Breaking Charge Transfer

In recent years, considerable attention has been directed toward symmetric donor–acceptor architectures, including quadrupolar A–D–A and D–A–D compounds, octupolar star-shaped systems such as $D(-A)_3$ and $A(-D)_3$, as well as symmetric dimers. These molecular systems exhibit a diverse range of excited-state phenomena, most notably (SBCT) and excimer formation, which has become the subject of extensive experimental and theoretical investigation [36–47]. A key feature of these systems is their ability to undergo ultrafast charge separation without requiring thermal activation [48], as evidenced by characteristic charge transfer timescales on the order of a few picoseconds—comparable to solvent relaxation times in polar environments [49]. While highly relevant to the broader study of photoinduced ET, these topics lie outside the scope of this chapter and will not be discussed further.

7.1.5 Modeling Ultrafast Competitive Electron Transfer

Here, we present a unified semiclassical framework for modeling ultrafast multistage ET in molecular systems containing multiple redox centers embedded in polar environments. Building on the stochastic point-transition approach [50], the theory extends classical models, including those of Najbar & Tachiya [51], Zusman & Beratan [52], and Sumi & Marcus [53], by explicitly incorporating:

- multiple coupled ET steps across diabatic free energy surfaces;
- hot electron transfer proceeding from nonequilibrium nuclear configurations;
- complex multicomponent dynamics of environmental relaxation.

We demonstrate how this framework can be used for simulation of ultrafast charge separation, quantification of ET quantum yields, and identification of molecular design parameters—including reorganization energies, vibrational coupling strengths, donor–acceptor distances, and overall geometry—that govern ET efficiency. Applications are illustrated for donor–acceptor–acceptor (DA_1A_2) triads inspired by Zn-porphyrin–imide systems, with relevance to light-activated molecular switches and artificial photosynthetic assemblies.

7.2 General Theoretical Framework for Multistage Electron Transfer in Nonequilibrium Environments

Photoinduced ET between donor (D) and acceptor (A) molecules in solution is significantly influenced by the dielectric polarization of the surrounding medium. This interaction is typically characterized by a single parameter—the medium reorganization energy, λ. In the classical Marcus theory of outer-sphere ET [54], the activation free energy is given by $G^{\sharp} = (\lambda + \Delta G_{\mathrm{ET}})^2 / 4\lambda$, where ΔG_{ET} is the ET driving force. Together, λ and ΔG_{ET} define the barrier height and therefore the ET rate. In weakly coupled systems, the reaction is treated as a nonadiabatic transition between two parabolic diabatic free energy surfaces associated with the donor and acceptor states.

The Marcus framework assumes that the nuclear degrees of freedom remain in thermal equilibrium within the initial donor state before the ET event. Under such conditions, electron transfer is activated by thermally driven fluctuations of polarization. However, this equilibrium assumption is often violated in ultrafast photoreactions, where the initial Franck-Condon state is created by a short laser pulse, placing the system on an upper electronic free energy surface. In such cases, photoinduced ET can proceed on timescales comparable to or even shorter than the characteristic solvent and intramolecular relaxation times.

Electron transfer in polar solvents is often modeled using a single reaction coordinate, z, which represents the vertical free energy gap between the donor (reactant) and acceptor (product) electronic states: $z = \Delta G$. This coordinate effectively captures the influence of solvent fluctuations and enables the reduction of the many-body solvent environment to a one-dimensional representation. The resulting two-state, single-coordinate framework has proven useful in many foundational theoretical models [50, 54, 55].

However, in macromolecular systems where photoexcitation triggers a cascade of ultrafast ET steps among multiple redox centers, each individual ET event is associated with its own energy-gap coordinate, z_k. These reaction coordinates typically exhibit mutual dependence [16, 51, 56–59], resulting in correlations and nonorthogonality across the multidimensional space of ET processes. Such coupling arises in both discrete and continuum treatments of the environment and must be accounted for to accurately model multistage ET dynamics [60].

For instance, in a A_1DA_2 triad composed of a donor directly linked to two acceptor units (see Fig. 7.1), competitive ET from D to A_1 and A_2 proceeds along distinct energy-gap coordinates, z_1 and z_2. These coordinates are often significantly correlated, with the degree of correlation strongly dependent on the spatial geometry of the

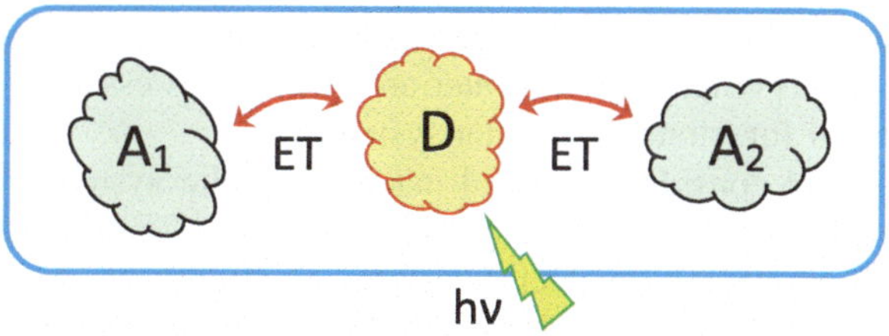

Fig. 7.1 Schematic representation of competitive photoinduced electron transfer in A_1DA_2 compounds. Optical excitation promotes the electron donor (D) to an excited state, followed by subsequent ET pathways to the acceptors A_1 and A_2

system and the polarity of the surrounding medium [51]. A similar situation arises in DA_1A_2 triads, where charge separation toward the $|D^+A_1A_2^-\rangle$ state is governed not only by individual ET rates but also by the correlation between the reaction coordinates associated with the $D \to A_1$ and $A_1 \to A_2$ steps [16, 58–62]. These examples highlight the role of solvent polarization, established during the earliest stages of ultrafast ET, in controlling the dynamics of subsequent electron transfer events.

Naturally, the correlation between the energy-gap coordinates z_1 and z_2 of two sequential ET steps manifests itself only when the second ET occurs on an ultrafast timescale, comparable to or shorter than the characteristic solvent relaxation time. In such cases, the second ET step proceeds from the nonequilibrium nuclear configuration established by the preceding charge transfer event. If z_1 and z_2 are orthogonal, deviations from equilibrium along z_1 do not influence the distribution along z_2, resulting in independent kinetics for the two steps. However, when z_1 and z_2 are non-orthogonal, the rate of the secondary ET depends on the degree of nuclear nonequilibrium along the z_2 coordinate, which, in turn, is determined by the angle between z_1 and z_2. This mechanism explains the influence of the reaction coordinate correlation on the kinetics of multistage reactions. A similar effect is observed in systems where multiple ET reactions occur in parallel [63].

It is important to emphasize that this effect represents a notable exception to the Ostwald principle of the independence of elementary reactions. This principle, foundational to classical chemical kinetics, holds that the rate of an elementary step depends solely on the species directly involved and is unaffected by other concurrent reactions within the same system. However, in the context of ultrafast ET, this assumption may break down. When ET occurs on timescales comparable to or faster than solvent and intramolecular relaxation, the system deviates from equilibrium conditions, and the polarization established during earlier ET steps can significantly influence subsequent reaction pathways [26, 53]. By contrast, in slower multistage reactions where the nuclear environment equilibrates between steps, the independence of elementary reactions remains a valid approximation.

The mutual influence of consecutive ultrafast ET steps arising from correlations between their respective energy-gap reaction coordinates has been recognized and analyzed in previous studies [16, 51, 56–60]. However, these investigations have primarily focused on three-center systems involving only two ET steps. Extending such models to more complex, multicenter architectures is essential for advancing the understanding of functional molecular systems.

In this section, we consider a general macromolecular system composed of N fixed electron donor and acceptor subunits (redox centers). The rigid spatial arrangement of these sites allows us to neglect slow, large-amplitude intramolecular motions and focus instead on the dominant role of electrostatic interactions between the transferred charge and the polar environment in governing reaction energetics. Nonetheless, it is well established that high-frequency, small-amplitude intramolecular vibrations can significantly influence ultrafast ET by modulating activation barriers and affecting the quantum yield of nonequilibrium transitions [64–67]. Although the present formulation emphasizes solvent dynamics, the theoretical framework admits a natural extension to include ET-active quantum vibrational modes, which will be developed in later sections of this chapter.

The construction of diabatic free energy surfaces (FESs) corresponding to different electronic states of the system is an important component of any ET model. In multistage systems, the full set of FESs defines not only the activation barriers for individual ET steps but also governs the pathways of solvent relaxation between successive electronic transitions. As a foundation, we adopt the linear response approximation for the dielectric response of the medium. Within this framework, the FESs for electronic states are represented as quadratic functions of the energy-gap reaction coordinates associated with individual ET events [16, 51]. These coordinates can be linearly transformed to a new basis in which the quadratic form becomes diagonal, yielding a set of independent generalized solvent coordinates. This diagonalization procedure has been explicitly demonstrated for two-center ET models [51] and is generalized here for multistage systems.

We show that the dimensionality of the generalized configuration space is determined not by the number of possible ET transitions between redox sites, but rather by the total number of redox-active centers in the molecular system. To facilitate modeling, we introduce a systematic algorithm for constructing diabatic FESs by incrementally expanding the configuration space. The algorithm enables efficient and scalable simulation of multistage ET dynamics within the stochastic point-transition framework, which extends beyond the applicability of the standard Golden Rule formalism and its perturbative variants [68]. The stochastic approach has previously been applied to describe ET in systems with multiple solvent relaxation modes as well as those involving ET-active high-frequency intramolecular vibrations [26, 69], and is well suited for addressing nonequilibrium effects in complex environments.

7.2.1 Electronic States and Polarization Coordinates of Environment

Consider N redox centers C_n $(n = \overline{1, N})$ within a macromolecule in a polar environment, as illustrated in Fig. 7.2A. At $t = 0$, the chromophoric center C_1 absorbs

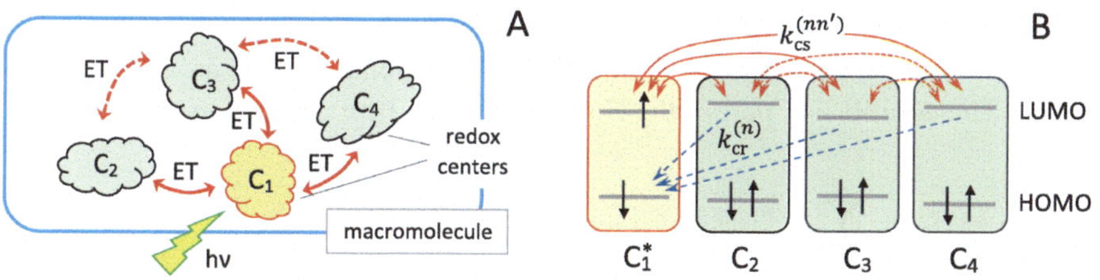

Fig. 7.2 **A** Schematic representation of a macromolecular compound containing multiple redox centers C_n, where C_1 serves as the photoactive unit initiating a cascade of competitive electron transfer steps. **B** Energy-level diagram illustrating charge separation (CS) and recombination (CR) transitions between HOMO and LUMO orbitals of individual redox units. The quantities $k_{cs}^{(nn')}$ and $k_{cr}^{(n)}$ denote the rate constants of CS and CR transitions in the quasi-equilibrium regime

a photon, initiating a sequence of nonadiabatic electronic transitions, including forward ET (charge separation) and backward ET (charge recombination). The specific pathway of the photoreaction is primarily determined by the characteristics of individual ET steps, particularly the rate constants for charge separation (CS) and recombination (CR) processes. However, during the initial nonequilibrium stage of the photoreaction the standard concept of the ET rate constant becomes inadequate. This transient regime, typically spanning several to tens of picoseconds, is marked by incomplete solvent and intramolecular vibrational relaxation.

Figure 7.2B illustrates the electronic energy levels associated with the HOMO and LUMO orbitals of the redox centers C_n. Immediately after photoexcitation, the system is promoted to an electronically excited state denoted by $|\psi_1\rangle$. Subsequent charge separation results in configurations in which the electron is localized on one of the redox centers C_n $(n \geq 2)$, yielding charge-separated states $|\psi_n\rangle$. The relevant electronic configurations can be defined as

$$\begin{aligned} |\psi_0\rangle &= |C_1 C_2 \dots C_N\rangle, \quad \text{ground state} \\ |\psi_1\rangle &= |C_1^* C_2 \dots C_N\rangle, \quad \text{excited state} \\ |\psi_n\rangle &= |C_1^+ \dots C_n^- \dots C_N\rangle, \quad (n = \overline{2, N}), \quad \text{CS states.} \end{aligned} \tag{7.1}$$

Charge separation, charge recombination, and internal conversion processes are defined in terms of the $|\psi_n\rangle$ states as

$$\begin{aligned} \text{CS} &: |\psi_n\rangle \to |\psi_{n'}\rangle, \quad (n, n' = 1 \dots N, n \neq n', n' \neq 1) \\ \text{CR} &: |\psi_n\rangle \to |\psi_0\rangle, \quad (n = 2 \dots N) \\ \text{IC} &: |\psi_1\rangle \to |\psi_0\rangle. \end{aligned} \tag{7.2}$$

It is convenient to express the system Hamiltonian $\hat{H}$ as the sum of two components

$$\hat{H} = \hat{H}_{eq} + \hat{H}_{ne}, \tag{7.3}$$

where $\hat{H}_{\text{eq}}$ describes the macromolecule in equilibrium with the environment, while $\hat{H}_{\text{ne}}$ accounts for nonequilibrium configurations of a polar medium around the redox centers. The equilibrium part of the Hamiltonian can be expressed as

$$\hat{H}_{\text{eq}} = \sum_{n=0}^{N} \check{G}^{(n)} |\psi_n\rangle\langle\psi_n| + \sum_{n=1}^{N} V_{\text{cr}}^{(n)} \left(|\psi_n\rangle\langle\psi_0| + \text{h.c.}\right) + \\ + \sum_{n=2}^{N} \sum_{n'>n} V_{\text{cs}}^{(nn')} \left(|\psi_n\rangle\langle\psi_{n'}| + \text{h.c.}\right). \tag{7.4}$$

Here, $\check{G}^{(n)}$ denotes the free energy of the system in the electronic state $|\psi_n\rangle$ when it is in equilibrium with the polarization of the medium. The quantities $V_{\text{cr}}^{(n)}$ and $V_{\text{cs}}^{(nn')}$ are the electronic coupling energies governing charge recombination (CR) and charge separation (CS) transitions, respectively. Within the one-electron approximation, these couplings are expressed through orbital overlaps as

$$V_{\text{cr}}^{(n)} = \langle \text{LUMO}_n | \hat{\mathcal{H}} | \text{HOMO}_1 \rangle, \\ V_{\text{cs}}^{(nn')} = \langle \text{LUMO}_n | \hat{\mathcal{H}} | \text{LUMO}_{n'} \rangle, \tag{7.5}$$

where $\hat{\mathcal{H}}$ is the one-electron Hamiltonian. The LUMO orbitals appear in these matrix elements because they host the photoexcited electron and therefore directly participate in recombination with the donor HOMO as well as in electron transfer between different acceptor sites, as illustrated in Fig. 7.2B.

The nonequilibrium part of the system Hamiltonian, $\hat{H}_{\text{ne}}$, accounts for unbalanced dielectric polarization and can be expressed in terms of independent polarization coordinates q_k, following the approach developed in Ref. [70]. Assuming a linear response of the medium to charge redistribution within the macromolecule, $\hat{H}_{\text{ne}}$ takes the form of a parabolic function in the q_k coordinates

$$\hat{H}_{\text{ne}} = \sum_{n=0}^{N} |\psi_n\rangle\langle\psi_n| \sum_{k=1}^{K} \left(q_k - \check{q}_k^{(n)}\right)^2 = \sum_{n=0}^{N} |\psi_n\rangle\langle\psi_n| \left|\boldsymbol{q} - \check{\boldsymbol{q}}^{(n)}\right|^2. \tag{7.6}$$

Here, $\boldsymbol{q} = (q_1, q_2, \ldots q_K)^{\text{T}}$ and $\check{\boldsymbol{q}}^{(n)} = (\check{q}_1^{(n)}, \check{q}_2^{(n)}, \ldots \check{q}_K^{(n)})^{\text{T}}$ are vectors in the K-dimensional configuration space, and $\check{\boldsymbol{q}}^{(n)}$ represents the equilibrium polarization associated with the electronic state $|\psi_n\rangle$. The operator $|\psi_n\rangle\langle\psi_n|$ acts as a projector onto the state ψ_n, ensuring that the corresponding polarization contribution is applied only when the system occupies this state. Finally, $|\boldsymbol{q}|$ denotes the Euclidean norm of the vector $\boldsymbol{q}$, $|\boldsymbol{q}| = (q_1^2 + q_2^2 + \cdots + q_K^2)^{1/2}$.

The $\boldsymbol{q}$-space is complete in the sense that it fully characterizes the state of the environment with respect to the CS/CR transitions. The number of the q_k coordinates is in general determined by the number of redox centers N and follows the relation [70]

$$K = N - 1. \tag{7.7}$$

The free energy surface associated with the diabatic electronic state $|\psi_n\rangle$ is defined as the diagonal matrix element of the total Hamiltonian, $G^{(n)} \equiv \langle\psi_n|\hat{H}|\psi_n\rangle$. This matrix element represents the expectation value of the Hamiltonian when the system is constrained to remain in the state $|\psi_n\rangle$, and thus provides the energy surface governing polarization coordinates in that electronic state. Substituting Eqs. (7.4) and (7.6) yields

$$G^{(n)}(\boldsymbol{q}) = \check{G}^{(n)} + \sum_{k=1}^{K} \left(q_k - \check{q}_k^{(n)}\right)^2 = \check{G}^{(n)} + \left|\boldsymbol{q} - \check{\boldsymbol{q}}^{(n)}\right|^2. \tag{7.8}$$

Each diabatic FES $G^{(n)}$ thus corresponds to a K-dimensional paraboloid centered at the equilibrium polarization configuration $\check{\boldsymbol{q}}^{(n)}$, with identical curvature along all q_k coordinates and a vertical energy offset $\check{G}^{(n)}$ (see Fig. 7.3).

The energetic cost of reorganizing the medium during an electron transfer event $|\psi_n\rangle \rightarrow |\psi_{n'}\rangle$ is quantified by the reorganization free energy $\lambda^{(nn')}$, defined as the energy required to move the system from its equilibrium configuration in state n to the equilibrium configuration of state n', while remaining on the FES of state n

$$\lambda^{(nn')} \equiv G^{(n)}\left(\check{\boldsymbol{q}}^{(n')}\right) - G^{(n)}\left(\check{\boldsymbol{q}}^{(n)}\right). \tag{7.9}$$

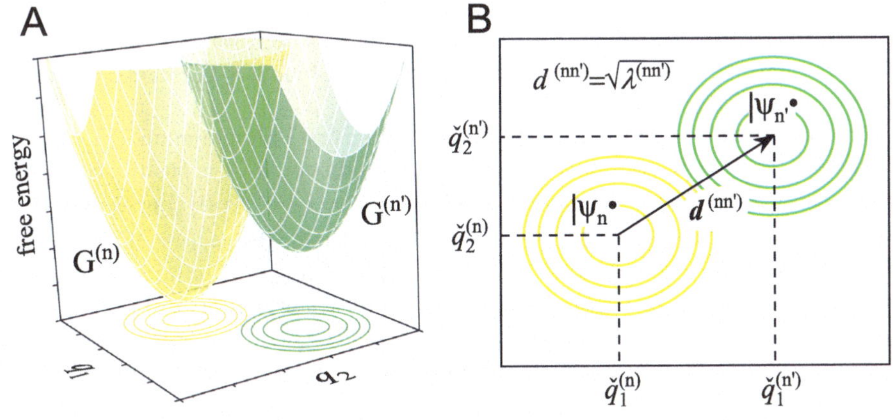

Fig. 7.3 **A** Diabatic free energy surfaces (FESs) for multistage ET, shown in the three-dimensional space of polarization coordinates (q_1, q_2) and free energy. Each surface $G^{(n)}(\boldsymbol{q})$ corresponds to a distinct diabatic state $|\psi_n\rangle$. **B** Geometric relationship between the FES minima: the distance between any two equilibrium positions $\check{\boldsymbol{q}}^{(n)}$ and $\check{\boldsymbol{q}}^{(n')}$ is equal to $\sqrt{\lambda^{(nn')}}$, establishing a direct connection between reorganization free energy and spatial configuration of the two FESs

Using Eq. (7.8), this becomes

$$\lambda^{(nn')} = G^{(n)}\left(\breve{q}^{(n')}\right) - G^{(n)}\left(\breve{q}^{(n)}\right) = \breve{G}^{(n)} + \sum_{k=1}^{K}\left(\breve{q}_k^{(n')} - \breve{q}_k^{(n)}\right)^2 -$$
$$-\left[\breve{G}^{(n)} + \sum_{k=1}^{K}\left(\breve{q}_k^{(n)} - \breve{q}_k^{(n)}\right)^2\right] = \sum_{k=1}^{K}\left(\breve{q}_k^{(n')} - \breve{q}_k^{(n)}\right)^2 = \left|\boldsymbol{d}^{(nn')}\right|^2, \quad (7.10)$$

where $\boldsymbol{d}^{(nn')} \equiv \breve{\boldsymbol{q}}^{(n')} - \breve{\boldsymbol{q}}^{(n)}$ is the displacement vector between the FES minima corresponding to states n and n'. Thus, the reorganization energy $\lambda^{(nn')}$ is directly related to the squared distance between the equilibrium polarization configurations in $\boldsymbol{q}$-space. Given a set of $\lambda^{(nn')}$ values, one can calculate the pairwise distances $d^{(nn')}$ between FES minima as $d^{(nn')} = \sqrt{\lambda^{(nn')}}$. These distances provide the necessary input for constructing the diabatic FESs.

From a geometrical standpoint, constructing the diabatic free energy surfaces $G^{(n)}$ reduces to the problem of embedding N points in an $(N-1)$-dimensional configuration space, such that the pairwise distances correspond to the known $\lambda^{(nn')}$. When the matrix of reorganization energies, $\hat{\lambda}$, is non-degenerate, this embedding problem has a well-defined solution: a unique set of N distinct points $\breve{\boldsymbol{q}}^{(n)}$ in $\mathbb{R}^{N-1}$, determined up to isometric (distance-preserving) transformations. These transformations include global translations, rotations, and reflections of the coordinate frame. Importantly, this ambiguity does not influence the ET model, as isometric transformations leave the ET energetics unchanged and therefore have no impact on observable quantities.

The spatial configuration of the FES minima $\breve{\boldsymbol{q}}^{(n)}$ can be constructed using the iterative embedding algorithm proposed in Ref. [70]. This procedure incrementally builds the multidimensional coordinate system by sequentially introducing each point $\breve{\boldsymbol{q}}^{(n)}$ and increasing the dimensionality of the $\boldsymbol{q}$-space as necessary. Figure 7.4 illustrates this algorithm for a representative case involving four redox centers embedded in a three-dimensional space.

The construction begins with the placement of the first FES minimum, $G^{(1)}$, at the origin: $\breve{\boldsymbol{q}}^{(1)} = (0, 0, 0)$ (Fig. 7.4A). The second minimum, $\breve{\boldsymbol{q}}^{(2)}$, is positioned along the q_1 axis at a distance $d^{(12)} = \sqrt{\lambda^{(12)}}$, yielding $\breve{\boldsymbol{q}}^{(2)} = \left(d^{(12)}, 0, 0\right)$ (Fig. 7.4B). The third point, $\breve{\boldsymbol{q}}^{(3)}$, is placed in the (q_1, q_2) plane such that the distances to the previous points match $d^{(13)} = \sqrt{\lambda^{(13)}}$ and $d^{(23)} = \sqrt{\lambda^{(23)}}$ (Fig. 7.4C). Finally, the fourth point, $\breve{\boldsymbol{q}}^{(4)}$, is located in full three-dimensional space, consistent with its distances to the first three points: $d^{(14)}$, $d^{(24)}$, and $d^{(34)}$ (Fig. 7.4D).

In low-dimensional cases, the construction of diabatic FESs using the proposed algorithm can be carried out analytically. An explicit analytical solution for the $N = 3$ case is presented in Sect. 7.2. For systems with larger numbers of redox centers ($N > 3$), the positions of the FES minima $\breve{\boldsymbol{q}}^{(n)}$ must typically be obtained numerically. A widely used approach is classical multidimensional scaling (MDS),

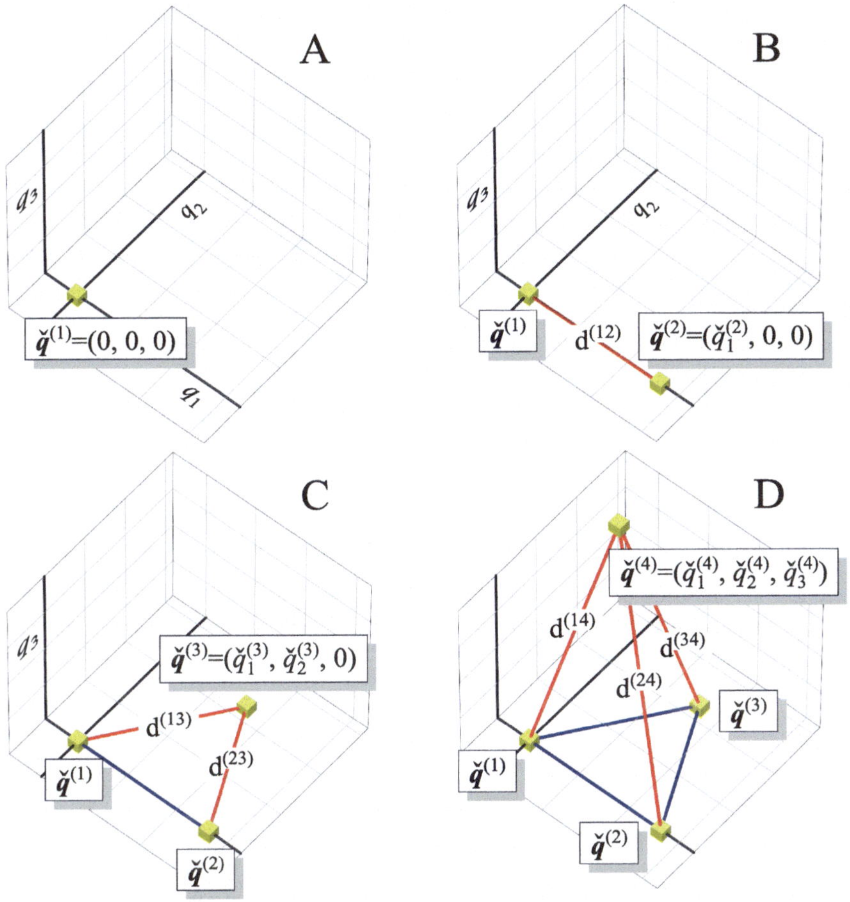

Fig. 7.4 Step-by-step construction of the FES minima $\check{q}^{(n)}$ for a macromolecular system with $N = 4$ redox centers. Because the dimensionality of the polarization coordinate space is $N - 1$, the minima are placed in a three-dimensional space (q_1, q_2, q_3). Panels **A**–**D** illustrate the sequential placement of the $\check{q}^{(n)}$ points. Their spatial arrangement is determined by the inter-state distances $d^{(nn')}$, which are directly related to the corresponding reorganization free energies $\lambda^{(nn')}$

which embeds a set of points in a Euclidean space such that the pairwise distances match a given dissimilarity matrix—in this context, derived from the reorganization free energies $\lambda^{(nn')}$. MDS minimizes a stress function that quantifies the discrepancy between the target and realized distances, yielding an optimal configuration in the least-squares sense [71].

7.2.2 Relaxation Components and Extended Coordinate Space

The system's motion along the q_k coordinates is primarily governed by the dynamic properties of the environment, particularly its response to charge redistribution during ET. Key contributors to this response include the electronic and dipolar components of medium polarization, as well as large-scale conformational intramolecular reorganization, such as shifts in the positions of redox centers within the macromolecule. This chapter focuses on ultrafast photochemical CS processes occurring on timescales of up to several tens of picoseconds. At such timescales, large-scale conformational modes can be considered effectively frozen and are therefore excluded from the present model.

The dynamic response of the medium is incorporated into the model through the energy-gap autocorrelation function,

$$\Gamma(t) = \frac{\langle \Delta E(0) \Delta E(t) \rangle}{\langle \Delta E(0)^2 \rangle},$$

where $\Delta E(t)$ denotes the instantaneous energy gap between the ground state with neutral reactants and the charge-separated excited state, and $\langle \ldots \rangle$ indicates ensemble averaging. The function $\Gamma(t)$ is an experimentally accessible quantity and is commonly represented as a sum of several exponentials [72, 73]:

$$\Gamma(t) = \sum_{i=1}^{L} \Gamma_i(t) = \sum_{i=1}^{L} \gamma_i e^{-t/\tau_i}. \tag{7.11}$$

Here, $\Gamma_i(t)$ represents the i-th relaxation component of the medium, with γ_i and τ_i denoting its weight and relaxation timescale. The weights satisfy the normalization condition $\sum \gamma_i = 1$. Polar solvents typically exhibit two to three relaxation components, with τ_i values often varying by an order of magnitude. In mixtures and structured environments, the range of τ_i values can be even wider. The early-time behavior of $\Gamma(t)$ is often studied using specialized experimental approaches that combine ultrafast spectroscopic techniques with computational methods [74].

The exponential form of $\Gamma_i(t)$ in Eq. (7.11) suggests an inertialess diffusive motion of the system along the energy-gap coordinate ΔE. Consequently, the sum of L exponentials corresponds to L distinct energy-diffusion relaxation processes in the environment, each characterized by a diffusion coefficient $D_i = 2k_B T/\tau_i$ [26, 67, 75].

Owing to the linearity of the medium response, the decomposition of $\Gamma(t)$ in Eq. (7.11) applies not only to the energy-gap coordinate ΔE, but also to each polarization coordinate of the system. Following the approach of Ref. [76], we generalize the single-coordinate representation of q_k by introducing the L-dimensional vector

$$\boldsymbol{y}_k = \boldsymbol{x} q_k = (x_1 q_k, x_2 q_k, \ldots, x_L q_k)^{\mathrm{T}}, \tag{7.12}$$

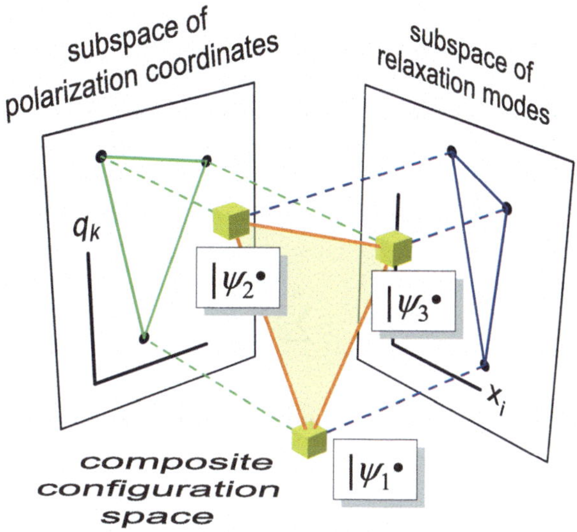

Fig. 7.5 Visualization of the coordinate space architecture used in the extended ET framework. Shown are the polarization coordinate subspace ($\boldsymbol{q}$), the relaxation component subspace ($\boldsymbol{x}$), and the composite configuration space ($\boldsymbol{y}$). Yellow objects indicate equilibrium positions $\check{\boldsymbol{y}}^{(n)}$ corresponding to different diabatic states $|\psi_n\rangle$

where $\boldsymbol{x} \equiv (x_1, x_2, \ldots, x_L)^{\mathrm{T}}$ is defined as

$$\boldsymbol{x} = \left(\sqrt{\gamma_1}, \sqrt{\gamma_2}, \ldots, \sqrt{\gamma_L}\right)^{\mathrm{T}}.$$

The vector $\boldsymbol{x}$ is normalized to unity, and its components are weighted by the relaxation amplitudes γ_i of the medium.

The transformation in Eq. (7.12) maps the polarization coordinates q_k into an extended configuration space $\boldsymbol{y}$, constructed as the outer product of the $\boldsymbol{x}$ and $\boldsymbol{q}$ subspaces (see Fig. 7.5):

$$\boldsymbol{y} = \mathrm{vec}(\boldsymbol{x} \otimes \boldsymbol{q}) = \mathrm{vec}\begin{bmatrix} x_1 q_1 & \ldots & x_1 q_K \\ \vdots & \ddots & \vdots \\ x_L q_1 & \ldots & x_L q_K \end{bmatrix}, \tag{7.13}$$

where $\otimes$ denotes the outer product and $\mathrm{vec}(\cdot)$ is the vectorization operation that reshapes a matrix into a column vector. The resulting vector $\boldsymbol{y}$ has dimension

$$M = LK = L(N-1). \tag{7.14}$$

Physically, this construction embeds the system into an enlarged configuration space that combines the K structural polarization coordinates with the L characteristic relaxation modes of the medium, thereby capturing both spatial and temporal aspects of solvent response.

Eq. (7.13) provides a higher-dimensional representation of the system, facilitating the analysis of multicomponent relaxation dynamics. As follows from this expression, the (ik)-th element of the extended configuration space corresponds to the i-th relaxation component of the q_k polarization coordinate. This explicit separation of relaxation components allows for a detailed description of the medium response and provides a framework for modeling complex relaxation processes [77].

Since $\boldsymbol{x}$ is a unit vector, $|\boldsymbol{x}| = 1$, it follows directly from Eq. (7.12) that $|\boldsymbol{y}| = |\boldsymbol{q}|$, indicating that the coordinate-splitting transformation preserves vector lengths. Consequently, the diabatic free energy surfaces $G^{(n)}$ in the $\boldsymbol{y}$-presentation take the form

$$G^{(n)}(\boldsymbol{y}) = \check{G}^{(n)} + \sum_{k=1}^{K} \sum_{i=1}^{L} \left(y_{ik} - \check{y}_{ik}^{(n)} \right)^2 = \check{G}^{(n)} + \left| \boldsymbol{y} - \check{\boldsymbol{y}}^{(n)} \right|^2, \tag{7.15}$$

where $\check{\boldsymbol{y}}^{(n)}$ defines the minimum of the FES and is given by

$$\check{\boldsymbol{y}}^{(n)} = \operatorname{vec}\left(\boldsymbol{x} \otimes \check{\boldsymbol{q}}^{(n)} \right) = \operatorname{vec} \begin{bmatrix} x_1 \check{q}_1^{(n)} & \dots & x_1 \check{q}_K^{(n)} \\ \vdots & \ddots & \vdots \\ x_L \check{q}_1^{(n)} & \dots & x_L \check{q}_K^{(n)} \end{bmatrix}. \tag{7.16}$$

By employing Eqs. (7.9) and (7.15), one derives the following expression for the reorganization free energy:

$$\lambda^{(nn')} = \sum_{i,k} \left(\check{y}_{ik}^{(n')} - \check{y}_{ik}^{(n)} \right)^2 = \left| \boldsymbol{D}^{(nn')} \right|^2, \tag{7.17}$$

where $\boldsymbol{D}^{(nn')} \equiv \check{\boldsymbol{y}}^{(n')} - \check{\boldsymbol{y}}^{(n)}$. This formulation mirrors the structure of Eq. (7.10), but is applicable to the higher-dimensional representation.

Together with the expansion of the coordinate space defined in Eq. (7.12), we introduce its inverse operation—a projection from the composite space $\boldsymbol{y}$ onto the subspaces $\boldsymbol{q}$ and $\boldsymbol{x}$. Projection onto the $\boldsymbol{q}$ subspace is performed by the operator $\hat{P}_q$

$$\hat{P}_q\, \boldsymbol{y} = |\boldsymbol{x}| \boldsymbol{q} = \boldsymbol{q}. \tag{7.18}$$

Similarly, projection onto the $\boldsymbol{x}$ subspace is defined by the operator $\hat{P}_x$

$$\hat{P}_x\, \boldsymbol{y} = |\boldsymbol{q}|\, \boldsymbol{x}. \tag{7.19}$$

Applying these operators to the displacement vector $\boldsymbol{D}^{(nn')}$ gives

$$\hat{P}_q\, \boldsymbol{D}^{(nn')} = \boldsymbol{d}^{(nn')}, \qquad \hat{P}_x\, \boldsymbol{D}^{(nn')} = \sqrt{\lambda^{(nn')}}\boldsymbol{x}. \tag{7.20}$$

These relations establish a direct correspondence between the system parameters—namely, the solvent reorganization energies and the relaxation amplitudes of the medium—and the spatial arrangement of the diabatic FESs in different configuration spaces.

7.2.3 ET Energy-Gap Coordinates and Equations of Motion

Nonadiabatic ET typically occurs when the free energies of the reactant and product states are equal, a condition realized in the vicinity of the intersection region of the corresponding diabatic FESs. The $|\psi_n\rangle \rightarrow |\psi_{n'}\rangle$ transitions can thus be described in terms of the energy gap between the two diabatic states, $\Delta G^{(nn')}(\boldsymbol{y}) \equiv G^{(n')}(\boldsymbol{y}) - G^{(n)}(\boldsymbol{y})$. Substituting Eq. (7.15), this gap can be expressed as

$$\Delta G^{(nn')}(\boldsymbol{y}) = \Delta \check{G}^{(nn')} + \lambda^{(nn')} - 2\sum_{k,i}\left(y_{ik} - \check{y}_{ik}^{(n)}\right)\left(\check{y}_{ik}^{(n')} - \check{y}_{ik}^{(n)}\right),$$

where $\Delta \check{G}^{(nn')} = \check{G}^{(n')} - \check{G}^{(n)}$ is the driving force of ET. Employing the definition $\boldsymbol{D}^{(nn')} \equiv \check{\boldsymbol{y}}^{(n')} - \check{\boldsymbol{y}}^{(n)}$, the expression simplifies to

$$\Delta G^{(nn')}(\boldsymbol{y}) = \Delta \check{G}^{(nn')} + \lambda^{(nn')} - 2\left(\boldsymbol{y} - \check{\boldsymbol{y}}^{(n)}\right) \cdot \boldsymbol{D}^{(nn')}.$$

The final term here represents a scalar product and can be reformulated as

$$\Delta G^{(nn')}(\boldsymbol{y}) = \Delta \check{G}^{(nn')} + \lambda^{(nn')} - 2\sqrt{\lambda^{(nn')}}z^{(nn')}, \tag{7.21}$$

where $z^{(nn')}$ is calculated as

$$z^{(nn')} \equiv \sum_{i,k}\left(y_{ik} - \check{y}_{ik}^{(n)}\right)\cos\theta_{ik}^{(nn')}, \tag{7.22}$$

and $\cos\theta_{ik}^{(nn')}$ are the directional cosines of the displacement vector $\boldsymbol{D}^{(nn')}$

$$\cos\theta_{ik}^{(nn')} = \frac{\check{y}_{ik}^{(n')} - \check{y}_{ik}^{(n)}}{\sqrt{\lambda^{(nn')}}}. \tag{7.23}$$

As follows from Eq. (7.21), the quantity $z^{(nn')}$ represents the energy-gap reaction coordinate for the $|\psi_n\rangle \rightarrow |\psi_{n'}\rangle$ ET transition, and Eq. (7.22) provides a straightforward computational recipe for evaluating this coordinate in terms of the directional cosines. This geometric construction is illustrated in Fig. 7.6, where the displacement vector $\boldsymbol{D}^{(nn')}$ is resolved into directional cosines that determine the energy-gap reaction coordinate $z^{(nn')}$.

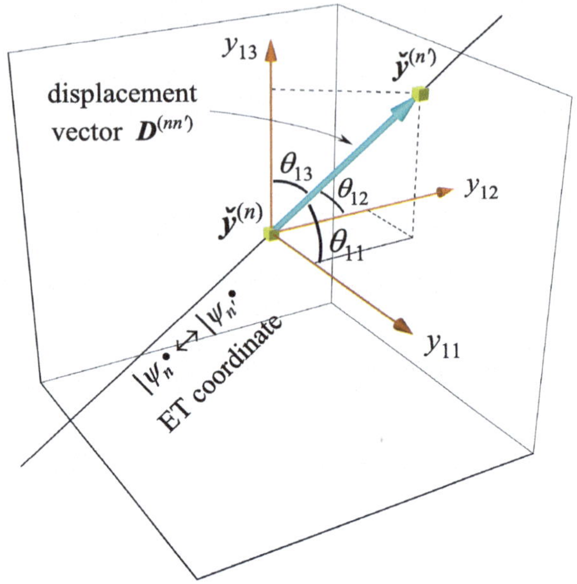

Fig. 7.6 Illustration of the displacement vector $\boldsymbol{D}^{(nn')}$ connecting the equilibrium positions $\check{\boldsymbol{y}}^{(n)}$ and $\check{\boldsymbol{y}}^{(n')}$ of two diabatic FESs in the extended configuration space $\boldsymbol{y}$. The components of $\boldsymbol{D}^{(nn')}$ along the coordinate axes define the directional cosines $\cos\theta_{ik}^{(nn')}$, which are used in the construction of the energy-gap reaction coordinate $z^{(nn')}$

We now apply this result to evaluate the intrinsic rates of ET transitions in the macromolecule. In the weak electronic coupling limit, the position-dependent rate $K_{\rm cs}^{(nn')}(\boldsymbol{y})$ is given by Fermi's Golden Rule

$$K_{\rm cs}^{(nn')} = \frac{2\pi}{\hbar}|V_{\rm cs}^{(nn')}|^2\,\delta\Big(G^{(n')} - G^{(n)}\Big) = \frac{\pi\,|V_{\rm cs}^{(nn')}|^2}{\hbar\sqrt{\lambda^{(nn')}}}\,\delta\Big(z^{(nn')} - \tilde{z}^{(nn')}\Big)\,, \qquad (7.24)$$

where $\delta(z)$ is the Dirac delta function, defined as a generalized function that vanishes everywhere except at $z = 0$ and satisfies

$$\int_{-\infty}^{\infty} f(z)\,\delta(z)\,dz = f(0)$$

for any smooth test function $f(z)$. In this context, the delta function enforces the energy-conservation condition at the crossing of diabatic FESs. The quantity

$$\tilde{z}^{(nn')} = \frac{\Delta\check{G}^{(nn')} + \lambda^{(nn')}}{2\sqrt{\lambda^{(nn')}}} \qquad (7.25)$$

is the value of the reaction coordinate $z^{(nn')}$ at the intersection point of the two diabatic surfaces. Analogous expressions can be written for $K_{\rm cr}^{(n0)}$ by replacing $V_{\rm cs}^{(nn')}$ with $V_{\rm cr}^{(n0)}$ in Eq. (7.24).

We now turn to the dynamic description of photochemical processes in macromolecular systems. The kinetics of multistage ET are formulated in terms of time-dependent probability density functions $\varrho_n(\boldsymbol{y}, t)$, which describe the distribution of the system over nuclear configurations $\boldsymbol{y}$ within a given diabatic electronic state $|\psi_n\rangle$. The complete state of the system is thus represented by an $(N + 1)$-dimensional vector

$$\boldsymbol{\varrho} = (\varrho_0, \varrho_1, \ldots, \varrho_N)^{\mathrm{T}},$$

whose temporal evolution is governed by the following kinetic master equation

$$\frac{\partial \boldsymbol{\varrho}(\boldsymbol{y}, t)}{\partial t} = \left(\hat{W} + \hat{T} + \hat{L}\right) \boldsymbol{\varrho}(\boldsymbol{y}, t). \tag{7.26}$$

In this formulation, $\hat{W}$ denotes the operator describing nonadiabatic electronic transitions, encompassing both charge separation and charge recombination processes, with transition rates $K_{\mathrm{cs}}^{(nn')}$ and $K_{\mathrm{cr}}^{(n0)}$ specified by Fermi's Golden Rule. The operator $\hat{T}$ accounts for irreversible deactivation of the photoexcited state, such as internal conversion or radiative relaxation. The operator $\hat{L}$ represents Smoluchowski diffusion (i.e., the overdamped limit of the Fokker–Planck equation) and describes relaxation of the nuclear degrees of freedom through interactions with the thermal environment.

The ET operator $\hat{W}$ is written as a summation over all allowed transition channels

$$\hat{W} = \sum_{n>1} K_{\mathrm{cr}}^{(n0)} \hat{M}^{(n0)} + \sum_{n\geq 1} \sum_{n'>n} K_{\mathrm{cs}}^{(nn')} \hat{M}^{(nn')}, \tag{7.27}$$

where $\hat{M}^{(nn')}$ is a sparse transition matrix. The matrix $\hat{M}^{(nn')}$ contains nonzero elements at four positions: the diagonal components (n, n) and (n', n') are set to -1, while the off-diagonal components (n, n') and (n', n) are assigned the value $+1$, thereby ensuring proper accounting of population depletion and accumulation between diabatic states.

The deactivation operator $\hat{T}$ describes the irreversible decay of the excited electronic state $|\psi_1\rangle$ to the ground state $|\psi_0\rangle$ with a first-order rate constant k_{d}

$$[\hat{T}]_{n,n'} = k_{\mathrm{d}}\, \delta_{n,0}\, \delta_{n',1} - k_{\mathrm{d}}\, \delta_{n,1}\, \delta_{n',1}, \tag{7.28}$$

where $\delta_{i,j}$ denotes the Kronecker delta.

The nuclear relaxation operator $\hat{L}$ governs the diffusive dynamics in the extended nuclear configuration space and incorporates the multicomponent nature of the environmental response. Given independence of the nuclear coordinates y_{ik}, the total operator $\hat{L}$ can be expressed as a sum over one-dimensional Smoluchowski operators

$$[\hat{L}]_{n,n'} = \delta_{n,n'} \sum_{i,k} \hat{L}_{ik}^{(n)}, \quad \hat{L}_{ik}^{(n)} = \frac{1}{\tau_i} \left[1 + \left(y_{ik} - \check{y}_{ik}^{(n)}\right) \frac{\partial}{\partial y_{ik}} + k_{\mathrm{B}} T \frac{\partial^2}{\partial y_{ik}^2}\right], \tag{7.29}$$

where τ_i is the relaxation time associated with the i-th mode, k_B is the Boltzmann constant, and T denotes the temperature.

Equations (7.26)–(7.29) constitute a comprehensive kinetic framework for modeling ultrafast multistage electron transfer dynamics. This formulation captures the essential features of nonadiabatic electronic transitions, nuclear relaxation, and irreversible deactivation of excited states. The initial condition for the probability density vector $\boldsymbol{\varrho}(\boldsymbol{y}, t)$ is determined by the physical conditions under which photoexcitation occurs. In particular, when excitation is induced by a femtosecond laser pulse, the initial nonequilibrium distribution reflects not only intrinsic molecular parameters and excitation frequency, but also the spectral width of the excitation pulse [18, 25, 78, 79].

Experimentally relevant observables include the time-dependent populations of diabatic states, defined by the configuration-space integrals of the corresponding probability densities

$$P_n(t) = \int \varrho_n(\boldsymbol{y}, t), d\boldsymbol{y}. \tag{7.30}$$

In the subsequent section, we address numerical techniques for solving the kinetic equation (7.26), with emphasis on computational approaches applicable to multistage ET cascades in complex molecular environments.

7.3 Numerical Method: Brownian Simulations with Surface Hopping

From a computational perspective, a distinctive feature of the kinetic equation (7.26) is the presence of multiple sink and source terms associated with nonadiabatic transitions. These terms are proportional to position-dependent transition rates, $K_{\mathrm{cs}}^{(nn')}(\boldsymbol{y})$ and $K_{\mathrm{cr}}^{(n0)}(\boldsymbol{y})$, whose sharply localized, delta-function-like structure follows directly from Eq. (7.24). This singular behavior poses a major challenge for traditional numerical approaches, such as finite-difference schemes and grid-based Monte Carlo methods, which typically require high spatial resolution to achieve sufficient accuracy.

This computational difficulty becomes even more pronounced when intramolecular quantum vibrational modes are included in the model. In these scenarios, the number of coupled electron–vibrational states increases substantially, leading to transition regions that are no longer confined to narrow domains in configuration space. As a result, the sink and source terms become distributed, further diminishing the efficiency of conventional discretization-based methods for solving the kinetic equations [30, 80–82].

To overcome the numerical challenges posed by delta-localized transition regions, we employ a trajectory-based simulation method that avoids the need for explicit spatial discretization of the coupling terms and has proven effective for solving the kinetic equations (7.26). The computational cost of this algorithm scales linearly with the dimensionality of the nuclear configuration space, i.e., with the number of independent coordinates y_{ik}, making it particularly well-suited for high-dimensional systems. This approach has previously been applied to simulate a variety of ultrafast ET processes, including intervalence charge transfer in mixed-valence metal complexes [83], photoinduced charge separation in zinc–porphyrin/imide dyads and triads [61], and fluorescence quenching via electron transfer in polar solvents [22, 84].

In the algorithm, the nuclear dynamics are treated as overdamped Brownian motion on diabatic free energy surfaces, representing stochastic fluctuations of the polarization coordinates driven by thermal interactions with the environment. Surface hopping is superimposed on this Brownian propagation: when a trajectory enters a region of near-degeneracy between two FESs, stochastic transition rules determine whether the system hops to the neighboring electronic state. Thus, Brownian motion accounts for the continuous, thermally induced evolution of nuclear coordinates, while surface hopping captures the discrete and probabilistic character of nonadiabatic electronic transitions.

The general Brownian surface-hopping algorithm proceeds according to the following sequence

1. Initialize an ensemble of particles representing the initial probability density $\varrho_0(\boldsymbol{y})$ on the excited-state free energy surface $G^{(1)}$.
2. For each particle residing in electronic state $|\psi_n\rangle$, propagate a stochastic Brownian trajectory on the corresponding diabatic surface $G^{(n)}$, performing the following operations at each time step:

   a. Detect whether the particle enters a transition region between $G^{(n)}$ and a neighboring surface $G^{(n')}$ by evaluating the energy-gap coordinate $z^{(nn')}$.
   b. For each potential transition, compute the probability of nonadiabatic transition and determine stochastically whether a surface hop occurs.
   c. If a hop occurs, update the electronic state to $|\psi_{n'}\rangle$ and continue trajectory propagation on $G^{(n')}$.
   d. If the particle remains in the photoexcited state $|\psi_1\rangle$, evaluate the probability of irreversible deactivation within the time step. If deactivation occurs, transfer the particle to the ground-state surface $G^{(0)}$ and continue propagation accordingly.

3. Compute the time-dependent populations $P_n(t)$ of each electronic state by averaging over the full ensemble of trajectories

$$P_n(t) = \langle \chi_n(t) \rangle ,$$

   where $\chi_n(t)$ is a binary indicator function that equals 1 if the particle is in state $|\psi_n\rangle$ at time t, and 0 otherwise.

In the following subsections, we describe the core components of this algorithm in detail, with particular attention to the stochastic integration of Brownian dynamics and the simulation of surface-hopping events, including both nonadiabatic charge transfer and irreversible excited-state decay.

7.3.1 Stochastic Propagation on Diabatic Free Energy Surfaces

Nonreactive Brownian trajectories on a given diabatic FES $G^{(n)}$ are numerically generated using Green's function formalism, which evaluates the propagator corresponding to the Smoluchowski operator $\hat{L}_{ik}^{(n)}$. Due to the orthogonality of the generalized coordinates y_{ik}, the multidimensional diffusion process decomposes into a set of independent one-dimensional stochastic processes. For clarity, we suppress the indices n, i, and k, and consider a single representative coordinate y. Green's function $F(y, t \mid y_0)$ is defined as the solution to the equation

$$\left(\frac{\partial}{\partial t} - \hat{L}\right) F(y, t \mid y_0) = \delta(t)\delta(y - y_0), \tag{7.31}$$

where $\hat{L}$ is the one-dimensional Smoluchowski operator associated with the harmonic potential $G(y) = (y - \check{y})^2$.

The corresponding Green's function has a closed-form Gaussian solution

$$F(y, t \mid y_0) = \frac{1}{\sqrt{2\pi\sigma^2(t)}} \exp\left[-\frac{(y - \bar{y}(t))^2}{2\sigma^2(t)}\right], \tag{7.32}$$

with mean and variance given by

$$\bar{y}(t) = \check{y} - (y_0 - \check{y})e^{-t/\tau}, \qquad \sigma^2(t) = k_B T \left(1 - e^{-2t/\tau}\right), \tag{7.33}$$

where τ denotes the relaxation time associated with the coordinate y.

As evident from Eq. (7.32), the propagator remains Gaussian at all times, with the mean relaxing exponentially toward the potential minimum $\check{y}$ and the variance growing from zero to its thermal equilibrium value. In the long-time limit ($t \to \infty$), the distribution approaches the stationary Boltzmann distribution corresponding to the harmonic potential.

Given the analytical form of Green's function, nonreactive Brownian trajectories can be generated using the following update rule

$$y(t + \Delta t) = \check{y} + \left(y(t) - \check{y}\right) e^{-\Delta t/\tau} + N\sqrt{k_B T \left(1 - e^{-2\Delta t/\tau}\right)}, \tag{7.34}$$

where Δt is the time step, and N is a normally distributed random variable with zero mean and unit variance, i.e., $\langle N \rangle = 0$ and $\langle N^2 \rangle = 1$. This update rule is exact with respect to the time step Δt, as it is derived directly from the closed-form solution of the Smoluchowski equation and does not rely on finite-difference approximations.

It is important to note that Eq. (7.34) does not generate a continuous physical trajectory of the particle on the $G(y)$ FES. In the mathematical sense, Brownian motion is nowhere differentiable, and thus the notion of a smooth trajectory is ill-defined. Rather, the update rule provides a stochastic sequence of positions at discrete time points $t_m = t_0 + m\Delta t$, which accurately reflect the statistical properties of the underlying diffusion process. The particle's path between consecutive time steps remains undefined, representing the inherent stochasticity of Brownian dynamics at short timescales.

As a consequence of this discretization, certain dynamical quantities cannot be directly obtained from the simulation. These include: (1) the exact time t^* at which a particle crosses the intersection between two diabatic FESs, (2) the instantaneous velocity v_y at the point of crossing, and (3) the number of repeated crossings that may occur within a single time step Δt. Nevertheless, the algorithm does allow for accurate estimation of the total residence time spent within an intersection region, and therefore yields statistically reliable rates for charge separation and recombination processes, in agreement with Fermi's Golden Rule.

A key component of the algorithm is the probabilistic evaluation of surface-hopping transitions, which is addressed in detail in the following subsection.

7.3.2 *Modeling Surface Hops at FESs Intersections*

Within the nonadiabatic framework, electronic transitions between diabatic states are permitted only when the system's nuclear configuration enters a region where two diabatic FESs become degenerate, $G^{(n)} = G^{(n')}$. Accurately detecting such intersection events is therefore essential for simulating surface hops.

The intersection region between surfaces $G^{(n)}$ and $G^{(n')}$ corresponds to a hyperplane defined by the condition $z^{(nn')} = \mathfrak{z}^{(nn')}$, where $z^{(nn')}$ is the scalar projection of the configuration vector $\boldsymbol{y}$ onto the displacement vector $\boldsymbol{D}^{(nn')}$ connecting the minima of the two surfaces. This hyperplane is oriented perpendicular to $\boldsymbol{D}^{(nn')}$ in the extended configuration space. To identify potential transitions, the algorithm monitors the sign of the function $z^{(nn')} - \mathfrak{z}^{(nn')}$ along the Brownian trajectory. A change in sign between two successive time steps indicates that the trajectory has crossed the corresponding intersection hyperplane, and thus entered the region where a nonadiabatic electronic transition may occur.

Let us consider a Brownian trajectory that intersects the degeneracy region between $G^{(n)}$ and $G^{(n')}$ within the time interval $[t, t + \Delta t]$. The probability for the particle to remain in the original diabatic state $|\psi_n\rangle$ throughout this interval is given by the survival function

$$\begin{aligned} S^* &= \exp\left\{-\int_t^{t+\Delta t} K^{(nn')}(\boldsymbol{y}(t))\,dt\right\} \\ &= \exp\left\{-\frac{\pi |V^{(nn')}|^2}{\hbar\sqrt{\lambda^{(nn')}}}\int_{t^*-0}^{t^*+0} \delta\left(z^{(nn')}(t) - \mathfrak{z}^{(nn')}\right) dt\right\}, \end{aligned} \tag{7.35}$$

where t^* denotes the (unknown) time at which the trajectory intersects the hyperplane, $y(t)$ is the particle's position in configuration space at time t, and $z^{(nn')}(t)$ is its projection onto the corresponding ET reaction coordinate.

This formulation is consistent with Fermi's Golden Rule for transitions between diabatic states. The presence of the Dirac delta function in Eq. (7.35) reflects the assumption that electronic transitions occur only at the exact crossing point between the diabatic surfaces, where their energies coincide. In practical simulations, the evaluation of this expression is replaced by a discrete probabilistic criterion for surface hopping.

The integral in Eq. (7.35) can be evaluated analytically, yielding an explicit expression for the survival probability of the particle in its initial diabatic state [70]

$$S^* = \exp\left\{-\frac{\pi|V^{(nn')}|^2}{\hbar|\dot{z}^{(nn')}|\sqrt{\lambda^{(nn')}}}\right\}, \tag{7.36}$$

where $\dot{z}^{(nn')}$ denotes the average velocity of the particle along the corresponding energy-gap coordinate, computed over the time interval $[t, t + \Delta t]$ as

$$\dot{z}^{(nn')} = \frac{z^{(nn')}(t + \Delta t) - z^{(nn')}(t)}{\Delta t}. \tag{7.37}$$

Equation (7.36) forms the basis for implementing electronic transitions within the surface-hopping algorithm. A transition event is simulated by drawing a uniform random number $\xi \in [0, 1)$ and comparing it to the calculated survival probability S^*. If $\xi > S^*$, the particle undergoes a nonadiabatic transition from $|\psi_n\rangle$ to $|\psi_{n'}\rangle$. Otherwise, the particle remains on the current FES $G^{(n)}$. This stochastic decision-making procedure ensures consistency with Fermi's Golden Rule and accurately captures the probabilistic nature of electronic transitions at surface intersections.

Figure 7.7 illustrates the surface hopping algorithm for a simple one-dimensional model consisting of two symmetric diabatic FESs, defined by $G_1(y) = (y + \sqrt{\lambda}/2)^2$ and $G_2(y) = (y - \sqrt{\lambda}/2)^2$. These surfaces intersect at $\tilde{y}^{(12)} = 0$. Figure 7.7A shows a Brownian trajectory calculated using Eq. (7.34) with a time step $\Delta t = 0.01$ ps. The inset highlights that only a part of the intersection events result in electronic transitions. Figure 7.7B provides a schematic representation of the surface hopping mechanism. Possible realizations of the Brownian trajectory near the intersection region are depicted, along with the associated survival and transition probabilities computed over a single time step Δt. The parameters used in the simulation are: $\lambda = 0.15$ eV, $k_B T = 0.025$ eV, $V = 0.006$ eV, $\tau = 1$ ps.

Unlike the update rule for nonreactive trajectory Eq. (7.34), the surface hopping algorithm based on Eq. (7.36) exhibits sensitivity to the time step Δt. Due to well-known characteristics of Brownian motion, the mean velocity $\dot{z}^{(nn')}$, calculated using Eq. (7.37), depends on Δt and even diverges as $\Delta t \to 0$. This dependence influences the single-crossing survival probability S^* and, consequently, the frequency of reaction events. Nevertheless, ensemble-averaged simulations reveal that the overall ET

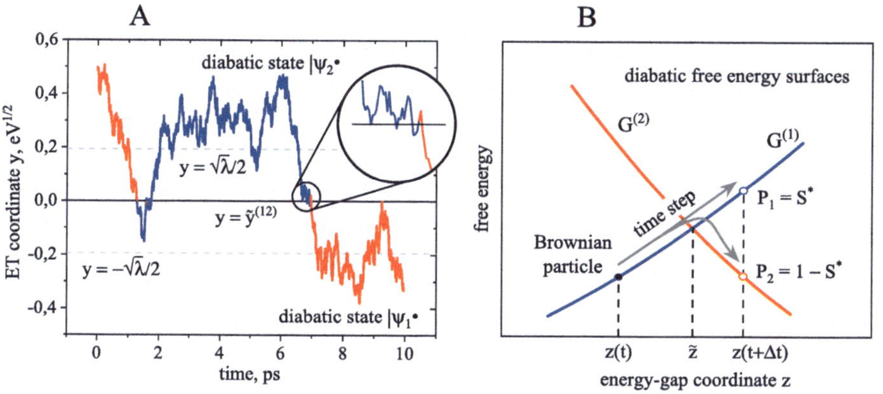

Fig. 7.7 **A** Simulated Brownian trajectory in a one-dimensional model with two intersecting diabatic FESs, $G_1(y)$ and $G_2(y)$. Dashed lines mark the FESs minima, the solid line at $y = 0$ indicates the intersection point. Surface hopping events are indicated by color changes: red corresponds to state $|\psi_1\rangle$, blue to $|\psi_2\rangle$. **B** Schematic illustration of the decision process for probabilistic surface hops during Brownian motion

kinetics remain invariant with respect to the choice of Δt, provided the following condition is met throughout the simulation [70]

$$\langle T^* \rangle \ll 1, \tag{7.38}$$

where $T^* \equiv 1 - S^*$ represents the probability of a transition during a single crossing event. This criterion ensures that the system operates in the weak-coupling regime and can be satisfied by selecting a sufficiently small time step Δt.

Assuming small time steps $\Delta t \ll \tau$, one can derive from Eqs. (7.34) and (7.37) the following estimate for the particle's velocity along the energy-gap coordinate

$$\dot{z}^{(nn')} = -\sum_{i,k} \frac{y_{ik} - \check{y}_{ik}^{(n)}}{\tau_i} \cos\theta_{ik}^{(nn')} + N\sqrt{\frac{2k_B T}{\Delta t}} \sum_{i,k} \frac{\cos\theta_{ik}^{(nn')}}{\sqrt{\tau_i}}.$$

The first term in this expression describes a regular drift of the particle toward the FES minimum, while the second term represents stochastic fluctuations arising from thermal noise. At $\Delta t \ll \tau k_B T/\lambda^{(nn')}$, the stochastic component dominates. In this regime, the mean absolute velocity $\langle|\dot{z}^{(nn')}|\rangle$ can be estimated as

$$\langle|\dot{z}^{(nn')}|\rangle \approx \frac{2}{\sqrt{\pi}}\sqrt{\frac{k_B T}{\Delta t}}\zeta^{(nn')}, \qquad \zeta^{(nn')} \equiv \sum_{i,k} \frac{\cos\theta_{ik}^{(nn')}}{\sqrt{\tau_i}}. \tag{7.39}$$

The quantity $\zeta^{(nn')}$ characterizes the effective diffusive mobility of a particle along the energy-gap coordinate $z^{(nn')}$ on short time scales. For convenience, this result can be recast in terms of an effective relaxation time

$$\langle|\dot{z}^{(nn')}|\rangle \approx \frac{2}{\sqrt{\pi}}\sqrt{\frac{k_{\rm B}T}{\tau_{\rm eff}^{(nn')}\Delta t}}, \tag{7.40}$$

where the effective relaxation time $\tau_{\rm eff}^{(nn')}$ is defined as

$$\tau_{\rm eff}^{(nn')} = \left(\zeta^{(nn')}\right)^{-2}. \tag{7.41}$$

By inserting Eq. (7.40) into Eq. (7.36), one obtains an explicit relation between the time step Δt and the average single-crossing transition probability $\langle T^*\rangle$. The condition $\langle T^*\rangle \ll 1$, required for the validity of the algorithm, is satisfied if the time step obeys the inequality

$$\Delta t \ll \frac{4\hbar^2\lambda^{(nn')}k_{\rm B}T}{\pi^3\tau_{\rm eff}^{(nn')}\left|V^{(nn')}\right|^4}. \tag{7.42}$$

This expression highlights the strong dependence of the allowed time step on the electronic coupling $V^{(nn')}$. As a result, the algorithm is particularly efficient for simulating nonadiabatic (weak-coupling) ET processes, where $V^{(nn')}$ is small.

It is important to note that Eq. (7.42) is a local condition, meaning it only needs to be satisfied in the vicinity of the FESs intersection region where transitions occur. This locality enables the use of efficient adaptive time-stepping schemes, in which smaller time steps are applied near term-crossing regions, while larger steps can be used elsewhere to reduce computational cost [70].

Although the above approach is particularly useful in the weak-coupling regime, alternative methods exist for stronger electronic coupling, as discussed in the following subsection.

7.3.3 Comparison with Reactive Green's Function Method

As an alternative to the surface hopping algorithm described above, a distinct numerical technique was developed in Refs. [21, 83, 85] for treating electron transfer in the strong-coupling (solvent-controlled) regime, where $V^{(nn')}$ is relatively large. This method is based on the evaluation of Green's functions for coupled diffusive equations that include delta-function sink-source terms to model ET transitions.

In contrast to the nonadiabatic surface hopping algorithm, the trajectories generated by the reactive Green's function method are inherently reactive–electronic transitions are incorporated directly into the propagation of each Brownian path. This intrinsic reactivity makes the method especially efficient for simulating solvent-

Table 7.1 Comparison of the two numerical schemes for simulating ET dynamics in the diabatic representation: the nonadiabatic surface hopping algorithm and the reactive Green's function method from Ref. [85]

Feature	Surface hopping algorithm	Reactive green's function method
ET regime applicability	Best suited for nonadiabatic (weak coupling) regime	Best suited for diffusion-controlled (strong coupling) regime
Trajectory type	Nonreactive; surface hops occur probabilistically at the FESs intersection regions	Reactive; transitions are incorporated within the trajectory
Time step sensitivity	High sensitivity; requires small Δt to maintain accuracy	Less sensitive; transitions handled within Green's function
Treatment of FESs crossings	Requires detection and evaluation of survival probability	Handled analytically through delta-localized terms
Computational cost per step	Low (simple updates and random number generation)	Higher (requires evaluation of the inverse Laplace transforms)

controlled ET, as confirmed by prior applications [85]. However, in the weak-coupling regime, the approach becomes less computationally efficient due to the increased numerical effort required to resolve rare transition events. Table 7.1 summarizes the key characteristics of both simulation strategies, highlighting their respective strengths and domains of applicability.

We now apply the general theoretical and computational framework to a representative three-center molecular system. Specifically, we consider D–A_1–A_2 triads, which serve as minimal models for investigating competitive ultrafast ET processes in multiredox architectures.

7.4 Competitive ET in Donor–Acceptor(1)–Acceptor(2) Compounds: General Formulation of the Model

Building on the theoretical framework developed in Sect. 7.2, we now consider a model molecular triad of the form DA_1A_2, where D is a photoactive electron donor, and A_1 and A_2 are electron acceptors. This DA_1A_2 motif serves as a minimal yet representative architecture for a broad class of molecular assemblies employed in optoelectronics, photocatalysis, and solar energy conversion, where the efficiency of photoinduced charge separation is essential [4, 6, 8, 9, 11, 86–93]. Moreover, natural photosynthetic reaction centers in plants and bacteria embody analogous multicenter systems that exploit ultrafast ET to achieve highly efficient conversion of solar energy into long-lived charge-separated states [35, 94].

Photoexcitation of the DA_1A_2 triad initiates a sequence of nonadiabatic electronic transitions, as depicted schematically in Fig. 7.8. Some of these transitions may take place under nonequilibrium conditions, during the system's evolution on diabatic

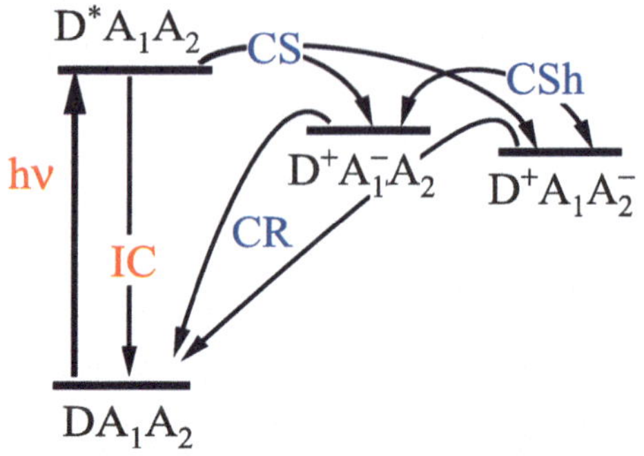

Fig. 7.8 Photochemical processes in the DA_1A_2 compound. CS, CR, CSh and IC denote charge separation, charge recombination, charge shift and internal conversion, respectively

FESs prior to full thermal relaxation. To account for these effects, we employ a general model incorporating a multicomponent relaxation function $\Gamma(t)$, as introduced in Eq. (7.11). A key advantage of this three-center/multicomponent formulation is that the corresponding diabatic FESs can be derived in closed analytical form. This property is important for validation of the theoretical framework developed in this work.

We assume that photoexcitation does not produce a substantial redistribution of electron density within the triad, thereby justifying the neglect of environmental reorganization at the excitation stage. Accordingly, the minima of the ground- and excited-state FESs are taken to coincide. The system is described in a diabatic basis of states with fixed electronic configurations

$$|\varphi_0\rangle = |DA_1A_2\rangle, \quad |\varphi_1\rangle = |D^*A_1A_2\rangle, \quad |\varphi_2\rangle = |D^+A_1^-A_2\rangle, \quad |\varphi_3\rangle = |D^+A_1A_2^-\rangle.$$

The equilibrium free energies $\check{G}^{(n)}$ associated with diabatic states $|\varphi_n\rangle$, as well as the corresponding reorganization free energies $\lambda^{(nn')}$, are treated as known parameters of the model.

For a three-center molecular system ($N = 3$) embedded in an L-mode environment, the extended configuration space $\boldsymbol{y}$ contains $(N-1)L = 2L$ independent nuclear coordinates (see Eq. (7.14)). It is convenient to represent the vector $\boldsymbol{y}$ in matrix form as

$$\boldsymbol{y} = \boldsymbol{x}^{\mathrm{T}} \otimes \begin{pmatrix} q_1 \\ q_2 \end{pmatrix} = \begin{pmatrix} y_{11} & y_{21} & \cdots & y_{L1} \\ y_{12} & y_{22} & \cdots & y_{L2} \end{pmatrix}, \tag{7.43}$$

where (q_1, q_2) are the polarization coordinates that span the two-dimensional configuration space, and $\boldsymbol{x}^{\mathrm{T}} = (x_1, x_2, \ldots, x_L)$ contains the corresponding relaxation amplitudes. The coefficients $x_i = \sqrt{\gamma_i}$ are normalized such that $\sum x_i^2 = 1$.

To compute the coordinates of the FES minima $\check{\boldsymbol{q}}^{(n)}$ (in polarization space) and $\check{\boldsymbol{y}}^{(n)}$ (in the full configuration space), we apply the general algorithm described in Sect. 7.2. Since arrangement of the $\check{\boldsymbol{q}}^{(n)}$ points depends solely on the relative displacement vectors $\boldsymbol{d}^{(nn')}$ (see Eq. (7.10)), the absolute positions of the FES minima in the $\boldsymbol{q}$-space are defined only up to isometric transformations, i.e., global translations and rotations. This invariance allows us to fix one of the FES minima arbitrarily. For convenience, we place the minimum of the excited state at the origin

$$\check{\boldsymbol{q}}^{(1)} = \begin{pmatrix} 0 \\ 0 \end{pmatrix}. \tag{7.44}$$

This choice simplifies the construction of the remaining minima.

The corresponding minimum in the extended space $\boldsymbol{y}$ is then obtained by applying the outer product structure

$$\check{\boldsymbol{y}}^{(1)} = \boldsymbol{x}^{\mathrm{T}} \otimes \check{\boldsymbol{q}}^{(1)} = \begin{pmatrix} 0\ 0 \dots 0 \\ 0\ 0 \dots 0 \end{pmatrix}. \tag{7.45}$$

To determine $\check{\boldsymbol{y}}^{(2)}$, we use the known reorganization energy $\lambda^{(12)}$ associated with the $|\varphi_1\rangle \to |\varphi_2\rangle$ transition. According to Eq. (7.10), the displacement between $\check{\boldsymbol{q}}^{(1)}$ and $\check{\boldsymbol{q}}^{(2)}$ is $\sqrt{\lambda^{(12)}}$. At this stage, it is sufficient to consider a displacement along a single polarization coordinate, such as q_1. This yields

$$\check{\boldsymbol{q}}^{(2)} = \begin{pmatrix} \sqrt{\lambda^{(12)}} \\ 0 \end{pmatrix}, \tag{7.46}$$

and the corresponding FES minimum in the composite space

$$\check{\boldsymbol{y}}^{(2)} = \boldsymbol{x}^{\mathrm{T}} \otimes \check{\boldsymbol{q}}^{(2)} = \sqrt{\lambda^{(12)}} \begin{pmatrix} x_1\ x_2 \dots x_L \\ 0\ \ 0\ \dots\ 0 \end{pmatrix}. \tag{7.47}$$

Similar geometric approach is used to evaluate $\check{\boldsymbol{y}}^{(3)}$. Given the reorganization energies $\lambda^{(13)}$ and $\lambda^{(23)}$, the distances from $\check{\boldsymbol{q}}^{(3)}$ to $\check{\boldsymbol{q}}^{(1)}$ and $\check{\boldsymbol{q}}^{(2)}$ are $d^{(13)} = \sqrt{\lambda^{(13)}}$ and $d^{(23)} = \sqrt{\lambda^{(23)}}$, respectively. The point $\check{\boldsymbol{q}}^{(3)}$ therefore lies in the plane spanned by the two polarization coordinates q_1 and q_2. The explicit form is given by

$$\check{\boldsymbol{q}}^{(3)} = \begin{pmatrix} \sqrt{\lambda^{(13)}} \cos\theta \\ \sqrt{\lambda^{(13)}} \sin\theta \end{pmatrix}, \tag{7.48}$$

with the angle θ determined by the law of cosines

$$\cos\theta = \frac{\lambda^{(12)} + \lambda^{(13)} - \lambda^{(23)}}{2\sqrt{\lambda^{(12)}\lambda^{(13)}}}. \tag{7.49}$$

This result can be verified by direct evaluation of $d^{(nn')}$ from Eqs. (7.44)–(7.48), and comparing them to the expected values $\sqrt{\lambda^{(nn')}}$.

Using Eq. (7.48) one obtains the coordinates of the $G^{(3)}$ FES minimum in $\boldsymbol{y}$ space

$$\check{\boldsymbol{y}}^{(3)} = \boldsymbol{x}^{\mathrm{T}} \otimes \check{\boldsymbol{q}}^{(3)} = \sqrt{\lambda^{(13)}} \begin{pmatrix} x_1 \cos\theta \; x_2 \cos\theta \; \ldots \; x_L \cos\theta \\ x_1 \sin\theta \; x_2 \sin\theta \; \ldots \; x_L \sin\theta \end{pmatrix}. \tag{7.50}$$

From Eqs. (7.15) and (7.45)–(7.50), the diabatic FESs of the DA_1A_2 triad are expressed in terms of the y_{ik} coordinates as follows

$$\begin{aligned} G^{(0)}(\boldsymbol{y}) &= \sum_i (y_{i1}^2 + y_{i2}^2) + \check{G}^{(0)}, \\ G^{(1)}(\boldsymbol{y}) &= \sum_i (y_{i1}^2 + y_{i2}^2) + \check{G}^{(1)}, \\ G^{(2)}(\boldsymbol{y}) &= \sum_i \left[\left(y_{i1} - x_i \sqrt{\lambda^{(12)}} \right)^2 + y_{i2}^2 \right] + \check{G}^{(2)}, \\ G^{(3)}(\boldsymbol{y}) &= \sum_i \left[\left(y_{i1} - x_i \cos\theta \sqrt{\lambda^{(13)}} \right)^2 + \left(y_{i2} - x_i \sin\theta \sqrt{\lambda^{(13)}} \right)^2 \right] + \check{G}^{(3)}, \end{aligned} \tag{7.51}$$

where $\check{G}^{(n)}$ denotes the equilibrium free energy of diabatic state $|\varphi_n\rangle$.

This set of FESs fully specifies the energetic landscape for all ET processes within the triad, including activation barriers for both charge separation and recombination transitions. As an illustrative example, consider electron transfer from the photoexcited donor D^* to the primary acceptor A_1, corresponding to the transition $|\varphi_1\rangle \to |\varphi_2\rangle$. The relevant diabatic FESs, $G^{(1)}$ and $G^{(2)}$, intersect along a $(2L-1)$-dimensional hyperplane, defined in terms of the energy-gap coordinate by

$$z^{(12)} = \sum_{i=1}^{L} y_{i1} x_i = \frac{\lambda^{(12)} + \Delta\check{G}^{(21)}}{2\sqrt{\lambda^{(12)}}} \equiv z^{(12)}_{\ddagger}. \tag{7.52}$$

This result gives the ET activation free energy

$$G^{\sharp} = \frac{\left(\lambda^{(12)} + \Delta\check{G}^{(21)} \right)^2}{4\lambda^{(12)}}, \tag{7.53}$$

which exactly recovers the classical Marcus result for nonadiabatic electron transfer [95].

Similar expressions can be derived for the activation barriers of other elementary transitions in the triad, confirming that the proposed model reproduces the correct equilibrium-limit behavior for all quasi-thermal ET reactions.

In the following section, we extend the analysis beyond the equilibrium framework to demonstrate that the model also captures nonequilibrium effects in ultrafast

ET. Specifically, we examine two limiting cases of the DA_1A_2 model for which exact analytical solutions are available in the literature, providing benchmarks for validating the theoretical approach.

7.4.1 Model Validation via Subspace Projection Analysis

The general theoretical framework for DA_1A_2 systems developed in the preceding subsection can be validated by comparison with two well-established models that represent limiting cases of our formulation. Specifically, we consider: (1) the Najbar–Tachiya model, which describes sequential ET in a three-center system coupled to a single-mode Debye solvent [51], and (2) the extended Sumi–Marcus model, which treats single-step ET in a two-center system embedded in a multicomponent environment [53, 96].

These models differ in both system dimensionality and the structure of their configuration spaces. The Najbar–Tachiya model incorporates three redox centers ($N = 3$) but restricts the environmental dynamics to a single collective relaxation mode ($L = 1$). In contrast, the extended Sumi–Marcus model accommodates an arbitrary number of relaxation modes (L), while limiting the electronic subsystem to two redox centers ($N = 2$).

A central distinction between the two lies in the choice of coordinate representation: the Najbar–Tachiya model is formulated in polarization coordinates, whereas the extended Sumi–Marcus model adopts relaxation coordinates. These coordinate sets span different subspaces of the full multidimensional configuration space employed in our generalized approach, as schematically illustrated in Fig. 7.5. As our framework generalizes both limits, it should correctly reproduce their results under appropriate parameter constraints, consistent with the correspondence principle.

To establish agreement with the Najbar–Tachiya model, we impose uniform solvent relaxation times, $\tau_1 = \tau_2 = \cdots = \tau_L = \tau$, thereby collapsing the relaxation dynamics to a single effective mode with a time scale τ. We then project the diabatic FESs from the full $\boldsymbol{y}$-space onto the reduced polarization subspace $\boldsymbol{q}$, effectively eliminating the relaxation coordinates $\boldsymbol{x}$ by folding the $\boldsymbol{x}$-space. This yields the following coordinates for the FES minima

$$\begin{aligned}
\check{\boldsymbol{q}}^{(0)} &= \check{\boldsymbol{q}}^{(1)} = \hat{P}_q\, \check{\boldsymbol{y}}^{(1)} = \begin{pmatrix} 0 \\ 0 \end{pmatrix}, && (|DA_1A_2\rangle \text{and} |D^*A_1A_2\rangle \text{states}) \\
\check{\boldsymbol{q}}^{(2)} &= \hat{P}_q\, \check{\boldsymbol{y}}^{(2)} = \begin{pmatrix} \sqrt{\lambda^{(12)}} \\ 0 \end{pmatrix}, && (|D^+A_1^-A_2\rangle \text{ state}) \\
\check{\boldsymbol{q}}^{(3)} &= \hat{P}_q\, \check{\boldsymbol{y}}^{(3)} = \begin{pmatrix} \sqrt{\lambda^{(13)}}\cos\theta \\ \sqrt{\lambda^{(13)}}\sin\theta \end{pmatrix}, && (|D^+A_1A_2^-\rangle \text{ state})
\end{aligned} \tag{7.54}$$

where $\hat{P}_q$ denotes the projection operator defined by Eq. (7.18).

The corresponding free energy surfaces in the polarization coordinate space (q_1, q_2) are then given by

$$G^{(0)}(q_1, q_2) = q_1^2 + q_2^2 + \check{G}^{(0)}, \qquad G^{(1)}(q_1, q_2) = q_1^2 + q_2^2 + \check{G}^{(1)},$$
$$G^{(2)}(q_1, q_2) = \left(q_1 - \sqrt{\lambda^{(12)}}\right)^2 + q_2^2 + \check{G}^{(2)}, \tag{7.55}$$
$$G^{(3)}(q_1, q_2) = \left(q_1 - \cos\theta\sqrt{\lambda^{(13)}}\right)^2 + \left(q_2 - \sin\theta\sqrt{\lambda^{(13)}}\right)^2 + \check{G}^{(3)}.$$

The system's time evolution in this representation is governed by a set of coupled Smoluchowski equations describing diffusion over the two-dimensional parabolic potentials $G^{(n)}(q_1, q_2)$. The diffusion coefficients along both polarization coordinates are equal, $D_1 = D_2 = k_B T/\tau$, reflecting the single-mode relaxation assumption. This formulation recovers the structure of the Najbar–Tachiya model for three-center electron transfer in a Debye solvent [51, 52, 59]. The two-dimensional description of diabatic FESs in terms of polarization coordinates, as expressed in Eq. (7.55), has been widely employed in models that incorporate electron–vibrational interactions in DA_1A_2 systems (e.g., [26, 61, 62, 97]).

Figure 7.9 depicts the diabatic free energy surfaces projected into the $\boldsymbol{q}$-space, constructed using Eqs. (7.55). The parameters are set to: $\lambda^{(12)} = 0.9$ eV, $\lambda^{(13)} = 1.0$ eV, and $\lambda^{(23)} = 0.4$ eV, resulting in an inter-coordinate angle $\theta \approx 40°$ between the energy-gap directions $z^{(12)}$ and $z^{(13)}$. Colored lines indicate regions of FES intersection corresponding to different ET channels: red denotes charge separation (CS), blue marks charge recombination (CR), and green represents the charge shift (CSh)

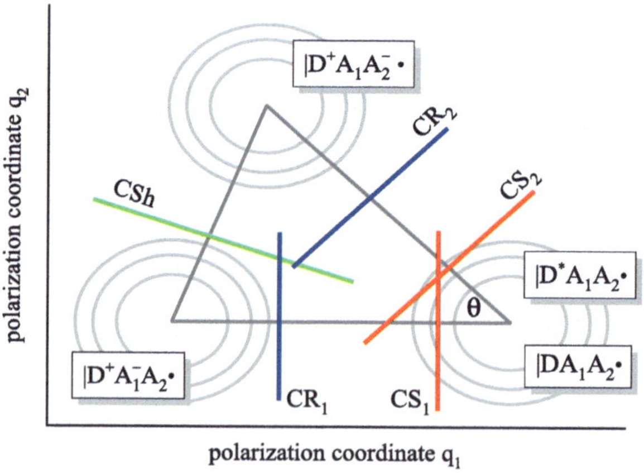

Fig. 7.9 Diabatic FESs of the DA_1A_2 compound in the polarization coordinate space $\boldsymbol{q}$. Colored lines indicate electron transfer regions: red (CS), blue (CR), and green (CSh), corresponding to charge separation, recombination, and shift, respectively. The position and orientation of each intersection line relative to the FES minima are governed by the values of $\Delta\check{G}^{(nn')}$ and $\lambda^{(nn')}$ for the corresponding ET transition

between ionic configurations. These processes compete during the early stages following photoexcitation. Within this reduced, two-dimensional representation, the FES topology fully captures the energetics and competition between ultrafast ET pathways in the DA_1A_2 triad embedded in a single-mode polar environment.

We now examine the correspondence between the general DA_1A_2 framework and the extended Sumi–Marcus model by projecting the diabatic FESs from the full configuration space $\boldsymbol{y}$ onto the relaxation subspace $\boldsymbol{x}$. The resulting FESs take the form

$$G^{(n)}(\boldsymbol{X}) = \sum_i \left(X_i - \check{X}_i^{(n)}\right)^2 + \check{G}^{(0)}, \tag{7.56}$$

where the new variables are defined as $X_i \equiv (\hat{P}_x \boldsymbol{y})_i = |\boldsymbol{q}|x_i$, and $\hat{P}_x$ denotes the projection operator onto the relaxation coordinate subspace. The equilibrium positions of the FES minima in this representation are given by

$$\check{X}_i^{(n)} \equiv \left(\hat{P}_x \check{\boldsymbol{y}}^{(n)}\right)_i = |\check{\boldsymbol{q}}^{(n)}|x_i, \tag{7.57}$$

with $|\check{\boldsymbol{q}}^{(n)}|$ representing the Euclidean norm of the polarization coordinate vector associated with state $|\varphi_n\rangle$.

Applying this projection to the state vectors $\check{\boldsymbol{y}}^{(n)}$ derived previously (Eqs. (7.45), (7.47), (7.50)) yields

$$\check{\boldsymbol{X}}^{(0)} = \check{\boldsymbol{X}}^{(1)} = \boldsymbol{0}, \quad \check{\boldsymbol{X}}^{(2)} = \sqrt{\lambda^{(12)}}\boldsymbol{x}, \quad \check{\boldsymbol{X}}^{(3)} = \sqrt{\lambda^{(13)}}\boldsymbol{x},$$

where $\boldsymbol{0}$ is the L-dimensional zero vector. The corresponding free energy surfaces, expressed in terms of the X_i coordinates, are

$$\begin{aligned} G^{(0)}(\boldsymbol{X}) &= \sum_i X_i^2 + \check{G}^{(0)}, \qquad G^{(1)}(\boldsymbol{X}) = \sum_i X_i^2 + \check{G}^{(1)}, \\ G^{(2)}(\boldsymbol{X}) &= \sum_i \left(X_i - \sqrt{\lambda^{(12)}}x_i\right)^2 + \check{G}^{(2)}, \\ G^{(3)}(\boldsymbol{X}) &= \sum_i \left(X_i - \sqrt{\lambda^{(13)}}x_i\right)^2 + \check{G}^{(3)}. \end{aligned} \tag{7.58}$$

We now demonstrate that the present framework reproduces the well-established multicomponent model of electron transfer for any pair of diabatic states. As a representative example, consider ET between the photoexcited and charge-separated configurations, $|D^*A_1A_2\rangle$ and $|D^+A_1^-A_2\rangle$, associated with the reorganization free energy $\lambda^{(12)}$. To facilitate comparison with standard models, we introduce a rescaled set of relaxation coordinates defined as $Q_i = X_i/2\sqrt{\lambda^{(12)}}$. In terms of these variables, the diabatic FESs corresponding to the excited and charge-separated states are given by

$$G^{(1)}(Q) = \sum_i \frac{Q_i^2}{4\lambda_i^{(12)}} + \check{G}^{(1)}, \qquad G^{(2)}(Q) = \sum_i \frac{(Q_i - 2\lambda_i^{(12)})^2}{4\lambda_i^{(12)}} + \check{G}^{(2)}, \quad (7.59)$$

where $\lambda_i^{(12)} = \lambda^{(12)} x_i^2$ denotes the contribution of the i-th environmental relaxation mode to the total reorganization energy. These expressions exactly reproduce the canonical form of the two-state, multicomponent ET model frequently employed in the theoretical treatment of single-step ET processes in non-Debye solvents (see, e.g., [67, 75, 98]).

Figure 7.10 illustrates the arrangement of the diabatic FESs $G^{(n)}$ in the relaxation coordinate space $\boldsymbol{X}$, computed according to Eqs. (7.58). The model parameters used are: $\lambda^{(12)} = 0.4$ eV, $\lambda^{(13)} = 0.9$ eV, with relaxation mode weights $\gamma_1 = 0.6$, $\gamma_2 = 0.4$, which define the respective contributions of the two environmental modes to the total reorganization energies. As implied by Eq. (7.57), the minima of the diabatic FESs in $\boldsymbol{X}$-space lie along a common line directed along the unit vector $\boldsymbol{x}$. Explicitly, these positions satisfy the linear relation $X_2 = \sqrt{\gamma_2/\gamma_1}X_1$, reflecting the anisotropic relaxation contributions encoded by γ_1 and γ_2.

The intersection hyperplane between any two diabatic surfaces $G^{(n)}(\boldsymbol{X})$ and $G^{(n')}(\boldsymbol{X})$ can be derived analytically by equating the corresponding expressions in Eq. (7.58). This yields the condition

$$\sum_i x_i X_i = \boldsymbol{x} \cdot \boldsymbol{X} = -\frac{\lambda^{(nn')} + \Delta\check{G}^{(nn')}}{2\sqrt{\lambda^{(nn')}}}, \quad (7.60)$$

where $\Delta\check{G}^{(nn')} \equiv \check{G}^{(n')} - \check{G}^{(n)}$ is the equilibrium free energy gap between the diabatic states. The resulting hyperplane, orthogonal to $\boldsymbol{x}$, defines the reaction zone for each ET process (see Fig. 7.10).

Equation (7.60) provides the condition for resonance between two diabatic states: it specifies the set of points $\boldsymbol{X}$ in the relaxation coordinate space where the free energies of $|\psi_n\rangle$ and $|\psi_{n'}\rangle$ are equal. Geometrically, this relation defines a hyperplane perpendicular to the vector $\boldsymbol{x}$, which encodes the relative contributions of the environmental relaxation modes. The position of the hyperplane is determined by two energetic quantities: (i) the reorganization free energy $\lambda^{(nn')}$, which sets the distance between the minima of the diabatic surfaces, and (ii) the thermodynamic driving force $\Delta\check{G}^{(nn')}$, which shifts the relative vertical alignment of the surfaces. Thus, Eq. (7.60) directly links the system's energetic parameters to the spatial location of the ET reaction zone in $\boldsymbol{X}$-space.

In nonequilibrium ET, the system does not propagate from the minimum of the donor FES but instead starts from a displaced configuration, typically generated by direct photoexcitation or a preceding ET step. As a result, the system's trajectory in $\boldsymbol{X}$-space may enter the reaction region before reaching thermal equilibrium—a process known as non-thermal or hot electronic transition [26, 79, 99, 100]. In the following section, we explore how hot transitions in molecular triads can suppress hot charge recombination, thereby enhancing the efficiency of photoinduced charge separation in DA_1A_2 compounds.

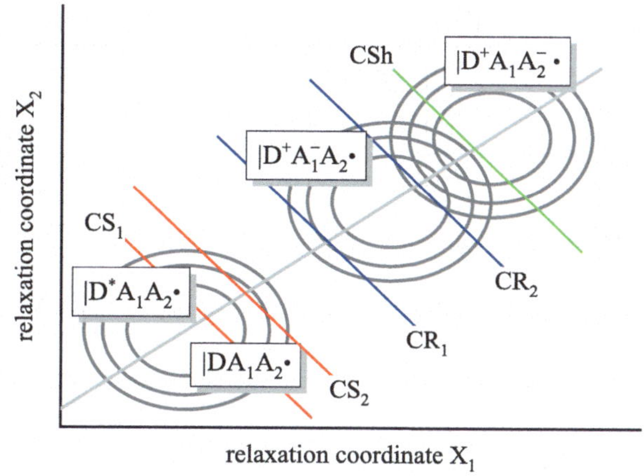

Fig. 7.10 Diabatic FESs of the DA_1A_2 triad in the relaxation coordinate space $\boldsymbol{X}$, evaluated using Eqs. (7.58). The gray line, defined by $X_2 = \sqrt{\gamma_2/\gamma_1}X_1$, represents the direction of the energy-gap ET coordinate. Intersection lines for CS, CR, and CSh processes are highlighted in color

7.5 Efficiency of Charge Separation from the Second Excited State in DA_1A_2 Compounds: Impact of Hot Electron Transfer

A promising direction in the development of photovoltaic and optoelectronic materials involves designing macromolecular systems that facilitate efficient photochemical charge separation through cascaded ultrafast ET steps, leading to the formation of long-lived charge-separated states. A broad class of such systems has been explored, frequently employing porphyrins as photoactive electron donors in combination with fullerenes or imide-based moieties as electron acceptors [87, 101–106]. These donor–acceptor assemblies emulate the architecture of natural photosynthetic reaction centers, where multistep ET enables spatial separation of charges and suppression of recombination, thereby stabilizing the CS state.

Recent experimental investigations have shown that molecular triads of the type A_L–D–A_R, in which the electron donor possesses two distinct locally excited states, can exhibit wavelength-dependent charge separation. In such systems, the direction of ET can be selectively modulated by tuning the excitation frequency. A representative example is the NDI–ZnP–NI triad, in which ZnP (zinc porphyrin) functions as the central photoactive donor, while NI (naphthaleneimide) and NDI (naphthalenediimide) serve as electron-accepting units on opposite sides of the porphyrin ring [87]. Upon excitation to the lowest singlet excited state (S_1, Q-band), electron transfer predominantly proceeds toward the NDI acceptor (A_L). Conversely, excitation to the higher-lying singlet state (S_2, Soret band) initiates ET to the NI acceptor (A_R), enabling photoselective control over the CS direction.

Such wavelength-dependent ET behavior opens opportunities for optoelectronic applications, including optical molecular switches whose dipole orientation and CS character depend on the excitation frequency. However, the CS efficiency of ZnP-based molecular switches remains limited, primarily due to ultrafast charge recom-

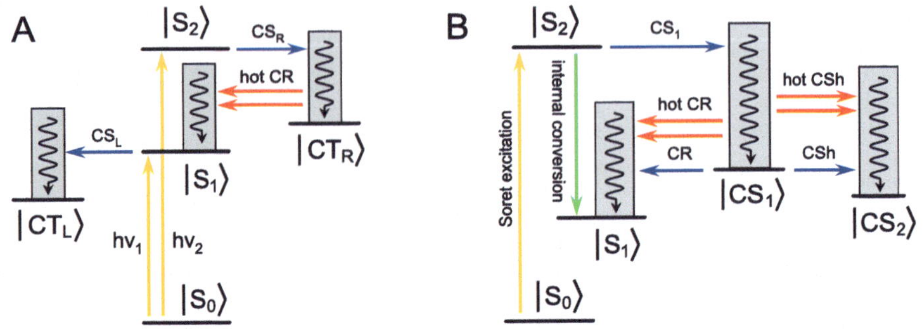

Fig. 7.11 **A** Wavelength-selective photoinduced charge separation in the NDI–ZnP–NI triad, as reported in Ref. [87]. The two charge-separated states, $|CT_L\rangle$ and $|CT_R\rangle$, correspond to NDI^-–ZnP^+–NI and NDI–ZnP^+–NI^- configurations, respectively. Red arrows denote nonequilibrium back electron transfer (hot charge recombination), which significantly reduces the CS efficiency in the right-hand ET branch. **B** Schematic representation of photochemical processes in DA_1A_2 compound, which models the right-hand ET branch of the NDI–ZnP–NI triad but includes an additional secondary acceptor covalently attached to the NI moiety. This design aims to suppress hot CR by enabling ultrafast charge shift (CSh) to A_2

bination in the $D^+A_R^-$ state generated from the S_2 excitation [87]. The electronic structure of these systems promotes fast back electron transfer from the CS state $|CT_R\rangle$ to the lower-lying neutral excited state $|S_1\rangle$, driven by solvent and vibrational relaxation (see Fig. 7.11A). This recombination process, which occurs before full thermal equilibration, constitutes hot electron transfer [17]. Time-resolved spectroscopic studies reveal that only 10–20% of the initially formed $D^+A_R^-$ pairs avoid hot recombination within the first few picoseconds following excitation.

The issue of hot charge recombination in the right-hand ET branch of the NDI–ZnP–NI triad may be addressed by incorporating an additional electron acceptor covalently linked to the NI moiety. In this extended four-center configuration, A_L–D–A_{R1}–A_{R2}, the initially formed charge-separated state $D^+A_{R1}^-$ can undergo ultrafast subsequent electron transfer to A_{R2}. This sequential process competes kinetically with charge recombination and effectively diverts the system away from back electron transfer to the lower-lying neutral state $|S_1\rangle$. Such intramolecular competition offers a strategy for suppressing hot recombination and enhancing the overall yield of photoinduced charge separation.

In this section, we examine the potential of suppressing hot CR via ultrafast sequential ET in DA_1A_2 triads. We assume that photoinduced charge separation is initiated from the second locally excited state $|S_2\rangle$, consistent with the mechanism illustrated in Fig. 7.11B. To focus on the relevant dynamics, we restrict our analysis to the "right" ET branch of the system, neglecting the "left" branch on the premise that it exerts minimal influence on the overall photochemical behavior under the conditions considered. The primary objective of this analysis is to quantify the effect of the secondary acceptor A_2 on the CS efficiency in triads.

7.5.1 Model Equations

To describe the photochemical dynamics of the DA_1A_2 system following excitation to the Soret band, we employ the general theoretical framework presented in Sect. 7.4. For notational clarity, we define the relevant diabatic electronic states as follows

$$|\psi_0\rangle = |S_1\rangle = |D^*A_1A_2\rangle, \qquad |\psi_1\rangle = |S_2\rangle = |D^{**}A_1A_2\rangle,$$
$$|\psi_2\rangle = |CS_1\rangle = |D^+A_1^-A_2\rangle, \qquad |\psi_3\rangle = |CS_2\rangle = |D^+A_1A_2^-\rangle. \tag{7.61}$$

The electronic ground state, $|S_0\rangle = |DA_1A_2\rangle$, is excluded from subsequent analysis, as recombination to this state is negligible on the timescale considered. This simplification is justified by the substantial energy gap between the $|CS_1\rangle$ and $|S_0\rangle$ states in ZnP–imide systems, expressed as $\Delta G_{\mathrm{cr0}} = \check{G}_{\mathrm{CS1}} - \check{G}_{\mathrm{S0}} \ll -\lambda^{(12)}$, which renders this recombination pathway significantly slower than the picosecond-scale dynamics of interest.

In a single-component polar environment, ultrafast competitive CS and CR processes in DA_1A_2 triads are frequently modeled using a two-dimensional polarization coordinate space [16, 51, 58, 107]. Since the present analysis focuses on the sequential two-step ET pathway $|S_2\rangle \to |CS_1\rangle \to |CS_2\rangle$, it is convenient to characterize the geometry of the diabatic FESs in terms of the angle between the displacement vectors $\boldsymbol{d}^{(21)}$ and $\boldsymbol{d}^{(23)}$ corresponding to the respective ET transitions. This angle, denoted $\theta^{(23)}$, is given by the geometric relation

$$\cos\theta^{(23)} = \frac{\lambda^{(12)} + \lambda^{(23)} - \lambda^{(13)}}{2\sqrt{\lambda^{(12)}\lambda^{(23)}}}, \tag{7.62}$$

where $\lambda^{(12)}$, $\lambda^{(23)}$, and $\lambda^{(13)}$ are the reorganization free energies associated with ET between the D–A_1, A_1–A_2, and D–A_2 pairs, respectively.

Using the notation introduced in Sect. 7.2, the diabatic FESs in the two-dimensional polarization coordinate space (q_1, q_2) are expressed as

$$\begin{aligned}
G_{\mathrm{S2}}^{(0)} &= (q_1 - \sqrt{\lambda^{(12)}})^2 + q_2^2,\\
G_{\mathrm{CS1}}^{(n)} &= q_1^2 + q_2^2 + n\hbar\Omega_{\mathrm{v}} + \Delta G_{\mathrm{CS1}},\\
G_{\mathrm{CS2}}^{(m)} &= (q_1 - \sqrt{\lambda^{(23)}}\cos\theta^{(23)})^2 + (q_2 - \sqrt{\lambda^{(23)}}\sin\theta^{(23)})^2 + m\hbar\Omega_{\mathrm{v}} + \Delta G_{\mathrm{CS2}},\\
G_{\mathrm{S1}}^{(l)} &= (q_1 - \sqrt{\lambda^{(12)}})^2 + q_2^2 + l\hbar\Omega_{\mathrm{v}} + \Delta G_{\mathrm{S1}},
\end{aligned} \tag{7.63}$$

where n, m, and l denote vibrational quantum numbers, and $\hbar\Omega_{\mathrm{v}}$ represents the characteristic vibrational energy. The free energy gaps ΔG_{CS1}, ΔG_{CS2}, and ΔG_{S1} are defined relative to the reference state $|S_2\rangle$ as

$$\Delta G_{\mathrm{CS1}} = \check{G}_{\mathrm{CS1}} - \check{G}_{\mathrm{S2}}, \quad \Delta G_{\mathrm{CS2}} = \check{G}_{\mathrm{CS2}} - \check{G}_{\mathrm{S2}}, \quad \Delta G_{\mathrm{S1}} = \check{G}_{\mathrm{S1}} - \check{G}_{\mathrm{S2}}.$$

In Eq. (7.63), the influence of ET-active high-frequency intramolecular vibrations is incorporated by introducing vibrational sublevels for each electronic state. It is assumed that these vibrations possess the same frequency Ω_v across all electronic states [61]. The notation $G_i^{(n)} = G_i^{(n)}(q_1, q_2)$ denotes the diabatic FES associated with the n-th vibrational sublevel of the i-th electronic state.

The time evolution of the system is governed by a set of coupled kinetic equations describing the probability density functions $\rho_i^{(n)} = \rho_i^{(n)}(q_1, q_2, t)$ corresponding to each electron-vibrational state. The governing equations are

$$
\begin{aligned}
\frac{\partial \rho_{S2}^{(0)}}{\partial t} &= \hat{L}_{S2}\rho_{S2}^{(0)} - \sum_n K_{cs}^{(0n)}\left(\rho_{S2}^{(0)} - \rho_{CS1}^{(n)}\right) - k_d \rho_{S2}^{(0)}, \\
\frac{\partial \rho_{CS1}^{(n)}}{\partial t} &= \hat{L}_{CS1}\rho_{CS1}^{(n)} + K_{cs}^{(0n)}\left(\rho_{S2}^{(0)} - \rho_{CS1}^{(n)}\right) - \sum_m K_{csh}^{(nm)}\left(\rho_{CS1}^{(n)} - \rho_{CS2}^{(m)}\right) \\
&\quad - \sum_l K_{cr}^{(nl)}\left(\rho_{CS1}^{(n)} - \rho_{S1}^{(l)}\right) - \frac{1}{\tau_v^{(n)}}\rho_{CS1}^{(n)} + \frac{1}{\tau_v^{(n+1)}}\rho_{CS1}^{(n+1)}, \\
\frac{\partial \rho_{CS2}^{(m)}}{\partial t} &= \hat{L}_{CS2}\rho_{CS2}^{(m)} + \sum_n K_{csh}^{(nm)}\left(\rho_{CS1}^{(n)} - \rho_{CS2}^{(m)}\right) - \frac{1}{\tau_v^{(m)}}\rho_{CS2}^{(m)} + \frac{1}{\tau_v^{(m+1)}}\rho_{CS2}^{(m+1)}, \\
\frac{\partial \rho_{S1}^{(l)}}{\partial t} &= \hat{L}_{S1}\rho_{S1}^{(l)} + \sum_n K_{cr}^{(nl)}\left(\rho_{CS1}^{(n)} - \rho_{S1}^{(l)}\right) - \frac{1}{\tau_v^{(l)}}\rho_{S1}^{(l)} + \frac{1}{\tau_v^{(l+1)}}\rho_{S1}^{(l+1)} + \delta_{ll_0} k_d \rho_{S2}^{(0)}.
\end{aligned}
\tag{7.64}
$$

Here, $\hat{L}_i$ denotes the Smoluchowski diffusion operator for the i-th electronic state, defined by Eq. (7.29), and k_d is the rate constant for irreversible internal conversion from the second to the first singlet excited state, $|S_2\rangle \to |S_1\rangle$. The functions $K_{cs}^{(0n)}$, $K_{csh}^{(nm)}$, and $K_{cr}^{(nl)}$ are position-dependent rate constants for nonadiabatic electron-vibrational transitions associated with charge separation, charge shift, and charge recombination, respectively. These rates are evaluated as

$$
\begin{aligned}
K_{cs}^{(0n)}(q_1, q_2) &= \frac{2\pi}{\hbar}|V_{cs}|^2 F_{0n}\, \delta\left(G_{S2}^{(0)} - G_{CS1}^{(n)}\right), \\
K_{cr}^{(nl)}(q_1, q_2) &= \frac{2\pi}{\hbar}|V_{cr}|^2 F_{nl}\, \delta\left(G_{CS1}^{(n)} - G_{S1}^{(l)}\right), \\
K_{csh}^{(nm)}(q_1, q_2) &= \frac{2\pi}{\hbar}|V_{csh}|^2 F_{nm}\, \delta\left(G_{CS1}^{(n)} - G_{CS2}^{(m)}\right),
\end{aligned}
\tag{7.65}
$$

where V_{cs}, V_{cr}, and V_{csh} are the corresponding electronic coupling elements.

The Franck–Condon factors F_{nm} in Eq. (7.65) quantify the overlap between the n-th and m-th vibrational levels and are calculated using the standard expression

$$
F_{nm} = e^{-S} n! m! \left[\sum_{r=0}^{\min(n,m)} \frac{(-1)^{n-r} (\sqrt{S})^{n+m-2r}}{r!(n-r)!(m-r)!} \right]^2, \tag{7.66}
$$

where $S = \lambda_{vib}/\hbar\Omega_v$ is the Huang–Rhys factor representing the strength of electron–vibrational coupling, and λ_{vib} is the reorganization energy associated with a high-frequency intramolecular mode.

The Franck–Condon factors F_{nm} quantify the vibrational overlap between the nuclear wavefunctions of the initial and final electronic states. Physically, they represent the probability that an electronic transition between two diabatic states is accompanied by a change in vibrational quantum number from n to m. In the displaced harmonic oscillator model adopted here, F_{nm} depends on the Huang–Rhys factor S, which measures the strength of electron–vibrational coupling. Large values of S correspond to strong vibronic displacement and hence significant population of higher vibrational levels after an electronic transition, while small S values indicate that transitions predominantly occur without vibrational excitation (the so-called zero-phonon line). Thus, the Franck–Condon factors provide the link between molecular geometry changes upon electronic excitation and the distribution of accessible vibronic states during charge separation and recombination.

In Eq. (7.64), vibrational relaxation is modeled as a single-quantum cascade process, wherein each transition from level n to $n-1$ occurs with a rate constant $1/\tau_v^{(n)}$. The relaxation times are assumed to follow the relation $\tau_v^{(n)} = \tau_v^{(1)}/n$, consistent with previously established treatments of high-frequency intramolecular vibrational dissipation.

The initial conditions for the kinetic system (7.64) are specified by assuming that, immediately following excitation, the system is thermally equilibrated on the lowest vibrational sublevel of the excited-state FES $G_{S2}^{(0)}(q_1, q_2)$. This leads to the initial probability distributions

$$\rho_{S2}^{(0)}(q_1, q_2, t=0) = \frac{1}{\pi k_B T} \exp\left(-\frac{q_1^2 + q_2^2}{k_B T}\right), \tag{7.67}$$
$$\rho_{CS1}^{(n)}(q_1, q_2, t=0) = \rho_{CS2}^{(m)}(q_1, q_2, t=0) = \rho_{S1}^{(l)}(q_1, q_2, t=0) = 0.$$

The time-dependent population $P_i(t)$ of each diabatic electronic state $|\psi_i\rangle$ is obtained by integrating the corresponding vibrationally resolved probability densities over the configuration space and summing over all vibrational sublevels

$$P_i(t) = \sum_n \int \rho_i^{(n)}(q_1, q_2, t)\, dq_1\, dq_2. \tag{7.68}$$

These populations provide a quantitative description of the kinetics of photoinduced CS and CR under nonequilibrium conditions. They also offer insight into the role of vibrational relaxation and solvent dynamics in modulating the efficiency of ultrafast ET processes.

To quantitatively characterize the efficiency of charge separation in DA_1A_2 triads during the nonequilibrium (hot) stage of the photoreaction, we define the hot product yields Y_i for all electronic states (i = S2, CS1, CS2, S1) as the population values $Y_i = P_i(t')$ at a characteristic time t' that marks the end of the transient regime and

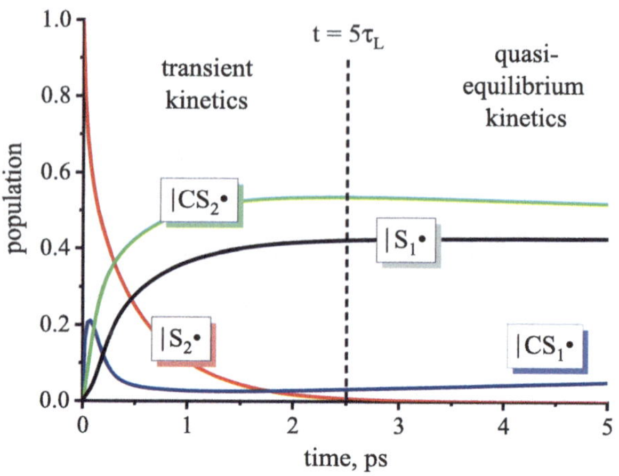

Fig. 7.12 Time evolution of diabatic state populations $P_i(t)$ in DA_1A_2 triad following Soret-band excitation. The ultrafast photochemical processes are complete by $t' = 2.5$ ps $= 5\tau_L$ (dashed vertical line). The populations at this time, $P_i(t')$, define the hot product yields Y_i

precedes the onset of quasi-equilibrium dynamics. This time point should be chosen such that fast hot transitions are complete, but thermally activated ET processes have not yet significantly occurred.

Figure 7.12 illustrates the time evolution of the state populations $P_i(t)$ over a few picoseconds following photoexcitation. The early-time dynamics ($t < 2.5$ ps) are dominated by nonequilibrium transitions: the population of the initially excited state $|S_2\rangle$ (red curve) decays sharply within the first 1–2 ps, primarily due to charge separation into the $|CS_1\rangle$ state (blue curve), which quickly rises and subsequently feeds into the $|CS_2\rangle$ state (green curve) through hot charge shift. Simultaneously, the $|S_1\rangle$ population (black curve) increases due to internal conversion and hot charge recombination from $|CS_1\rangle$.

These processes are essentially complete by $t' = 5\tau_L = 2.5$ ps, as indicated in the figure by the vertical line. Beyond this point, population changes proceed slowly, governed by thermally activated transitions within the quasi-equilibrium regime. Accordingly, we adopt $t' = 2.5$ ps as the cutoff time for evaluating the hot product yields Y_i and analyzing the efficiency of ultrafast, nonequilibrium charge separation in the triad.

The numerical simulations shown in Fig. 7.12 were performed using the following model parameters. The lifetime of the second excited state was set to $\tau_d = k_d^{-1} = 2$ ps, a representative value for ZnP-based systems [108]. The characteristic medium relaxation time was chosen as $\tau_L = 0.5$ ps, corresponding to low-viscosity polar solvents such as acetonitrile. The thermal energy was fixed at $k_B T = 0.025$ eV. Reorganization energies for low-frequency (solvent) modes were set to $\lambda^{(12)} = 0.9$ eV, $\lambda^{(23)} = 0.4$ eV, and $\lambda^{(13)} = 0.8$ eV, while the contribution from high-frequency intramolecular vibrations was neglected ($\lambda_{vib} = 0$). Free energy gaps ΔG_{CS1}, ΔG_{CS2}, and ΔG_{S1} were taken equal to -0.8 eV. Electronic coupling parameters for charge separation, charge shift, and charge recombination transitions were uniformly set to $V_{cs} = V_{csh} = V_{cr} = 0.05$ eV.

To characterize the sensitivity of CS efficiency to the underlying energetic parameters of the DA_1A_2 system, the following section presents numerical simulations analyzing how the hot product yields Y_i depend on key factors, including (i) the angle $\theta^{(23)}$ between the energy-gap coordinates associated with the two sequential ET steps (CS and CSh), (ii) the low-frequency and high-frequency reorganization free energies, $\lambda^{(nn')}$ and λ_{vib}, respectively, and (iii) the electronic coupling parameter V_{cr}, which governs the rate of hot charge recombination from the intermediate charge-separated state $|CS_1\rangle$ to the first excited state $|S_1\rangle$.

7.5.2 *Numerical Results: Effect of the System Energetics*

To analyze the influence of nonequilibrium charge recombination and charge shift processes on the photochemical dynamics of DA_1A_2 triads, we performed a series of numerical simulations under the condition of zero high-frequency vibrational reorganization energy ($\lambda_{vib} = 0$). The electronic coupling strength for the CR transition, V_{cr}, was systematically varied to examine its impact on the CS efficiency. The resulting hot product yields Y_i for the $|CS_1\rangle$, $|CS_2\rangle$, and $|S_1\rangle$ states are shown in Fig. 7.13A–C as functions of the angle $\theta^{(23)}$ between the two sequential reaction coordinates.

To facilitate interpretation of the simulation results, Fig. 7.13D presents a geometric visualization of the intersection regions between diabatic FESs associated with charge separation, charge recombination, and charge shift transitions. As previously illustrated in Fig. 7.9, the spatial arrangement of these intersections, serving as reactive sinks, along the relaxation trajectory of the nuclear wave packet on the G_{CS1} surface is a key determinant of the photoreaction outcome.

Notably, the relative locations of the CR and CSh sinks with respect to the initial CS region vary systematically with the angle $\theta^{(23)}$, thereby modulating the extent of their dynamic overlap with the evolving wave packet. This geometric sensitivity underscores an important principle for optimizing the CS efficiency: by tuning the angle $\theta^{(23)}$ through rational molecular design, one may reduce the undesired recombination and selectively enhance the forward ET pathway.

To evaluate the role of geometric factors, we first examine the red curves in Fig. 7.13A–C, which correspond to vanishing charge recombination, $V_{cr} = 0$. In this limit, Fig. 7.13C shows that the hot yield of the $|S_1\rangle$ state, Y_{S1}, remains constant across all values of $\theta^{(23)}$. This outcome has a clear physical interpretation: in the absence of recombination, Y_{S1} is solely determined by the competition between charge separation $|S_2\rangle \rightarrow |CS_1\rangle$ and internal conversion $|S_2\rangle \rightarrow |S_1\rangle$ processes that are independent of $\theta^{(23)}$.

A similar trend is observed for the yields of the $|CS_1\rangle$ and $|CS_2\rangle$ states, which exhibit no dependence on $\theta^{(23)}$ within a certain angular range. Specifically, this independence persists as long as the nonequilibrium wave packet formed on the G_{CS1} surface lies above the CSh transition region. This condition is fulfilled for

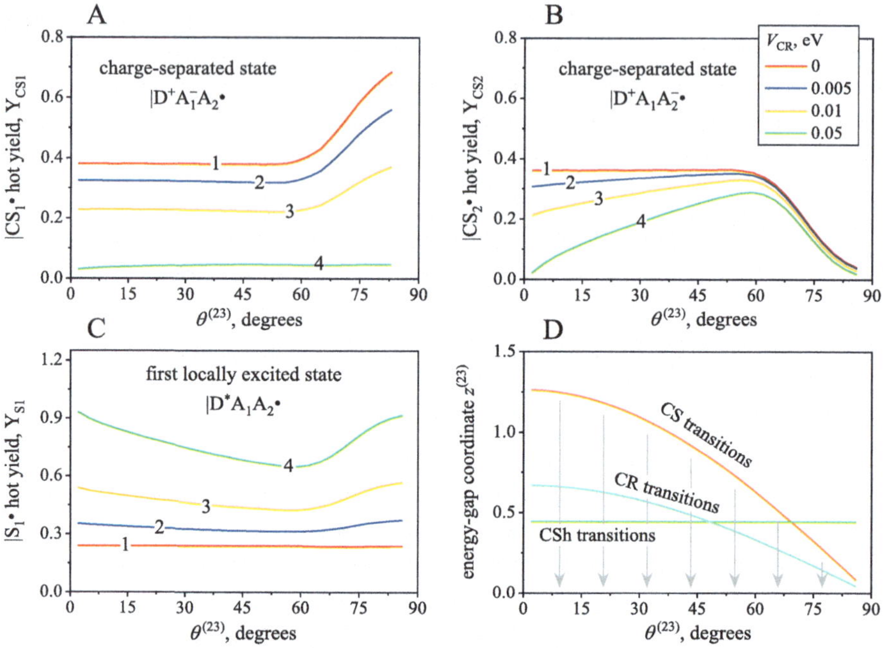

Fig. 7.13 Influence of nonequilibrium (hot) charge recombination on CS efficiency in DA_1A_2 triads. **A**–**C** Hot product yields for the charge-separated states $|CS_1\rangle$, $|CS_2\rangle$ and the first excited state $|S_1\rangle$ are shown as functions of the angle $\theta^{(23)}$ between the CS and CSh energy-gap coordinates. Results are presented for several values of the electronic coupling V_{cr}, which governs the rate of hot recombination: 0 (1), 0.005 (2), 0.01 (3), and 0.05 eV (4). The CR driving force is fixed at $\Delta G_{cr} = -0.3$ eV. All other parameters are as given in the caption to Fig. 7.12. **D** Geometric interpretation of the transition regions in $\boldsymbol{q}$-space: intersection points $\tilde{z}^{(23)}$ for CS, CR, and CSh processes, plotted as functions of $\theta^{(23)}$. Vertical arrows illustrate the trajectory of a nonequilibrium wave packet on the G_{CS1} surface, highlighting how spatial overlap with different transition regions is controlled by $\theta^{(23)}$

$\theta^{(23)} < \theta_c^{(23)}$, where $\theta_c^{(23)}$ denotes the critical angle at which the CSh transition sink intersects the path of the CS-generated wave packet (see the red and green lines in Fig. 7.13D).

This insensitivity to geometry arises because the rate of a hot transition is primarily determined by the gradient (slope) of the diabatic FESs at their intersection, which remains independent of $\theta^{(23)}$ under the assumptions of the model [17]. When $\theta^{(23)}$ exceeds this critical value, Y_{CS2} decreases significantly: the CS-generated wave packets then fall below the CSh reaction sink, reducing the likelihood of hot charge shift to the secondary acceptor. This behavior emphasizes the geometric control that $\theta^{(23)}$ exerts over ET pathway selectivity.

As the electronic coupling V_{cr} increases, the population yield of the $|S_1\rangle$ state correspondingly rises, while the yields of the charge-separated states $|CS_1\rangle$ and $|CS2\rangle$ decrease, reflecting the enhanced efficiency of charge recombination. At higher val-

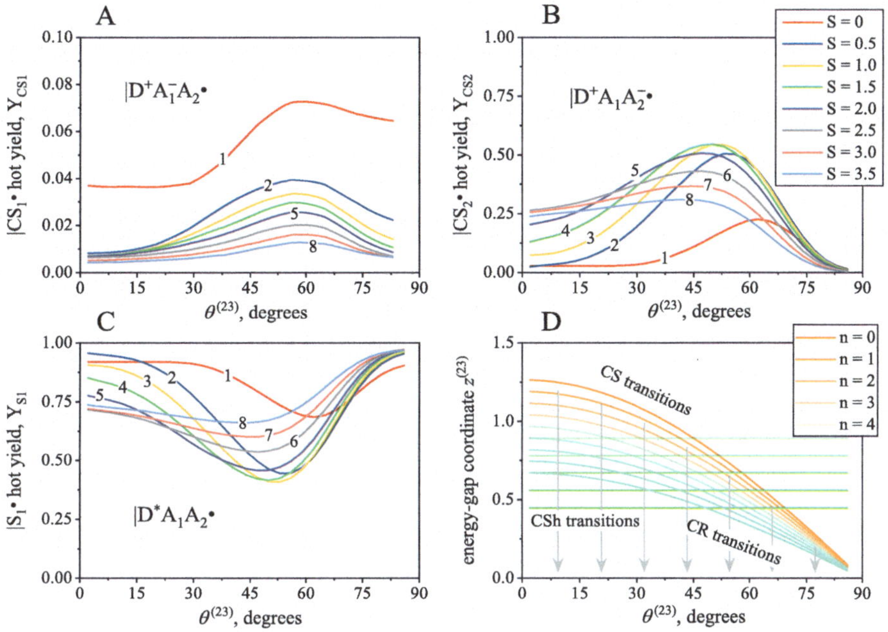

Fig. 7.14 Effect of ET-active high-frequency intramolecular vibrations on the efficiency of charge separation in DA_1A_2 triads. **A–C** Hot product yields for $|CS_1\rangle$, $|CS_2\rangle$, and $|S_1\rangle$ states plotted as functions of the angle $\theta^{(23)}$, shown for different values of the Huang–Rhys factor S. Curve labels correspond to $S = 0$ (1), 0.5 (2), 1.0 (3), 1.5 (4), 2.0 (5), 2.5 (6), 3.0 (7), and 3.5 (8). **D** Arrangement of the CS, CSh, and CR transition regions in the energy-gap coordinate $z^{(23)}$, taking into account vibrational structure of diabatic FESs

ues of V_{cr}, the $|CS_2\rangle$ yield displays a pronounced peak near the critical angle $\theta_c^{(23)}$, accompanied by a corresponding dip in the $|S_1\rangle$ yield. This behavior is attributable to the interaction between the CR and CSh pathways. This interplay between competitive ET channels has been thoroughly analyzed in Ref. [61].

The effect of high-frequency vibrational reorganization on hot product yields is illustrated in Fig. 7.14, which presents the calculated dependencies $Y_i(\theta^{(23)})$ for several values of the Huang–Rhys parameter, $S = \lambda_{vib}/\hbar\Omega_v$. At small $\theta^{(23)}$, the hot yield of the $|CS_2\rangle$ state increases monotonically with S, reflecting the growing contribution of vibrationally assisted charge shift transitions.

As $\theta^{(23)}$ approaches the region associated with maximum yield, the population of $|CS_2\rangle$ exhibits a pronounced peak whose magnitude increases with S up to approximately $S \approx 1.0$–1.5, beyond which a gradual decline is observed. Furthermore, the position of this maximum shifts systematically toward smaller values of $\theta^{(23)}$ as S increases, indicating that vibrational reorganization facilitates more efficient charge separation at reduced angular displacements between the CS and CSh transition vectors.

Reorganization of intramolecular high-frequency vibrational modes opens additional reaction channels by enabling transitions into vibrationally excited product states. This effect initially enhances the efficiency of hot electronic transitions, as previously discussed in Ref. [80]. As a result, the hot yields of both the $|S_1\rangle$ and $|CS_2\rangle$ states are expected to increase with growing vibrational reorganization energy λ_{vib}, or equivalently, with the Huang–Rhys factor S. This trend is clearly reflected in Fig. 7.14. However, for sufficiently large values of λ_{vib}, this enhancement becomes non-monotonic and eventually declines. This reversal in behavior has been explained in earlier studies [109] and is attributed to the redistribution of transition probability among an increasing number of vibrational levels, which effectively weakens the strength of individual reaction sinks. Despite this limitation, ET-active high-frequency vibrational modes can significantly enhance the overall yield of the final charge-separated state, $|CS_2\rangle$, with values reaching up to 0.6 under optimal conditions. In contrast, the hot yield of the intermediate charge-separated state $|CS_1\rangle$ remains consistently low (below 0.04) when λ_{vib} is nonzero, indicating fast ET progression toward the terminal acceptor.

To analyze the efficiency of ultrafast charge separation in DA_1A_2 triads, we define the total yield of charge-separated states as

$$Y_{cs} = Y_{CS1} + Y_{CS2},$$

where Y_{CS1} and Y_{CS2} denote the hot yields of the intermediate and terminal charge-separated states, respectively. Figure 7.15 presents the computed Y_{cs} values on the $(\lambda^{(12)}, \lambda^{(23)})$ plane for several combinations of Huang–Rhys parameter S and angular displacement $\theta^{(23)}$.

The results reveal that high quantum yields ($Y_{cs} > 0.7$) are predominantly observed when two conditions are met: (i) the angle $\theta^{(23)}$ is relatively small ($\lesssim 45°$), and (ii) the intramolecular vibrational reorganization is strong ($S \geq 1$). These findings are consistent with previous analyses (e.g., [61]), which identified that maximum CS efficiency occurs when the relaxation trajectory of the nuclear wave packet on the G_{CS1} surface first intersects the CSh transition region before reaching the CR transition region.

Furthermore, the location of the Y_{cs} maximum on the $(\lambda^{(12)}, \lambda^{(23)})$ landscape is sensitive to $\theta^{(23)}$: for $\theta^{(23)} = 30°$, the yield is maximized at $\lambda^{(23)} \approx 1.2$ eV, whereas at $\theta^{(23)} = 75°$, the optimal value shifts to $\lambda^{(23)} \approx 0.15$ eV. This shift reflects the changing geometry of the FES intersection regions with respect to the relaxation direction. In contrast, the position of the maximum is relatively insensitive to the Huang–Rhys parameter S, although the absolute yield increases with S in most cases.

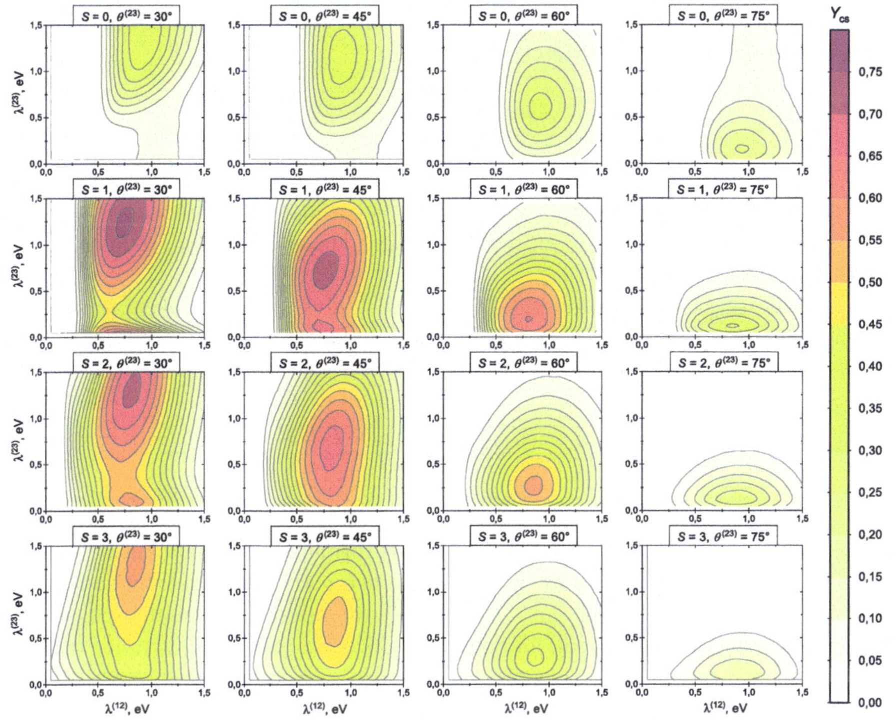

Fig. 7.15 Quantum yield of ultrafast charge separation, Y_{cs}, as a function of the reorganization energies $\lambda^{(12)}$ and $\lambda^{(23)}$, for different values of the Huang–Rhys factor S and angular displacement $\theta^{(23)}$. Each panel corresponds to a specific $(S, \theta^{(23)})$ pair, as indicated. Other model parameters are consistent with those used in Figs. 7.13 and 7.14

7.5.3 *Numerical Results: Effect of Molecular Geometry*

We now examine how the geometric structure of the DA_1A_2 triad influences the efficiency of ultrafast charge separation. Specifically, we investigate the impact of donor and acceptor effective radii, center-to-center ET distances, and the triad bending angle on the hot CS yield Y_{cs}.

The sensitivity of Y_{cs} to structural parameters arises from their direct influence on key quantities governing electron transfer: the reorganization free energies $\lambda^{(12)}$, $\lambda^{(23)}$, and $\lambda^{(13)}$, as well as the electronic coupling strengths V_{cs}, V_{cr}, and V_{csh}. Previous studies have demonstrated that Y_{cs} in DA_1A_2 systems is particularly responsive to ET energetics, especially during ultrafast charge shift from A_1 to A_2 [61, 62, 97, 110].

To quantify these effects, we adopt a classical electrostatic model in which the donor (D) and acceptors (A_1, A_2) are represented as spherical cavities embedded in a dielectric medium. The respective radii of the redox centers are denoted by R_1, R_2, and R_3, and the center-to-center distances between sites n and n' are indicated as $R_{nn'}$ $(n, n' = 1, 2, 3)$. The bending angle of the triad is designated by Φ.

Within this spherical-cavity model, the solvent reorganization energies can be evaluated using the classical Marcus formula [54]

$$\lambda^{(nn')} = \frac{c_0 e^2}{2}\left(\frac{1}{R_n} + \frac{1}{R_{n'}} - \frac{2}{R_{nn'}}\right), \tag{7.69}$$

where e is the elementary charge, and c_0 is the Pekar factor, given by $c_0 = 1/\epsilon_\infty - 1/\epsilon_0$. For acetonitrile, a prototypical polar solvent, we adopt $\epsilon_\infty = 1.806$ and $\epsilon_0 = 36.64$.

By substituting Eq. (7.69) into Eq. (7.62), one obtains an analytical expression for the angle $\theta^{(23)}$ between the ET energy-gap coordinates associated with the $|S_2\rangle \rightarrow |CS_1\rangle$ and $|CS_1\rangle \rightarrow |CS_2\rangle$ transitions [97]

$$\cos\theta^{(23)} = \frac{R_2^{-1} - R_{23}^{-1} - R_{12}^{-1} + R_{13}^{-1}}{\sqrt{\left(R_1^{-1} + R_2^{-1} - R_{12}^{-1}\right)\left(R_2^{-1} + R_3^{-1} - R_{23}^{-1}\right)}}. \tag{7.70}$$

This result demonstrates that, within the Marcus formalism, the angular correlation $\theta^{(23)}$ is determined exclusively by the molecular geometry and is independent of the dielectric properties of the medium. Numerical evaluations indicate that physically reasonable variations in the radii and inter-site distances yield a wide range of $\theta^{(23)}$ values, typically between 40° and 85°.

In addition to the geometric dependence of the reorganization free energies, the electronic coupling elements V_{cs}, V_{cr}, and V_{csh} are also strongly influenced by molecular structure. A widely employed approach to estimate these couplings is the exponential distance-dependence model, which assumes that the electronic interaction between two redox centers decays exponentially with their center-to-center separation. Within this model, the coupling constant V_i for the i-th ET process (i = cs, cr, csh) is expressed as

$$V_i = V_i^{(0)} \exp\left(-\frac{R_{nn'} - R_n - R_{n'}}{L_i}\right), \tag{7.71}$$

where $V_i^{(0)}$ denotes the maximum coupling at van der Waals contact ($R_{nn'} = R_n + R_{n'}$), and L_i is the characteristic tunneling decay length for the corresponding ET process [110].

To identify structural motifs that enhance ultrafast charge separation in DA_1A_2 triads, we adopt the following two-step strategy. First, we analyze a simpler DA_1 dyad to determine the geometric parameters that maximize the initial CS efficiency. Subsequently, we introduce a secondary acceptor (A_2), and systematically vary its spatial configuration to estimate its effect on the overall CS efficiency. This methodology provides a clearer understanding of how individual geometrical features influence the dynamics of multistep ET.

As a starting point, we consider the photophysical behavior of the DA_1 system following Soret-band excitation at $t = 0$. The resulting CS efficiency is determined by

Table 7.2 Optimized geometrical parameters for donor–acceptor dyads obtained from numerical simulations. $Y_{cs}^{(max)}$ denotes the maximum quantum yield of ultrafast charge separation, corresponding to the free energy gap ΔG_{cs}. R_1, R_2, and R_{12} represent the effective radii and the donor–acceptor center-to-center distance, respectively

ΔG_{cs}, eV	R_1 and R_2, Å	R_{12}, Å	$Y_{cs}^{(max)}$
−1.4	4	10.4	0.54
−1.2	5	12.5	0.50
−1.0	6	14.7	0.47
−0.8	7	17.0	0.40

competition between two primary processes: (1) internal conversion from the second excited state $|S_2\rangle$ to the lower-lying state $|S_1\rangle$, and (2) hot charge recombination from the intermediate CS state $|D^+A_1^-\rangle$ back to $|S_1\rangle$ during solvent and vibrational relaxation. High efficiency of charge separation requires the following conditions to be met:

1. Forward ET from $|S_2\rangle$ to $|D^+A_1^-\rangle$ must be faster than internal conversion to $|S_1\rangle$.
2. Hot CR from $|D^+A_1^-\rangle$ to $|S_1\rangle$ must be minimized.

Efficient forward ET is promoted by strong donor–acceptor coupling, which typically requires a short inter-site distance R_{12}. Additionally, ultrafast charge separation is most favorable when the reaction is nearly activationless, i.e., when the total reorganization energy satisfies $\lambda^{(12)} + \lambda_{vib} \approx -\Delta G_{cs}$. In contrast, suppression of hot CR favors larger donor–acceptor separations, which reduce wavefunction overlap and diminish the back-transfer coupling.

These competing requirements give rise to a non-monotonic dependence of Y_{cs} on R_{12}. Specifically, there exists an optimal intermediate distance that balances forward ET efficiency and recombination suppression. This behavior is confirmed by numerical simulations, shown in Fig. 7.16, where Y_{cs} is plotted as a function of R_{12} for various values of the effective radii ($R_1 = R_2$) and the driving force ΔG_{cs}. In all scenarios, Y_{cs} exhibits a distinct maximum at a separation exceeding the contact radius. The position of this maximum depends on the system parameters, confirming the existence of a geometry-dependent CS efficiency.

A summary of the optimal yields and corresponding structural parameters is provided in Table 7.2, which compiles peak Y_{cs} values and associated geometric configurations across a range of representative cases.

The plots presented in Fig. 7.16 also demonstrate a pronounced dependence of Y_{cs} on the driving force ΔG_{cs} associated with the $|S_2\rangle \rightarrow |CS_1\rangle$ transition. This sensitivity arises from the requirement that the activation barrier for forward electron transfer remain sufficiently small to ensure fast charge separation. The corresponding activation free energy, derived from the Marcus theory, is given by

$$G_{cs}^{\sharp} \equiv \frac{\left(\lambda^{(12)} + \lambda_{vib} + \Delta G_{cs}\right)^2}{4\lambda^{(12)}} \lesssim k_B T. \tag{7.72}$$

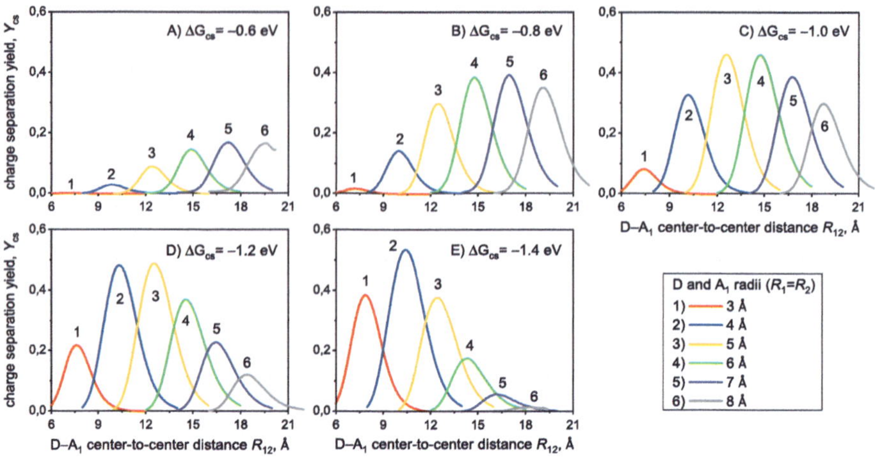

Fig. 7.16 Hot charge separation yield Y_{cs} in DA_1 dyads as a function of the donor–acceptor center-to-center distance R_{12}, for selected values of the driving force ΔG_{cs} and donor/acceptor radii $R_1 = R_2$ (indicated in each panel). A pronounced non-monotonic dependence $Y_{cs}(R_{12})$ reflects the competition between efficient forward ET and suppression of ultrafast CR

As follows from this expression, larger exergonicity $|\Delta G_{cs}|$ of the CS transition requires a proportionally larger reorganization energy $\lambda^{(12)}$ to maintain a low activation barrier. Given that $\lambda^{(12)}$ is inversely related to the effective donor and acceptor radii via Eq. (7.69), this implies that smaller sizes R_1, R_2 are advantageous under strongly exergonic conditions. Consequently, the optimal molecular geometry for maximizing Y_{cs} depends not only on spatial arrangement but also on the underlying thermodynamics of the ET process.

Despite optimization of donor–acceptor distance, the calculated hot CS yields in Fig. 7.16 do not exceed approximately 50% for the parameters considered. This upper limit primarily reflects the competition from internal conversion of $|S_2\rangle$ to $|S_1\rangle$, occurring on a characteristic timescale of $\tau_d = 2$ ps in zinc–porphyrin-based systems. Within the current model, this deactivation pathway cannot be readily suppressed.

Nevertheless, there exist molecular systems in which the $|S_2\rangle$ lifetime is substantially longer. Notably, xanthione-derivative compounds have been shown to possess the $|S_2\rangle$ state lifetimes on the order of 100 ps and longer [111]. In such systems, the extended excited-state lifetime may allow for significantly more efficient forward ET and reduced probability of hot charge recombination, thereby enhancing overall ET efficiency.

Figure 7.17 illustrates the dependence of Y_{cs} on the timescale τ_d of internal conversion from the second to the first excited singlet state. The results are presented for $\tau_d = 2$, 10, and 50 ps, and for several values of the effective donor and acceptor radii, $R_1 = R_2 = R$. In each case, Y_{cs} is plotted as a function of the donor–acceptor center-to-center distance R_{12}. As shown in panels A–C, an increase in τ_d leads to a marked enhancement in CS efficiency. For $\tau_d = 50$ ps, the quantum yield approaches

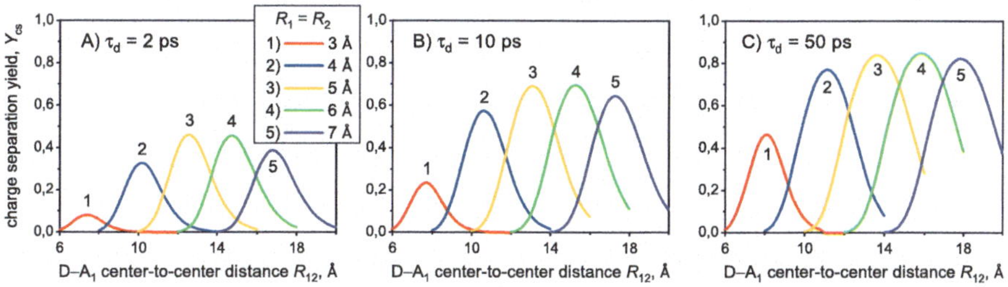

Fig. 7.17 Influence of the second excited-state lifetime τ_d on the efficiency of ultrafast charge separation in DA_1 dyads. The hot CS yield Ycs is shown as a function of donor–acceptor distance R_{12} for different values of the effective radii $R_1 = R_2 = R$ (indicated in each panel). Panels **A–C** correspond to $\tau_d = 2$, 10, and 50 ps, respectively, demonstrating how slower internal conversion enhances the CS efficiency

0.85 under optimal geometric conditions, as compared to less than 50% for $\tau_d = 2$ ps. This enhancement reflects the increased probability that the $|S_2\rangle$ state undergoes charge separation before internal conversion to $|S_1\rangle$ occurs.

Having established the structural and kinetic factors that govern CS efficiency in DA_1 dyads, we now extend the analysis to DA_1A_2 triads. In this context, the inclusion of a secondary acceptor A_2 can significantly improve the overall Y_{cs} yield if two conditions are satisfied: (i) charge shift from A_1 to A_2 proceeds as ultrafast (hot) electron transfer, and (ii) the charge shift occurs predominantly before the system undergoes hot charge recombination. These requirements impose specific constraints on the relationship between reorganization energies and free energy gaps of the relevant ET steps, as previously discussed in Ref. [62].

To quantitatively assess the influence of the triad geometry on quantum yield of hot CS, we systematically explore how variations in the effective radius of the secondary acceptor A_2 and the distance between the primary and secondary acceptors affect Y_{cs}. This analysis is based on the optimized donor–acceptor configurations identified for DA_1 dyads (see Table 7.2), wherein R_1, R_2, and R_{12} are fixed according to the CS driving force ΔG_{cs}. Focusing on the representative case $\Delta G_{cs} = -1.4$ eV, we perform simulations over a two-dimensional parameter space defined by the effective radius of the secondary acceptor, R_3, and the surface-to-surface separation $\Delta R_{23} \equiv R_{23} - R_2 - R_3$.

As shown in Fig. 7.18, the incorporation of A_2 can substantially enhance the CS yield, provided that the secondary acceptor is both sufficiently large in size ($R_3 \sim 8$–9 Å) and spatially proximate to the primary acceptor ($\Delta R_{23} \lesssim 1$ Å). Under these conditions, Y_{cs} increases by approximately 0.12 relative to the dyad baseline, yielding a total CS efficiency of approximately 0.64 for the full triad.

These results highlight a clear structural optimization strategy: high-efficiency ultrafast CS requires compact donor and primary acceptor units (small R_1, R_2) with large reorganization energies ($\lambda^{(12)} \gtrsim 1$ eV) conducive to ultrafast forward ET. Simultaneously, the reorganization free energy associated with the secondary ET step ($\lambda^{(23)}$) should remain relatively small ($\lambda^{(23)} \lesssim 0.5$ eV), a requirement that

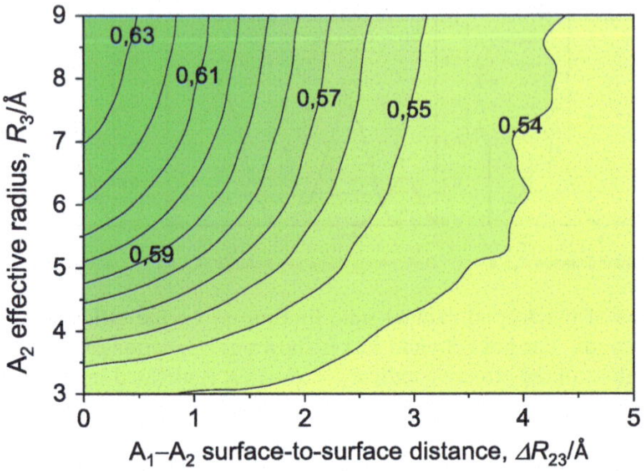

Fig. 7.18 Influence of the secondary acceptor's geometric configuration on efficiency of ultrafast charge separation in DA_1A_2 triads. Contour plots show the dependence of Y_{cs} on the effective radius of the secondary acceptor A_2 and the surface-to-surface distance between A_1 and A_2. Simulations were performed for $\Delta G_{cs} = -1.4$ eV. Parameters R_1, R_2, and R_{12} are fixed according to the optimized DA_1 geometry (Table 7.2)

can be satisfied by selecting a large-radius A_2 moiety and minimizing its separation from A_1. This interplay between the structural and energetic parameters underscores the importance of precise molecular design in engineering multistep ET systems for high-efficiency photochemical applications.

Another important structural parameter in donor–acceptor triads is the bending angle Φ, defined as the angle between the D–A_1 and A_1–A_2 segments of the molecular framework. This parameter significantly influences the dielectric polarization response of the environment during the two sequential ET events, namely, the initial charge separation and the subsequent charge shift. The effect arises from the fact that changes in Φ modify the spatial separation between the donor D and the secondary acceptor A_2, thereby altering the R_{13} distance. As indicated by Eqs. (7.69) and (7.70), such variations impact both the total reorganization energy $\lambda^{(13)}$ and the angle $\theta^{(23)}$ between the energy-gap vectors associated with the two ET steps.

Sensitivity of CS efficiency to triad geometry, particularly to the bending angle, has also been reported in systems exhibiting symmetry-breaking charge separation in excited states. For instance, recent studies of perylenediimide dimers revealed that the CS yield depends strongly on conformational flexibility and the twist angle between the constituent subunits [39, 40].

To quantitatively evaluate the influence of the bending angle on CS efficiency, a series of numerical simulations was carried out. Figure 7.19 presents the resulting dependencies of Y_{cs} on Φ for several values of the charge shift driving force ΔG_{csh}. The simulations were performed using the following fixed parameters: donor and primary acceptor radii $R_1 = R_2 = 4$ Å, secondary acceptor radius $R_3 = 8$ Å, donor–acceptor distance $R_{12} = 10.4$ Å, and primary–secondary acceptor distance $R_{23} =$

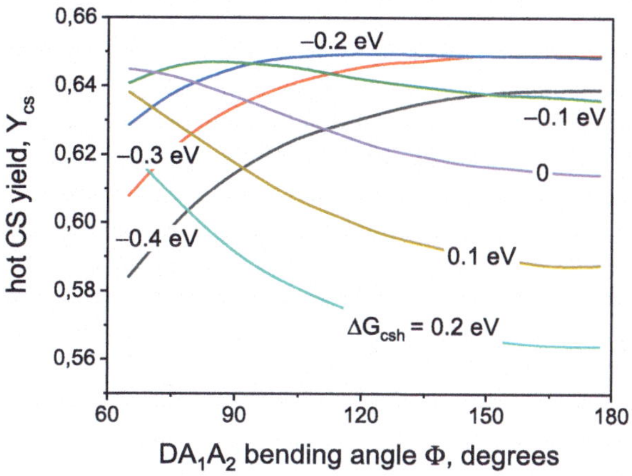

Fig. 7.19 Dependence of the ultrafast charge separation yield Y_{cs} on the bending angle Φ of the DA_1A_2 triad. Curves correspond to different values of the charge shift free energy gap ΔG_{csh}, as indicated. Fixed parameters: $R_1 = R_2 = 4$ Å, $R_3 = 8$ Å, $R_{12} = 10.4$ Å, $R_{23} = 12$ Å

12 Å. All other model parameters are consistent with those used in the preceding sections.

As shown in Fig. 7.19, the dependence of Y_{cs} on the bending angle Φ is strongly influenced by the thermodynamic driving force for the charge shift (CSh) process:

- For strongly exergonic CSh transitions ($\Delta G_{csh} = -0.4$ eV), Y_{cs} increases monotonically with increasing Φ, indicating that in this limit more linear DA_1A_2 geometries promote more efficient sequential ET to the secondary acceptor.
- For endergonic charge shifts ($\Delta G_{csh} = +0.2$ eV), Y_{cs} exhibits a monotonic decrease with Φ, reflecting the diminished probability of uphill CSh transitions at larger bending angles.
- For intermediate driving forces ($\Delta G_{csh} = -0.2$ eV), the $Y_{cs}(\Phi)$ curve is relatively insensitive to variations in Φ, indicating that the system maintains robust CS performance across a wide conformational range.

These findings suggest that triad architectures with moderately exergonic charge shift energetics and intermediate bending angles can represent an effective balance between efficient secondary electron transfer and suppression of hot charge recombination. Such configurations offer a practical design strategy for enhancing the photochemical efficiency of multiredox molecular assemblies.

7.6 Concluding Remarks

7.6.1 General Formalism

Many distinctive features of ultrafast photoinduced electron transfer (ET) in macromolecular systems originate from the nonequilibrium character of the nuclear environment. This environment—comprising both solvent and intramolecular vibrational degrees of freedom—is driven out of equilibrium either by optical excitation or by rapid charge redistribution through electronic tunneling between redox centers. A consistent theoretical description of multistage ET processes must therefore incorporate the time-dependent evolution of the nuclear subsystem, including its dynamics during and between successive ET steps.

In this chapter, we have summarized the main results of our recent work on the development of a general semiclassical framework for modeling ultrafast ET in multiredox molecular assemblies. The proposed formalism describes complex ET pathways in macromolecules embedded in polar environments with multicomponent relaxation by employing a multidimensional coordinate space that explicitly incorporates both polarization and relaxation coordinates. This approach captures essential features of nonequilibrium ET dynamics and provides a unified theoretical basis for a broad range of ET regimes, spanning from single-step relaxation-mediated processes to cascaded and hot electron transfer events.

7.6.2 Applications

Application of the framework to donor–acceptor–acceptor (DA_1A_2) triads has revealed the crucial role of hot electron transfer pathways in enhancing charge-separation efficiency. Numerical simulations demonstrated how molecular and environmental parameters—including reorganization energies, electron–vibrational couplings, donor–acceptor separations, and triad bending angles—influence ultrafast ET kinetics and ultimately determine the quantum yield of the charge-separated state.

7.6.3 Outlook

Taken together, these results elucidate how molecular architecture and environmental relaxation dynamics shape the efficiency and directionality of photoinduced electron transfer. The developed framework provides not only a conceptual foundation but also a predictive tool for guiding the rational design of photochemical systems, including artificial photosynthetic complexes, molecular optoelectronic devices, and organic photovoltaic materials.

Acknowledgements This work was supported by Russian Science Foundation (project 22-13-00180-P), https://rscf.ru/en/project/22-13-00180/.

References

1. Ponseca CSJ, Chábera P, Uhlig J, Persson P, Sundström V. Chem Rev. 2017;117(16):10940. https://doi.org/10.1021/acs.chemrev.6b00807.
2. Lee YM, Nam W, Fukuzumi S. Chem Sci. 2023;14:4205. https://doi.org/10.1039/D2SC07101K.
3. Lawrence J, Egan R, Hoefer T, Scarampi A, Shang L, Howe CJ, Zhang JZ. Nat Rev Bioeng. 2023;1:887–905. https://doi.org/10.1038/s44222-023-00093-x.
4. Scattergood PA, Delor M, Sazanovich IV, Bouganov OV, Tikhomirov SA, Stasheuski AS, Parker AW, Greetham GM, Towrie M, Davies ES, Meijer AJHM, Weinstein JA. Dalton Trans. 2014;43:17677. https://doi.org/10.1039/C4DT01682C.
5. Fukuzumi S, Ohkubo K, Suenobu T. Acc Chem Res. 2014;47(5):1455. https://doi.org/10.1021/ar400200u.
6. Bottari G, de la Torre G, Guldi DM, Torres T. Coord Chem Rev. 2021;428: 213605. https://doi.org/10.1016/j.ccr.2020.213605.
7. Ballabio M, Cánovas E, Nanosci ACS. Au. 2022;2(5):367. https://doi.org/10.1021/acsnanoscienceau.2c00015.
8. Machín A, Cotto M, Ducongé J, Márquez F. Biomimetics. 2023;8(3):298. https://doi.org/10.3390/biomimetics8030298.
9. Wang L, Zhu W. Adv Sci. 2024;11(10):2307227. https://doi.org/10.1002/advs.202307227.
10. Vauthey E. J Photochem Photobiol, A. 2006;179(1–2):1. https://doi.org/10.1016/j.jphotochem.2005.12.019.
11. D'Souza, F, Ito, O. Chem. Commun. pp. 4913–4928 (2009). https://doi.org/10.1039/B905753F.
12. Ostroverkhova O. Chem Rev. 2016;116:13279. https://doi.org/10.1021/acs.chemrev.6b00127.
13. Santos E, Schmickler W. Chem Rev. 2022;122(12):10581. https://doi.org/10.1021/acs.chemrev.1c00583.
14. Xin N, Guan J, Zhou C, Chen X, Gu C, Li Y, Ratner MA, Nitzan A, Stoddart JF, Guo X. Nat Rev Phys. 2019;1(3):211. https://doi.org/10.1038/s42254-019-0022-x.
15. Fisher JM, Williams ML, Palmer JR, Powers-Riggs NE, Young RM, Wasielewski MR. J Am Chem Soc. 2024;146(14):9911. https://doi.org/10.1021/jacs.4c00201.
16. Cho M, Silbey RJ. J Chem Phys. 1995;103:595. https://doi.org/10.1063/1.470094.
17. Ivanov AI, Potovoi VV. Chem Phys. 1999;247(2):245. https://doi.org/10.1016/S0301-0104(99)00197-4.
18. Feskov SV, Ionkin V, Ivanov AI, Hagemann H, Vauthey E. J Phys Chem A. 2008;112(4):594. https://doi.org/10.1021/jp709587x.
19. Zimmermann C, Willig F, Ramakrishna S, Burfeindt B, Pettinger B, Eichberger R, Storck W. J Phys Chem B. 2001;105:9245. https://doi.org/10.1021/jp011106z.
20. Wan C, Xia T, Becker HC, Zewail AH. Chem Phys Lett. 2005;412:158. https://doi.org/10.1016/j.cplett.2005.06.101.
21. Feskov SV, Gladkikh V, Burshtein AI. Chem Phys Lett. 2007;447:162.
22. Feskov SV, Burshtein AI. J Phys Chem A. 2009;113(48):13528. https://doi.org/10.1021/jp901863t.
23. Kumpulainen T, Lang B, Rosspeintner A, Vauthey E. Chem Rev. 2017;117:10826. https://doi.org/10.1021/acs.chemrev.6b00491.
24. Torieda H, Nozaki K, Yoshimura A, Ohno T. J Phys Chem A. 2004;108(22):4819. https://doi.org/10.1021/jp037259z.
25. Barykov VY, Ionkin VN, Ivanov AI. J Phys Chem C. 2015;119(6):2989. https://doi.org/10.1021/acs.jpcc.5b00005.
26. Feskov SV, Mikhailova VA, Ivanov AI. J Photochem Photobiol, C. 2016;29:48. https://doi.org/10.1016/j.jphotochemrev.2016.11.001.
27. Kundu M, He TF, Lu Y, Wang L, Zhong D. J Phys Chem Lett. 2018;9(11):2782. https://doi.org/10.1021/acs.jpclett.8b00882.

28. Matyushov DV. J Mol Liq. 2018;266:361. https://doi.org/10.1016/j.molliq.2018.06.087.
29. Lu Y, Kundu M, Zhong D. Nat Comm. 2020;11(1):2822. https://doi.org/10.1038/s41467-020-15535-y.
30. Siplivy NB, Feskov SV, Ivanov AI. J Chem Phys. 2020;153(4): 044301. https://doi.org/10.1063/5.0013708.
31. Santoro A, Bella G, Cancelliere AM, Serroni S, Lazzaro G, Campagna S. Molecules. 2022;27(9): https://doi.org/10.3390/molecules27092713.
32. Martín C, Ziółek M, Douhal A. J Photochem Photobiol, C. 2016;26:1. https://doi.org/10.1016/j.jphotochemrev.2015.12.001.
33. LeBard DN, Matyushov DV. J Phys Chem B. 2009;113(36):12424. https://doi.org/10.1021/jp904647m.
34. Blankenship R. Molecular Mechanisms of Photosynthesis. Wiley; 2021.
35. Cherepanov DA, Semenov AY, Mamedov MD, Aybush AV, Gostev FE, Shelaev IV, Shuvalov VA, Nadtochenko VA. Biophys Rev. 2022;14(4):805. https://doi.org/10.1007/s12551-022-00983-1.
36. Young RM, Wasielewski MR. Acc Chem Res. 2020;53(9):1957. https://doi.org/10.1021/acs.accounts.0c00397.
37. Verma P, Tasior M, Roy P, Meech SR, Gryko DT, Vauthey E. Phys Chem Chem Phys. 2023;25:22689. https://doi.org/10.1039/D3CP02810K.
38. Swathi K, Sujith M, Divya PS, Delledonne A, Phan Huu DKA, Di Maiolo F, Terenziani F, Lapini A, Painelli A, Sissa C, Thomas KG. Chem Sci. 2023;14:1986. https://doi.org/10.1039/D2SC05206G.
39. Dereka B, Balanikas E, Rosspeintner A, Li Z, Liska R, Vauthey E. J Phys Chem Lett. 2024;15(32):8280. https://doi.org/10.1021/acs.jpclett.4c01694.
40. Mazumder A, Vinod K, Thomas AC, Hariharan M. J Phys Chem Lett. 2025;16(19):4819. https://doi.org/10.1021/acs.jpclett.5c00372.
41. Ivanov AI. J Phys Chem C. 2018;122:29165. https://doi.org/10.1021/acs.jpcc.8b10985.
42. Antipov IF, Ivanov AI. J Chem Phys. 2022;157(22): 224104. https://doi.org/10.1063/5.0129697.
43. Siplivy NB, Ivanov AI. J Chem Phys. 2024;160(19): 194302. https://doi.org/10.1063/5.0211030.
44. Mikhailova TV, Ivanov AI. J Chem Phys. 2024;160: 054302. https://doi.org/10.1063/5.0193532.
45. Mikhailova TV, Mikhailova VA, Ivanov AI. J Chem Phys. 2024;161(15): 154303. https://doi.org/10.1063/5.0237870.
46. Ivanov AI. J Photochem Photobiol, C. 2024;58: 100651. https://doi.org/10.1016/j.jphotochemrev.2024.100651.
47. Ivanov AI. J Chem Phys. 2025;162(2): 024303. https://doi.org/10.1063/5.0243375.
48. Nazarov AE, Ivanov AI. J Phys Chem B. 2020;124(47):10787. https://doi.org/10.1021/acs.jpcb.0c07612.
49. Dereka B, Vauthey E. J Phys Chem Lett. 2017;8(16):3927. https://doi.org/10.1021/acs.jpclett.7b01821.
50. Zusman LD. Chem Phys. 1980;49(2):295. https://doi.org/10.1016/0301-0104(80)85267-0.
51. Najbar J, Tachiya M. J Phys Chem. 1994;98:199. https://doi.org/10.1021/j100052a033.
52. Zusman LD, Beratan DN. J Chem Phys. 1999;110(21):10468. https://doi.org/10.1063/1.478976.
53. Sumi H, Marcus RA. J Chem Phys. 1986;84:4894. https://doi.org/10.1063/1.449978.
54. Marcus RA. J Chem Phys. 1956;24(5):966. https://doi.org/10.1063/1.1742723.
55. Bixon M, Jortner J. Adv Chem Phys. 1999;106:35. https://doi.org/10.1002/9780470141656.ch3.
56. Marchi M, Gehlen JN, Chandler D, Newton M. J Am Chem Soc. 1993;115:4178. https://doi.org/10.1021/ja00063a041.
57. Tang J, Norris JR. J Chem Phys. 1994;101:5615. https://doi.org/10.1063/1.467348.
58. Ando K, Sumi H. J Phys Chem B. 1998;102:10991.

59. Newton MD, Isr J. Chem. 2004;44:83. https://doi.org/10.1560/LQ06-T9HQ-MTLM-2VC1.
60. Newton MD. J Phys Chem B. 2015;119(46):14728. https://doi.org/10.1021/acs.jpcb.5b07456.
61. Feskov SV, Ivanov AI. J Phys Chem A. 2013;117(45):11479. https://doi.org/10.1021/jp408516q.
62. Feskov SV, Ivanov AI. Russ. J Phys Chem. 2016;90(1):144. https://doi.org/10.1134/S0036024416010106.
63. Feskov SV, Yudanov VV. Russ. J Phys Chem A. 2017;91(9):1816. https://doi.org/10.1134/S0036024417090102.
64. Jortner J, Bixon M. J Chem Phys. 1988;88(1):167. https://doi.org/10.1063/1.454632.
65. Akesson E, Johnson AE, Levinger NE, Walker GC, DuBruil TF, Barbara PF. J Chem Phys. 1992;96:7859.
66. Barbara PF, Walker GC, Smith TP. Science. 1992;256:975.
67. Bagchi B, Gayathri N. Adv Chem Phys. 1999;107:1. https://doi.org/10.1002/9780470141663.ch1.
68. Barzykin AV, Frantsuzov PA, Seki K, Tachiya M. Adv Chem Phys. 2002;123:511.
69. Ivanov AI, Mikhailova VA. Russ. Chem Rev. 2010;79(12):1047. https://doi.org/10.1070/RC2010v079n12ABEH004167.
70. Feskov SV, Ivanov AI. J Chem Phys. 2018;148(10): 104107. https://doi.org/10.1063/1.5016438.
71. Borg I, Groenen PJF. Modern Multidimensional Scaling: Theory and Applications (Springer. N Y. 2005. https://doi.org/10.1007/0-387-28981-X.
72. Jimenez R, Fleming GR, Kumar PV, Maroncelli M. Nature. 1994;369:471.
73. Maroncelli M, Kumar VP, Papazyan A. J Phys Chem. 1993;97:13.
74. Nazarov AE, Ivanov AI, Rosspeintner A, Angulo G. J Mol Liq. 2022;360: 119387. https://doi.org/10.1016/j.molliq.2022.119387.
75. Zusman LD. Chem Phys. 1988;119(1):51. https://doi.org/10.1016/0301-0104(88)80005-3.
76. Feskov SV, Russ J. Phys Chem B. 2024;18:1. https://doi.org/10.1134/S1990793124010081.
77. Feskov SV. Int J Mol Sci. 2022;23(24): https://doi.org/10.3390/ijms232415793.
78. Nicolet O, Banerji N, Pagès S, Vauthey E. J Phys Chem A. 2005;109:8236. https://doi.org/10.1021/jp0532216.
79. Fedunov RG, Feskov SV, Ivanov AI, Nicolet O, Pagès S, Vauthey E. J Chem Phys. 2004;121(8):3643. https://doi.org/10.1063/1.1772362.
80. Feskov SV, Ionkin VN, Ivanov AI. J Phys Chem A. 2006;110(43):11919. https://doi.org/10.1021/jp063280z.
81. Feskov SV, Kichigina AO, Ivanov AI. J Phys Chem A. 2011;115(9):1462. https://doi.org/10.1021/jp108607t.
82. Nazarov AE, Barykov VY, Ivanov AI. J Phys Chem B. 2016;120(12):3196. https://doi.org/10.1021/acs.jpcb.6b00539.
83. Feskov SV, Gladkikh V, Burshtein AI. Chem Phys Lett. 2008;458(1–3):71. https://doi.org/10.1016/j.cplett.2008.04.063.
84. Feskov SV, Rogozina MV, Ivanov AI, Aster A, Koch M, Vauthey E. J Chem Phys. 2019;150(2):024501. https://doi.org/10.1063/1.5064802.
85. Feskov SV. Comput Theor Chem. 2018;1145:15. https://doi.org/10.1016/j.comptc.2018.10.007.
86. Guldi DM. Chem Soc Rev. 2002;31:22. https://doi.org/10.1039/B106962B.
87. Wallin S, Monnereau C, Blart E, Gankou JR, Odobel F, Hammarström L. J Phys Chem A. 2010;114(4):1709. https://doi.org/10.1021/jp907824d.
88. Wróbel D, Graja A. Coord Chem Rev. 2011;255(21):2555. https://doi.org/10.1016/j.ccr.2010.12.026.
89. Kirner SV, Henkel C, Guldi DM, Megiatto JD Jr, Schuster DI. Chem Sci. 2015;6:7293. https://doi.org/10.1039/C5SC02895G.
90. Phelan BT, Zhang J, Huang GJ, Wu YL, Zarea M, Young RM, Wasielewski MR. J Am Chem Soc. 2019;141(31):12236. https://doi.org/10.1021/jacs.9b06166.

91. Loong H, Zhou J, Jiang N, Feng Y, Xie G, Liu L, Xie Z. J Phys Chem B. 2022;126(12):2441. https://doi.org/10.1021/acs.jpcb.2c00965.
92. Brown PJ, Qiu Y, Latawiec EI, Phelan BT, Tcyrulnikov NA, Palmer JR, Krzyaniak MD, Kopp SM, Huang Y, Young RM, Wasielewski MR. J Phys Chem A. 2024;128(43):9371. https://doi.org/10.1021/acs.jpca.4c05015.
93. Wang X, Wang H, Zhang M, Pullerits T, Song P. J Phys Chem C. 2023;127(6):2805. https://doi.org/10.1021/acs.jpcc.2c07364.
94. Nguyen HH, Song Y, Maret EL, Silori Y, Willow R, Yocum CF, Ogilvie JP. Sci. Adv. 9(18), eade7190 (2023). https://doi.org/10.1126/sciadv.ade7190.
95. Marcus RA, Sutin N. Biochim Biophys Acta. 1985;811(3):265. https://doi.org/10.1016/0304-4173(85)90014-X.
96. Bicout DJ, Szabo A. J Chem Phys. 1997;109:2325. https://doi.org/10.1063/1.476800.
97. Feskov SV, Ivanov AI. Chem Phys. 2016;478:164. https://doi.org/10.1016/j.chemphys.2016.03.013.
98. Feskov SV, Malykhin RE, Ivanov AI. J Phys Chem B. 2020;124(46):10442. https://doi.org/10.1021/acs.jpcb.0c07978.
99. Canton SE, Kjær KS, Vankó G, van Driel TB, Adachi SI, Bordage A, Bressler C, Chabera P, Christensen M, Dohn AO, Galler A, Gawelda W, Gosztola D, Haldrup K, Harlang T, Liu Y, Møller KB, Németh Z, Nozawa S, Pápai M, Sato T, Sato T, Suarez-Alcantara K, Togashi T, Tono K, Uhlig J, Vithanage DA, Wärnmark K, Yabashi M, Zhang J, Sundström V, Nielsen MM. Nat Commun. 2015;6(1):6359. https://doi.org/10.1038/ncomms7359.
100. Gladkikh V, Burshtein AI, Feskov SV, Ivanov AI, Vauthey E. J Chem Phys. 2005;123: 244510. https://doi.org/10.1063/1.2140279.
101. Rizzi AC, van Gastel M, Liddell PA, Palacios RE, Moore GF, Kodis G, Moore AL, Moore TA, Gust D, Braslavsky SE. J Phys Chem A. 2008;112(18):4215. https://doi.org/10.1021/jp712008b.
102. Rego LGC, Hames BC, Mazon KT, Joswig JO. J Phys Chem C. 2014;118(1):126. https://doi.org/10.1021/jp408955e.
103. Manna AK, Balamurugan D, Cheung MS, Dunietz BD. J Phys Chem Lett. 2015;6(7):1231. https://doi.org/10.1021/acs.jpclett.5b00074.
104. Sun X, Zhang P, Lai Y, Williams KL, Cheung MS, Dunietz BD, Geva E. J Phys Chem C. 2018;122(21):11288. https://doi.org/10.1021/acs.jpcc.8b02697.
105. Hou Y, Zhang X, Chen K, Liu D, Wang Z, Liu Q, Zhao J, Barbon A. J Mater Chem C. 2019;7:12048. https://doi.org/10.1039/C9TC04285G.
106. Hu Z, Tong Z, Cheung MS, Dunietz BD, Geva E, Sun X. J Phys Chem B. 2020;124(43):9579. https://doi.org/10.1021/acs.jpcb.0c06306.
107. Motylewski T, Najbar J, Tachiya M. Chem Phys. 1996;212:193.
108. Mataga N, Taniguchi S, Chosrowjan H, Osuka A, Yoshida N. Chem Phys. 2003;295:215.
109. Ivanov AI, Ionkin VN, Feskov SV. Russ. J Phys Chem. 2008;82(2):303. https://doi.org/10.1134/S0036024408020295.
110. Bazlov SV, Feskov SV, Ivanov AI. Russ. J Phys Chem B. 2017;11(2):242. https://doi.org/10.1134/S1990793117020026.
111. Burdzinski G, Kubicki J, Maciejewski A, Steer R, Velate S, Yeow E. Organic Photochemistry and Photophysics (Taylor & Francis. N Y. 2005. https://doi.org/10.1201/9781420036992.

Chapter 8
Exciton Storage in Symmetry-Protected Open Quantum Batteries

Zohreh Khodadad and Gabriel Hanna

Abstract The global surge in energy demand, combined with rapid advancements in quantum technology, highlights the need for efficient energy storage systems. While traditional batteries, which rely on chemical reactions, continue to improve, quantum thermodynamics offers a promising alternative. By harnessing the unique principles of quantum mechanics that govern microscopic systems, quantum-based energy storage solutions present new opportunities for enhanced performance and efficiency. Several models of quantum batteries (QBs) have been proposed, aiming to optimize charging speed, storage capacity, and discharge efficiency. Although much of this work remains theoretical, recent experimental breakthroughs have demonstrated the feasibility of manipulating quantum systems to store and release energy. One particularly compelling model is the symmetry-protected excitonic QB (SPEQB), which leverages symmetry-protected dark states for long-term exciton storage. A key advantage of this model is its ability to suppress energy leakage to the environment under symmetry-preserving conditions, even in open systems. The exciton can then be released on demand by intentionally breaking the symmetry. This chapter begins with an introduction to QBs and a review of major models. We highlight notable experimental advances, explain the theoretical framework of SPEQBs, and present simulation results that illustrate their behavior and potential in future energy applications.

Keywords Quantum battery · Dark state · Exciton dynamics · Open quantum systems · Quantum technologies · Symmetry protection · Energy storage · Quantum thermodynamics · Quantum coherence · Decoherence

Z. Khodadad
University of Alberta, Edmonton, Canada
e-mail: zkhodada@ualberta.ca

G. Hanna (✉)
University of Alberta, Edmonton, Canada
e-mail: gabriel.hanna@ualberta.ca

T. Onishi (ed.), *Advances in Quantum Science I*,
https://doi.org/10.1007/978-981-95-9717-8_8

8.1 Introduction

8.1.1 Foundational Models and the Role of Quantum Effects in Quantum Batteries

A quantum battery (QB) is a quantum system designed to store and release energy by exploiting quantum mechanical effects. Its operation consists of three main phases: charging, where external energy is coherently transferred into the system; storage, during which the energy is preserved in quantum states with minimal loss; and discharge, where the stored energy is extracted to perform work (see Fig. 8.1). The concept of a QB was first introduced by Alicki and Fannes [1], who investigated how quantum entanglement—nonclassical correlations between subsystems characterized by inseparable joint quantum states—can enhance the extractable work from a quantum system. Their study considered systems composed of multiple identical copies of a battery and, using the concept of passivity, defined the maximum extractable work, known as ergotropy [2]. They demonstrated that collective, entangling unitary operations significantly increase energy extraction efficiency, with performance improving as the number of battery copies grows.

Building on this foundational work, Binder et al. [3] investigated whether quantum effects provide an operational advantage over their classical counterparts. To address this question, they analyzed the charging process for both single qubits—two-level systems that can exist in superpositions of energy eigenstates $|0\rangle$ and $|1\rangle$—and arrays of qubits. For multi-qubit systems, they demonstrated that allowing global entangling

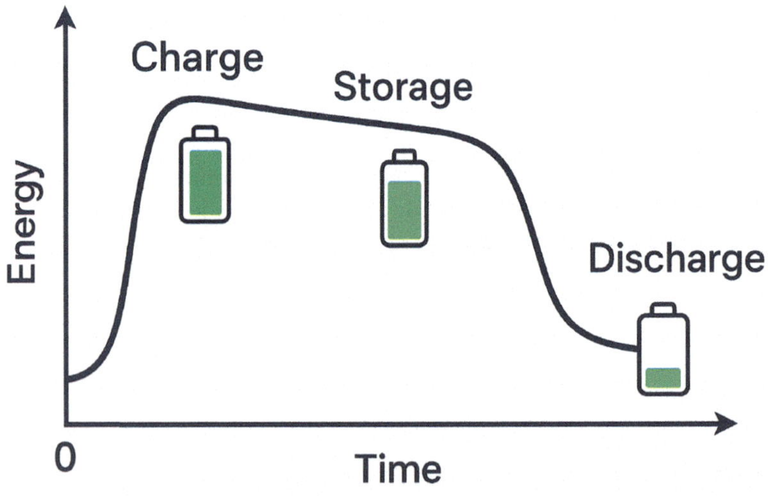

Fig. 8.1 Schematic representation of the three operational phases of a QB: (i) charge, where energy is supplied to the system; (ii) storage, during which the energy is maintained within the quantum states; and (iii) discharge, where the stored energy is extracted for use

operations leads to a significant enhancement in power, in contrast to the limited performance of local operations. This quantum advantage arises from the entanglement, offering insights into the practical benefits of quantum effects in QB design.

To explore whether quantum effects beyond entanglement can be exploited in QB models, Campaioli et al. [4] expanded the scope by investigating the role of collective quantum operations in enhancing thermodynamic processes. They demonstrated that a significant advantage can still be achieved even in the absence of entanglement during the charging process.

Ferraro et al. [5] further advanced the earlier works [3, 4] by introducing the "Dicke QB" model, where N two-level systems (TLSs) are coupled to a single photonic mode in a cavity. By contrasting this model with the "Rabi QB", [5] where each TLS is individually coupled to its own cavity mode, they demonstrated a quantum advantage in charging power achieved through collective interactions. The results showed that the Dicke QB outperforms the linear scaling observed in non-collective charging schemes, with the enhancement attributed to quantum entanglement generated by the shared photonic mode. This study provided a realistic and experimentally feasible framework for leveraging quantum collective effects in solid-state QBs.

Building on the idea of leveraging quantum effects beyond entanglement, Kamin et al. [6] explored an alternative approach that emphasizes quantum coherence—the property of a quantum system to exist in superpositions of states, reflected in the presence of non-zero off-diagonal elements in its density matrix—as a key resource for QB performance. To address whether entanglement is necessary for optimal charging, they considered a two-qubit QB system and compared the charging dynamics under both collective and non-collective (parallel) protocols. Their findings revealed that efficient charging can occur without generating entanglement, and that coherence generation can drive optimal charging performance.

Beyond coherence, entanglement, and collective effects, Le et al. [7] explored an alternative mechanism for enhancing QB performance. The study explored the use of a one-dimensional spin chain with intrinsic two-body interactions as a QB. Unlike models that rely on global operations for charging, this approach focused on locally charging the spin chain while leveraging the system's internal interactions to enhance performance. The authors found that anisotropic spin–spin interactions significantly improve charging power, particularly for long-range couplings. Interestingly, they concluded that this enhancement arises from interaction energetics rather than quantum correlations or entanglement.

8.1.2 Open Quantum Batteries

While most initial studies on QBs were limited to closed quantum systems—isolated systems that do not interact with their environment—more recent models have been developed to address open quantum systems [8–16]. These new models are more realistic, as they account for the unavoidable interactions between a QB and its environment in real-world scenarios. While some models exploit system–environment interactions to facilitate charging [9, 10] or stabilize QBs [11, 12], such interactions can also lead to detrimental effects, including dissipation and dephasing, which degrade QB performance. By incorporating these environmental factors, these advanced models provide deeper insights into the dynamics, stability, and practical implementation of QBs under realistic conditions.

One approach to preventing energy leakage in open QBs during both the charging and storage phases is to utilize dark states [13, 14, 17]. Dark states are quantum states that are not optically accessible, making them robust against radiative decay and other dissipative processes. By leveraging these states, QBs can retain their charge over extended periods with minimal energy loss. For example, Santos et al. [17] used the Stimulated Raman Adiabatic Passage (STIRAP) [18] protocol in a qutrit-based QB to achieve robust charging by steering the system's evolution through a dark state, thereby avoiding the spontaneous discharging caused by coherent oscillations typical of two-level QBs.

Quach and Munro [13] exploited dark states in a Dicke QB to enhance its stability. In contrast to the approach proposed by Santos et al. [17], the dark states in this work arise from subradiance—a phenomenon in which collective emission from quantum emitters is significantly suppressed compared to that of isolated emitters. This suppression occurs when the emitters are arranged so that their emitted fields interfere destructively. As a result, the subradiant mechanism effectively protects the stored energy from environmental dissipation, making the Dicke QB a strong candidate for robust and long-lived quantum energy storage.

Another example of exploiting dark states in QBs is the symmetry-protected excitonic QB (SPEQB) proposed by Liu et al. [14]. In their model, the dark states emerge as a consequence of the system's topological symmetries [19]. The authors demonstrated that by preparing the system in such a state, the QB remains completely isolated from dissipative processes. This QB model will be the main focus of this chapter and will be discussed in detail in the next section.

8.1.3 Experimental Realizations of Quantum Batteries

It is also worth highlighting some experimental realizations of QBs that have successfully demonstrated quantum advantages in energy storage and transfer. One notable example is a QB based on superabsorption in a Dicke QB [20], a cooperative quantum phenomenon where energy absorption scales superextensively with system size. This study used a microcavity containing organic dye molecules forming an ensemble of two-level systems. Using ultrafast transient absorption spectroscopy, the experiment revealed femtosecond charging dynamics, confirming both superextensive energy storage and enhanced charging power. This realization provided strong experimental evidence of how quantum collective effects can significantly improve QB performance.

Another significant experimental realization involves QBs implemented using nuclear magnetic resonance techniques in star–topology spin networks [21]. In this setup, a central spin (the battery) is surrounded by multiple charger spins, enabling collective charging. By experimentally characterizing battery energy and ergotropy through quantum state tomography, the authors demonstrated quantum speedup in energy storage due to collective interactions. Moreover, they successfully implemented a charger–battery–load circuit, demonstrating the QB's ability to transfer stored energy even after storage durations of up to two minutes.

8.2 Symmetry-Protected Excitonic Quantum Battery Model

The theoretical and experimental advancements to date represent significant progress toward the practical implementation of QBs, showcasing how quantum effects such as entanglement, collective interactions, and quantum coherence can enhance energy storage and transfer. While these developments provide valuable insights, QB models still struggle with energy dissipation, which limits their long-term stability. As previously mentioned, a promising approach to overcoming these challenges is exploiting topological symmetry in a quantum system. This method enables long-term energy storage by protecting the stored energy from environmental interactions. In this section, we provide a detailed description of the SPEQB, followed by an analysis of the results that demonstrate its stability and efficiency.

The SPEQB model was originally formulated as a cyclic open quantum network (OQN) composed of six sites [14], where each site is occupied by a TLS, as depicted in Fig. 8.2a. Sites 1 and 4 are directly coupled to thermal baths and are referred to as surface sites (SSs), while the remaining four sites (2, 3, 5, and 6) are referred to as bulk sites (BSs), which interact indirectly with the baths.

The model incorporates Coulombic interactions between adjacent sites, facilitating the transfer of a Frenkel exciton (localized on one site) to its neighboring sites. It is restricted to the single-excitation manifold, meaning that the system can only host one exciton (although the model could be extended to a multi-excitation manifold). As a result, the Hilbert space of the OQN is spanned by single-excitation basis states. In this case, the network possesses the following Frenkel Hamiltonian:

$$\hat{H}_N = \sum_{n=1}^{6} E_n |n\rangle\langle n| + \sum_{n \neq m} J_{n,m} |n\rangle\langle m| \tag{8.1}$$

where $|n\rangle$ is a state in which the system at site n is excited while the systems at all other sites remain in their ground states. Here, E_n is the on-site energy of site n, and $J_{n,m}$ is the coupling strength between sites n and m. To date, studies using this QB model have considered only nearest-neighbor couplings between sites, although there is no fundamental obstacle to including non-nearest-neighbor couplings as well.

The OQN is coupled to two thermal baths at temperature T through sites 1 and 4, where each bath is represented by a set of harmonic oscillators. The Hamiltonian

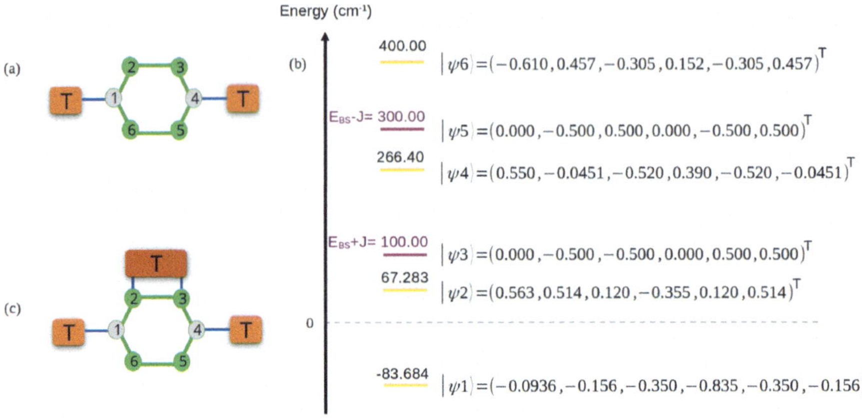

Fig. 8.2 **a** Architecture of the SPEQB during the storage phase, where the exciton population remains confined within the BSs (depicted in green circles), with zero population on the surface sites (represented by gray circles). The SQEQB is coupled to two thermal baths at temperature T. **b** Energy level diagram illustrating the eigenstates and energies (in cm^{-1}) of the OQN Hamiltonian in Eq. 8.1 with $E_1 = 250$ cm^{-1}, $E_{BS=2,3,5,6} = 200$ cm^{-1}, $E_4 = 0$ cm^{-1}, and all the non-zero inter-site couplings set equal to $J = -100$ cm^{-1}. The dark states are highlighted in purple. **c** Architecture of the SPEQB during the discharge phase, where the exciton population is transferred to the SSs. To break the symmetry, the SQEQB is coupled to an additional bath at temperature T through sites 2 and 3

describing these thermal baths ($\hat{H}_B$) and their interaction with OQN ($\hat{H}_{NB}$) is given by

$$\hat{H}_B + \hat{H}_{NB} = \frac{1}{2}\sum_{n=1,4}\sum_{j}^{M}\left[\hat{P}_{n,j}^2 + \omega_{n,j}^2\left[\hat{R}_{n,j} - \frac{C_{n,j}}{\omega_{n,j}^2}|n\rangle\langle n|\right]^2\right], \tag{8.2}$$

where M represents the number of harmonic oscillators in each bath; $\hat{P}_{n,j}$, $\hat{R}_{n,j}$, and $\omega_{n,j}$ represent the mass-weighted momentum, position, and frequency of the jth oscillator, respectively; $C_{n,j}$ is the coupling strength between the jth oscillator and SS n. The system–bath interaction can be characterized by a spectral density function, e.g., a Debye–Drude spectral density, $J_\omega = 2\lambda_b \frac{\omega\omega_c}{\omega^2+\omega_c^2}$, where λ_b represents the reorganization energy and ω_c the cut-off frequency of the bath.

In the Frenkel Hamiltonian, if $E_2 = E_6$, $E_3 = E_5$, $J_{1,2} = J_{1,6}$, $J_{2,3} = J_{5,6}$, and $J_{3,4} = J_{4,5}$, an exchange symmetry emerges, viz., the system is invariant under the exchange of sites 2 and 6 and of sites 3 and 5 (see Fig. 8.2a). This symmetry allows for the definition of the following unitary symmetry operator:

$$\hat{\Pi} = |1\rangle\langle 1| + |4\rangle\langle 4| + (|2\rangle\langle 6| + |3\rangle\langle 5| + H.C.). \tag{8.3}$$

It commutes with the network Hamiltonian, i.e., $[\hat{\Pi}, \hat{H}_N] = 0$, and also with the interaction Hamiltonian, i.e., $[\hat{\Pi}, |n\rangle\langle n|] = 0$ for n= 1, 4. As a result, $\hat{\Pi}$ and $\hat{H}_N$

share a common basis, which could be used to decompose the system's Hilbert space and its corresponding operator space. In this case, it could be shown that there exists a one-dimensional diagonal subspace $|\psi_\alpha^{(1)}\rangle\langle\psi_\alpha^{(1)}|$ within the operator space, which remains invariant and maps onto itself under the influence of the total Hamiltonian, $\hat{H}_N + \hat{H}_B + \hat{H}_{NB}$ [22]. Here, α indexes the eigenvalues of $\hat{\Pi}$, and superscript 1 denotes the one-dimensional nature of the subspace. These subspaces are referred to as decoherence-free subspaces and the states $|\psi_\alpha^{(1)}\rangle$, residing in these subspaces, are dark states. These states have garnered significant attention in the development of quantum technologies due to their robust protection against environmental decoherence [22–26]. In the context of this QB model, the necessary conditions for the existence of these dark states are satisfied, allowing them to be used for exciton storage.

To identify the dark states of a particular OQN, one must diagonalize the system Hamiltonian in Eq. 8.1. The eigenstates that exhibit zero amplitude on the SSs (sites 1 and 4) are classified as dark. As an example, Fig. 8.2b shows the eigenenergies and corresponding eigenstates of $\hat{H}_N$ for a specific choice of on-site energies and inter-site couplings. The parameters are set as $E_1 = 250\ \mathrm{cm}^{-1}$, $E_{BS=2,3,5,6} = 200\ \mathrm{cm}^{-1}$, $E_4 = 0\ \mathrm{cm}^{-1}$, and all non-zero inter-site couplings as $J = -100\ \mathrm{cm}^{-1}$. Under these conditions, the OQN possesses six eigenstates. Each eigenstate can be expressed as a linear combination of site basis states $\{|n\rangle\}$, where the coefficients quantify the contribution of each site to that eigenstate. From the figure, we can identify states 3 and 5 as the dark states. When the OQN is initially prepared in one of its dark states, the exciton population remains constant, ensuring that the exciton remains confined within the BSs. This confinement defines the storage phase of the QB.

The discharge phase of the QB can be initiated by introducing an additional thermal bath at temperature T (referred to as the probe), which breaks the exchange symmetry of the system. One possible implementation involves attaching the probe to sites 2 and 3, as depicted in Fig. 8.2c. This symmetry breaking enables exciton population transfer from the BSs to the SSs. Additionally, by setting the on-site energy of site 4 lower than that of site 1, a bias potential is created, directing the exciton to accumulate predominantly at site 4. This site, referred to as the exit site, therefore serves as the primary site for exciton extraction during the discharge phase. The Hamiltonian governing this probe ($\hat{H}_P$) and its interaction with the OQN ($\hat{H}_{NP}$) is given by

$$\hat{H}_P + \hat{H}_{NP} = \frac{1}{2}\sum_k^M \left[\hat{p}_k^2 + \Omega_k^2\left[\hat{r}_k - \frac{\gamma_k}{\Omega_k^2}(|2\rangle\langle 2| + |3\rangle\langle 3|)\right]^2\right], \quad (8.4)$$

where M is the number of harmonic oscillators constituting the probe; $\hat{p}_k$, $\hat{r}_k$, and Ω_k correspond to the mass-weighted momentum, position, and frequency of each oscillator, respectively; and γ_k is the strength of the coupling between the probe oscillator and sites 2 and 3, where the probe is attached. The system–probe interaction can be characterized by a spectral density function, e.g., a Debye–Drude spectral density.

Taking all of the above into account, the total Hamiltonian can be expressed as

$$\hat{H} = \hat{H}_N + \hat{H}_B + \hat{H}_{NB} + \chi(\hat{H}_P + \hat{H}_{NP}), \tag{8.5}$$

where the parameter χ takes on a value of either 0 during the storage phase or 1 during the discharge phase.

8.3 Simulation of Exciton Dynamics

To simulate the exciton dynamics of the OQN, Liu et al. used the DECIDE approach [27], an approximate quantum dynamics method that treats the OQN fully quantum mechanically while describing the bath oscillators in a classical-like manner. Despite its approximate nature, DECIDE has demonstrated high accuracy across a range of model systems, including the spin–boson model, a photo-induced electron transfer model, and the Fenna–Matthews–Olson complex model. Compared to phenomenological approaches such as the Lindblad master equation [28, 29], DECIDE achieves greater accuracy and explicitly incorporates the environmental degrees of freedom into the equations of motion.

In the DECIDE approach, formulated within the partial Wigner representation [30], one specifies the coordinates of both the quantum subsystem and the classical-like environment that are to be evolved [27]. In this case, the OQN coordinates are taken to be the set of $\hat{\mathcal{P}}_{nm} = |n\rangle\langle m|$ operators and the bath coordinates the set of positions and momenta of all oscillators, i.e., $X = (\{P_{n,j}\}, \{R_{n,j}\}, \chi\{p_k\}, \chi\{r_k\})$, as defined below Eqs. 8.2 and 8.4. As in Eq. 8.5, χ takes on a value of either 0 (during the storage phase) or 1 (during the discharge phase). The DECIDE equations of motion for the coupled network–bath evolution are

$$\begin{aligned}
\frac{d}{dt}\mathcal{P}^{\beta\beta'}_{nm}(t) = i &\left[\sum_{l=1}^{6} V_{ln}(t)\hat{\mathcal{P}}_{lm}(t) - \sum_{v=1}^{6} V_{mv}(t)\hat{\mathcal{P}}_{nv}(t)\right]^{\beta\beta'} \\
&- \frac{i}{2}\sum_{j} C_{n,j}(R_{n,j}(t)\hat{\mathcal{P}}_{nm}(t) + \hat{\mathcal{P}}_{nm}(t)R_{n,j}(t))^{\beta\beta'}(\delta_{n,1} + \delta_{n,4}) \\
&+ \frac{i}{2}\sum_{j} C_{m,j}(R_{m,j}(t)\hat{\mathcal{P}}_{nm}(t) + \hat{\mathcal{P}}_{nm}(t)R_{m,j}(t))^{\beta\beta'}(\delta_{m,1} + \delta_{m,4}) \\
&- \chi\frac{i}{2}\sum_{k} C_k(R_k(t)\hat{\mathcal{P}}_{nm}(t) + \hat{\mathcal{P}}_{nm}(t)R_k(t))^{\beta\beta'}(\delta_{n,2} + \delta_{n,3}) \\
&+ \chi\frac{i}{2}\sum_{k} C_k(R_k(t)\hat{\mathcal{P}}_{nm}(t) + \hat{\mathcal{P}}_{nm}(t)R_k(t))^{\beta\beta'}(\delta_{m,2} + \delta_{m,3}),
\end{aligned}$$

$$
\begin{aligned}
&\frac{d}{dt}R_{n,j}^{\beta\beta'}(t) = P_{n,j}^{\beta\beta'}(t),\\
&\frac{d}{dt}P_{n,j}^{\beta\beta'}(t) = -\omega_{n,j}^2 R_{n,j}^{\beta\beta'}(t) + C_{n,j}\mathcal{P}_{nn}^{\beta\beta'}(t)(\delta_{n,1}+\delta_{n,4}),\\
&\frac{d}{dt}r_k^{\beta\beta'}(t) = p_k^{\beta\beta'}(t),\\
&\frac{d}{dt}p_k^{\beta\beta'}(t) = -\Omega_k^2 r_k^{\beta\beta'}(t) + \chi\gamma_k(\hat{\mathcal{P}}_{22}+\hat{\mathcal{P}}_{33})^{\beta\beta'}(t),
\end{aligned}
\tag{8.6}
$$

where $\beta = \{|1\rangle, |2\rangle, ..., |6\rangle\}$, $V_{nn}(t) = E_n(t) + \sum_{j=1}^{M} C_{n,j}^2/(2\omega_{n,j}^2)(\delta_{n,1}+\delta_{n,4}) + \chi\sum_{k=1}^{M}\gamma_k^2/(2\Omega_k^2)(\delta_{n,2}+\delta_{n,3})$, and $V_{nm}(t)$ is the time-dependent (in general) value of the nearest-neighbor electronic coupling when $n \neq m$. In total, there are $36 \times [35 + 2(2+\chi)M]$ coupled first-order differential equations for the network, side thermal baths, and probe [14, 27].

The expressions for the system–bath couplings and oscillator frequencies, obtained by discretizing the Debye–Drude spectral density, are given by [31]

$$
\begin{aligned}
C_{n,j} &= 2\sqrt{\lambda_b \arctan(\omega_{max}/\omega_c)/(\pi M)}\,\omega_{n,j}\\
\omega_{n,j} &= \tan(j\ \arctan(\omega_{max}/\omega_c)/M)\omega_c\\
C_k &= 2\sqrt{\lambda_p \arctan(\omega_{max}/\omega_p)/(\pi M)}\,\Omega_k\\
\Omega_k &= \tan(j\ \arctan(\omega_{max}/\omega_p)/M)\omega_p,
\end{aligned}
\tag{8.7}
$$

where the first and second equations provide the system–bath couplings and frequencies for the oscillators in the side thermal baths, respectively, while the third and fourth equations provide those in the probe bath.

The exciton population localized on site n at time t can be calculated according to

$$
\langle \hat{P}_n(t)\rangle = \sum_{\beta\beta'}\int dX(0)\,\mathcal{P}_n^{\beta\beta'}(t)\rho_N^{\beta'\beta}(0)\rho_B(0)\rho_P(0),
\tag{8.8}
$$

where $\rho_N^{\beta'\beta}(0)$ is the network's initial density matrix element and $\rho_B(0)\rho_P(0)$ is the initial state of the thermal environments. The initial state of the network is taken to be one of its dark states, so $\hat{\rho}_N(0) = |DS\rangle\langle DS|$. The initial state of the thermal side baths is taken to be the thermal equilibrium distribution for a system of harmonic oscillators [32]

$$
\rho_B(0) = \prod_{n=1,4}\prod_{j=1}^{M}\frac{\tanh(\beta\omega_{n,j}/2)}{\pi}\times\exp\left[-\frac{2\tanh(\beta\omega_{n,j}/2)}{\omega_{n,j}}\left[\frac{P_{n,j}^2}{2}+\frac{\omega_{n,j}^2 R_{n,j}^2}{2}\right]\right],
\tag{8.9}
$$

where $\beta = 1/k_B T$ is the inverse temperature. (The corresponding expression for the probe bath possesses the same form.)

DECIDE is a trajectory-based approach. Thus, to compute $\mathcal{P}_n^{\beta\beta'}(t)$ along a given trajectory, one first samples the harmonic bath coordinates $X(0)$ from their respective thermal equilibrium distributions and sets the initial values of $\{\mathcal{P}_n^{\beta\beta'}(0)\}$ according to the network's initial state. Then, the DECIDE equations of motion are integrated in time to obtain $\mathcal{P}_n^{\beta\beta'}(t)$. Finally, the population of site n, $\langle \hat{P}_n(t) \rangle$, is obtained by averaging over the ensemble of trajectories in time.

8.4 Exciton Dynamics During the Storage and Discharge Phases

The charging, storage, and discharge dynamics of the SPEQB have been examined in several previous studies [14, 33–35]. A detailed understanding of the exciton dynamics under different conditions is crucial for optimizing both the performance and stability of the SPEQB. In this chapter, we focus on how structural and environmental factors shape exciton storage and discharge. We begin with the idealized case without disorder and noise, then assess the influence of periodic oscillations on the population dynamics, and finally evaluate the effects of both diagonal and off-diagonal disorder.

8.4.1 No Disorder

The disorder-free model, as presented in Sect. 8.2, serves as a useful benchmark for understanding the core features of the SPEQB model. Figure 8.3a and b show its population dynamics during the storage and discharge phases, respectively. These results were obtained using the parameter set: $E_1 = 250$ cm^{-1}, $E_{BS} = 200$ cm^{-1}, $E_4 = 0$ cm^{-1}, with all non-zero inter-site couplings set to $J = -100$ cm^{-1}. In the storage phase, the BS populations remain constant at 0.25 each, while the SS populations remain zero. Upon attaching the probe and breaking the symmetry, the exciton population is discharged from the BSs to the SSs, with site 4 accumulating the majority of the population due to the bias potential between site 1 and site 4. These results demonstrate the ability of this open quantum system to store excitonic energy without dissipation, showcasing its potential for loss-free energy storage and controlled discharge in a QB.

8.4.2 Inter-site Oscillations

To account for the possibility that the inter-site couplings can vary in time due to changes in inter-site distances, which in turn modify the dipole–dipole interactions, Khodadad et al. introduced a sinusoidal variation of the electronic couplings [34]. This variation, which is valid for small-amplitude motions, can be expressed as [36]

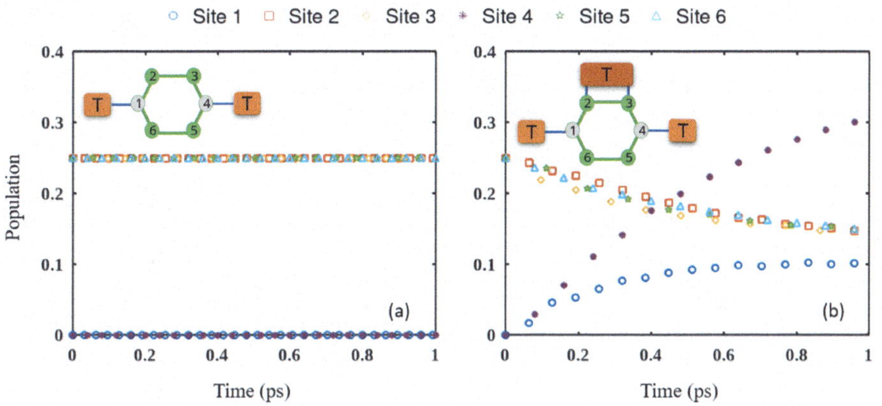

Fig. 8.3 Site population dynamics of the QB in the **a** storage phase and **b** discharge phase, obtained using the following parameter values: $T = T_p = 300$ K, $\lambda_b = 35$ cm^{-1}, $\lambda_p = 10$ cm^{-1}, $\omega_c = \omega_p = 106$ cm^{-1}, $E_1 = 250$ cm^{-1}, $E_{BS} = 200$ cm^{-1}, $E_4 = 0$ cm^{-1}, and all non-zero inter-site couplings set to $J = -100$ cm^{-1}. For each simulation, a time step of 1 fs, 10^4 trajectories, and $M = 35$ oscillators in each bath, was used. Each inset schematically depicts the architecture specific to the corresponding phase

$$J(t) = \frac{J_0}{[1 - 2A\sin(\omega t + \phi)]^3}, \tag{8.10}$$

where J_0 is a reference coupling value, A is the oscillation amplitude, ω is the oscillation frequency, and ϕ is the phase. In Ref. [34], the variations of all non-zero couplings were taken to be equal and $J_0 = -100$ cm^{-1} was chosen to be the value of J that maximizes the exit site (site 4) population during the discharge phase of the disorder-free model with the system parameters given in the caption of Fig. 8.3.

Figure 8.4a and b show the effects of the oscillation frequency and oscillation amplitude, respectively, on the exciton discharge efficiency during the first picosecond of the discharge phase. The discharge efficiency is defined as

$$\eta(t) = 2\kappa \int_0^t \langle e|\hat{\rho}_N(t')|e\rangle dt'. \tag{8.11}$$

This expression assumes a sink coupled to the exit site, where the population is given by $\langle e|\hat{\rho}_N(t')|e\rangle$, and the exciton transfer into the sink occurs at a rate κ [14, 37, 38]. In the simulations, a transfer rate of $\kappa = 1$ ps^{-1} was used, consistent with typical values observed in exciton transfer dynamics [37]. In Fig. 8.4a, it is seen that the $\omega = 10$ cm^{-1}, 20 cm^{-1}, and 80 cm^{-1} results exhibit discharge efficiencies that are comparable to the case without oscillations, while the 2 cm^{-1} result exhibits a lower discharge efficiency. These results suggest that there exists a window of frequencies that do not significantly affect the discharge dynamics. In Fig. 8.4b, it is seen that a sufficiently large oscillation amplitude (viz., $A = 0.19$) leads to a substantial reduction in the

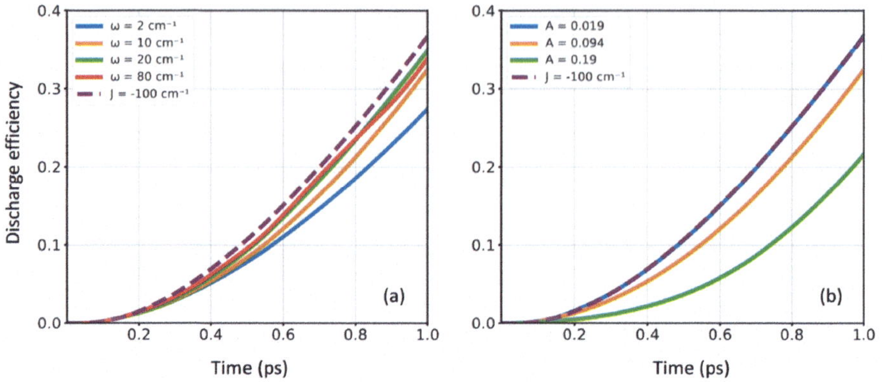

Fig. 8.4 Effect of the periodic variation of the inter-site couplings on the exciton discharge efficiency for **a** $A = 0.094$ and different ω values, and **b** $\omega = 10$ cm^{-1} and different A values. The purple curve represents the case with a constant value of $J = -100$ cm^{-1}. The remaining values of the parameters are the same as in the caption of Fig. 8.3 with $\phi = \pi/2$. For each simulation, a time step of 0.32 fs, 10^4 trajectories, and $M = 35$ oscillators in each bath, was used

discharge efficiency after 1 ps, while the smaller amplitudes (viz., $A = 0.019$ and 0.094) have little to no impact.

The differences between the cases with and without inter-site oscillations can be attributed to the fact that oscillations cause the coupling strength to deviate from its optimal value of -100 cm^{-1}, thereby reducing the population transfer to the exit site and, consequently, to the sink. At lower J values, the sites are more weakly coupled, leading to less population transfer. At higher J values, the stronger inter-site coupling causes the system's eigenstates to be highly delocalized, leading to a more uniform population distribution across all sites and a reduction in the discharge efficiency.

8.4.3 *Diagonal and Off-diagonal Disorder*

In addition to periodic modulation of the inter-site coupling strength, real systems are often subject to various sources of disorder that can affect both the coupling magnitude and on-site energies. Unlike periodic modulations, disorder can break the system's intended symmetry, potentially degrading the storage efficiency of the QB [14, 34]. For instance, static disorder often arises from material impurities or fabrication defects, while dynamic disorder typically results from interactions with a thermal environment, such as electron–phonon coupling [39].

Structural defects can be modeled by adding static Gaussian disorder, while thermal fluctuations can be modeled by adding dynamic Gaussian (white or colored) disorder. Disorder can be added to the diagonal and/or off-diagonal matrix elements of the Hamiltonian. In the case of static Gaussian disorder, random numbers are sampled from a Gaussian distribution, $P(x) = \frac{1}{\sqrt{2\pi\sigma^2}} \exp(\frac{-(x-\mu)^2}{2\sigma^2})$, where the standard deviation (SD) σ characterizes the noise strength. These values are added to the relevant matrix elements at the start of each trajectory. In the case of dynamic Gaussian disorder, random numbers are sampled from a Gaussian distribution and added to the relevant matrix elements at each time step of a trajectory. For white noise, the noise values are uncorrelated in time, whereas for colored noise, the noise values are correlated in time according to the following noise correlation function:

$$\langle \zeta(t)\zeta(s)\rangle = \sigma^2 \exp(-\lambda(t-s)), \tag{8.12}$$

where $\zeta(t)$ is the noise value at time t, λ^{-1} is the noise correlation time, and σ is the standard deviation of the Gaussian distribution governing the noise strength. Colored noise can be generated according to the algorithm described in Ref. [40].

To gain insight into how these different types of disorder impact the QB performance, Khodadad et al. simulated the exciton dynamics of disordered QBs during both the storage and discharge phases [34]. Figure 8.5 shows the effects of different types of off-diagonal Fig. 8.5a–c and diagonal disorders Fig. 8.5d–f on the storage efficiency, defined as the sum of the BS populations, during the storage phase. For all disorder types, the reduction in storage efficiency is negligible when the disorder strength is very weak (viz., 2 cm^{-1}). However, for higher disorder strengths, the

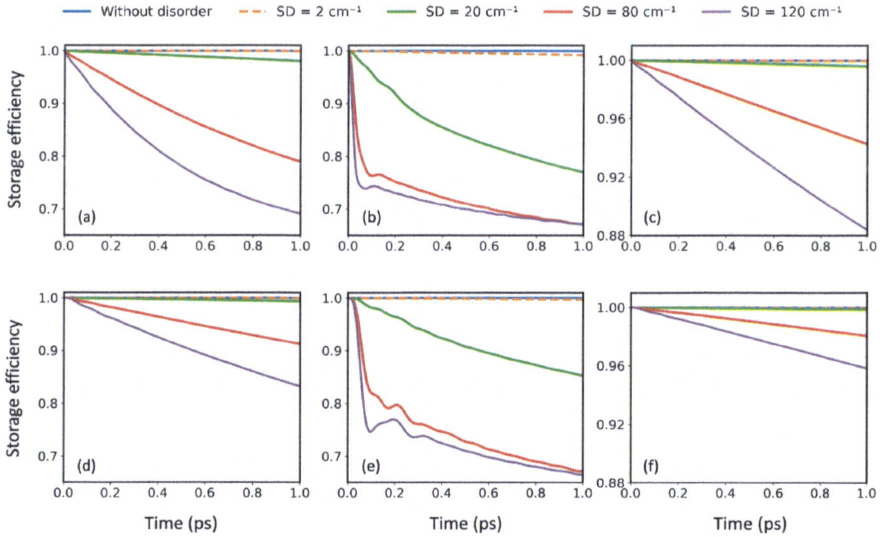

Fig. 8.5 Storage efficiency under **a** off-diagonal Gaussian colored noise, **b** off-diagonal Gaussian static disorder, **c** off-diagonal Gaussian white noise, **d** diagonal Gaussian colored noise, **e** diagonal Gaussian static disorder, and **f** diagonal Gaussian white noise. Results for the following standard deviations are shown: 0 cm^{-1} (blue), 2 cm^{-1} (orange), 20 cm^{-1} (green), 80 cm^{-1} (red), and 120 cm^{-1} (purple). For each simulation, a time step of 0.32 fs, 10^5 trajectories, and $M = 35$ oscillators in each bath, was used. The remaining values of the parameters are the same as in the caption of Fig. 8.3

reduction becomes significant, with its extent depending on the type of noise. For example, in the off-diagonal disorder case, focusing on the SD = 80 cm^{-1} curves (red), it can be observed that static disorder induces the highest leakage, followed by colored noise, and finally white noise, with storage efficiency values after 1 ps of $\approx$ 0.68, 0.79, and 0.94, respectively. The same trend is observed in the diagonal disorder case (Fig. 8.5d, e, and f).

The results in Fig. 8.5 also suggest that the same type of disorder can have a stronger impact when applied to the off-diagonal elements. For example, comparing colored noise with SD = 120 cm^{-1} in panels (a) and (d) (off-diagonal vs. diagonal colored noise), the storage efficiency values after 1 ps are approximately 0.7 and 0.84, respectively, indicating a more pronounced effect in the off-diagonal case. However, in the case static disorder, the impact remains the same for SD = 80 cm^{-1} and SD = 120 cm^{-1} (see panels (b) and (e)), regardless of whether the disorder is diagonal or off-diagonal. This suggests that under sufficiently strong disorder, static disorder affects the system in a similar manner, irrespective of the matrix elements it perturbs. The higher leakage observed in the presence of off-diagonal disorder, compared to diagonal disorder, is mainly due to its stronger disruption of system symmetry and the coherences required to maintain it. As seen in the equations of motion in Eq. 8.6, when the site indices are equal, the diagonal on-site energy terms cancel out, leaving the off-diagonal couplings as the main drivers of population dynamics. As a result, population leakage is more sensitive to variations in coupling strengths than to fluctuations in on-site energies.

Figure 8.6 shows the effects of different types of off-diagonal Fig. 8.6a–c and diagonal disorders Fig. 8.6d–f on the exciton discharge efficiency. Compared to the storage phase, static disorder leads to smaller changes in the discharge efficiency after 1 ps across the same range of disorder strengths (2 to 120 cm^{-1}), indicating that the discharge phase is more robust to static disorder. Over this range, colored noise causes only minor changes in the discharge efficiency, while white noise has a negligible effect.

8.5 Conclusion and Outlook

This chapter examined the operation of the SPEQB as a model for storing excitation energy. It demonstrated how symmetry and the resulting dark states can be leveraged for efficient energy storage. The importance of preserving symmetry during the storage phase—where it protects the exciton from environmental losses—and intentionally breaking it during the discharge phase to enable controlled energy release was discussed.

The influence of inter-site coupling fluctuations and on-site energy disorder on exciton dynamics in the SPEQB model was also discussed. Three types of disorder—static disorder, white noise, and colored noise—were considered. During the storage phase, these disturbances induced exciton leakage, with static disorder having the

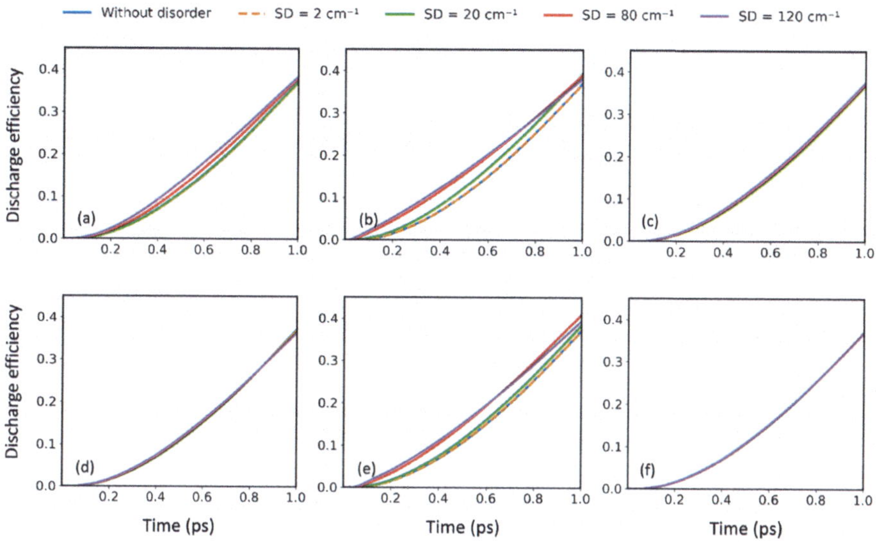

Fig. 8.6 Discharge efficiency under **a** off-diagonal Gaussian colored noise, **b** off-diagonal Gaussian static disorder, **c** off-diagonal Gaussian white noise, **d** diagonal Gaussian colored noise, **e** diagonal Gaussian static disorder, and **f** diagonal Gaussian white noise. Results for the following standard deviations are shown: 0 cm^{-1} (blue), 2 cm^{-1} (orange), 20 cm^{-1} (green), 80 cm^{-1} (red), and 120 cm^{-1} (purple). For each simulation, a time step of 0.32 fs, 4×10^5 trajectories, and $M = 35$ oscillators in each bath, was used. The remaining values of the parameters are the same as in the caption of Fig. 8.3

greatest impact and white noise the least. In contrast, the discharge phase was generally robust against disorder.

Future work will investigate the effects of alternative network architectures, including variations in ring size and topology. In parallel, efforts will focus on developing an optical protocol for preparing the system in the dark state (charging phase) and optimizing the discharge process. These studies are essential for advancing the design of efficient SPEQBs and will offer valuable insights into the feasibility and performance of this theoretical model in realistic molecular systems under experimentally relevant conditions.

Acknowledgements G. H. acknowledges funding from the Natural Sciences and Engineering Research Council of Canada (NSERC).

Competing Interests The authors have no conflicts of interest to declare that are relevant to the content of this chapter.

References

1. Alicki R, Fannes M. Phys Rev E. 2013;87:1.
2. Allahverdyan AE, Balian R, Nieuwenhuizen TM. Europhys Lett. 2004;67:565.
3. Binder FC, Vinjanampathy S, Modi K, Goold J. New J Phys. 2015;17: 075015.
4. Campaioli F, Pollock FA, Binder FC, Céleri L, Goold J, Vinjanampathy S, Modi K. Phys Rev Lett. 2017;118: 150601.
5. Ferraro D, Campisi M, Andolina GM, Pellegrini V, Polini M. Phys Rev Lett. 2018;120: 117702.
6. Kamin FH, Tabesh FT, Salimi S, Santos AC. Phys Rev E. 2020;102: 052109.
7. Le TP, Levinsen J, Modi K, Parish MM, Pollock FA. Phys Rev A. 2018;97: 022106.
8. Zhao F, Dou F-Q, Zhao Q. Phys Rev A. 2021;103: 033715.
9. Carrega M, Crescente A, Ferraro D, Sassetti M. New J Phys. 2020;22: 083085.
10. Tabesh FT, Kamin FH, Salimi S. Phys Rev A. 2020;102: 052223.
11. Bai S-Y, An J-H. Phys Rev A. 2020;102: 060201.
12. Gherardini S, Campaioli F, Caruso F, Binder FC. Phys Rev Res. 2020;2: 013095.
13. Quach JQ, Munro WJ. Phys Rev Appl. 2020;14: 024092.
14. Liu J, Segal D, Hanna G. J Phys Chem C. 2019;123:18303.
15. Ghosh S, Chanda T, Mal S, Sen A. Phys Rev A. 2021;104: 032207.
16. Zakavati SS, Tabesh FT. Phys Rev E. 2021;104: 054117.
17. Santos AC, Çakmak B, Campbell S, Zinner NT. Phys Rev E. 2019;100: 032107.
18. Vitanov NV, Rangelov AA, Shore BW, Bergmann K. Rev Mod Phys. 2017;89: 015006.
19. Thingna J, Manzano D, Cao J. Sci Rep. 2016;6:28027.
20. Quach JQ, McGhee KE, Ganzer L, Rouse DM, Lovett BW, Gauger EM, Keeling J, Cerullo G, Lidzey DG, Virgili T. Sci. Adv. 8, eabk3160 (2022).
21. Joshi J, Mahesh TS. Phys Rev A. 2022;106: 042601.
22. Manzano D, Hurtado PI. Adv Phys. 2018;67:1.
23. Bacon D, Lidar DA, Whaley KB. Phys Rev A. 1999;60:1944.
24. Buča B, Prosen T. New J Phys. 2012;14: 073007.
25. Lidar DA, Chuang IL, Whaley KB. Phys Rev Lett. 1998;81:2594.
26. Manzano D, Hurtado PI. Phys Rev B. 2014;90: 125138.
27. Liu J, Hanna G. J Phys Chem Lett. 2018;9:3928.
28. Gorini V, Kossakowski A, Sudarshan ECG. J Math Phys. 1976;17(5):821.
29. Lindblad G. Commun Math Phys. 1976;48:119.
30. Kapral R. J Chem Phys. 1999;110(18):8919.
31. Wang H, Thoss M, Miller WH. J Chem Phys. 2001;115:2979.
32. Imre K. özizmir E, Rosenbaum M, Zweifel PF. J Math Phys. 1967;8:1097.
33. Liu Z, Hanna G. Molecules. 2024;29:889.
34. Khodadad Z, Tarabishi J, Hanna G. J Chem Phys. 2025;162: 074309.
35. Khodadad Z, Mahdian M, Hanna G. Adv. Quantum Technol. p. e00211 (2025).
36. Asadian A, Tiersch M, Guerreschi GG, Cai J, Popescu S, Briegel HJ. New J Phys. 2010;12: 075019.
37. Allegra M, Giorda P, Lloyd S. Phys Rev A. 2016;93: 042312.
38. Chin AW, Datta A, Caruso F, Huelga SF, Plenio MB. New J Phys. 2010;12: 065002.
39. May V, Kühn O. 3rd edn. (John Wiley & Sons, Weinheim, 2023).
40. Fox F, Gatland IR, Roy R, Vemuri G. Computer. 1988;38:5938.

Chapter 9
Recent Advances in Relativistic Coupled-Cluster Studies of P,T-Odd Properties in Ultracold Molecules for Precision Measurements

Kaushik Talukdar, Malaya K. Nayak, and Sourav Pal

Abstract Ultracold molecules containing heavy metals promise many exciting applications in cutting-edge science and technology. Heavy, polar diatomic and triatomic molecular systems, in particular, are of great importance for exploring new physics beyond the Standard Model through precision measurements. In this context, highly accurate relativistic electronic-structure calculations are essential for determining parameters associated with parity (P) and time-reversal (T) violating (P,T-odd) interactions. Herein, we review recent calculations of P,T-odd parameters for diatomic molecules within the fully relativistic single-reference coupled-cluster framework. We focus especially on results obtained using the Z-vector (energy-derivative) method and the extended coupled-cluster approach as employed by our group. Both methods are reliable for calculating P,T-odd parameters, with the Z-vector approach emerging as computationally more efficient.

Keywords Ultracold molecules · Precision measurement · Spectroscopy · Electron correlation · Relativistic effects · Coupled-cluster theory

---

K. Talukdar
Department of Chemical & Biological Sciences, National Institute of Technology Meghalaya, Sohra, India

M. K. Nayak
Theoretical Chemistry Section, Bhabha Atomic Research Centre, Mumbai, India

Homi Bhabha National Institute, BARC Training School Complex, Mumbai, India

S. Pal (✉)
Department of Chemistry, Ashoka University, Sonipat, India
e-mail: sourav.pal@ashoka.edu.in

T. Onishi (ed.), *Advances in Quantum Science I*,
https://doi.org/10.1007/978-981-95-9717-8_9

9.1 Introduction

9.1.1 Parity (P) and Time-Reversal (T) Symmetry Violation in Ultracold Molecules

Molecular systems cooled to temperatures near absolute zero, typically in the nanokelvin range, are known as ultracold molecules. These molecules have exciting applications in fundamental physics, ultracold chemistry, quantum science, and technology [1–4]. In particular, ultracold molecules provide a powerful low-energy platform for probing fundamental physics. To search for "new physics" beyond the Standard Model of elementary particles, a number of experiments intended for precision measurement on ultracold molecules have already been performed or are in progress. Such experiments can help to discover new sources of violation of discrete symmetries in physics: charge conjugation (C), parity (P), and time-reversal (T). Violation of charge conjugation symmetry means that the physical laws are not invariant under the replacement of particles with their antiparticles. Similarly, violation of parity symmetry indicates that the laws of physics are not invariant under spatial inversion, whereas violation of T-symmetry signifies non-invariance under reversal of the direction of time. According to the CPT theorem, the universe is invariant under the combined CPT transformation. Consequently, any violation of T-symmetry is equivalent to a violation of CP symmetry. It is important to mention here that the CP-violation can explain the existing mysteries of the universe such as the matter–antimatter asymmetry [5, 6].

Permanent electric dipole moment (EDM) and magnetic quadrupole moment (MQM) of a species arise due to the simultaneous violations of P and T invariance. In paramagnetic molecules, three dominant P,T-violating (i.e., P,T-odd) interactions, *viz.*, the intrinsic EDM of the electron (eEDM, d_e), the scalar–pseudoscalar (S-PS) nucleon–electron neutral current coupling, and the nuclear MQM induce permanent molecular EDM, which can be measured through high-precision spectroscopic experiments on ultracold molecules. However, molecular sensitivity parameter for each P,T-violating interaction is necessary for the analysis of the experimental results. The sensitivity parameters, also known as P,T-odd interaction constants, cannot be measured from any experiment. These can be determined only through the highly accurate electronic-structure calculations.

9.1.2 Experimental Measurements of P,T-Violating Electric Dipole Moments (EDMs)

In recent decades, significant experimental efforts have been devoted to measuring atomic and molecular electric dipole moments [8, 9]. Owing to their large polarizability and strong internal electric fields, molecules offer enhanced sensitivity compared to atoms and are particularly well suited for EDM searches. As mentioned

Table 9.1 Experimental upper bound limit on the electron electric dipole moment (EDM) using open-shell diatomic molecules

Molecule	Electronic state	$\|d_e\|$ ($e$cm)	Year of experiment
^{174}YbF [10]	$^2\Sigma_{1/2}$	$< 10.5 \times 10^{-28}$	2011
^{232}ThO [11]	$^3\Delta_1$	$< 1.1 \times 10^{-29}$	2018
^{180}HfF$^+$ [12]	$^3\Delta_1$	$< 4.1 \times 10^{-30}$	2023

above, in paramagnetic molecules, the observed EDM is primarily sensitive to the electron electric dipole moment, the scalar–pseudoscalar nucleon–electron neutral current interaction, and the nuclear magnetic quadrupole moment. Precision measurements performed on open-shell molecules such as YbF [10], ThO [11], and HfF$^+$ [12, 13] have led to stringent upper limits on the eEDM (see Table 9.1). Most notably, recent experiments by the JILA group using trapped metastable HfF$^+$ ions reported an improved limit of $|d_e| < 4.1 \times 10^{-30} e$cm at 90% confidence [12], surpassing the earlier result obtained by the ACME collaboration with ThO [11] by more than a factor of two. Both the ThO and HfF$^+$ experiments exploited the first excited metastable electronic state, which offers important advantages such as Ω-doublet structure and strong suppression of systematic errors. While the ACME collaboration employed a molecular beam of metastable ThO, the JILA group utilized trapped molecular ions, illustrating complementary experimental approaches to EDM measurements. Beyond these, alternative strategies have been proposed, including the use of polar molecules embedded in cryogenic inert-gas matrices [14] and linear triatomic molecules such as YbOH and LuOH$^+$ [15–17]. In the latter case, the presence of l-doubling in excited bending modes allows complete molecular polarization using relatively weak external electric fields, making these systems especially attractive for future experiments. Consequently, several experimental efforts involving BaF [18, 19], YbF [14, 20], ThF$^+$ [21], YbOH [15, 16], LuOH$^+$ [17], HgX (X=F, Cl, Br, I) [14], RaF [14], and WC [14] are currently underway or under development.

9.1.3 Role of Electronic Structure Theory in Precision Measurements

Knowledge of P,T-odd molecular parameters is necessary to analyze the results of precision experiments with ultracold molecules, and these parameters can only be reliably obtained through highly accurate quantum-chemical calculations. In addition, since the electric dipole moment of a paramagnetic molecule can arise from multiple P,T-violating interactions, measurements on different molecular systems with distinct sensitivity ratios are essential for disentangling the individual contributions. In this context, accurate molecular electronic-structure calculations play a critical role in extracting fundamental limits on the eEDM and scalar–pseudoscalar coupling constants [22–25], as well as in guiding the design of next-generation eEDM experiments. Note that the larger the magnitude of the P- and T-odd parameters, the better the candidate molecule will be. It is also noteworthy that the accuracy of such calculations is highly sensitive to the quality of the molecular electronic wavefunction in the near-nuclear region. Consequently, a method capable of efficiently treating both electron correlation and relativistic effects is required for the reliable computation of P- and T-odd parameters. Four-component relativistic coupled-cluster theory meets these criteria. Recently, we reviewed numerous analytic and numerical methods based on relativistic single-reference coupled-cluster (SRCC) theory that have emerged as reliable tools for determining the sensitivity parameters W_d, W_M, and W_S for the P,T-odd properties such as the eEDM, the nuclear MQM, and S-PS nucleus–electron interaction, respectively [26]. We also provided a systematic survey on recent experimental and relativistic SRCC investigations of potential molecular candidates in search of new physics beyond the Standard Model, conducted by various research groups over the globe.

In this chapter, we review recent advances in computational investigations of P,T-violating interactions in paramagnetic molecules, however, with a primary focus on relativistic coupled-cluster-based Z-vector and extended-coupled-cluster (ECC) predictions of molecular sensitivity parameters in potential ultracold systems. We highlight the theory of these two analytic approaches, along with their applications in P,T-violating physics. It is worth noting that the Z-vector technique is a non-variational approach, whereas the ECC approach is variational. Therefore, this chapter offers special focus on determining molecular sensitivity parameters in ultracold molecules using the variational coupled-cluster approach within the four-component relativistic framework.

9.2 P,T-odd Sensitivity Parameters

9.2.1 Effective Electric Field for the Electron Electric Dipole Moment (eEDM) Interaction

The effective electric field experienced by the unpaired electron within an open-shell molecule can be defined as

$$E_{\text{eff}} = W_d|\Omega| = \langle\Psi_\Omega|\sum_j^n \frac{\hat{H}_{\text{d}}(j)}{d_e}|\Psi_\Omega\rangle, \tag{9.1}$$

where W_d is known as the P,T-odd eEDM interaction parameter and Ω denotes the projection of total electronic angular momentum of the molecule (with n number of electrons) on the molecular axis (say, z-axis). Ψ_Ω refers to the coupled-cluster wavefunction for the Ωth state and $\hat{H}_{\text{d}}$ represents the effective Hamiltonian [27] for the eEDM interaction, given by

$$\hat{H}_{\rm d} = 2icd_e\gamma^0\gamma^5\hat{\mathbf{p}}^2, \tag{9.2}$$

with c as the speed of light, γs and $\hat{\mathbf{p}}$ as the Dirac matrices and the momentum operator, respectively.

9.2.2 Scalar–Pseudoscalar (S-PS) Nucleus–Electron Neutral Current Interaction Parameter

The S-PS interaction coefficient, W_S, can be defined in terms of the following matrix element:

$$W_S = \frac{1}{\Omega}\langle\Psi_\Omega|\sum_j^n \frac{\hat{H}_{\rm SP}(j)}{k_{\rm s}}|\Psi_\Omega\rangle. \tag{9.3}$$

Here k_s, a dimensionless constant, is known as the fundamental scalar–pseudoscalar nucleon–electron neutral current coupling constant quantifying the strength of the S-PS interaction, and $\hat{H}_{\rm SP}$, the effective Hamiltonian for this interaction [28] reads

$$\hat{H}_{\rm SP} = i\frac{G_F}{\sqrt{2}}Zk_s\gamma^0\gamma^5\rho_N(r), \tag{9.4}$$

where G_F and Z refer to the Fermi constant and the nuclear charge, respectively, and $\rho_N(r)$ is known as the nuclear charge density normalized to unity.

9.2.3 Nuclear Magnetic Quadrupole Moment (MQM) Interaction Parameter

The sensitivity parameter for the nuclear magnetic quadrupole moment interaction, W_M [29–31], reads

$$W_M = \frac{3}{2\Omega}\cdot\langle\Psi_\Omega|\sum_j^n\left(\frac{\vec{\alpha}_j\times\vec{r}_j}{r_j^5}\right)_z r_z|\Psi_\Omega\rangle, \tag{9.5}$$

where $\vec{\alpha}_j$ is the vector representing Dirac's matrices, i.e., $\vec{\alpha}_j = (\alpha_x, \alpha_y, \alpha_z)$ and $\vec{r}_j$ refers to the position vector for the jth electron.

9.3 Relativistic Coupled-Cluster Method

The coupled-cluster (CC) method is a post-Hartree–Fock quantum chemistry method approach that accurately describes electron correlation effects. The CC electronic wavefunction [32–35] is expressed as

$$|\Psi_{cc}\rangle = e^{\hat{T}}|\Phi_0\rangle. \tag{9.6}$$

Here, $|\Phi_0\rangle$ refers to a reference state (usually the Dirac–Hartree–Fock determinant to describe the relativistic effects) and $\hat{T} = \hat{T}_1 + \hat{T}_2 + \cdots + \hat{T}_K = \sum_{n=1}^{K} \hat{T}_n$ is known as the cluster (excitation) operator. In second quantization formalism, the n-tuple excitation operator $\hat{T}_n$ is defined as

$$\hat{T}_n = \frac{1}{(n!)^2} \sum_{ij\ldots ab\ldots} t_{ij\ldots}^{ab\ldots} a_a^\dagger a_b^\dagger \ldots a_j a_i, \tag{9.7}$$

where $i, j \ldots$ and $a, b \ldots$ denote occupied (hole) and unoccupied (particle) spinors, respectively; the cluster amplitudes $t_{ij\ldots}^{ab\ldots}$ can be determined solving the CC amplitude equation. For coupled-cluster singles doubles (CCSD) approximation, $\hat{T} = \hat{T}_1 + \hat{T}_2$; that is, only single- and double-excitation operators are included. Using these operators, the singly- and doubly-excited determinants can be written as $|\Phi_i^a\rangle = \hat{T}_1|\Phi_0\rangle$ and $|\Phi_{ij}^{ab}\rangle = \hat{T}_2|\Phi_0\rangle$.

The CCSD wavefunction is expressed as

$$|\Psi_{cc}\rangle = e^{\hat{T}_1+\hat{T}_2}|\Phi_0\rangle. \tag{9.8}$$

The corresponding amplitudes for the excitation operators ($\hat{T}_1$ and $\hat{T}_2$) can be evaluated from the following equations:

$$\begin{aligned} \langle\Phi_i^a|(\hat{H}_N e^{\hat{T}})_c|\Phi_0\rangle &= 0, \\ \langle\Phi_{ij}^{ab}|(\hat{H}_N e^{\hat{T}})_c|\Phi_0\rangle &= 0. \end{aligned} \tag{9.9}$$

Here, $\hat{H}_N$ means the normal-ordered Hamiltonian operator (i.e., $\hat{H}_N = \hat{H} - \langle\Phi_0|\hat{H}|\Phi_0\rangle$) and the subscript c denoting refers to connectedness, which means that only the connected terms survive in the contraction between $\hat{H}_N$ and $\hat{T}$ operators. In other words, only connected diagrams in the exponential cluster ansatz are included, ensuring linked clusters that guarantee size-extensivity of the coupled-cluster method. For more insights into the coupled-cluster theory, readers can see references [34] and [35].

Finally, the coupled-cluster correlation energy reads

$$E_{corr} = \langle\Phi_0|(\hat{H}_N e^{\hat{T}})_c|\Phi_0\rangle. \tag{9.10}$$

9.3.1 Z-vector Approach within Relativistic Coupled-Cluster Framework

The Z-vector method [36, 37], which is an energy-derivative approach, uses a perturbation independent linear (de-excitation) operator, $\hat{\Lambda}$, to evaluate the energy derivative. Within the CCSD framework, the operator $\hat{\Lambda} = \hat{\Lambda}_1 + \hat{\Lambda}_2$ is defined as follows:

$$\hat{\Lambda}_m = \frac{1}{(m!)^2} \sum_{ij...ab...} \lambda_{ab...}^{ij...} a_i^{\dagger} a_j^{\dagger} \ldots a_b a_a, \tag{9.11}$$

where the amplitude $\lambda_{ab...}^{ij...}$ is obtained by solving the following equations:

$$\langle \Phi_0 | [\hat{\Lambda}(\hat{H}_N e^{\hat{T}})_c]_c | \Phi_i^a \rangle + \langle \Phi_0 | (\hat{H}_N e^{\hat{T}})_c | \Phi_i^a \rangle = 0,$$

$$\langle \Phi_0 | [\hat{\Lambda}(\hat{H}_N e^{\hat{T}})_c]_c | \Phi_{ij}^{ab} \rangle + \langle \Phi_0 | (\hat{H}_N e^{\hat{T}})_c | \Phi_{ij}^{ab} \rangle + \langle \Phi_0 | (\hat{H}_N e^{\hat{T}})_c | \Phi_i^a \rangle \langle \Phi_i^a | \hat{\Lambda} | \Phi_{ij}^{ab} \rangle = 0. \tag{9.12}$$

Using the amplitudes of $\hat{T}$ and $\hat{\Lambda}$, one can easily evaluate the energy derivative as follows:

$$\Delta E' = \langle \Phi_0 | (\hat{O}_N e^{\hat{T}})_c | \Phi_0 \rangle + \langle \Phi_0 | [\hat{\Lambda}(\hat{O}_N e^{\hat{T}})_c]_c | \Phi_0 \rangle, \tag{9.13}$$

where $\hat{O}_N$ is the derivative of the normal-ordered perturbed Hamiltonian with respect to the external perturbing field, herein it is the P,T-odd interaction Hamiltonian in normal-ordered form. We extended the Z-vector method into the four-component relativistic coupled-cluster framework [38].

9.3.2 Relativistic Extended Coupled-Cluster Approach

In contrast to the traditional coupled-cluster theory, the extended coupled-cluster (ECC) approach is variational. In general, the ECC functional [39, 40] of the operator $\hat{O}$ reads

$$\langle \hat{O} \rangle = \frac{\langle \Phi_0 | e^{\hat{T}'} \hat{O} e^{\hat{T}} | \Phi_0 \rangle}{\langle \Phi_0 | e^{\hat{T}'} e^{\hat{T}} | \Phi_0 \rangle}, \tag{9.14}$$

where $\hat{T}'$ denotes the hole–particle (h-p) destruction operator, and $\hat{T}$ is the usual cluster (h-p creation) operator. Arponen [39] introduced a new h-p destruction operator, $\hat{T}''$, that satisfies the following equation:

$$\frac{\langle \Phi_0 | e^{\hat{T}'} e^{\hat{T}}}{\langle \Phi_0 | e^{\hat{T}'} e^{\hat{T}} | \Phi_0 \rangle} = \langle \Phi_0 | e^{\hat{T}''}. \tag{9.15}$$

Thus, the ECC functional can be defined using the $\hat{T}''$ operator as follows:

$$\langle \hat{O} \rangle = \langle \Phi_0 | e^{\hat{T}''} e^{-\hat{T}} \hat{O} e^{\hat{T}} | \Phi_0 \rangle = \langle \Phi_0 | e^{\hat{T}''} (\hat{O} e^{\hat{T}})_c | \Phi_0 \rangle. \tag{9.16}$$

Despite the fact that $\hat{O}(e^{\hat{T}})_c$ leads to a terminating series, the full contraction between the h-p destruction operator $\hat{T}''$ and excitation operator $\hat{T}$ often yields disconnected terms in the amplitude equation. To get rid of this problem, the ECC functional is re-defined using two sets of amplitude, *viz.*, t and λ, as follows:

$$\langle \hat{O} \rangle = \langle \Phi_0 | e^{\hat{\Lambda}} (\hat{O} e^{\hat{T}})_L | \Phi_0 \rangle_{DL}, \tag{9.17}$$

where the subscripts L and DL mean that $\hat{T}$ is linked to the $\hat{O}$ vertex and $\hat{\Lambda}$ is connected to either $\hat{O}$ or at least two $\hat{T}$ operators, respectively. The $\hat{\Lambda}$ operator of the ECC functional is mathematically same as that defined in the previous section (see Eq. 9.11).

The ECC energy functional reads

$$E = \langle \hat{H} \rangle = \langle \Phi_0 | e^{\hat{\Lambda}} (\hat{H} e^{\hat{T}})_L | \Phi_0 \rangle_{DL}. \tag{9.18}$$

In presence of an external field of perturbation (say, $\hat{W}$), this functional can be written as

$$E(\omega) = \langle \Phi_0 | e^{\hat{\Lambda}(\omega)} [\hat{H}(\omega) e^{\hat{T}(\omega)}]_L | \Phi_0 \rangle_{DL}. \tag{9.19}$$

Here, ω is a small perturbation parameter, and $\hat{H}(\omega)$ is the field-dependent Hamiltonian, defined as $\hat{H}(\omega) = \hat{H} + \omega \hat{W}$.

As a variational method, the extended coupled-cluster (ECC) framework satisfies the $2n + 1$ rule, which states that knowledge of cluster amplitudes up to the nth order is sufficient to evaluate the $2n + 1$th-order energy derivative. Consequently, the first-order derivative of energy can be determined from the zeroth-order m-body cluster amplitudes, solving the following equations:

$$\frac{dE^{(0)}}{dt_m^{(0)}} = 0, \quad \frac{dE^{(0)}}{d\lambda_m^{(0)}} = 0. \tag{9.20}$$

We implemented the ECC method within the fully relativistic framework to compute various first-order properties of molecules [41–43].

9.4 Computational Prediction of P,T-odd Molecular Parameters for Precision Measurement

The accurate theoretical evaluation of parity and time-reversal-violating (P,T-odd) molecular sensitivity parameters has become an essential component in the search for physics beyond the Standard Model. In particular, open-shell diatomic and polyatomic molecules have attracted considerable attention due to their enhanced sensitivity to CP-violating effects. Recent developments in relativistic coupled-cluster theory have enabled systematic and high-precision investigations of such effects, with substantial contributions from the research groups of Pal, Das, Borschevsky, Titov, and Skripnikov. This section provides an overview of these efforts, reviewing key numerical findings, with particular emphasis on contributions from our group. Toward the end of this section, we also highlight recent works by Das, Borschevsky, Skripnikov, and others to give readers a flavor of their research; for a detailed discussion, readers are referred to Ref. [26] or to the references therein.

9.4.1 Z-Vector Results

A major methodological advance for precise calculations of P,T-odd properties was introduced by us through the implementation of the Z-vector method within the four-component CCSD framework. This analytical energy-derivative approach allows direct evaluation of molecular properties without relying on numerical differentiation, thereby improving both accuracy and computational efficiency. The key results using the Z-vector approach with triple-zeta (TZ) or quadruple-zeta (QZ) quality basis sets are presented in Table 9.2. Using the Z-vector method, accurate values of E_{eff} of 38.1 GV/cm and W_S of 88.4 kHz were obtained for PbF [44, 45] and E_{eff} of 123.2 GV/cm and W_S of 284.2 kHz were obtained for HgH [46], highlighting the crucial role of all-electron correlation for properties sensitive to the electronic

Table 9.2 P,T-odd sensitivity parameters—E_{eff}(GV/cm), W_S (kHz), and W_M (10^{33}Hz/ecm^2)—in open-shell diatomic molecules within the Z-vector/CCSD framework

Molecule	Basis	$\lvert E_{\mathrm{eff}}\rvert$	$\lvert W_S\rvert$	$\lvert W_M\rvert$
SrF	TZ	2.10	1.87	0.20
BaF	QZ	6.52	8.35	0.39
RaF	QZ	52.5	141.2	1.60
HgF	QZ	115.9	266.4	3.59
PbF	QZ	38.1	88.4	0.87
SrH	TZ	2.61	2.33	0.24
CdH	QZ	12.2	14.0	0.82
HgH	QZ	123.2	284.2	3.22

wavefunction near the nucleus. The Z-vector formalism was also applied to RaF [47], and high-precision calculations employing quadruple-zeta basis sets identified RaF as a highly promising candidate for probing P, T-violating interactions. Comparisons with expectation-value approaches showed that the Z-vector method provides improved accuracy for core-dependent properties, including hyperfine structure constants. Another important investigation using the Z-vector approach was performed on the laser cooled molecule, BaF [48]. This molecule is an exciting system for the precision measurement and an experiment is underway. The sensitivity parameters—E_{eff} and W_S—were reported as 6.52 GV/cm and 8.35 kHz, respectively, within an error of 8%.

Beyond electron EDM and scalar–pseudoscalar interactions, attention has also been directed toward P, T-violating effects arising from the nuclear magnetic quadrupole moment. The first high-accuracy relativistic coupled-cluster calculations of MQM sensitivity parameters were performed by us using the Z-vector approach in HgH [49]. The sensitivity parameter W_M was reported to be 3.22×10^{33}Hz/ecm^2 with an estimated uncertainty of 6%. Z-vector approach was further employed to determine the P,T-odd parameters in HgF [50] and the group-12 monohydrides [51]. These studies showed that proper treatment of core-electron correlation effects is key to accurate prediction of the sensitivity parameters and the higher-energy virtual energy functions are important for all-electron correlation.

MQM sensitivity parameters of metal monofluorides [52] such as MgF, CaF, SrF, BaF, RaF, and PbF were reported by us from the four-component relativistic CCSD calculations, and RaF, with the deformed ^{223}Ra nucleus, emerged as a promising candidate for probing physics in the hadronic sector. RaF exhibits large sensitivity parameters (W_M=1.60×10^{33}Hz/ecm^2) for the nuclear magnetic quadrupole moment interaction. The study revealed that the magnetic quadrupole moment interaction constants receive substantial contributions from electron correlation, and that the MQM sensitivity parameter exhibits an approximate Z^2 scaling with the nuclear charge of the metal atom within the group-2 of the periodic table, highlighting the importance of heavy-element systems. In other words, the magnitude of P,T-odd effects is approximately proportional to Z^2, where Z is the nuclear charge of the group-2 metal atoms. Electron correlation effects were shown to contribute significantly, accounting for about 20–30% of the total MQM interaction, while mean-field corrections for higher-order relativistic effects such as the Gaunt interaction were found to be minor. Additional benchmark studies on SrF [52, 53] and SrH [54] were reported recently by Pal.

Table 9.2 shows that the Z-vector method predicts that the P,T-odd effect sizes in the monohydrides of Sr and Hg are larger than those in the corresponding monofluorides, a trend that was also demonstrated by Das using the expectation-value approach. Among the systems studied, the HgH molecule exhibits the largest sensitivity parameters ($|E_{\text{eff}}|$=123.2 GV/cm, $|W_S|$=284.2 kHz, $|W_M|$=3.22×10^{33}Hz/ecm^2), followed by HgF. It is noteworthy that the P,T-odd effects generally increase with the nuclear charge of the metal atom within the same group of the periodic table due to relativistic enhancement effects; however, this trend does not necessarily hold within the same period. This behavior arises from differences in the electronic structures of molecules formed by metals within the same period, whereas within a given group the electronic structures of metal diatomics are typically similar. Nevertheless, the P,T-odd parameters in diatomic molecules such as SrF, BaF, RaF, HgF, PbF, SrH, CdH, and HgH are sufficiently large to make them promising candidates for precision measurements.

9.4.2 Dependence on Basis Set Quality

The sensitivity of the P,T-odd parameters to basis-set quality was assessed for the RaF molecule using the Z-vector approach. A systematic study of the basis-set dependence, using TZ and QZ basis sets, is summarized in Table 9.3, with the number of correlated electrons varied (61, 79, and 97). Increasing the number of correlated electrons from 61 to 97 leads to an increase in both the effective electric field strength (from 53 GV/cm to 55 GV/cm in the TZ basis) and the scalar–pseudoscalar (S-PS) interaction parameter (from approximately 143 to 148 kHz in the TZ basis), highlighting the importance of inner-core electron correlation. In contrast, the corresponding increase in the QZ basis set is not significant, because virtual spinors above 20 a.u. are excluded from the correlation treatment. Proper correlation of inner-core electrons requires high-energy virtual spinors, whose absence limits the impact of additional correlated electrons in the QZ calculations. One can expect significant increase in the magnitude of P,T-odd parameters in QZ basis if high-energy unoccupied spinors are included in the molecular calculation.

Table 9.3 P,T-odd sensitivity parameters of RaF [47] at different basis sets: QZ (dyall.cv4z for Ra and cc-pCVQZ for F) and TZ (dyall.cv3z for Ra and cc-pCVTZ for F). The energy cutoffs for virtual spinors in QZ and TZ are 20 and 500 a.u., respectively

Correlated Electrons	Basis	$\|E_{\text{eff}}\|$ (GV/cm)	$\|W_S\|$ (kHz)
61	TZ	53.5	143.6
79	TZ	54.5	146.3
97	TZ	55.1	148.1
61	QZ	52.3	140.4
79	QZ	52.5	141.1
97	QZ	52.5	141.2

Table 9.4 Dependence of P,T-odd parameters of CdH on internuclear distance [51] (Dyall.ae2z and cc-pCVDZ basis sets are used for Cd and H, respectively, with 500 a.u. as the energy cutoff for virtual spinors)

Internuclear Distance (Å)	$\lvert E_{\text{eff}} \rvert$ (GV/cm)	$\lvert W_S \rvert$ (kHz)	$\lvert W_M \rvert$ 10^{33}Hz/ecm^2
1.580	10.43	11.45	0.760
1.680	10.43	11.41	0.766
1.728	10.39	11.35	0.765
1.780	10.32	11.25	0.762
1.834	10.21	11.13	0.756
1.880	10.09	10.99	0.749
1.980	9.76	10.61	0.726

9.4.3 Dependence on Internuclear Distance

Dependence of P,T-odd parameters on the bond length of a diatomic system was also understood from the study on CdH (see Table 9.4). A series of calculations was performed using a double-zeta-quality basis set, in which the internuclear distance was varied around the equilibrium bond length (1.780 Å). It was observed that a change of 3% in the internuclear separation from the equilibrium bond distance results in a change of less than 1% in the magnitude of the sensitivity parameters. This implies that the P,T-odd parameters do not have very strong dependence on the internuclear distance in the vicinity of equilibrium bond length.

9.4.4 Dependence on Relativistic Hamiltonian

To explore the dependence of the P,T-odd properties on the relativistic Hamiltonian in SrF [53], four different Hamiltonians were used in the molecular calculations, *viz.*, the four-component Hamiltonian: Dirac–Coulomb (DC) and Dirac–Coulomb–Gaunt (DCG), and the exact two-component Hamiltonian: X2C (includes mean-field treatment of the two-electron spin-same-orbit coupling) and X2C-G (includes both spin-same-orbit and spin-other-orbit coupling treatments). Our results are shown in Table 9.5, which demonstrates that the X2C Hamiltonian reproduces the four-component DC results with good accuracy, and that the inclusion of the Gaunt interaction in the X2C-G Hamiltonian yields results comparable to those obtained with the DCG Hamiltonian. Additionally, incorporating the Gaunt interaction does not result in any significant change in the calculated properties, implying that higher-order relativistic effects contribute only marginally.

Table 9.5 Dependence of P,T-odd parameters of SrF on the relativistic Hamiltonian, at dyall.cv3z for Sr, cc-pCVTZ for F, and energy cutoff for virtual spinors as 500 a.u

Hamiltonian	$\lvert E_{\text{eff}} \rvert$ (GV/cm)	$\lvert W_S \rvert$ (kHz)
X2C	2.10	1.88
DC	2.10	1.88
X2C-G	2.10	1.87
DCG	2.04	1.88

9.4.5 Relativistic Extended Coupled-Cluster Results: Dependence on Virtual Spinors

We also introduced a variational coupled-cluster approach for predicting P,T-odd properties [43]. Specifically, we implemented the extended coupled-cluster (ECC) method within a relativistic framework and reported the first calculations of P,T-odd properties in diatomic molecules using this approach. Sensitivity parameters were computed for four laser-coolable diatomic molecules, SrH, SrF, CaH, and CaF, in their ground electronic states. We further assessed the sensitivity of the molecular calculations to the basis-set size, the treatment of high-energy virtual spinors, and the choice of relativistic Hamiltonian; for additional details, the reader is referred to Ref. [43].

It is not always feasible to include all virtual (unoccupied) spinors in correlation calculations; therefore, virtual spinors above a chosen energy cutoff are often excluded. However, this raises important questions regarding the influence of high-energy virtual spinors on molecular properties, and understanding this aspect is crucial for advancing molecular computational techniques. Accordingly, we performed a set of calculations on SrH as a test case using a double-zeta basis set with different energy cutoffs for virtual spinors, and the results are presented in Table 9.6. The inclusion of high-energy virtual spinors in the correlation treatment increases the magnitude of the P,T-odd parameters; in particular, virtual spinors with energies above 50 a.u. and 500 a.u. contribute approximately 4% and 2%, respectively. This

Table 9.6 Dependence of P,T-odd sensitivity parameters of SrH on the virtual spinors. Basis set used for the calculation: dyall.cv2z for Sr and cc-pVDZ for H

Cutoff (a.u.)	Spinors	$\lvert E_{\text{eff}} \rvert$ (GV/cm)	$\lvert W_S \rvert$ (kHz)	$\lvert W_M \rvert$ (10^{33} Hz/ecm^2)
50	131	2.29	2.06	0.234
500	167	2.35	2.12	0.240
2000	187	2.37	2.13	0.241
No cutoff	219	2.38	2.15	0.243

indicates that the use of a low-energy cutoff for virtual spinors is not recommended for precise calculations of P,T-odd parameters.

9.4.6 Comparison of Extended Coupled-Cluster (ECC), Z-Vector, and Linear Expectation-Value (LECC) Results

We also bench-marked the ECC results with those reported using the Z-vector and linear expectation-value (LECC) methods within the relativistic CCSD framework (see Table 9.7). One can see that the magnitudes of the P,T-odd sensitivity parameters for CaF and CaH are approximately an order of magnitude smaller than those for SrF and SrH. This implies that CaF and CaH are not promising candidates for P,T-odd EDM searches. Nonetheless, the close agreement between ECC results and those obtained using these two methods indicates that the ECC method is a reliable tool for computing P,T-odd properties. Moreover, the property expression within the ECC approach contains additional terms compared to the Z-vector method, and likewise the Z-vector expression includes additional terms relative to the LECC formulation. Consequently, from the standpoint of electron correlation, the ECC method is expected to provide a more complete description than the other two approaches. Therefore, the ECC method is expected to yield more reliable values for P,T-odd parameters in ultracold molecules, and these properties should be explored in molecules such as BaF, RaF,

Table 9.7 Comparison of P,T-odd sensitivity parameters computed by using extended-coupled-cluster (ECC), Z-vector, and linear expectation-value (LECC) approaches for CaH, CaF, SrH, SrF at the triple-zeta basis (dyall.cv3z for metal, cc-pVTZ for non-metal, with virtual energy cutoff: 500 a.u.)

Molecule	Method	$\lvert E_{\text{eff}} \rvert$ (GV/cm)	$\lvert W_S \rvert$ (kHz)	$\lvert W_M \rvert$ (10^{33} Hz/ecm^2)
CaH	ECC	0.32	0.24	0.065
	Z-vector	0.33	0.24	0.065
	LECC	0.33	0.25	0.066
CaF	ECC	0.27	0.20	0.054
	Z-vector	0.27	0.20	0.055
	LECC	0.28	0.21	0.055
SrH	ECC	2.60	2.32	0.247
	Z-vector	2.62	2.33	0.248
	LECC	2.65	2.35	0.251
SrF	ECC	2.10	1.87	0.201
	Z-vector	2.11	1.88	0.202
	LECC	2.13	1.90	0.205

HgF, PbF, HgH, and others using the ECC approach with larger, high-quality basis sets.

9.4.7 Studies by Other Research Groups

In addition to contributions from our group, some of the earliest comprehensive relativistic coupled-cluster studies in this field were carried out by Das and collaborators, who employed coupled-cluster methods to investigate P,T-violating effects in atoms and molecules using the linear expectation-value (LE) formulation and the finite-field (numerical) approach. Uncontracted relativistic basis sets of double-, triple-, and quadruple-zeta quality were adopted to ensure a reliable treatment of relativistic and correlation effects. These investigations produced benchmark values of the effective electric field E_{eff} and the scalar–pseudoscalar interaction constant W_S for several molecules of experimental relevance. For example, YbF was studied using a four-component LECC framework, yielding values of E_{eff} and W_S with uncertainties below 10% [55, 56]. The importance of correlating semi-core electrons was clearly demonstrated, as their contributions were found to be essential for obtaining accurate P, T-odd parameters. Subsequent applications of the same methodology revealed large effective electric fields in weakly polar molecules, such as YbH and HgH, highlighting the role of relativistic effects in heavy-element hydrides [57]. These findings challenged the traditional assumption that strong molecular polarity is a perquisite for large P, T-odd enhancements. Further CSSD-level studies extended to mercury mono-halides (HgF, HgCl, HgBr, and HgI) where Prasanna et al. [58, 59] showed that HgBr and HgI possess particularly favorable properties for eEDM searches. Detailed analyses of the coupled-cluster wavefunction indicated that both mean-field and higher-order correlation terms contribute significantly to the large values of Eeff observed in these molecules. Investigations of alkaline-earth and group-12 monofluorides revealed systematic trends across the periodic table [60]. In particular, group-12 monofluorides were found to exhibit larger P, T-odd sensitivity parameters than their group-2 counterparts. This behavior was attributed to stronger relativistic contraction of valence orbitals in heavier elements. Notably, these studies showed that while the absolute magnitude of E_{eff} does not increase monotonically with nuclear charge Z, the ratio W_S/E_{eff} does display a monotonic dependence on Z. In addition, Hg-alkali metal system [61] and Ra-containing molecules such as RaH [62], RaX (X=F, Cl, Br, I, and At) [63], and RaZ (Z=Cu, Ag, and Au) [63] were also explored using the LECC method. These systems were predicted to exhibit exceptionally large effective fields and scalar–pseudoscalar interaction constants, making them promising candidates for future experimental investigations. Expectation-value approach was further employed to predict the P,T-odd parameters in YbOH [64], HgOH [65], superheavy Lr-systems [66], and TlF^+ [67]. Similarly, E_{eff} and W_S for the alkaline-earth monofluorides, the mercury mono-halides (HgX), and PbF [68] were investigated by Das and co-workers through the application of the finite-field approach within the relativistic CCSD framework.

To incorporate higher-order correlation effects, Borschevsky and co-workers [69, 70] employed finite-field coupled-cluster methods at the CCSD(T) (i.e., CCSD with partial triples) level, along with multireference Fock-space coupled-cluster theory, where necessary. These approaches were applied not only to diatomic molecules [71] such as LaS, LaO, and LuO but also to polyatomic systems such as BaOH [69, 70], YbOH [69, 70], $BaCH_3$ [72], and $YbCH_3$ [72] which are of increasing experimental interest. Their results showed that polyatomic molecules can retain large P, T-odd sensitivity parameters comparable to those of their diatomic analogues, while offering additional experimental advantages. Another interesting observation from their studies is that enhanced covalent character in metal–ligand bonds was found to further increase P,T-odd sensitivity, providing valuable chemical insights into molecular design strategies for future experiments.

High-precision calculations were also carried out by Skripnikov and co-workers using the GRECP/Restoration approach combined with four-component CCSD(T) [24, 73, 74]. These studies focused on experimentally relevant systems such as ThO and HfF^+, for which detailed decompositions of contributions from inner-core electrons, relativistic Gaunt interactions, and higher-order correlation effects were provided. Their investigations re-confirmed the importance of inner-core electron correlation for accurate prediction of P,T-odd parameters. Such findings underscore the need for comprehensive relativistic and correlation treatments for the quantitative interpretation of high-precision experiments.

Collectively, these studies indicate that while CCSD-level calculations provide valuable qualitative and quantitative insights, the inclusion of perturbative triple excitations is essential for achieving higher predictive accuracy. Expectation-value methods, though computationally efficient, often suffer from truncation-related limitations, particularly for core-sensitive properties. Energy-derivative approaches, such as finite-field and Z-vector methods, offer improved robustness and systematic convergence. In addition, the relativistic ECC approach stands out as reliable as the Z-vector technique; however, the former is computationally more demanding. Among the available techniques, the Z-vector method emerges as a particularly attractive framework for evaluating multiple P, T-odd molecular parameters simultaneously. When combined with CCSD(T)-level wavefunctions, it offers a balanced compromise between computational feasibility and accuracy, making it a promising approach for future high-precision studies of CP-violating effects in molecules.

9.5 Conclusion

Recent theoretical studies have shown that fully relativistic four-component coupled-cluster methods provide a powerful framework for evaluating molecular parameters associated with P,T-violating phenomena. Molecular parameters for the electron-electric dipole moment (EDM) interaction, the scalar–pseudoscalar nucleon–electron coupling, and the nuclear magnetic quadrupole moment have been computed for a range of open-shell species of interest for precision EDM measurements. These molecular calculations play a crucial role in guiding and interpreting ongoing and planned high-resolution experimental studies. Several relativistic coupled-cluster variants have been tested for their ability to describe P,T-odd interactions, and despite their methodological differences, they generally yield compatible and reliable results. These findings also emphasize that achieving accurate P,T-odd parameters requires more than a sophisticated correlation method alone. High-quality basis sets, inclusion of high-energy virtual spinors, and proper treatment of inner-core electrons are all essential for obtaining quantitatively reliable predictions. Among the available computational approaches, the Z-vector/CCSD formulation introduced by our group has proven to be an efficient and reliable tool for computing multiple P,T-odd properties. Variational approaches, such as the extended coupled-cluster (ECC) method, are also promising for evaluating these properties with comparable accuracy. Overall, advancements in four-component relativistic coupled-cluster theory have enabled the identification of promising molecular candidates such as BaF, RaF, HgF, HgH, YbF, PbF, ThO, HfF^+, SrF, YbOH, and BaOH for electron EDM experiments and the prediction of their sensitivities with high confidence. As experimental EDM searches and theoretical techniques continue to evolve, molecular systems are expected to remain central to probing physics beyond the Standard Model.

Glossary

- **Electric Dipole Moment:** The electric dipole moment (EDM) measures the separation of positive and negative charges in a system.
- **P,T-odd Interactions:** Such interactions violate both parity (P) and time-reversal (T) symmetries and are collectively referred to as P,T-violating or P,T-odd interactions.
- **Magnetic Quadrupole Moment:** The magnetic quadrupole moment (MQM) is a second-rank tensor and is the lowest P,T-odd magnetic moment of a system.
- **Scalar–Pseudoscalar (S-PS) Interaction:** It refers to scalar–pseudoscalar nucleon–electron neutral current coupling in beyond-Standard-Model physics that violates P- and T-symmetry.
- **Coupled-Cluster Singles Doubles:** Coupled-cluster singles and doubles (CCSD) is a correlated method that approximates the electronic wavefunction using an exponential ansatz with single- and double-excitation cluster operators.

References

1. DeMille D, Hutzler NR, Rey AM, et al. Nat Phys. 2024;20:741.
2. Cornish SL, Tarbutt MR, Hazzard KRA. Nat Phys. 2024;20:730.
3. Hutzler NR. Quantum. Sci Technol. 2020;5: 044011.
4. Ye J, Zoller P. Phys Rev Lett. 2024;132: 190001.

5. Dine M, Kusenko A. Rev Mod Phys. 2003;76:1.
6. Sakharov AD. JETP Lett. 1967;5:24.
7. Christenson JH, Cronin JW, Fitch VL, Turlay R. Phys Rev Lett. 1964;13:138.
8. Safronova MS, Budker D, DeMille D, Kimball DFJ, Derevianko A, Clark CW. Rev Mod Phys. 2018;90: 025008.
9. Chupp TE, Fierlinger P, Ramsey-Musolf MJ, Singh JT. Rev Mod Phys. 2019;91: 015001.
10. Hudson JJ, Kara DM, Smallman IJ, Sauer BE, Tarbutt MR Hinds EA, (2011) Nature (London) 473:493.
11. The ACME Collaboration, Andreev V et al. (2018) Nature (London) 562:355.
12. Roussy TS, et al. Science. 2023;381:46.
13. Cairncross WB, et al. Phys Rev Lett. 2017;119: 153001.
14. Vutha AC, Horbatsch M, Hessels EA. Atoms. 2018;6:3.
15. Kozyryev I, Hutzler NR. Phys Rev Lett. 2017;119: 133002.
16. Isaev TA, Zaitsevskii AV, Eliav E. J Phys B: At Mol Opt Phys. 2017;50: 225101.
17. Maison DE, Skripnikov LV, Penyazkov G, Grau M, Petrov AN. Phys Rev A. 2022;106: 062827.
18. Aggarwal P, Bethlem HL, Borschevsky A, Denis M, Esajas K, Haase PAB, Hao Y, Hoekstra S, Jungmann K, Meijknecht TB, et al. Eur Phys J D. 2018;72:197.
19. Vutha A, Horbatsch M, Hessels E. Phys Rev A. 2018;98: 032513.
20. Fitch NJ, Lim J, Hinds EA, Sauer BE, Tarbutt MR. Quantum. Sci Technol. 2020;6: 014006.
21. Gresh DN, Cossel KC, Zhou Y, Ye J, Cornell EA. J Mol Spectrosc. 2016;319:1.
22. Meyer ER, Bohn JL, Deskevich MP. Phys Rev A. 2006;73: 062108.
23. Petrov A, Mosyagin N, Isaev T, Titov A. Phys Rev A. 2007;76: 030501.
24. Skripnikov LV. J Chem Phys. 2017;147: 021101.
25. Fleig T. Phys Rev A. 2017;96: 040502.
26. Talukdar K, Avasare V, Pal S. (2025) in: Pal S, Srivastava V, Avasare V, Maruani J (eds) Theoretical Methods, Algorithms, and Applications of Quantum Systems in Chemistry, Physics, and Biology. QSCP 2023. Progress in Theoretical Chemistry and Physics, vol 35. Springer, Cham.
27. Kozlov MG, Labzowsky LN. J Phys B: At Mol Opt Phys. 1995;30:1933.
28. Hunter LR. Science. 1991;252:73.
29. Kozlov MG, Fomichev V, Dmitriev YY, Labzovsky LN, Titov AV. J Phys B: At Mol Opt Phys. 1987;20:4939.
30. Skripnikov LV, Titov AV, Flambaum VV. Phys Rev A. 2017;95: 022512.
31. Ginges J, Flambaum VV. Phys Rep. 2004;397:63.
32. Cizek J. J Chem Phys. 1966;45:4256.
33. Cizek J. Adv Quantum Chem. 1969;14:35.
34. Bartlett RJ, Musiał M. Rev Mod Phys. 2007;79:291.
35. Crawford TD, Schaefer HF III. Rev Comp Chem. 2000;14:33.
36. Handy NC, Schaefer HF. J Chem Phys. 1984;81:5031.
37. Salter EA, Trucks GW, Bartlett RJ. J Chem Phys. 1989;90:1752.
38. Sasmal S, Pathak H, Nayak MK, Vaval N, Pal S. Phys Rev A. 2015;91:030503(R).
39. Arponen J. Ann Phys (NY). 1983;151:311.
40. Bishop R, Arponen J, Pajanne P. Aspects of Many-Body Effects in Molecules and Extended Systems. Berlin: Springer; 1989.
41. Sasmal S, Pathak H, Nayak MK, Vaval N, Pal S. Phys Rev A. 2015;91: 022512.
42. Buragohain H, Talukdar K, Nayak MK. Theor Chem Acc. 2024;143:5.
43. Talukdar K, Nayak MK, Vaval N, Pal S. J Chem Theory Comput. 2025;21(11):5481.
44. Sasmal S, Pathak H, Nayak MK, Vaval N, Pal S. J Chem Phys. 2015;143: 084119.
45. Sasmal S, Talukdar K, Nayak MK, Vaval N, Pal S. Mol Phys. 2017;115:2807.
46. Sasmal S, Pathak H, Nayak MK, Vaval N, Pal S. J Chem Phys. 2016;144: 124307.
47. Sasmal S, Pathak H, Nayak MK, Vaval N, Pal S. Phys Rev A. 2016;93: 062506.
48. Talukdar K, Nayak MK, Vaval N, Pal S. J Phys B: At Mol Opt Phys. 2020;53: 135102.
49. Talukdar K, Nayak MK, Vaval N, Pal S. Phys Rev A. 2019;99: 032503.
50. Talukdar K, Nayak MK, Vaval N, Pal S. J Chem Phys. 2019;150: 084304.

51. Talukdar K, Nayak MK, Vaval N, Pal S. Phys Rev A. 2020;101: 032505.
52. Talukdar K, Nayak MK, Vaval N, Pal S. J Chem Phys. 2020;153: 184306.
53. Talukdar K, Buragohain H, Nayak MK, Vaval N, Pal S. Theor Chem Acc. 2023;142:15.
54. Talukdar K, Nayak MK, Pal S. J Chem Sci. 2025;137:96.
55. Abe M, Gopakumar G, Hada M, Das BP, Tatewaki H, Mukherjee D. Phys Rev A. 2014;90: 022501.
56. Sunaga A, Abe M, Hada M, Das BP. Phys Rev A. 2016;93: 042507.
57. Sunaga A, Abe M, Hada M, Das BP. Phys Rev A. 2017;95: 012502.
58. Prasannaa VS, Vutha AC, Abe M, Das BP. Phys Rev Lett. 2015;114: 183001.
59. Prasannaa VS, Abe M, Bannur VM, Das BP. Phys Rev A. 2017;95: 042513.
60. Sunaga A, Prasannaa VS, Abe M, Hada M, Das BP. Phys Rev A. 2018;98: 042511.
61. Sunaga A, Prasannaa VS, Abe M, Hada M, Das BP. Phys Rev A. 2019;99:040501(R).
62. Fazil NM, Prasannaa VS, Latha KVP, Abe M, Das BP. Phys Rev A. 2019;99: 052502.
63. Sunaga A, Abe M, Hada M, Das BP. Phys Rev A. 2019;99: 062506.
64. Prasannaa VS, Shitara N, Sakurai A, Abe M, Das BP. Phys Rev A. 2019;99: 062502.
65. Mitra R, Prasannaa VS, Sahoo BK, Hutzler NR, Abe M, Das BP. Atoms. 2021;9:7.
66. Mitra R, Prasannaa VS, Ruiz RFG, Sato TK, Abe M, Sakemi Y, Das BP, Sahoo BK. Phys Rev A. 2021;104: 062801.
67. Bala R, Prasannaa VS, Abe M, Das BP. Eur Phys J Plus. 2023;138:478.
68. Abe M, Prasannaa VS, Das BP. Phys Rev A. 2018;97: 032515.
69. Denis M, Haase PAB, Timmermans RGE, Eliav E, Hutzler NR, Borschevsky A. Phys Rev A. 2019;99: 042512.
70. Denis M, Hao Y, Eliav E, Hutzler NR, Nayak MK, Timmermans RGE, Borschesvky A. J Chem Phys. 2020;152: 084303.
71. Chamorro Y, Flambaum V, Ruiz R, Borschevsky A, Paesteka LF. 2024.arXiv preprint.
72. Chamorro Y, et al. Phys Rev A. 2022;106: 052811.
73. Skripnikov LV. J Chem Phys. 2016;145: 214301.
74. Skripnikov LV, Titov AV, Flambaum VV. Phys Rev A. 2017;95: 022512.

The manufacturer's authorised representative in the EU is Springer
Nature Customer Service Centre GmbH, Europaplatz 3, 69115 Heidelberg,
Germany. If you have any concerns regarding our products, please
contact ProductSafety@springernature.com

Printed and bound by CPI Group (UK) Ltd, Croydon, CR0 4YY
07/10/2026
02229349-0014